People of the Desert and Sea

People of the Desert and Sea

ETHNOBOTANY OF THE SERI INDIANS

Richard Stephen Felger
and
Mary Beck Moser

THE UNIVERSITY OF ARIZONA PRESS
Tucson, Arizona

About the Authors

RICHARD S. FELGER has published extensively on desert plants, their uses, and new food crops around the world. His field of expertise is the flora and economic botany of the Sonoran Desert region. He has studied the relationship between the Seri Indians and the plants and animals in their region of northwestern Mexico for more than two decades. He earned his Ph.D. at the University of Arizona and in 1980 was appointed a research scientist at the Office of Arid Lands Studies, University of Arizona.

MARY BECK MOSER began studying in the Seri language in the early 1950s when she and her husband, the late Edward W. Moser, made their home in the Seri village of El Desemboque. Working with the Summer Institute of Linguistics, she wrote a number of linguistic and anthropological papers and completed her husband's translation of the New Testament into Seri. In the early 1980s Mrs. Moser continued her work with the Seri in various areas of study, including an extensive dictionary of the language. She is a graduate of the University of North Dakota.

The University of Arizona Press
www.uapress.arizona.edu

© 1985 by The Arizona Board of Regents
All rights reserved. Published 1985
First paperback edition 1991
Century Collection edition 2016

Printed in the United States of America
21 20 19 18 17 16   8 7 6 5 4 3

ISBN-13: 978-0-8165-0818-1 (cloth)
ISBN-13: 978-0-8165-1267-6 (paper)
ISBN-13: 978-0-8165-3475-3 (Century Collection paper)

This book was set in 11/13 Linotron 202 Sabon

Library of Congress Cataloging-in-Publication Data
Felger, Richard Stephen.
  People of the desert and sea.

  Bibliography: p.
  Includes index.
1. Seri Indians—Ethnobotany.   2. Indians of Mexico—
Sonora (State)—Ethnobotany.   3. Ethnobotany—Mexico—
Sonora (State) 1. Moser, Mary Beck.   II. Title.
F1221.S43F45   1985   581.6'1'097217   84-16357

∞ This paper meets the requirements of ANSI/NISO Z39.48-1992
(Permanence of Paper).

For the Comcáac
Edward William Moser
Alexander Russell
Jean Straub Russell

For the Comcáac . . . *Xica quih quiistox quih
hant quih iti coomii hiz cocoom quih comcaac
hantx mocat com hehe an com ziix quih ano
quiih hehe quih hant quiti quiih com ziix quih
ihasiijim quih chaa quih ipacta quih ocoaaj taax
ipacta z ihaaj cömiimcaj quih haa siihca caha.
Taax ziix quih ihasiijim ta taax cmique quih
issactim tama toc cöcahcaaha. Taaxah ox
tpactama hapaspoj hipquih it hant toiima hant
quih iti tahca hiz cöcahca ha taax mizj pata
cömisiiyaj ac quih oo cömatcmaai hehe quih
ziix quih ihasiijim quih hant quih iti coom
hantx mocat coii ocoaajc.*

Contents

Maps

Preface

The hunter-gatherer way of life has spanned the time of human existence. Within recent millennia it has declined as agriculture and population have increased. The transition has been a long one, and even into modern times many societies have engaged in mixed agricultural and hunter-gatherer economies. Since the mid-twentieth century the hunter-gatherer way of life has nearly vanished from the world.

This book is about the Seri Indians, traditionally a hunting and gathering and seafaring people, and their knowledge and usage of plants. Although known to outsiders as the Seri, they call themselves the *Comcáac* "the People." They live along the extremely hot and arid coast of the Gulf of California in Sonora, Mexico. Like most hunting and gathering societies surviving into the twentieth century, the Seri live where agriculture is generally not feasible.

Although for several centuries the Seri had varying degrees of interaction with the outside world, their traditional culture and knowledge of the environment remained remarkably intact until the second half of the twentieth century. The older people with whom we worked spent their earlier adulthood living off the desert and sea. Rapidly accelerating acculturation began in the 1950s. However, many traditional practices and some hunting and gathering were still evident in 1983, even though modern houses and pickup trucks had become a part of Seri life.

We have concentrated our studies on traditional practices and information, and in our opinion most Seri knowledge of the plants, animals, and their environment is their own.

Although hunting and gathering society covered most of human history, it remained poorly understood until relatively recently. With pioneer works, such as those of Lee and Devore (1968), there has been a renewed interest in and better understanding of hunter-gatherer life. However, detailed information of hunter-gatherer knowledge of plants and animals has remained elusive, partly because of loss of the information due to acculturation—particularly for North American peoples.

This book is primarily an ethnobotany—the botany of a culture other than ours. It is a book which addresses several basic questions: What knowledge did hunting and gathering people have of natural history and the plant world? More specifically, what did desert people, the Seri, know of the natural world? How were plants used, classified and named, and incorporated into their lives?

Even during our first exposures to Seri culture it became apparent that Seri knowledge of their cultural history, including the local plant world, was vast. It was not possible to delve into this body of ethnobotanical information without touching upon almost every other aspect of Seri life. Especially in Parts I and II of this book we have summarized and condensed these other aspects of Seri culture in order to place the ethnobotanical information, mostly in Part III, into a meaningful context. By concentrating on our own areas of expertise and training—botany and linguistics—we have been able to record and interpret a slice of the world's information which otherwise might have slipped into oblivion.

The information was obtained mostly in the Seri language, and over the years we have been able to verify the data repeatedly. Information obtained in Spanish almost always was verified in the Seri language. For the most part what we report here was common knowledge among the older people, and was often discussed by them. Additionally, many traditional practices were still being carried out and we were able to learn by observation and in some cases by participation.

At one point early in the study we attempted a standardized approach to data gathering, but it was a disaster. As one Seri woman said, "Do you want to do this your way or ours?" As much as possible we tried to listen as any good student does, and not ask too many questions, or ask open-ended questions. However, during the later years of the study we were able to ask meaningful, specific questions about things which we should have learned much earlier. We certainly have had our share of the classical problem of the ethnologist learning something important at a late date because earlier, "You didn't ask."

At first we learned Seri names of plants and animals, and observed the many still extant uses the people made of them. We often went on outings with several people at a time who gave us names, uses and other information. Elderly people were pleased that we were writing down information, stating that they felt it would otherwise die with them. A plant was often brought to us because someone thought that we did not yet have information about it. In addition, over the years we were able to observe repeatedly the use of plants and animals in the context of everyday life.

Whatever success we have had is owed to the Seri and, in no small degree, to the fact that the data gathering was fun for all concerned. However, it should be clearly understood that in Seri culture, at least during the time of our studies, a price was paid for any service rendered except within the usual kinship sharing system. Payment was proportional to the task performed, but generosity did not interfere with obtaining accurate information.

The Seri orthography in this book conforms as nearly as possible to Spanish conventions. It is the same as that used by the Seri to read and write in their own language, with the exception that we have added hyphens in compound words and written stress accents as a pronunciation guide for non-Seri speakers. An accent (ˊ) is shown when the stress is not on the first syllable of the word or on the first syllable of the second word of a hyphenated compound.

For the sake of convenience, we have also revised quoted passages and words from earlier works to conform to this style. Unless otherwise indicated, all translations are ours, or based on those of Edward Moser or Stephen Marlett.

Throughout the book the gloss, or word-for-word translation, is indicated by single quotation marks (' . . .'), and the free translation by double quotation marks (" . . ."). In order to broaden understanding of the information, we have sometimes amplified or modified the gloss. For example, *tis* is an unanalyzable word for catclaw (*Acacia greggii*) or for harpoon point. We gloss *tis* as 'catclaw' or 'harpoon point,' depending upon intended meaning. In many cases we provide the scientific name in the gloss and the common name in the free translation. If there is no gloss or free translation for a Seri term in this book, it is an unanalyzable term or one which we were unable to translate.

In different parts of the book, particularly in the species accounts and tables, plants not native to the original region of Seri occupation are indicated by an asterisk (*). This category includes: plants naturalized in the region since post-contact times, e.g., tree tobacco (*Nicotiana glauca*); plants cultivated in adjacent regions, even since ancient times, e.g., corn (*Zea mays*) and domesticated tepary (*Phaseolus acutifo-*

lius); and exotic crops such as wheat (*Triticum aestivum*), and store-bought vegetables, e.g., potato (*Solanum tuberosum*). We omit repeated references to our own earlier publications; these are listed in the literature cited. In general only primary published sources are cited.

Botanical nomenclature and taxonomy are based on our interpretation and understanding of the existing literature and regional flora. For the most part synonyms given by Wiggins (1964) are not repeated here. In many cases only one species of a polytypic genus is covered in this book. In such cases only the generic name might be used in the text. For example, *Pachycereus* refers to *P. pringlei*, since it is the only species in that genus listed in the species accounts. Likewise, we often omit the specific epithet for a monotypic genus, e.g., *Carnegiea*.

In Part III the species accounts for the marine algae, non-vascular land plants, and ferns and fern relatives are listed separately (Chapter 16), and arranged alphabetically by genus and species. The species accounts for the flowering plants (Chapter 17) are arranged alphabetically by family, genus, and species (with the exception of the Cactaceae, in which the columnar cacti are treated separately from the rest of the cacti; otherwise, the cacti are also arranged alphabetically by genus and species). Scientific names for plant families are very conservative for the benefit of non-botanists. For example, we place the Mimosaceae, Caesalpinaceae, and Papilionaceae into a single family listed as Leguminosae rather than Fabaceae.

Some of the dates given in the text are approximate. Many dates were determined by relating events told to us by the Seri to known dates of various other incidents. We were also usually able to ascertain if a given event in their oral history occurred before or after a recorded historical date. We compared ages of individuals, since usually a person knew that he was younger or older than another. By these means the Seri helped us date many older photographs; they also helped us identify the people in the photographs. The Seri often indicated a size for a plant or other object with their hands,

and we have given these in approximate metric measurements.

Throughout the book we have generally used the past tense, even though many practices and customs were still in use at the time of writing. We have done this because in many cases it would be difficult to state which customs and practices were for certain no longer resorted to, at least on occasion. Conversely, it was impossible to predict when extant practices might cease, if at all.

Some artifacts illustrated in the text are replicas, or objects made for us or at our request, or items made to be sold to tourists. Others were made and used by the Seri themselves. However, these distinctions were not always clear. Using Seri material culture as a model, Schindler (1981) provided a methodology to distinguish artifacts and practices of indigenous origin from those affected by outside influence.

One major point needs to be emphasized. The Seri are highly individualistic, and different people often have different versions and ways of doing things. This may even be a characteristic of hunting and gathering people in general. For example, there are different and sometimes contradictory versions and information concerning the origin myth and the leatherback fiesta. This does not necessarily mean that one version is correct and the other wrong. However, there are overall patterns and in many cases the information is amazingly uniform. The Seri are highly pragmatic, and use of a given plant is usually more a matter of practicality than cultural preference or custom. In many cases we report that a given species was used for a certain purpose. This information does not mean that other species were not also similarly used. Negative ethnobotanical information is generally inconclusive. A given species or material usually was chosen because it was available and served the intended purpose. However, some variation in information or in a given plant name or term may be related to different geographic and dialect backgrounds of the people.

Herbarium voucher specimens other than marine algae are at the University of Arizona and the Instituto Politécnico Nacional, with duplicates in other herbaria in Mexico and the U.S. Marine algae are deposited in the Algal Collections, United States National Herbarium, Smithsonian Institution, with duplicates distributed to herbaria in Mexico and elsewhere.

In many cases our studies confirmed earlier reports, and these have been cited in the text. One of the classic works of the Seri dates from 1692, when Adamo Gilg, a Jesuit priest, wrote a long letter from the Seri mission of Pópulo. This letter, expertly translated from German by Daniel Matson (Di Peso and Matson 1965), contains the earliest detailed descriptions of the Seri, including some valuable ethnobotanical information. We have found Gilg's descriptions to be accurate, at least for those cases in which information could still be confirmed.

Between August 9 and August 13, 1826, Lieutenant Robert William Hale Hardy, an Englishman surveying Mexico for pearl fisheries, made several brief visits to a Seri camp at Tecomate, on the north shore of Tiburón Island. The encounter was friendly and his first-hand observations are well described in the account of his extensive travels in Mexico, first published in 1829.

The first anthropological study of the Seri was done by W. J. McGee, who visited the Seri region from November 1 to November 4, 1894, and again from December 1, 1895, to January 2, 1896 (McGee 1894–1896; Fontana and Fontana 1983). During his first visit he spent less than two days with some Seri camped at Rancho Costa Rica. These were defeated people, who had suffered great losses during the previous decades. During his second visit the Seri eluded McGee's party and he never made contact with them, although he went to Tiburón Island, visited Seri camps, collected many artifacts, and had harrowing adventures (Carmony and Brown 1983). The eminent anthropologist Alfred Kroeber had this to say about McGee:

The Seri . . . are often considered the wildest and most primitive tribe surviving in North America. They owe this repute partly to McGee's monograph about them. . . . It is easy to read between the lines of this description that McGee leaned toward a romantic and imaginative interpretation of the Seri (Kroeber 1931:3).

However, there is a great deal of valuable information if one can separate McGee's direct observations from his wordy opinions, preconceived ideas, and second-hand information.

Charles Sheldon, a hunter, naturalist, and talented writer, made an unprecedented trip to the region between December 1, 1921, and January 10, 1922. His journal, published in 1979, gives an unbiased first-hand account of the Seri. He camped with the Seri on Tiburón Island without a guide, and communicated with them in Spanish. His account of the Seri is the most accurate one available for that time period in Seri history.

Alfred Kroeber, an astute, professional anthropologist, visited the Seri for six days in spring, 1930. His 1931 monograph is the first truly scientific account of the Seri, and, by and large, it contains accurate but necessarily fragmentary information. However, he concentrated on social and religious aspects and, unfortunately, he was not able to communicate adequately. His principal source of information was a Seri man named Chico Romero.

Between 1922 and 1939, Edward H. Davis made six visits to the Seri. During this time he collected many artifacts for the Heye Foundation in New York. Also at the Heye Foundation are his diary and more than 450 photographs taken in the Seri region. Portions of his diary and some of the photographs were published by Quinn and Quinn (1965). Davis's journal contains valuable first-hand observations. However, like McGee before him, Davis relied far too heavily on second-hand, local, prejudicial information. At the time no outsider spoke Seri. We have generally referred to the published Quinn and Quinn excerpt of Davis's journal because it is much more accessible. However, the Quinns did not follow the original chronological order of the journal exactly, and it is sometimes difficult to ascertain time and place without reference to the original journal. By courtesy of the Heye Foundation, with the help of S. A. "Alex"

Alexandride, a number of the Davis photos are reproduced here, some for the first time.

William Griffen's 1959 monograph of the Seri provided the first broadly based modern study of Seri culture. Thomas Bowen (1983) has provided a concise overview of the Seri and the best general account in Spanish was given by Nolasco (1967). In 1976 Edward Moser published an extensive bibliographic account of the Seri.

Acknowledgments

Many Seri have contributed information for this book, and in this regard we especially thank Victoria Astorga, Lolita Blanco, Ramona Casanova, the late María Luisa Chilión, Carlota Colosio, María Antonia Colosio, Pedro Comito, Rosa Flores, Roberto Herrera, Rosita Méndez, the late Jesús Morales, the late Chico Romero, Elvira Valenzuela Félix, and the late Sara Villalobos.

We have examined Seri ethnographic materials in various museum collections, including: the Arizona State Museum at the University of Arizona (Tucson); Amerind Foundation (Dragoon); Heard Museum (Phoenix); Instituto Nacional de Antropología e Historia, Centro Regional del Noroeste (Hermosillo); Lowie Museum, University of California (Berkeley); Museo Nacional de Antropología (Mexico City); Museum of Navajo Ceremonial Art (Santa Fe); Peabody Museum, Harvard University (Cambridge); Southwest Museum (Los Angeles); Taylor Museum (Colorado Springs); Natural History Museum, Smithsonian Institution (Washington, D.C.); and the National Museum of Ethnology (Osaka). We thank the personnel of these museums for making the collections available to us.

Our research was greatly facilitated by grants from the National Science Foundation (SOC 75-13-628 and BNS 77-08-582). We are grateful for support and facilities provided by the Office of Arid Lands Studies and the Environmental Research Laboratory of the University of Arizona. Our special thanks to Bernard L. Fontana.

We have relied heavily on the generous assistance and encouragement of the late Edward Moser, Cathy Moser Marlett, Stephen Marlett, the late Alexander Russell, and Jean Straub Russell. Many others have helped us in our research, and in this regard we particularly thank the following people: Christopher H. Bailey, Thomas Bowen, Richard Brusca, Katina Bucher, Stephen Buchmann, Martha Chidester, Otis Chidester, Beatriz Braniff, Kim Cliffton, Dennis O. Cornejo, Stephanie Daniel, Sherry Dashiel, Imogene Davis, Irving Davis, Mahina Drees, Richard Ford, Rodney G. Engard, Robert Gasser, Lucretia Brezeale Hamilton, Julian Hayden, James Henrickson, James Hills, Ellen Horn, Jane Harrison Ivancovich, Donald Johnston G., Michael Mahar, Patrick Manley, Robert C. Moser, Gary Paul Nabhan, Carlos Nagel, Thomas Naylor, E. Tad Nichols, Nancy Nicholson, James N. Norris, Michael Owens, H. H. Patterson, Hilda Patterson, Doris Potwin, Edward Potwin, Adrian Rankin, Lynn Ratener, Amadeo Rea, Philip J. Regal, Frances Runyan, Cherie Ryerson, Scott Ryerson, Daphne Scott, Thomas Sheridan, the late Edward H. Spicer, Charles Stigers, Barbara Tapper, Donald A. Thomson, Alfredo Topete, Oscar Topete, Richard S. White, Harold Walton, and Vendla Walton.

RICHARD STEPHEN FELGER
MARY BECK MOSER

Abbreviations

The talent of many photographers and illustrators and the cooperation of several institutions have contributed to the illustrations for this book. Each is identified in individual credit lines below the illustrations as follows:

Illustrators
CMM—Cathy Moser Marlett
FR   —Frances Runyan
KM   —Kay Mirocha
LBH —Lucretia Brezeale Hamilton
NEW —Nancy Evans Weaver
NLN —Nancy L. Nicholson
SM   —Susan Manchester

Photographers
AHH —Alison Hyde Habel
BM   —Borys Malkin
CS   —Charles Sheldon
DLB —David L. Burckhalter
EHD —Edward H. Davis
ETN —E. Tad Nichols
EWM —Edward W. Moser
FFD —F. Faurest Davis
GHX —Gwyneth Harrington Xavier
HT   —Helga Teiwes
JDH —Julian D. Hayden
JWM —James W. Manson
LMH —Laurence M. Huey
MBM —Mary Beck Moser

MM   —Maron Meeks
NLN —Nancy L. Nicholson
RJH —R. James Hills
RMT —Raymond M. Turner
RSF —R. S. Felger
TGB —Thomas Gerald Bowen
WCS —W. Charles Swett
WD   —William Dinwiddie

Archives and Institutions
AF   —Amerind Foundation, Dragoon, Arizona
AH   —Allan Hancock Foundation, University of Southern California, Los Angeles
ARS —Archivum Romanum Societatis Jesu, Rome
ASM —Arizona State Museum, University of Arizona, Tucson
HF   —Museum of the American Indian, Heye Foundation, New York
NAA —National Anthropology Archives, Smithsonian Institution, Washington, D.C.
SD   —San Diego Natural History Museum, San Diego
UAL —University of Arizona Library, Special Collections

PART I

The People and the Setting

Figure 1.1. A group of Seri in September, 1958. From left
to right, the adults are Roberto Camposano, María Burgos,
María Morales, Amalia Burgos, María de la Luz Moreno,
and Adolfo Burgos. *JWM; ASM-25114*.

1. The People

The Seri are gregarious, outgoing and aggressive, and have a sharp sense of humor (Figure 1.1). They tend to be highly independent, nonconforming, and quick to adjust for the sake of convenience. In their harsh and highly variable environment such versatile and pragmatic behavior was undoubtedly of major adaptive value. The total population of the various Seri groups was probably never more than several thousand (see McGee 1898:135). Due to warfare and disease their territory and population dwindled until there were fewer than two hundred by the 1930s. Since then their population has increased and reached five hundred in 1982. These hunting and gathering, seafaring people call themselves the *Comcáac* 'the People.' They have long lived along the coastal desert of Sonora, Mexico (Map 1.1).

Most Seri had access to marvelous marine resources which the world will never again know. Others had little access to marine resources and depended on the desert for their subsistence. However, the single overriding problem was the scarcity of drinking water. Social groups had to be small and mobile to survive.

The traditional settlement pattern generally consisted of temporary camps of several or more extended families. It seems to have been commonplace to occupy a campsite for a month or more. Camps were relocated as local food and fresh water resources went out of season or were depleted, or after a rain so that there would be sufficient water en route. Other reasons also influenced choice of campsite. For example, El Desemboque was a favorite summer location because of the relative absence of biting insects. Since climate and resources vary from year to year in this desert region, the same camps were not necessarily reoccupied each year. There are more than four hundred Seri names for campsites, water holes, and other local geographic features.

Like most people with very low population density, the Seri had essentially no political organization other than a local, temporary war chief. There were no leaders or spokesmen until interactions with Spanish and Mexican authorities created a need for them.

Culture and Society

The extended family formed the center of Seri social life. The complex kinship system (Di Peso and Matson 1965:49), with more than sixty distinct terms, shows relatively close affinity with Yuman systems (Kroeber 1931:9). Several obligatory customs associated with the Seri kinship system placed strict social controls on the members of the extended family. Each person had an obligation to share one of two classes of goods: material goods or food. When available, these were to be shared with specified members of the extended family, who were obliged, in turn,

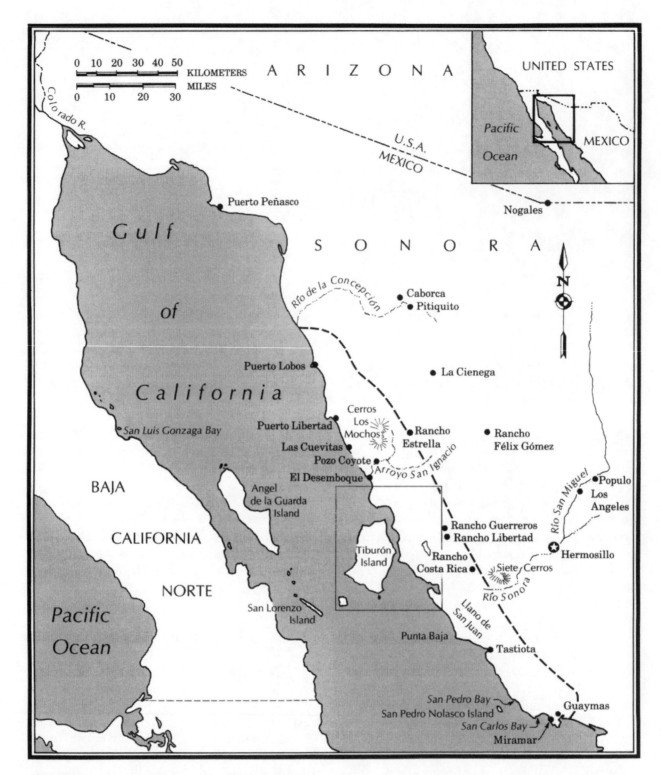

Map 1.1. The Seri Region in northwestern Mexico. The original area of Seri occupation closely approximates the Gulf Coast of Sonora subdivision of the Sonoran Desert as defined by Shreve (1951), which stretches from near Puerto Lobos to Guaymas (area west of dashed line). The enlargement on the right shows details of Tiburón Island and the mainland, which are separated by the Infiernillo Channel. KM

4

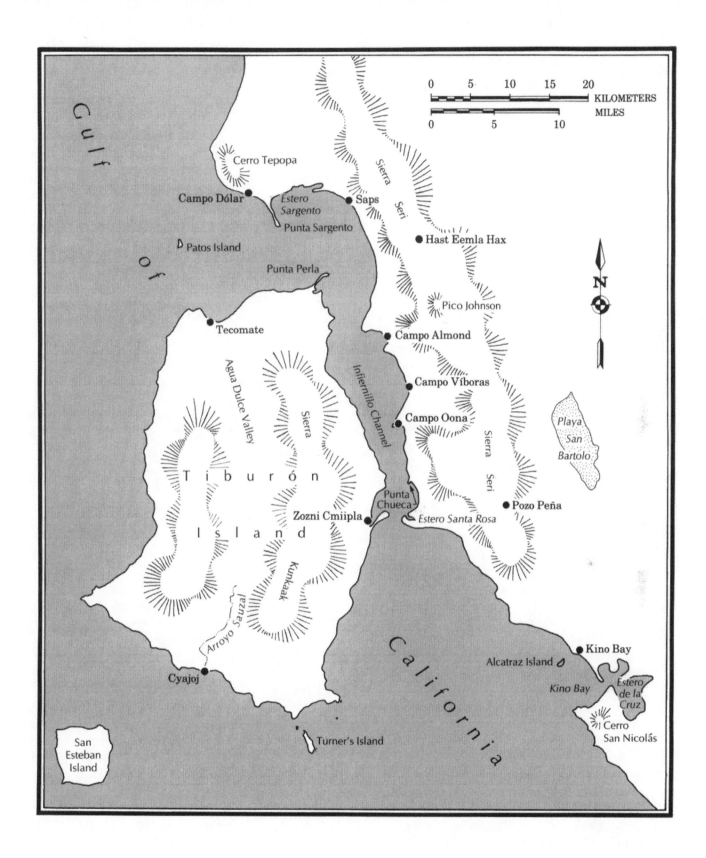

to share the opposite class of goods when available. Although this custom insured distribution, women bore the greater burden since they were obliged to share items such as meat and other food more often than men. In the past a buildup of individual wealth would be tantamount to acknowledging stinginess, a transgression of which no one wanted to be guilty.

Another custom providing social control involved nonspeaking relationships within the extended family. Each person was prohibited from speaking directly to certain relatives. In this matter men were more restricted than women. A man, for example, was not to speak directly to his father, uncles, brothers, or children after they reached puberty, nor with his parents-in-law or most of his other in-laws. In 1692 Adamo Gilg wrote that "no father-in-law may talk with his son-in-law, nor the latter with the former, nor may the one come near the other" (Di Peso and Matson 1965:52). Kroeber (1931:8) likewise added that "Neither husband nor wife looks at nor addresses either parent-in-law." Such behavior undoubtedly reduced the chance of friction among family members. However, as with so many Seri cultural ideals, the rules were often broken, at least since about the 1930s (Kroeber 1931:8). When communication between nonspeaking members of the family was necessary, it was accomplished by several means. For example, a man might tell his wife something he wanted his son to know, and she could convey the message. Since family members lived in close contact, most conversations were heard by everyone present.

Ceremonial fiestas or social gatherings were held to celebrate a number of events. These events were non-calendric and non-predictable, effectively preventing a large gathering at any one time and place. The meager local freshwater resources could not support large gatherings. In the traditional fiesta held after the capture of a leatherback sea turtle, one of the first functions to be performed by the hunter upon his return to camp was to set out to carry back water for the celebration. Like most native North American peoples, the Seri regarded four as a sacred number. Fiestas lasted four days, and most dances and songs were performed in multiples of four.

The puberty fiesta was the most common celebration. It included dancing, singing, games, and feasting. Since the early part of the twentieth century, it has been held only for the girl, who remained in seclusion, her face painted with a traditional design by her fiesta sponsor. She abstained from eating meat and remained awake during the final night. Before dawn female relatives took the girl to the beach and ceremonially purified her by washing her hair with sea water. She was then of marriageable age. The last puberty fiesta for a boy was held about 1923.

Marriage among the Seri is monogamous, but once included polygyny, at least during the period of severe hostilities in the nineteenth century. There was a taboo against marriage among family members, including cousins (Griffen 1959:27; Sheldon 1979:98). Marriage was usually initiated by the parents of the two young people and in the past a shaman was often consulted. After the marriage proposal was accepted by the girl's family, a period of six months to a year or more elapsed during which the young man's family presented a series of gifts to her family (Sheldon 1979:98). The bride price in the 1970s often included a boat, radio, money, food, yardgoods, basket-making materials, ironwood carvings to be sold, or even a pickup truck. The marriage took place when the girl's family gave permission. Since about the 1960s the ceremony usually has taken place in the local Apostolic church and has included formal vows. However, in earlier times no such vows were exchanged. The young couple usually began married life in a house built near the home of the boy's parents. The husband was obliged to help maintain his parents-in-law as long as they lived. Divorce was rare.

The *hamác* system consisted of "several roles that are called into play in certain contexts where supernatural danger is involved (Griffen 1959: 42–46)." Each family had another family which was its *hamác*. The *hamác* family might or might not be related, and one's *hamác* was apparently inherited, and basically patrilineal. Duties of the *hamác* family included sponsoring the puberty fiestas and burials. The fiesta sponsor was simply called *hamác*, while the burial sponsor was called *hamác hacátol* (the latter term is related to *cacátol*, the word for dangerous).

In earlier times, a returning warrior handed over booty taken from an enemy to his burial sponsor. The warrior was thus protected from the dangerous supernatural powers that such objects possessed. The burial sponsor, in turn, was obligated to trade back other goods "safe" to the warrior in compensation for the items of plunder (Griffen 1959:43).

Burial was performed by the deceased's burial sponsor, who was a specified male relative. He painted his hands, either black or white, to avoid the dangers involved in handling the corpse (Figure 1.2). A few of the deceased's most-used possessions were buried with him. Sometimes cholla (*Opuntia*, e.g., *O. fulgida*) was put in a partially filled grave, and then more cholla, rocks, shells, and thorny brush were heaped over it (Figure 1.3). McGee (1898:288) reported graves with heaps of cholla pinned together with ocotillo

Figure 1.2. Burial sponsor, Fernando Romero, with his hands painted to avoid danger. *ETN, El Desemboque, March 1951.*

Figure 1.3. A grave at Kino Bay with thorny brush to protect it from coyotes. *JDH, 1941; ASM-21017.*

(*Fouquieria splendens*). This practice discouraged coyotes (Quinn and Quinn 1965:211).

The house of the deceased, traditionally a temporary structure usually constructed of ocotillo poles and brush, was burned. If a deceased man had owned a balsa (a reed boat, see *Phragmites*), it too was burned. The remaining possessions became the property of the burial sponsor, who, in turn, was obliged to give all of his equivalent possessions to the bereaved family. This exchange eliminated the spirit power that was said to contaminate the deceased's possessions from the time of death. However, since the middle of the twentieth century the influences of Christianity, the local church, and changing economic conditions have affected practices associated with death (Griffen 1959:43–46).

Language

In earlier times there were six major geographic groups of Seri speaking three mutually intelligible dialects (E. Moser 1963a; see also discussion of the Seri regions in Chapter 6). Members of the various groups were generally on friendly terms, and marriages and other social exchanges took place. However, there was also frequent violence between neighboring groups, as well as with non-Seri neighbors. Since the latter part of the nineteenth century, the Seri, living mostly on Tiburón Island and nearby mainland regions, have been an amalgamation of survivors from the various geographic regions.

The Seri language is classified as a member of the group of language families known as the Hokan stock (Langdon and Silver 1976; Marlett 1981). Since modern Seri does not bear close affinity to other Hokan languages, it is considered to be, together with extinct Serian dialects, a language family in itself. The closest linguistic family is believed to be that of the Yuman groups of northwestern Mexico and the southwestern United States, which includes such languages as Cocopa, Diegueño, Havasupai, Kiliwa, Maricopa, Mojave, Paipai, Quechan (Yuma), Yavapai,

and Walapai. Seri is an agglutinative language, highly inflected and phonologically complex, characterized by verbs with many prefixes and suffixes. There are frequent glottal stops and consonant clusters. A simplified key to pronunciation and the corresponding international phonetic symbols for Seri are presented in Table 1.1. The first language of the Seri people is still their native one, and linguistic borrowing remains minimal. Spanish is their second language.

TABLE 1.1.
Key to Pronunciation for Seri Words

Seri Orthographic Symbol	International Phonetic Association Symbol	English Approximation
a	a	*father*
c	k	*cup*
e	æ	*cat*
f	ɸ	*fin*
h	ʔ	"*oh* oh!" (glottal stop)
i	i	mach*i*ne
j	x	voiceless velar fricative (as in Scottish 'Lo*ch*')
l	ɬ	voiceless fricative *l* (as Welsh *ll*)
m[1]	m	*map*
n	n̪	*never*
o	o	*no*
ö	w̥	*which*
p	p	*pin*
qu	k	*cup*
r	r	(as in Spanish 'pe*ro*')
s	s	*sit*
t	t̪	*tip*
x	χ	voiceless uvular fricative (as in French 'let*tre*')
y	j	*yet*
z	ʂ	*ship*

Note: Double vowels and consonants represent long sounds. Stress in Seri words is usually on the first syllable.

1. *m* in unstressed syllables assimilates to the point of articulation of the following consonant. For example, the *m* in com*cáac* 'people' is pronounced as *ng* in the English word *sing*.

m following *c* or *cö* is pronounced as a nasalized *w*, as in cmi*que* |kw̄ikæ| 'person.'

Prehistory

According to Bowen (1976:94), the first people who lived along the central coast of Sonora

> . . . arrived long before the introduction of pottery, and it appears that there may have been a succession of distinct preceramic peoples, including the bearers of the Clovis, San Dieguito, and Amargosa cultures. . . . The entire ceramic occupation of the central coast constitutes a single continuous tradition culminating in the contemporary culture of the Seri.

Prehistoric Seri pottery, called Tiburón Plain, was a hard, well-fired plain ware made without organic temper. This unusually thin pottery, also known as "eggshell pottery," ranged from 2 to 5 mm in thickness, and averaged about 3 mm. It was made thin by scraping the clay with a clamshell while it was still damp (Bowen and Moser 1968:99–100). The ones used for water transport were among the largest and thinnest utilitarian pottery vessels in the world (see Chapter 6). By 1983 there was only a single radiocarbon date of an archaeological site associated with Seri culture. A burial site found along the Infiernillo coast in 1974 contained eelgrass and Tiburón Plain pottery. The eelgrass was dated at nearly 2,000 years of age (see *Zostera*).

The Seri undoubtedly had cultural contacts with the Cochimí and other Baja California people. On clear days mountains in Baja California are visible from the central coast of Sonora, and the greatest open-water distance, from San Lorenzo Island to Baja California, is only 19 km. The Seri regularly crossed the 10-km channel between San Esteban and Tiburón islands. Spanish colonial documents and Seri oral tradition indicate that the Seri made trans-Gulf voyages in their reed boats, or balsas (Bowen 1976:95–97). The vegetation, physical environment, and resource bases of the Seri and Cochimí regions are more similar to each other than to those of adjacent areas, and, according to Shreve (1951), both regions constitute the Gulf Coast subdivision of the Sonoran Desert.

Neighboring Indian groups consisted of Piman and Cahitan (Yaqui and Mayo) peoples. Piman-speaking neighbors of the Seri were the Papago to the north and northeast and the Lower Pima to the east and southeast. On the south and southeast, Seri territory abutted lands held by the Yaqui, and according to Spicer (1980:167 and personal communication) the distinctive Cerro Tetas de Cabra at San Carlos Bay was the mutually recognized boundary. In addition, there was probably some contact, and even trade, with the Opata living along the San Miguel River near the eastern edge of Seri territory (see *Zea*). In prehistoric times various peoples—such as those from southern Arizona and Casas Grandes, Chihuahua, or middlemen traders—came to the shores of the Gulf to obtain sea shells and salt (Di Peso 1974, III:500–505, 627–629; Hayden 1972; Haury 1976:305–308), and they most likely encountered the Seri. However, Bowen (1976:66) stated that "the near absence of foreign ceramics other than Trincheras pottery at central coast sites, in conjunction with the scarcity of Tiburón Plain outside the central coast, suggests that the only significant contact maintained with outsiders was with the Trincheras people." The Trincheras people lived in Sonora, north of the Seri region.

The Seri distinguished between two kinds of ancient Giants. "The *hant ihiyáxi comcáac* 'land its-edge people' were of enormous stature and are said to have lived in Baja California. Apart from one legend recorded by Griffen (1959:19), the Seri know little about these beings (E. Moser in Bowen 1976:105)." The second and more important group of Giants were the *xica coosyatoj* 'things singers.' They were undoubtedly the ancestors of the contemporary Seri.

According to Seri oral tradition, these Giants lived in Baja California as well as on Tiburón Island and the nearby mainland region when the Seri first appeared on the scene. These Giants continued to live on the Baja California peninsula and in Seri territory for some time afterward. Many met their demise through gambling, in which they bet their lives, or in duels. Some died in a great flood or they were transformed into rocks and various animals or plants (see boojum tree, *Fouquieria columnaris*). In another version, the survivors intermarried with the Seri or retired to Baja California (Griffen 1959:19; Moser and White 1968:146; Bowen 1976:105).

The Seri were able to describe the *xica coosyatoj* Giants and their culture in some detail (Bowen 1976:105–106). Despite their superhuman size, they were said to have eaten the same foods as the Seri and used many of the same implements. However, they also made certain artifacts and practiced certain customs that the Seri claimed they did not share. Some of the disclaimed artifacts are known archaeologically and even historically. These include gyratory crushers (see *Prosopis*), disk beads (Bowen 1976:87), Tiburón Plain or classic Seri "eggshell" pottery (Bowen and Moser 1968:127–128), nose ornaments and lip plugs (Och in McGee 1898:78; Hardy 1829:286; Pfefferkorn 1949:81; Di Peso and Matson 1965:53), and ceramic figurines (Moser and White 1968:47–48).

The association of outmoded aspects of Seri culture to the Giants also applied to language (Bowen 1976:106; E. Moser 1963a:17–18; Moser and White 1968:147). A limited amount of "Giants' speech" known to the Seri was actually archaic Seri speech. The Seri described Giants' speech, as well as the extinct San Esteban dialect, as "musical" and referred to both as "singing talk." The extinct dialect of the Sargento people was characterized by extreme intonational contours, and that of the San Esteban people by sharp pitch contours and heavy stress. The contemporary Seri regarded the San Esteban people as backward and primitive (E. Moser 1963a:16).

Spanish Colonial Period

During the Spanish explorations of the sixteenth century, Spaniards made contact with neighbors of the Seri and may have briefly encountered the Seri themselves. Some of the Seri occasionally journeyed to the *rancherías* of settled agricultural people, such as the Eudeve and the Opata, to trade salt and hides for agricultural produce. It is possible that a few of the Seri also made their way to Spanish settlements in Sinaloa. In any case, the sixteenth-century record clearly demonstrates that the Spaniards knew of people living along the coast who did not practice agriculture, were nomads, tall in stature, and had reed boats or rafts called balsas (see *Phragmites*). These characteristics fit the Seri.

In December, 1535, Alvar Núñez Cabeza de Vaca and his companions, near the end of their epic wandering across the continent (Oviedo 1972:258), heard about people who were probably the Seri. He reported:

> . . . *la costa no tiene maiz y comen poluo de bledo y de paja y de pescado que toman en la mar con balsas, porque no alcáçan canoas* (Núñez 1555:45).

> . . . on the coast there is no maize, and they eat powder of "*bledo*" and straw, and fish that they catch in the sea from balsas, because they do not have canoes.

The reference to non-agricultural people ("no maize") and the balsa is clear. The phrase "*polvo de bledo y de paja*," which might be translated as "flour of greens and grass," may refer to flour made from wild plants, especially eelgrass (*Zostera marina*).

In the fall of 1539, Francisco de Ulloa, captain of one of the three ships of exploration sent out by Cortez, sailed up the Sonoran coast to the Colorado River and then southward along the east side of Baja California. Ulloa saw no people from Guaymas Bay northward to the Colorado River and did not believe that such a land could be inhabited. Since he apparently sailed up the relatively barren west side of Tiburón Island, it is possible that he would not have encountered the Seri along such a course. Nevertheless, he and his men did come across what were apparently Cochimí Indians along the coast of Baja California in the vicinity of San Luis Gonzaga Bay (Wagner 1929:20, 306–308).

Ulloa's description of their material culture, along with subsequent missionary descriptions of the Indians living in the Central Desert of Baja California (Aschmann, 1959; Clavigero 1937; del Barco 1973), indicates similar cultural adaptations to the desert and sea on both shores of the Gulf of California. However, the Baja California Indians had "hooks of tortoiseshell (Wagner 1929:22)." Although the Seri utilized tortoiseshell (hawksbill turtle) there is no evidence that they had fishhooks (McGee 1898:194; Kroeber 1931:19). Ulloa's description of a Baja California balsa matches that of McGee (1898) and others for the Seri three and a half centuries later (see *Phragmites*). Although the Spaniards did not speak to the Baja California Indians, Ulloa wrote, "We judged these people to be nomads, of little intelligence (Wagner 1929:22)."

Since the earliest Spanish chronicles of the sixteenth century, one of the prominent features of the reports was vivid accounts of a dreaded arrow poison. This was almost certainly the arrow poison made from a common desert shrub, *hierba de la flecha* (*Sapium*).

The first record of a variation of the name "Seri" is that of the Jesuit missionary and administrator Padre Andrés Pérez de Ribas, in his classic *Triunfos de nuestra santa fe*, originally published in 1645. Although Pérez de Ribas probably never visited the Seri, as a missionary to the Yaqui he certainly heard about them. The southernmost Seri group, the Guaymas, had already become missionized in Yaqui territory. However, the Spaniards undoubtedly considered the Guaymas people to be different from other Serian groups. According to Pérez de Ribas (1944, II:148):

> . . . *hay noticias de gran gentío de otra nación, que llaman Heris: es sobremanera bozal, sin pueblos, sin casas ni sementeras. No tienen ríos, ni arroyos y beben de algunas lagunillas y charcos de agua: sustentándose de caza; aunque al tiempo de cosecha de maíz, con cueros de venados, y sal que recogen de la mar, van a rescatarlo a otras naciones. Los más cercanos destos a la mar, también se sustentan de pescado y dentro de la misma mar, en isla, se dice, que habitan otros de la misma nación, cuya lengua se tiene por dificilísima sobremanera.*

. . . there are notices of a great multitude of another nation, called the Heris. This nation is excessively wild, without pueblos, houses, or fields. They have no rivers or arroyos and drink from small ponds or *charcos* of water. They sustain themselves by hunting, although during the time of the maize harvest, they go with deer hides and salt, which they gather from the sea, to trade with other nations. Those closest to the sea also sustain themselves with fish. Within the same sea, on an island it is said, live others of the same nation, whose language is held to be extremely difficult.

Pérez de Ribas (1944, I:128) also described the use of a marine grass seed (see *Zostera*). This knowledge indicates that the Spaniards were in possession of relatively detailed information about certain aspects of Seri life by the early seventeenth century (Sheridan and Felger 1977), although contacts between the Seri and the Spanish remained infrequent through the middle of that century.

By the 1660s, however, groups of Seri had already adopted a pattern of petty livestock raiding, which brought them into hostile contact with Spaniards living in the San Miguel drainage. In 1662 several hundred Seri were killed in the desert west of Ures (Spicer 1962:105). During the seventeenth and eighteenth centuries many Seri augmented their hunting and gathering subsistence by raiding Spanish settlements and missions. European livestock became new and less wary types of game animals for the Seri. Raiding and rustling were an example of Seri opportunism, a flexible economic response to the introduction of new resources along their frontier. They were able to take advantage of these resources without radically changing either their social organization or their settlement patterns.

Some Seri groups had little or no contact with Europeans, and were not involved in raiding. This situation made little difference to the Spanish authorities, who neither knew how to deal with nor understood the anarchistic Seri. Within northwestern New Spain the Seri became an enclave of resistance to Spanish encroachment. Only the Apache to the north gave the Spaniards more trouble.

The Guaymas region was the only part of Seri territory ever permanently occupied by the Spaniards. The Guaymas people apparently welcomed the Spanish, and seemed to adapt to European life more quickly than the other Seri groups. By the late 1620s the Guaymas people seem to have taken up mission life with Lower Pimas and Yaquis in the multi-ethnic village of Belem (Spicer 1962:106).

The other Seri groups proved more intractable. In the late seventeenth and eighteenth centuries the Jesuits established missions for the Seri along the San Miguel River, more than one hundred kilometers—and a journey of two to four days—east of the arid coastal desert (Griffen 1961). These missions, on a permanent stream, attracted a minority of the Seri, who temporarily adopted agricultural mission life. The first mission, Santa María del Pópulo, was established by Father Juan Fernández in 1679. Four years later Fernández was transferred, the mission was devastated by an epidemic, and the survivors fled into the coastal desert. Adamo Gilg, a Moravian Jesuit, reestablished the mission in 1688, and it survived until the collapse of the Jesuit Seri mission system in 1748 (Di Peso and Matson 1965; Sheridan 1979). In 1692 Padre Gilg wrote to a fellow Jesuit at Brunn, Moravia. In the letter he included a map with an inset drawing (Figure 1.4), which is the earliest illustration of the Seri.

Jesuit missionaries and the Spanish frequently complained about the "wild" Seri, who drifted in and out of the missions to steal food and material goods. Gilg, in fact, claimed that the Seri were originally attracted to the missions along the San Miguel and Sonora Rivers for the purpose of raiding the missionized Lower Pimas:

> . . . [they] attached themselves to the Pimas, not from love for these their ancient neighbors, who also had moved southeast into newly constructed villages at the instigation of the Missionaries, but out of desire for booty that they could obtain by plundering the newly converted Pimas. Because of these attacks they drew down upon their necks Spanish soldiers under whose protection the Pimas stood . . . (Di Peso and Matson 1965:41).

The involvement of the opportunistic Seri with mission populations probably included kinship sharing, as well as outright pilfering and raiding. While a few Seri did become farmers and Christians, most did not capitulate to Spanish customs or authority but remained as hunter-gatherers.

In 1748 action taken by the Spanish completely upset the program of missionization for the Seri and resulted in more than two decades of brutal guerrilla warfare. The Spanish presidio of Pitic (modern Hermosillo) was moved to San Miguel de Horcasitas, near Pópulo, where better farmland was available. The lands which the Seri at Pópulo and Los Angeles had been farming with moderate success were distributed among the Spanish residents of the relocated presidio. When the Seri protested, they were arrested and their women summarily deported to Guatemala and elsewhere in New Spain. As might be expected, these actions were too much for the Seri men to endure. They warned the missionaries, whom they did not wish to harm, and then attacked Spanish settlers throughout central Sonora (Spicer 1962; Sheridan 1979).

Prior to 1750 there had been numerous petty raids by the Seri, including revenge raids. By the 1750s, however, hostilities had escalated into open war, which then raged unabated until the 1770s. During this time there were several campaigns to exterminate the Seri, including an expensive and abortive invasion of Tiburón Island by General Diego Ortiz Parrilla in 1750 and the more effective campaigns of Colonel Domingo Elizondo in the late 1760s and early 1770s. No decisive battles were fought, but increasing Spanish pressure, Old World diseases, and the weariness of several decades of unrelenting warfare gradually wore down Seri resolve.

By the time of the expulsion of the Jesuits in 1767 and near the end of the Spanish colonial empire in the closing decades of the eighteenth century, it was clear that attempts to make the Seri accept an agricultural European lifestyle had failed. The bureaucracy-loving Spanish administered a global empire. The anarchistic Seri hated authority. The two cultures failed to communicate.

The Nineteenth Century

By the end of the eighteenth century the Seri had been pacified or driven into their desert refuges. Extensive Seri raiding did not recur until governmental authority was weakened along the northwestern frontier following Mexican independence. From the 1830s on, Seri and Apache raiders harassed Mexican ranches and settlements. Mexican settlers slowly established themselves on the margins of Seri territory and began to curtail Seri raids. From time to time groups of Seri were rounded up, brought into Pitic (Hermosillo) and forcibly settled at Villa de Seris

Figure 1.4. A group of seventeenth-century Seri on the move. Drawn by Padre Adamo Gilg in 1692 as an inset to a map of the Seri region. The man on the left may have an elaborate hairdo rather than a hat (Bowen 1983:236). He carries a single curve bow and a pottery vessel suspended in a net from a carrying yoke. The other man holds a double curve bow and wears animal pelts as a breechclout (described by Gilg as foxskins). The pelts are tied on with a waistcord. Some are wearing sandals and jewelry, including a nose ornament. One of the men and the child wear headbands, perhaps similar to the modern leafy wreaths. The woman carries a mat, probably woven from reed (*carrizo*) or palm. She balances a basket on a head-ring and on top of the basket is a baby on a cradleboard. Gilg gives "*himamas*" as the Seri name for the cactus fruit. The columnar cactus more closely resembles *echo*, *Pachycereus pecten-aboriginum*, rather than the ones from the coastal Seri region. *Echo* occurs near Pópulo, where Gilg lived, and bears edible fruit. The shape of the basket is very different from that of any known Seri work basket. *ARS (print courtesy of NAA).*

on the south side of the Sonora River. However, as had occurred time and again in the past, most of them eventually returned to their desert homeland.

In 1844 Pascual Encinas established the Costa Rica ranch in the Siete Cerros area (McGee 1898:109). Encinas attempted to improve relations with the Seri, thinking that by paying them wages to work on his ranch they would begin to change their ways. Encinas's stock drank from some of the same water holes and grazed some of the same range where the Seri hunted and gathered. Naturally, the cattle and horses were adopted as part of the Seri cuisine. Despite Encinas's economic plan, the Seri continued to feast on his stock and engage in petty thievery. He declared war. The Encinas Wars from 1855 until the late 1860s decimated the Seri. The cowboys had rifles and were mounted. The Seri never acquired horses. Pascual Encinas told McGee (1898:113) that during the dozen years of strife his men killed about half of the Seri.

The isolated San Esteban group of Seri, ignorant of European ways and innocent of raiding or hostility toward Spaniards and Mexicans, were wiped out by the military, possibly in the 1860s. With the total Seri population greatly reduced, the survivors held out in their desert homeland on Tiburón Island and the opposite Sonora coast. Mounted cavalry could scarcely follow them into their mountainous island refuge without knowledge of the location of remote water holes. The military commonly complained that when they did manage to find water holes, the water had been poisoned. However, by the end of the nineteenth century Seri resistance had been broken.

The earliest photographs of the Seri known to us were two taken in 1874 and one around 1880 (Figure 1.5). The 1874 photographs, taken by Henry von Bayer on board the U.S.S. Narragansett, show some Seri men, including a posed archer (see McGee 1898:opposite 201). William J. McGee arrived at the Costa Rica cattle ranch in 1894. Although he spent less than two days with the Seri, he was able to amass a sizable collection of artifacts and information. William

Dinwiddie, the expedition photographer, secured the most extensive nineteenth-century photographic series of the Seri. His 1894 photo of a Seri family is shown in Figure 1.6.

The Twentieth Century

The Seri were able to adapt to centuries of European pressure without radically changing their social structure or economy, and well into the twentieth century they were still huntergatherers. Agriculture and mining remained at the periphery of their still remote territory. Not until nearly the middle of the twentieth century, when isolation was no longer possible and large segments of their natural resources began to decline, did the Seri finally accept and have access to the conveniences and customs of the modern world.

Distrust and prejudice between Mexicans and the Seri remained strong during the first half of the twentieth century. In the late nineteenth and early twentieth centuries there were sporadic killings on both sides, a few of them involving American adventurers. In 1894 two North American adventurers were killed at Tecomate, at the north side of Tiburón Island (Robinson and Flavell 1894), and sensational accounts of an attack by savage Indians fanned the fires of hatred characteristic of the times. Outrageous popular accounts continued to feature the Seri as cannibalistic, immoral, and degenerate. Cannibalism has never been a part of Seri culture. The Seri may have killed outsiders, but they did not consume their victims.

The first Seri wooden boats were made at the turn of the century, and these small sail- and paddle-powered craft soon replaced the traditional reed balsas. In the 1940s outboard motors began to be used in conjunction with sails. Beginning in the 1920s, demands for marine resources—such as fish and, later, sea turtles—steadily increased from the growing towns and cities in Sonora and the adjacent United States. Kino Bay was established as a Mexican fishing village, and the Seri also began to engage in small-scale commercial fishing.

Figure 1.5. These Seri men at Guaymas were probably on a trading expedition. They wear cloth kilts and lounge on pelican pelts which they undoubtedly brought to trade. The men on the right wear the *coton* shirt. One man wears a leafy wreath. Note the gun barrel in center and liquor bottles. *The photograph is from the studio of Alfredo Laurent, Guaymas, c. 1880; ASM-53994.*

Partially as a result of nonhostile encounters, such as those with a small number of hardy tourists and field scientists, the Seri began opening their way of life to others during the 1920s and 1930s. Roberto Thomson Encinas, nephew of Pascual Encinas and resident of Hermosillo, became an important peacemaker and friend of the Seri (Spicer 1962:114). The Seri population, however, had sunk to less than two hundred.

North Americans built a sports lodge at Kino Bay in the 1930s, partially as an escape from Prohibition north of the border. Because of their commercial fishing activities, the Seri often camped at the periphery of Kino Bay. The Americans amused themselves by giving clothing and trinkets to the Seri, who quickly added begging to their subsistence repertoire. Shark fishing became locally important because the livers were purchased for the vitamin industry. Seri fishing techniques were modified for the new commercial efforts (Spicer 1962:115).

While the Seri adapted to some outside ways,

Figure 1.6. A Seri family camped at Rancho Costa Rica in November 1894. The man is dipping food from a basket; the woman on the left is Juana María. *WD; NAA-4276.*

by and large their traditional life continued as it had for centuries:

> During the 1930s nearly every Seri family acquired boats. Food, clothing, and tools available through the stores maintained by the Mexican fish traders began to be more and more widely used. Essentially Seri life changed little. The old form of brush shelter continued to be used and the traditional forms of social life and religion were maintained. Although Kino Bay became a much-frequented base for the whole group, families still spent periods on Tiburón Island and along the coast north and south of Kino Bay, roaming in the old nomadic way. Their new mobility by means of the plank boat gave them, in fact, a somewhat wider range than before (Spicer 1962 : 115).

In 1938 the Mexican government took an interest in the Seri, which for the first time did not involve attempts to resettle them, make them into farmers, or missionize them. A Seri fishing cooperative was established at El Desemboque and this ancient camp, known as *Haxöl Ihoom* 'clams their-resting-place,' or "The Place of the Clams," became a village. This action was supposed to remove the Seri from the "immoral" influences at Kino Bay, nearly 100 km to the south. With the synthesis of vitamins in the early 1940s, however, the shark liver boom and the Seri fishing cooperative collapsed. In the 1950s El Desemboque was still a small fishing village (Figure 1.7).

In 1951 Edward and Mary Beck Moser arrived at Kino Bay to begin their work with the Seri under the auspices of the Summer Institute of Linguistics. During the following year Richard Felger made his first field trip to Kino Bay to collect plant specimens. In 1952 the Mosers built their adobe house at El Desemboque. A paved road linked the Sonora capital of Hermosillo to

Arizona in the north and to the port city of Guaymas in the south. The 100-km-long road west from Hermosillo to Kino Bay was in the process of being paved, but at the time was still a dusty road requiring half a day to negotiate. Rich agricultural fields, irrigated with fossil ground water, were spreading across the coastal plain west of Hermosillo.

Kino Bay in 1952 was a fishing village with less than a hundred inhabitants, several bars, no school, and no electricity. The Seri frequently camped around the village and freely ranged northward to Puerto Libertad and also lived on Tiburón Island. The road northward from Kino Bay and the one over Noche Buena Pass were passable only by vehicles with high road clearance. The trip from Hermosillo to El Desemboque took all day.

Most Seri men still wore their hair long and had only recently added outboard motors to their sail-powered small wooden boats. Tecomate, at the north end of Tiburón Island, was a thriving Seri camp, and the desert and sea provided more than enough food. There was little need for store-bought food, although supplies such as flour, sugar, coffee, and outboard motor fuel were eagerly traded or purchased from Mexican store-keepers and fish buyers on the mainland. The coastal desert landscape was in primordial condition; the hand of man was scarcely visible. Clean beaches and desert stretched unbroken as far as one could see. Game was abundant and the sea teemed with fish and huge sea turtles.

The Mexican Apostolic Church, which began missionary work among the Seri in 1953, was successful in gaining a substantial number of Seri adherents in the late 1950s and early 1960s. The traditional, individualistic Seri religion, based largely on the vision quest and shaman-ism, nearly ceased to function in modern Seri society.

Increases in tourism, particularly after 1960, helped to create a shift to a cash economy for the Seri by the sale of their artwork. Through the

Figure 1.7. A portion of El Desemboque in March 1951. *ETN.*

Figure 1.8. Aerial views of the two Seri villages along the
coast of Sonora. A) El Desemboque. The trees are tamarisk
(*Tamarix aphylla*). B) Punta Chueca. *RJH, May 1977.*

work of the Mosers, Seri became a written language, and a substantial number of the people became literate in their own language. A medical clinic and federal school were established by the Mexican government in El Desemboque and later at Punta Chueca. Until the 1960s measles epidemics continued to periodically plague the Seri. Tuberculosis was also a serious problem. The water from one of the two wells near El Desemboque was almost brackish, and its fluoride content so high that children growing up drinking it developed discolored, mottled teeth. However, with better sources of water, improved housing, and health care provided by the Mexican government in the 1970s, the problems of nonindigenous disease began to abate.

During the 1970s the roads were improved and many Seri families obtained pickup trucks or automobiles. The government provided modern housing and power plants which supplied electricity for part of each evening. El Desemboque and Punta Chueca were transformed into villages with straight streets. Figure 1.8 shows aerial views of the two permanent Seri villages.

Vacation homes, motels, and condominiums sprang up on the beach dunes at Kino Bay as the region became a major resort. Sea turtles declined to the brink of extinction and other marine resources also became increasingly scarce. At the same time ironwood sculpture became a major and unique industry among the Seri. In fact, it was so successful that it was copied.

Ironwood sculpture and other Seri folk art, such as necklaces and baskets, provided a major source of cash for virtually every Seri family.

Tourists and commercial art buyers sought out Seri crafts. Many families moved to Punta Chueca in order to be close to the tourists at Kino Bay. However, the *jejena* or *ziix cocósi* "small biting things," tiny biting flies, made life miserable during the summer and fall months. Also, fresh water was often difficult to obtain in Punta Chueca during hot weather. After the tourist influx during Holy Week, many Seri returned to El Desemboque, where they had more permanent homes.

In few other places of the world has a nonagricultural, hunting and gathering way of life persisted so vigorously into the middle of the twentieth century. There were joyous and romantic aspects to such a life, but the older people who actually lived off the desert and sea had no desire to return to the old ways. Drinking water was in short supply and few of the luxuries of the modern world were available. In those days, one too old or infirm to keep up with the group was left behind.

Although much traditional knowledge vanished with the death of each older person, Seri cultural identity remained strong in 1983. Their language was intact and their population increasing. The Seri were meeting the challenges of the modern world in much the same way as they reacted to earlier Spanish-Mexican influences— by taking advantage of new resources, new markets, and new opportunities without radically altering their loose and flexible way of life. They were accepting what they could of the new culture and rejecting that part they did not want.

2. Vegetation

The Seri live in an extremely arid environment along the Sonoran coast of the Gulf of California. Chains of steep, rugged desert mountains extend through the region in a general north-south direction. Peak elevations are somewhat less than 1,000 m. These disjunct mountains are surrounded by broad bajadas and coastal plains; the plains are apparently of Recent alluvial and deltaic origin (Runsk and Fisher 1964). The region is part of the Basin and Range Province of Southwestern North America. The Gulf Coast of Sonora has a basement dominated by Paleozoic metamorphic rocks which were intruded by Mesozoic granitic rocks. Overlying this basement are Middle and Upper Tertiary volcanic sequences. The geologic history of western Sonora is transitional between that of adjacent central Sonora and that of Baja California (Gastil and Krummenacher 1977).

The high relief land masses surrounding the narrow Gulf of California produce continental, rather than oceanic, conditions, and there is a correspondingly large fluctuation of environmental parameters (Roden 1964). Sea water temperatures near the surface vary from 14° to 18°C in winter to 32°C in summer (Robinson 1973).

The intertidal habitats of the Gulf are among the most diverse in the world. Much of the shoreline consists of sea cliffs, wave-cut bluffs, and broad, sandy beaches with high and unstable dunes. Lower and more stable dunes of an older age often occur inland and parallel to the higher beach dunes.

There are many small bays and coves along the rocky portions of the coastline. Along the shoreline of the coastal plains and bajadas are various lagoons and inlets, locally known as *esteros*, which support mangroves and other halophytic vegetation. Freshwater drainage from the land surface into the esteros is negligible. The waters of the esteros drain with the tides and tend to be hypersaline, particularly during the hotter times of the year.

The Seri region is a hot, arid, coastal desert, meteorologically classified as Arid to Extremely Arid (Meigs 1953, 1966), with average annual rainfall of 100 to 250 mm (Hastings and Humphrey 1969, García 1981). Rainfall is extremely uneven from year to year, and is unpredictable in the driest portions of the region. No perennial rivers or streams flow into the sea and nonindustrial agriculture in such a region is not feasible.

The total annual rainfall along the coast of Sonora decreases from south to north and east to west, although secondarily it increases with elevation. The rainfall is bi-seasonal (summer and winter-spring). The short, monsoon-like, summer rainy season is most pronounced and dependable towards the south and east. In addition, occasional hurricane-fringe storms in late summer and early fall may bring significant quantities of rainfall over a short time span. Toward the northwestern reaches of the region winter-spring rains become increasingly important. There is a corresponding shift from plants and animals dependent upon hot season rainfall in the southern parts of the region to a biota adapted to cool season rains in the northwestern portion.

Summer rains are characterized by thunderstorms, and are often very local in extent. These violent storms may bring large quantities of rainfall in a few hours or less. In contrast, winter rainfall is derived from Pacific frontal storms, which are extensive in area of coverage and tend to deliver gentle rains over longer spans of time. There appears to be no correlation between the quantity of summer or winter rainfall and the total annual rainfall from one year to the next.

The summers are long, hot, and humid (see Hastings and Humphrey 1969). Temperatures any day from mid-June through September may exceed 38°C (100° F) and sometimes exceed 43° C (109°F). The Seri said that they suffered most from the heat during September.

Most years are frost free. However, in January occasional periods with several degrees of freezing temperatures can cause extensive damage to frost-sensitive species. The northern distribution of many Sonoran Desert or thornscrub organisms, particularly the perennial plants, is limited by the most severe freezing weather, which might occur only once in several decades. The western and northwestern distributional limits also tend to be defined by increasing aridity. Since both aridity and winter freezing temperatures become more severe northward, it is often difficult to segregate these as limiting factors. However, soil moisture is the most important, overriding factor in the distribution of arid land plants.

The original area of Seri occupation coincides with the Sonora segment of the Sonoran Desert vegetational region designated by Shreve (1951) as the Central Gulf Coast (see Map 1.1). This cultural and natural area is in the heartland of the Gulf of California (Felger 1976b). Both the Sonora and Baja California sides of the Gulf Coast region are strikingly similar in physical environment, as well as in vegetation and flora. The topography and vegetation are complex and highly varied. While the Seri region lies within the confines of the Sonoran Desert, there are fringes of mangroves and, at higher elevations, local areas of thornscrub.

The flora of the region is moderately rich in terms of number of species: the original area of Seri occupation probably includes about 500 species of vascular plants. As a means of comparison, the flora of the entire Sonoran Desert (Wiggins 1964) contains approximately 2,500 species. There are 290 species of vascular plants known from Tiburón Island, and the flora of the island probably contains an additional 15 percent. San Esteban Island supports a flora of 110 species (Felger and Lowe 1976).

In general, the species richness (number of species per area) shows positive correlation with soil moisture. Thus, species richness generally increases to the south, with higher elevation, and in the more favorable microclimates, such as north-facing slopes and riparian or semiriparian situations.

The vegetation or major biotic communities in the Seri region can be classified as follows:

Marine vegetation
  marine algae communities
  seagrass meadow
Littoral scrub
  mangrove scrub
  salt scrub
Desertscrub
  coast scrub (*Frankenia* scrub)
  cactus scrub
  mesquite scrub or *mezquital*
  creosotebush scrub
  mixed desertscrub
  riparian desertscrub
Thornscrub

Major references on the vegetation and flora of the region include Dawson (1966), Felger (1966), Felger and Lowe (1976), Gentry (1942, 1949), Hastings, Turner, and Warren (1972),

Johnson et al. (1970), I. M. Johnston (1924), Norris (in preparation), Shreve and Wiggins (1964), Standley (1920–1926), and Wiggins (1980).

Marine Vegetation

The submerged marine vegetation exhibits considerable seasonal variation. Greatest biomass development occurs during the cooler months and peaks during the spring before water temperatures become too hot.

Marine Algae Communities

Extensive sublittoral beds of seaweeds occur mostly on hard or rocky substrates. However, in terms of biomass, marine algae in the Gulf are not as extensive as on the Pacific Coast of the Californias. There are no very large seaweeds, such as *Macrocystis* and other kelp. In the Gulf there is considerable seasonal variation, with greatest development of the brown and red algae during the cooler seasons. Many of the larger seaweeds are ephemerals. In late winter and early spring considerable quantities of seaweeds are cast ashore in stormy, windy weather. Although there are about two hundred species of marine macro-algae in the region, these seaweeds played but a minor role in Seri culture.

Seagrass Meadow

In the Gulf of California isolated, although often extensive, seagrass meadows occur in protected waters with muddy-sandy substrates. These grow at the lower limits of tidal zones, but mostly in subtidal benthic zones to approximately three fathoms, and occasionally to approximately five fathoms. These meadows are comprised mostly of eelgrass (*Zostera marina*) and ditch-grass (*Ruppia maritima*). Another seagrass, *Halodule wrightii*, is also encountered in the region.

Seagrass beds occur at the bays of Guaymas, Bacochibampo, San Carlos, and Kino, and in Estero Tastiota and the Infiernillo Channel. The most extensive seagrass meadows in western Mexico are those in the Infiernillo Channel (Figure 2.1). Eelgrass is best developed during the cooler seasons of the year and is often replaced by ditch-grass during the hottest months, particularly in shallow water. During spring great masses of eelgrass wash ashore (Figure 2.2).

Although floristically this is the simplest vascular plant community in the region, it has been one of the most important to the Seri. Seagrasses are among the favorite foods of the green turtle (*Chelonia*), the single most important traditional food resource of the Seri (Felger, Cliffton, and Regal 1976; McGee 1898:214). Turtle hunters often sought their prey in seagrass meadows, and the seed of eelgrass was used by the Seri as a major food resource (Felger and Moser 1973; Felger, Moser, and Moser 1980). A generalized food web emphasizing eelgrass and ditch-grass is shown in Figure 2.3.

Littoral Scrub

Low scrub, consisting largely of evergreen, perennial halophytes, is common at the tidally inundated interface of the desert and the sea. There is a predominance of succulent and semi-succulent species. These communities are relatively simple and usually consist of only one to several common species of plants. There are less than two dozen common plant species that make up all of the various littoral scrub communities in the Seri region.

Littoral scrub has been important in Seri society due to the resources found there. Camps were often located on adjacent dunes or the coast scrub zone. Many shore plants are unarmed (without spines) and glabrous (without trichomes or hairs). Such plants were called "soft" plants by the Seri and were extensively utilized as roofing materials and to line baskets or sea turtle carapaces to keep food clean.

Mangrove Scrub

There are three species of mangrove in the Gulf of California: *Avicennia germinans* (black

Figure 2.1. The Infiernillo Channel separating Tiburón Island from the mainland (at Campo Víboras, looking west to the island). The most extensive seagrass meadows in western Mexico occur in these shallow waters. *ETN, March 1951.*

Figure 2.2. Eelgrass cast ashore along the Infiernillo Channel. *RSF, April 1983.*

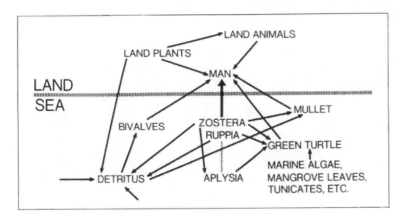

Figure 2.3. Diagrammatic representation of the seagrass–sea turtle–human ecosystem of the Infiernillo Channel.

mangrove), *Laguncularia racemosa* (white mangrove), and *Rhizophora mangle* (red mangrove). They occur in quiet bays and lagoons locally called esteros (Figures 2.4, 2.5). The major mangrove stands in the region are at Guaymas, Miramar, Tastiota, Punta Baja, Estero de la Cruz, in inlets along both sides of the Infiernillo Channel, and farther north at Puerto Lobos (where only the black mangrove occurs).

The dense, tangled mangrove roots are inundated and then exposed twice daily by the tide (Figure 2.6). Freezing weather once every few decades limits the northern distribution and extent of development of the mangroves. For example, the unusually severe freeze of January 1971 caused extensive damage to many of the mangroves at Estero Sargento, and many of them subsequently died. Freezing weather seems to be a major factor contributing to the relatively low stature of the mangrove colonies in the Gulf of California, where maximum heights seldom exceed 3 to 5 m.

At low tide oysters (*Ostrea columbiensis*) were collected from the stilt roots of the red

mangrove and from the inundated woody portions of the white mangrove. Clams (primarily *Chione*) were gathered in the draining channels and rivulets in the open areas between the mangroves, and swimming crabs (*Callenictes bellicosus*) were caught in quantity in shallow tide pools.

At high tide mullet (*Mugil* spp.) were poisoned with sand croton (see *Croton californicus*) or harpooned in the backwaters, and by the 1960s nets were employed to catch them in such places. Other fish and sea turtles entering the esteros were also hunted there. Numerous birds nesting or roosting in the mangroves were utilized for their eggs and meat.

The Seri often used the confusing maze of dense thickets and channels in the almost impenetrable mangroves to hide from the military. A tiny biting fly (*Culicoides sureus*), locally called *jején*, makes life miserable in the mangroves during the hotter months of the year. Although the mangroves in the Seri region are relatively dwarfed and restricted in area as compared with those of more tropical coasts, they yielded important ani-

Figure 2.4. Estero near Campo Víboras showing the extent of mangrove scrub (darker vegetation). *RJH, May 1977.*

Figure 2.5. Mangroves at low tide in the quiet waters of the estero near Campo Víboras. *NLN, April 1974.*

Figure 2.6. Exposed stilt roots of the red mangrove at low tide in the estero at Punta Santa Rosa. *RSF, April 1983.*

Figure 2.7. Salt scrub at the north end of the Infiernillo Channel, showing iodine bush (*Allenrolfea occidentalis*) and sea blite (*Suaeda moquinii*). RSF, April 1983.

mal resources and played a significant role in Seri culture.

Salt Scrub

This low, dense vegetation commonly consists of salt grasses, halophytic shrubs (salt bushes), and succulent forbs (Figure 2.7). Common salt grasses in the region are *Monanthochloe littoralis*, *Sporobolus virginicus*, and *Jouvea pilosa*. Halophytic shrubs include *Allenrolfea occidentalis* (iodine bush), *Atriplex barclayana* (coastal saltbush), *Suaeda moquinii* (sea blite), and *Maytenus phyllanthoides* (*mangle dulce*). Common succulent forbs are *Batis maritima* (saltwort), and *Salicornia* spp. (glasswort), and *Sesuvium verrucosum* (sea purslane).

The salt scrub zone adjoins the inland side of the mangroves; where mangroves are absent, it lies between the beach and coast scrub zone. Salt scrub characteristically occurs on low-lying flat terrain with saline soil and on upper beaches. Several minor food plants occur in salt scrub, e.g., *Allenrolfea*, *Batis*, and *Maytenus*.

Desertscrub

Most of the land surface of the region exhibits a complex intergrading array of desert plant communities. This desertscrub can conveniently be classified into six major community types.

Coast Scrub

Bordering the shore of much of the region, there is a band of low and monotonous coastal vegetation with few species. It is dominated by *Frankenia palmeri*, and occurs along the mainland south to the vicinity of Punta Baja and on Tiburón Island (Figure 2.8). Terrain and soil type are highly varied, often with a relatively high salt content. Frequent maritime dew seems to maintain the *Frankenia*. Width of this zone depends on the elevational grade; in steeper regions is may be only a hundred or so meters wide, but on flat terrain it may extend inland for about one kilometer.

The strategic location was much utilized for campsites, and most of the original vegetation at

Figure 2.8. Coast scrub. A) East side of Tiburón Island opposite Punta Chueca showing *Frankenia palmeri* with ocotillo (*Fouquieria splendens*) in the background.
B) West side of Tiburón Island, adjacent to the shore, showing widely scattered *Frankenia palmeri* and *Fagonia pachyacantha*. The west side of the island is much drier than the east side and the vegetation is correspondingly sparser. *RSF, March 1965.*

the settlements of Kino Bay, Punta Chueca, and El Desemboque consisted of coast scrub. *Frankenia palmeri* provided a convenient source of fast-burning fuel and brush for shelters, but otherwise there were few resources in this zone.

Cactus Scrub

Conspicuous, dense stands of giant columnar cacti occur scattered along the coast of Sonora south of Puerto Lobos (Figure 2.9) and on most of the islands in the Gulf of California. Common columnar cacti in the region are:

> *Carnegiea gigantea*, sahuaro
> *Lophocereus schottii*, senita
> *Pachycereus pringlei*, cardón
> *Stenocereus gummosus*, pitaya agria
> *S. thurberi*, organ pipe

Cactus scrub is often adjacent to the coast scrub zone, although expansive areas of it may also occur farther inland. Maximum development of *cardón* in this region characteristically is near the shore. Immediately inland it is often replaced by dense stands of sahuaro. Other common members of the cactus scrub community include:

> *Atriplex polycarpa*, desert saltbush
> *Bursera microphylla*, elephant tree
> *Cercidium floridum*, blue palo verde
> *Fouquieria splendens*, ocotillo
> *Jacquinia pungens*, San Juanico
> *Maytenus phyllanthoides*, mangle dulce
> *Olneya tesota*, ironwood
> *Opuntia fulgida*, jumping cholla

Columnar cacti played an important role in Seri culture, providing significant food, bev-

Figure 2.9. Cactus scrub at Punta Baja (about 1 km inland from shore). Prominent plants are cardón (*Pachycereus pringlei*), organ pipe (*Stenocereus thurberi*), senita (*Lophocereus schottii*), and jumping cholla (*Opuntia fulgida*). RSF, March 1965.

Figure 2.10. Aerial view of mesquite scrub (*mezquital*) alternating with alkaline flats supporting salt scrub on the coastal plain between Punta Baja and Tastiota. *RSF, July 1964.*

erage, and material resources. Ocotillo, particularly abundant in ecotonal regions between coast scrub, cactus scrub, and mixed desert-scrub, was of major importance for construction of Seri houses.

Mesquite Scrub (Mezquital)

Expansive areas of mesquite-dominated vegetation occur along the broad coastal plains between Tastiota and Kino Bay, and at scattered localities to the north, such as along the margins of Playa San Bartolo (Figure 2.10). Until recent decades the alluvial plain, Llano de San Juan Bautista, between Kino Bay and Tastiota supported vast *mezquital* composed of low, patchy forests or thickets of mesquite (*Prosopis glandulosa*).

The Llano is undoubtedly of deltaic origin (Shreve 1951), and mesquite seems well adapted to the fine-textured, deep, alkaline soils (see Simpson 1977). In addition to mesquite, there are numerous other important food resource species such as:

> *Bumelia occidentalis, bebelama*
> *Jacquinia pungens, San Juanico*
> *Lycium brevipes,* desert wolfberry
> *Opuntia fulgida,* jumping cholla
> *Vallesia glabra*

Wildlife must have been abundant in earlier times. By the early 1960s the few remnants of the original Llano vegetation which had not been replaced by irrigated fields were being seriously reduced by woodcutters, and were rapidly turning into wasteland flats. *Mezquital* was among the more important food-producing areas.

Creosotebush Scrub

Relatively simple communities dominated by *Larrea divaricata* (creosotebush) are widespread at the head of the Gulf of California (Figure

Figure 2.11. Creosotebush scrub in the Agua Dulce Valley in the middle of Tiburón Island. *RSF, April 1965.*

2.11) but are somewhat reduced in extent farther south in the region of Tiburón Island and the adjacent coast. This xerophytic vegetation is best developed in the more open, arid sites. Noteworthy stands occur on the expansive flats of the Agua Dulce Valley on Tiburón Island. In the Guaymas region, near its southern limit in Sonora, *Larrea* is limited to a small fraction of exposed flat ridges and other flatlands (mostly on disturbed sites).

Ephemerals (desert annuals) are seasonally common and some were utilized for food, e.g., *Amaranthus, Oligomeris,* and certain grasses. Creosotebush itself provided important non-food resources—such as the highly prized lac—and was an important medicinal plant.

Mixed Desertscrub

Most of the land surface of the region supports complex and highly varied communities which may be classified as mixed desertscrub (Figures 2.12–2.15). Included here is the majority of the regional flora. A more detailed analysis of the mixed desertscrub could differentiate it into additional units. Mixed desertscrub occupies open, desert flatlands, bajadas, and hill and mountain slopes.

Various desert shrubs and small desert trees make up many distinctive and usually local communities. There are numerous species of small-leaved and drought-deciduous shrubs and sub-shrubs, as well as other growth forms, such as succulents (e.g., cacti) and ephemerals. Each season in each year may bring a different species spectrum of ephemerals. Representative perennial species are:

Atamisquea emarginata, palo zorrillo
Bursera microphylla, elephant tree
Cercidium microphyllum, foothill palo verde
Colubrina viridis, palo colorado
Cordia parvifolia
Errazurizia megacarpa
Fouquieria splendens, ocotillo
Jatropha cinerea, ashy limberbush
J. cuneata, limberbush
Randia thurberi
Viscainoa geniculata

Although mixed desertscrub includes the majority of the regional flora and most of the land surface, it did not contribute a corresponding quantity of the resources used by the Seri. Common major subsistence species include:

Agave cerulata, century plant
A. subsimplex, century plant
Cercidium microphyllum, foothill palo verde
Opuntia fulgida, jumping cholla
Plantago insularis, desert plantago
Stenocereus thurberi, organ pipe

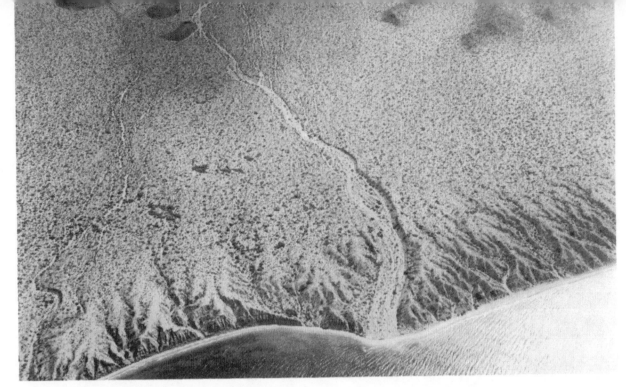

Figure 2.12. Mixed desertscrub on the mainland side of
the Infiernillo Channel. The dots are small desert trees and
shrubs. *RSF, 1965.*

Figure 2.13. Cactus scrub and mixed desertscrub in the
vicinity of Pozo Coyote. Prominent plants are *cardón*
(*Pachycereus pringlei*), a few organ pipe (*Stenocereus
thurberi*), and teddybear cholla (*Opuntia bigelovii*). *ETN,
March 1951.*

Figure 2.14. Mixed desertscrub on San Esteban Island. In the foreground is the San Esteban century plant, *Agave cerulata dentiens. RSF, March 1965.*

Figure 2.15. Mixed desertscrub on the mainland bajada on the west side of Sierra Seri, looking east from the vicinity of Campo Víboras. Conspicuous plants are *cardón* (*Pachycereus pringlei*), organ pipe (*Stenocereus thurberi*— the smaller columnar cactus), elephant tree (*Bursera microphylla*), and jojoba (*Simmondsia chinensis*). *ETN, March 1951.*

Two edaphic extremes—coastal dunes and sea cliffs—each support distinctive mixed desertscrub plant communities. High, unstable dunes facing the sea are developed along extensive stretches of the mainland and Tiburón Island shores. Proceeding inland there are often progressively smaller and more stable dunes positioned in parallel series. Those farther inland are older than the outer beach dunes. There is a gradual shift in dune flora from north to south and from east to west on the shores of Tiburón Island. However, in any one area the flora is highly predictable, depending on dune size, stability, and exposure to maritime wind.

Many dune species have deep roots, and plants with silvery or grey pubescent leaves are common. Certain dune species seldom occur elsewhere, e.g., *Astragalus magdalenae*, *Dicoria canescens*, *Euphorbia leucophylla*, and *Triteleiopsis palmeri*. Others are sandy-soil species with wide distributions in sandy non-dune habitats, e.g., *Abronia maritima*, *A. villosa*, *Aristida californica*, *Croton californicus*, and *Psorothamnus emoryi*. Other common species occur over a wider spectrum of soils, e.g., *Atriplex canescens*, *Jatropha cinerea*, *Lycium brevipes*, and *Palafoxia arida* (see Felger 1980).

The coastal dunes were important, favored sites for camps, both in modern times and in prehistory (Bowen 1976:23). There was good access to the sea, with excellent vantage points to spot approaching people, boats, and animals (such as the migrating *cooyam* green turtle) and to watch the sea surface and weather conditions across the water. Furthermore, it was pleasant to camp in the clean sand and enjoy the frequent breezes. However, resources on the dunes themselves were severely limited.

Sea cliffs relatively free of bird guano support a unique group of rock-holding plants comprising a very sparse mixed desertscrub. Prominent plants include *Ficus petiolaris*, *Hofmeisteria fasciculata*, *H. laphamioides*, and *Eucnide rupestris*. Cliff fig, *Ficus petiolaris*, was a fairly important species in Seri culture, but otherwise the cliff plants and habitat were not very important to the people.

Riparian Desertscrub

Desert canyons, arroyos, washes, and their floodplains support denser, taller, and more diverse plant communities than those found in the adjacent desert (Figure 2.16). Most of these watercourses are dry except for rare and usually brief occasions—perhaps only a few days each year or once every several years—when rainfall may be sufficient to produce runoff. In the drier regions, species which are elsewhere nonriparian are often restricted to the drainageways because

Figure 2.16. Riparian desertscrub along a dry watercourse in the Agua Dulce Valley, near the center of Tiburón Island. The tree is bebelama (*Bumelia occidentalis*), and the shrubs are ironwood (*Olneya tesota*) and desert hackberry (*Celtis pallida*). RSF, February 1965.

Figure 2.17. Freshwater plant communities in the Seri region are narrowly restricted to wet soil at the edge of water holes. The vegetation at this permanent water hole at Sauzal, Tiburón Island, includes tangled masses of reed-grass (*Phragmites australis*), salt cedar (*Tamarix ramosissima*), and mesquite (*Prosopis torreyana*) beyond the pool. Seep willow (*Baccharis salicifolia*) is in the foreground. *RJH, April 1977.*

of increased soil moisture. The canyons and other dry watercourses supporting riparian desertscrub provided some of the most productive and important hunting and gathering microenvironments.

Characteristic trees and shrubs occurring in riparian desertscrub in the region include:

Acacia greggii, catclaw
Ambrosia ambrosioides, canyon ragweed
Bumelia occidentalis, bebelama
Cercidium floridum, blue palo verde
Hymenoclea monogyra

Hyptis emoryi, desert lavender
Olneya tesota, ironwood
Prosopis glandulosa, mesquite
Stegnosperma halimifolium
Vallesia glabra

Permanent waters, such as springs and seeps feeding bedrock pools in canyons, locally called *tinajas*, provided the people with fresh water. Highly localized desert freshwater communities occur at these water holes (Figure 2.17) and most of these species seldom occur elsewhere in the region. For the most part, these plant species have cosmopolitan or widespread geographic distributions.

Throughout the original area of Seri occupation there are on the order of not more than a few dozen permanent freshwater sites supporting riparian vegetation. Although occupying but a small fraction of a percent of the land surface, these water holes were crucial to human survival in the area (see Chapter 6). Reedgrass (*Phragmites australis*) is almost entirely restricted to places of permanent water, and was an indispensable resource. Other characteristic species include *Cyperus elegans*, *Eleocharis genicularis*, *Salix gooddingii* (western willow), and *Typha domingensis* (cattail). Giant reed (*Arundo donax*) and salt cedar (*Tamarix ramosissima*), both non-native species, have invaded some of the water holes, even on Tiburón Island, where they were found to be well established in the 1960s.

Thornscrub

Limited areas of thornscrub and ecotones between desertscrub and thornscrub occur at higher elevations on Sierra Kunkaak on Tiburón Island and the mountains north of Guaymas and San Carlos Bay. These isolated pockets are mostly on north- and east-facing slopes and in the larger canyons. Such areas were apparently seldom frequented by the Seri—probably because of difficult access—except in riparian or semiriparian areas where canyons carry thornscrub or its ecotones into lower elevations.

These communities consist of complex, dry-deciduous subtropical vegetation with many small trees and large shrubs. Seasonal aspects are sharply distinct, with summer wet season perennial coverage on the order of 75 to 100 percent. Ephemerals and herbaceous perennials are seasonally common. Characteristic thornscrub or desertscrub-thornscrub ecotone species on Tiburón Island (T) and the San Carlos Bay–Guaymas region (G) include:

Bursera confusa, torote papalillo (G)
Elytraria imbricata (G and T)
Ficus petiolaris, cliff fig (G and T)
Fouquieria macdougalii, tree ocotillo (G)
Guaiacum coulteri, guayacán (G and T)
Jacquinia pungens, San Juanico (G and T)
Lysiloma divaricata, mauto (G and T)
Pachycereus pecten-aboriginum, echo (G)
Stenocereus thurberi, organ pipe (G and T)

3. Animal Life

The rich and diverse fauna of the region includes most terrestrial life-forms common to the Sonoran Desert region, as well as marine life characteristic of much of the west coast of North America and elsewhere in the warm seas of the world. Seri knowledge of the fauna and its role in their culture was as extensive as that of the plant kingdom. *Ziix ccam* 'thing that-is-alive,' or "living thing," is the term for any animal— invertebrate or vertebrate, marine or terrestrial. Examples of Seri names and information are given for selected animal groups and are not meant to be comprehensive.

In terms of subsistence and material culture, the most important faunal elements included various bivalves (e.g., clams and oysters), certain fish (including mullet), the green sea turtle, the desert tortoise, the San Esteban chuckwalla, various birds (including the brown pelican and other shore and sea birds), the mule deer, the desert bighorn sheep, the black-tailed jackrabbit, and the California sea lion.

Invertebrates

The invertebrate fauna in the region is rich and diverse (Brusca 1980). Most of the major or conspicuous forms were known to the Seri by name and there was considerable knowledge surrounding them. A few of the more conspicuous invertebrates are given here as examples.

As with the terrestrial flora and fauna, there is considerable variation in distributional patterns.

The marine mollusk fauna is diverse and abundant (Keen 1971). Numerous species have been used extensively as food, utensils, tools, and ornaments. They are also of major interest because the shells tend to be well preserved in archaeological sites and are biologically diagnostic. Shell-filled middens are a common feature along the shores of the Gulf of California (Figure 3.1; Bowen 1976). Shell middens occurring at some distance from the present shoreline, such as along the west side of Playa San Bartolo, may indicate changing sea levels or shorelines in late Pleistocene or Recent times.

Marine mollusks constituted a vital part of the diet of the people who lived along the coast. At low tide women selectively gathered clams and other shellfish from tidepools, tidal flats, and esteros. Oysters were easily pried off rocks, and from the southern end of the Infiernillo Channel and southward they were collected from the roots of the red mangrove (*Rhizophora*; Figure 3.2). Sometimes men, women, and children collected mollusks, such as octopus, which they extracted by hand from small, cave-like burrows beneath tidepool rocks. Shellfish were also collected and sold commercially.

Shellfish were steamed (see *Croton californicus*), placed around hot coals, put on a grill (a twentieth-century method), boiled, or eaten raw. Octopus was boiled. The large black murex was cooked by burning sticks, such as the dry ribs of organ pipe (*Stenocereus thurberi*), on top of it. Mollusks important in the diet included:

Class Pelecypoda (bivalves, such as clams)
    Arca pacifica, Pacific arc, *quiimosim xepe ano yaafc* 'he-who-begs(for food) sea in he-pounded-hard'
    Atrina tuberculosa, pen shell, *seten cmaam* 'seten female'
    Cardita affinis, *quiit*
    Chama mexicana, *imox*
    Chione californiensis, California chione, *spitjquim*
    C. fluctifraga, smooth chione, *haan*
    Dosinia ponderosa, *halit cahóoxp* 'head it-makes-it-white'
    Glycymeris gigantea, *xpanóis*
    Laevicardium elatum, giant egg cockle, *xtiip*
    Modiolus capax, fat horse mussel, *satoj*
    Ostrea columbiensis, mangrove oyster, *haxt*
    O. palmula, oyster, *stacj*
    Pinna rugosa, pen shell, *seten ctam* 'seten male'
    Pinctada mazatlanica, pearl oyster, *copas quictoj* 'copas with-pearls'
    Protothaca spp., littleneck clam, *haxöl*
    Spondylus calcifer, spiny oyster, *teexoj*
Class Gastropoda (univalves, such as snails)
    Crucibulum scutellatum, cup-and-saucer limpet, *caixona*
    Muricanthus nigritus, black murex, *nocat*
    Onchidella binneyi, *tamax*
    Strombus gracilior, *xica cotítzilca* 'things cotítzilca'
    Turbo fluctuosus, turban, *cotópis*
Class Cephalopoda
    Octopus sp., *hapaj*

Clamshells were indispensable as tools, utensils, and containers. Those from the various larger clams, such as *Laevicardium elatum* (Figure 3.3), served as all-purpose bowls, dippers for food, containers for pigments and paints (Hardy 1829:286; McGee 1898:186), and were used to scoop out dirt to dig or enlarge a well. The shells of medium-sized clams, especially of a

Figure 3.1. Shell midden along the shore of the Infiernillo Channel. *RSF, 1983.*

Figure 3.2. Oyster (*Ostrea columbiensis*) on a stilt root of red mangrove (*Rhizophora mangle*) in an estero at Punta Santa Rosa. *RSF, April 1983.*

Figure 3.3. Giant egg cockles (*Laevicardium elatum*) found at abandoned brush houses at *Saps*. The shell on the left is 13.5 cm long. *RSF, April 1983.*

surf clam (*Mactra dolabriformis*), or a small shell of the giant egg cockle (*Laevicardium elatum*), were used as spoons. The surf clam was called *haxöl icáai* 'haxöl to-make-with.' *Haxöl* is the littleneck clam, and *icáai* 'to-make-with' refers only to a woman making pottery. As the name indicates, the shell of this clam was used to scrape the damp clay of an unfinished pottery vessel of thin-walled Tiburón Plain or "eggshell ware" (Bowen and Moser 1968:127–128). A water-worn beach rock about 7 cm long, specifically called *iquémt* 'to rub with,' was held inside the vessel opposite the clamshell while the pot was being scraped. A large clamshell filled with water served as a mirror when placed in the shade. The thick shell of the Gulf cockle, *zacz* (*Trachycardium panamense*), was used to scrape out the pulp of the *siml* barrel cactus (*Ferocactus wislizenii*) for making emergency liquid and barrel cactus containers.

Clamshells, especially *Chione*, were stuck into columnar cacti and tree trunks (e.g., ironwood, *Olneya*) as good luck caches. *Olivella* and other seashells were extensively utilized for making necklaces and other personal ornaments (see Chapter 12). A tower shell, *xtapácaj* (*Turritella gonostoma*), was used in a game in which

the shell was flipped into the open end of a piece of reedgrass (see *Phragmites*).

When referring to the shell alone, without the fleshy parts, the term *ináil* 'its shell' was used together with the name of the mollusk. For example, one would speak of *cotópis ináil* or *satoj ináil*. The name for the mollusk used without the term *ináil* referred to the entire animal, both the shell and the body (soft parts).

Seri knowledge and use of other invertebrates were also extensive. Swimming crabs (*Callinectes*, mostly *C. bellicosus* [Figure 3.4A], probably some *C. arcuatus*), called *zamt*, were harpooned in shallow water along the shore and in the mangrove esteros (Sheldon 1979:123). The crab harpoon or *hacáaiz*, was made from a slender, lightweight pole about the thickness of a finger, with a nail as a point. The so-called turtle harpoon shown by McGee (1898:187) was obviously a crab harpoon. These crabs, esteemed for their sweet-tasting meat, were eaten in substantial quantities. The claws and carapaces were numerous in camp and house refuse (Figure 3.4B; McGee 1898:195). The claws were made into small dolls for girls. Crabs in several other genera were also eaten; examples include *oot izámt* 'coyote its-*zamt*,' or "coyote's swimming crab" (probably *Portunus*, a swimming crab similar in appearance to *Callinectes* but smaller in size) and *inl quixaz* 'arms tinkle' (probably *Eurytium affine* and/or *E. albidigitum*).

The Gulf of California spiny lobster, *Panulirus inflatus*, called *ptcamn*, was caught in tidepools along the southern shore of Tiburón Island. It was a favorite food of the people living in that region. Shrimp (*Penaeus* and related genera) were almost unknown to the Seri during traditional times, because the adult shrimp are seldom found in the intertidal zone.

A colorful mangrove spider, *Gasterocantha elipsoides*, was known as *coiz cahtxíma* 'spider rich.' It was said to be a rich spider because it was so beautiful. *Coiz*, the name of a certain common spider, was also a general term for spider. The harvestman or daddy longlegs (*Phalangida*) was called *hant cmaa tpaxi iti hacátax* 'land now finished on that-which-was-caused-to-go,' or "thing which went on the newly formed land." In the beginning, while the earth was still soft, the supernatural being known as *Hant Caai* "he who made the land" sent the daddy longlegs out to walk on the newly formed land to see if it had become firm enough for *Hant Caai* himself to walk on.

Like the Papago, the Seri gathered caterpillars of the white-lined sphinx moth (*Hyles lineata*), which feed on desert ephemerals during warm weather (Figure 3.5; see *Boerhaavia erecta*). This caterpillar was called *hehe icám* 'plant its-life,' or "plant's live thing." The head was twisted off, the viscera stripped out with the fingers, and the "skin" (actually mostly muscle or meat) cooked in oil in pottery vessels. The cooked caterpillars were often dried and stored in covered vessels. Gilg's letter of 1692 (Di Peso and Matson 1965 : 55) indicated that the Seri ate grasshoppers as well as caterpillars. *Caatc ipápl* "what grasshoppers are strung with," the name for two small shrubs in the mallow family (*Abutilon incanum* and *Horsfordia alata*), likewise shows that grasshoppers were probably eaten. Although at least eight kinds of ants were distinguished by name, they were not utilized.

Copni yamáax 'carpenter-bee its-wine,' the sweet "beebread" (pollen plus nectar) made by the carpenter bee (*Xylocopa*), was a highly esteemed food. It was described as being like cream, not honey, filling a half-finger-sized hole made in dead wood by the carpenter bee. The people ate a lot of it during the time of year when the adult *copni* was putting this food in the hole for its larva. The best place to collect *copni yamáax* was at the south shore of Tiburón Island, where the Arroyo Sauzal empties into the ocean. Strong ocean currents leave massive quantities of driftwood there, and large numbers of carpenter bees nest in this driftwood. Elsewhere there were relatively few carpenter bees and less of the *copni yamáax* to harvest.

The honeybee, introduced from the Old World, probably colonized the Sonora coast during the latter part of the nineteenth century. It was introduced into California in the 1850s (Woodward 1938), and hives were brought to

Figure 3.4. *Callinectes bellicosus* is a common swimming crab along the shore and in the esteros. A) The carapace of this crab at Estero Sargento is 15 cm wide. B) *C. bellicosus* was eaten in substantial quantities and the carapaces were abundant in camp refuse. The ones shown here were gathered for the photograph at *Saps*. The largest is 16.5 cm wide. Note harpoon holes. *RSF, April 1983.*

southern Arizona in 1872 (*Arizona Citizen* 1872, *Arizona Daily Star* 1897). Around 1900 Luis Torres (elder) collected the first honeycomb known to the Seri in their region. This well-remembered event occurred on a small mountain near Pozo Peña. Chico Romero told us,

Figure 3.5. Caterpillar of the white-lined sphinx moth. *RSF, near Estero Sargento, April 1983.*

"You think honey was a thing eaten by ancestors? No, it wasn't. It is a new food to us."

Dry branches of elephant tree (*Bursera microphylla*) were considered the best material to burn for smoking out bees. Honey was sometimes transported in a container made from barrel cactus (*Ferocactus wislizenii*). Although it is relatively new to the Seri, it came to be highly esteemed and eaten with many foods, including:

Amaranthus fimbriatus, fringed amaranth
Cercidium spp., palo verde
Cnidoscolus palmeri, *mala mujer*
Ferocactus wislizenii, barrel cactus
Opuntia fulgida, jumping cholla
Zostera marina, eelgrass

Honey was also used as a matrix for facepaint.

Fish

There are no freshwater fish in the Seri region. The marine fish fauna includes at least several hundred species (Thomson, Findley, and Kerstitch 1979:4), although only a small portion of these were important in Seri culture. Seri knowledge of fish was extensive (Findley 1971). The generic term for fish is *zixcám*, derived from *ziix ccam* 'thing that-is-alive.'

Seasonal changes in the fish fauna correspond with the relatively large seasonal fluctuations in water temperature. Larger sharks, sailfish, marlin, and swordfish migrate into the Gulf as the water warms in early summer and leave as it cools in fall. Others, such as groupers, migrate from deeper water into shallower water in winter, while the endangered *totoaba* (*Cynoscion macdonaldi*) migrates northward and from deeper to shallow water in winter.

Fish were an important part of Seri subsistence. They were generally cooked alongside hot coals, roasted on a grill, or boiled. Fish meat was also dried in the sun (the meat was not salted) and stored in sealed pottery vessels. Before the advent of commercial fishing (about the 1930s), the people living on Tiburón Island dried considerable quantities of meat from the larger fish, such as *cabrilla*, *mero*, and sea basses. In the 1980s fish continued to be an important part of the diet. Small fish caught by boys with hand lines from the beach were a commonplace breakfast.

A double-pronged harpoon was used for spearing many kinds of fish, such as mullet, triggerfish, groupers, sea basses, and snappers. It was used from boats and while wading in shallow water in pursuit of mullet (*Mugil* spp.). Mullet were also caught in the esteros with the aid of fish poison made from sand croton (*Croton californicus*). The single-pronged fish harpoon was used from the shore for reef fishes, such as *cochi* (triggerfish, *Balistes polylepis*), and sea basses (*cabrillas*) and groupers (family Serranidae)—including "*mero*" (probably the Gulf grouper, *Mycteroperca jordani*), "*pinta*" (probably *cabrilla de rocas*, *Paralabrax macula-*

Figure 3.6. Commercial fishing at Kino Bay, February 1935. Large sea basses and totoaba were abundant. The second man from left is Pascual Blanco; fourth from the left is Santo Blanco. *LMH. SD.*

tofasciatus), and *cabrilla* or *sardinera* (leopard grouper, *Mycteroperca rosacea*). For information on fish harpoons, see *Acacia greggii, Echinocereus grandis, Larrea,* and *Phragmites,* and the discussion on fishing equipment in Chapter 10.

Gulf grunion (*Leuresthes sardina*), called *caaha,* were clubbed with branched sticks when they came ashore to spawn at precise tidal phases from January to May. They were cooked in large numbers against coals, dried, and stored in sealed pottery vessels in caves. The bulls-eye puffer (*Sphoeroides annulatus*) was avoided because it is toxic, and stingrays and most kinds of sharks were generally not eaten. However, the guitarfish (*Rhinobatos productus,* a shark-like ray) was commonly eaten.

The Seri did not use fishhooks or nets until they began commercial fishing (Figure 3.6), starting more or less in the 1930s. Since then many Seri men have engaged in commercial fishing and turtle hunting. These new technologies and the depletion of certain near-shore commercial species have contributed to an increase in offshore fishing in deeper waters, yielding fish for which the Seri had no names. At various times from about the 1930s to the 1960s dynamite was used for fishing (Whiting 1951:9, 11) but was curtailed because it resulted in severe accidents and was illegal.

According to Seri oral history, earlier in the twentieth century there were several months of exceptional rainfall and considerable flooding. The usually dry lake at the San Bartolo Playa filled with water and had fish in it. Three Mexican men (Chico Sesma, Manuel Sesma, and Luis Saavedra) came with wagons, salt, food, and clothing. They paid the Seri with food and clothing to collect the fish and salt them. The Seri who worked in this venture were at a camp in the vicinity of a ranch near Pozo Peña. The Seri believed that the fish came down with the flood-waters from the north. This event may have taken place in 1926, when there was much flooding and the San Bartolo dry lake filled (Davis 1929:110, 115).

Amphibians

Over most of the Seri region the amphibian fauna is represented by only a few species of desert toads in the genera *Bufo* and *Scaphiopus* (spadefoot toad). They occur at scattered localities on the mainland, and *Bufo punctatus* and *Scaphiopus couchi* breed at the Sauzal water hole on Tiburón Island. They were inconsequential in Seri culture and the several species were not differentiated by name. Toads and frogs were called *otác* or *ziix hax ano quiij* "thing that sits in water."

Reptiles

The reptile fauna is diverse and formed an essential part of the culture. Seri knowledge of reptiles was extensive, and most of the species were distinguished by name. All five sea turtle species that exist on the Pacific coast of the Americas occur in the Seri region. For most Seri groups the green turtle (*Chelonia*) was the single most important food resource and was extensively utilized in their material culture. All sea turtles, especially the leatherback (*Dermochelys*), featured prominently in oral tradition and ritual (Cliffton, Cornejo, and Felger, 1982; Felger, Cliffton, and Regal, 1976; McGee 1898; Smith 1974).

The Seri distinguished by name fourteen kinds of sea turtles, eight of which were *Chelonia*. The term *moosni* was used distributively to indicate all sea turtles or specifically for the most common kind of green turtle. The Seri word is essentially identical to the Yaqui (*móosen*) and Mayo (*moósen*) terms for sea turtle (Felger, Moser, and Moser, 1983). While it may seem that the Seri

overclassified sea turtles, many of these folk taxa had biological significance. By the 1970s all sea turtles in North American waters were threatened with extinction. Some of the "micro-races" or stocks given taxonomic status by the Seri apparently became extinct by the mid–twentieth century (Cliffton, Cornejo, and Felger, 1982). In many respects Seri knowledge of sea turtle ecology was more extensive than that which has been published by biologists. Seri nomenclature of sea turtles is summarized in Table 3.1.

The leatherback (*Dermochelys coriacea*) is the largest turtle in the world. It has never been common in the Gulf of California (Cliffton, Cornejo, and Felger 1982). A four-day fiesta, similar to the girl's puberty fiesta, was held upon the capture of one of these giant turtles. It was a rare event; for example, Chico Romero, who was about 86 years old when he died in 1974, never witnessed one of these fiestas.

According to a prevalent version of the leatherback story, the sailfish, a certain large moth, and the leatherback were members of the same family. The leatherback was a female person in this family who had lost her sibling, and her face was spotted because she was crying, mourning for her dead relative. In another version, the leatherback was a mother who had lost her children, and she was crying for that reason. The fiesta was held to pacify her and make her happy. María Antonia Colosio explained concepts surrounding the leatherback in the following narrative:

In the beginning, when the earth was just finished, there were families, but no one had died. Then in a family someone died, it was said. The leatherback was a member of that family. The sailfish was one. The moth was one. They were in the same family. They saw the first dead person, when they were there. Then there came the flood. When the land was first new, the floods came regularly. Then at that time, when the flood came, that kind of turtle lived in the sea. The sailfish also lived in the sea. The moth didn't live in the sea, but they were all of one family. The first person that died, they saw it, that's when they were animals. That's the reason why on catching a leatherback, it is a different kind of thing, they said. Then if a fiesta isn't held,

TABLE 3.1
Seri Classification of Sea Turtles

Scientific and Common Names	Seri Names	Comments
Caretta caretta *caguama perica*, loggerhead	*xpeyo*	Apparently this is the population found north of Kino Bay. Rare at least since about 1940s; only juveniles seen since at least the 1950s. Not good eating; usually not harpooned.
	moosni ilítcoj caacöl 'sea-turtle their-heads large(plural)' "large-headed sea turtle"	Differentiated in part from *xpeyo* by its greener color, smaller head, and milder manner. Apparently this is the population found from Kino Bay southward. By the mid-twentieth century it was no longer seen.
Chelonia mydas *caguama negra*, *caguama prieta*, green turtle	*moosniáa* 'sea turtle real'	Usually referred to simply as *moosni*, this is the common green turtle in the region. It is probably the taxon known as *carrinegra*.
	moosnáapa 'sea turtle true'	This refers to the true or ultimate *moosni* eaten by the ancestors. It was generally not a very large turtle, and had a brownish grey carapace and flippers. It no longer existed by the mid-twentieth century or earlier.
	moosníil 'sea turtle its-blueness' "blue sea turtle"	The blue turtle was seen only by the ancestors. It occurred off the north shore of Tiburón Island and west from there into the open sea. It was huge and strong, and the harpoon point and mesquite rope line turned blue when the turtle was harpooned (see *Asparagopsis*).
	moosni ictoj 'sea-turtle its-redness' "red sea turtle"	This was probably an albino form and was always rare. It was found at the south end of the Infiernillo Channel at a certain place known as *moosni ictoj iime* "red sea turtle's home."
	quiquíi	This was a distinct kind of *moosni* noted as thin, with sunken eyes, and thin flippers which it could hardly move. There was much fat but very little meat. It was rare but still occasionally caught in the mid-twentieth century. The *quiquíi* was not sick; it was a turtle that had died but its spirit had been reborn.
	cooyam	This name referred to the young, migratory phase of the green turtle entering the region from the south for the first time.
	cooyam caacöl 'cooyam large'	As an adolescent turtle growing up in the region, the *cooyam* was reclassified as *cooyam caacöl*.
	ipxom haquíma 'its-fat the-most' "the fattest one"	This was the mature phase of the *cooyam caacöl*. It was found off the north and west coasts of Tiburón Islands and between San Esteban and Tiburón islands.
	moosni quimoja 'sea-turtle *quimoja*'	*Quimoja* is an archaic word for which the meaning has been lost. This was a giant *moosni*, perhaps weighing 100 kg. It fed primarily on sea grasses and tunicates. It was hunted and eaten by the people.
	moosni ctam hax ima 'sea-turtle male with not-be'	This was a male turtle which had a shorter tail than the regular *moosni*. This special type of *moosni* was never seen mating and reminded the Seri of a kind of adult mule deer, *hap imítjc* 'mule-deer without-scrotum,' in which the testicles do not develop. This turtle was always very fat.

TABLE 3.1
(continued)

Scientific and Common Names	Seri Names	Comments
Dermochelys coriacea siete filos, leatherback	*moosnípol* 'sea turtle its-blackness' "sea turtle's blackness"	It has never been common in the region and most of the ones known to the Seri were female. It is the largest turtle in the world, and had special ceremonial meaning in Seri culture.
Eretmochelys imbricata carey, hawksbill	*moosni quipáacalc* 'sea-turtle that-which-overlaps'	The name refers to the overlapping carapace plates, which are the source of commercial tortoiseshell. By the 1960s hawksbills were rare in the region.
	moosni sipoj 'sea-turtle osprey'	This turtle was differentiated from the other kind of hawksbill by its more pointed beak, whiter plastron (not as yellowish), and smooth carapace. It was no longer seen by the middle of the twentieth century.
Lepidochelys olivacea golfina, olive Ridley	*moosni otác* 'sea-turtle toad' "toad turtle"	The Ridley is the smallest of all sea turtles. The carapace is narrower at the anterior end than that of *Chelonia*, flared at the sides, and flattish. Once fairly common in the region, it has become rare since the 1960s or earlier.

it is a bad thing for the people who catch it, they who are there. That's why when a [leatherback] turtle is caught, then a kind of fiesta [is held], and happiness occurs. It is for that reason, when a leatherback is caught, no one causes harm [bothers it]. But if a fiesta isn't held, something bad will happen to the one who caught it. That's why a fiesta is held.

It was said that when the men found and harpooned a leatherback, she spoke to them and led them ashore. It was also said that the leatherback understood Seri, so the people whispered in her presence. *Moosnípol* 'sea turtle its-blackness,' the usual name for the animal, was not used in her presence; rather, the term *xica cmotómanoj* "weak things" was used instead.

Drinking water was supposed to be supplied by the harpooner, who made a trip for water as soon as the great turtle was beached. During the fiesta the turtle was kept in a specially built shelter. The ones we have seen resembled small brush houses with both ends open and only the upper two-thirds roofed (Figure 3.7A; also see Smith 1974). If not shaded the turtle would soon die.

Although the turtles often died during the fiesta, the people sometimes returned them to the sea alive. It was said that the ancestors ate the meat. It was also said that, upon eating the meat for the first time, one should close his eyes or else he would become blind. One man, who had tasted the meat without closing his eyes, said he had no ill effects from it. It may be that in earlier times the meat was eaten as a matter of course and that more recently it was only occasionally eaten or "tasted" (see *Lepidium*).

In March, 1981, two leatherbacks were captured two days apart. The first one died, and after two days the people felt compassion for the second one and released it. Sacred elephant tree branches were scattered over the turtle (see *Bursera microphylla*), and the head, flippers, and carapace were painted with powerful designs (Figure 3.7B). There were special leatherback songs and the same festivities—such as the *camóiilcoj* circle game, singing, and dancing—that were performed during the girl's puberty rites. In former times the larger bones were cleaned and painted, often with designs similar to those

Figure 3.7. Leatherback fiesta at *Saps* in March 1981.
A) A special shelter was constructed to shade the turtle,
sacred elephant tree branches were scattered over it, and
powerful designs were painted on the carapace and flippers. B) Detail of the designs. The paint, in a large clamshell (*Laevicardium elatum*), was applied with a twig
brush. *MBM*.

used for facepainting (Figure 3.8). A painted
bone was sometimes kept as a fetish (Bowen and
Moser 1968:111; Smith 1974:141).

The green turtle (*Chelonia mydas*) was hunted
by the men throughout the year, although it was
most abundant during warm weather (Figure 3.9). It provided a major source of protein
and fat. Sea turtles were harpooned from balsas
and later from wooden boats (see *Acacia greggii*,
Phragmites, *Prosopis*, and *Zostera*). These animals were so abundant (e.g., Hardy 1829:292)
that the hunters usually pursued only the larger
ones. McGee (1898:214) estimated that sea turtles constituted 25 percent of the Seri diet. It
many cases this percentage may have been much
higher.

Commercial hunting of sea turtles began mod-
estly around the 1930s and escalated sharply
after the 1940s (Figure 3.10). For several decades following the 1940s the Seri earned a considerable portion of their income by commercial

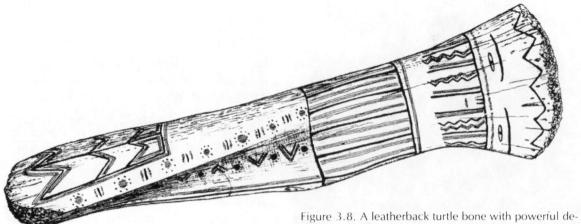

Figure 3.8. A leatherback turtle bone with powerful designs painted by Aurora Colosio. This bone, from a fiesta held in 1967, is 22.5 cm long and painted with red, blue, and green designs. *CMM.*

Figure 3.9. Hunting the green turtle. A) Pedro Comito with harpoon poised, ready to strike. B) A small turtle is boated. The light-colored plastron and small size indicate that it might be a *cooyam*. *EHD, Infiernillo Channel, 1936; HF-23939, HF-23940.*

A

B

Figure 3.10. Seri fishermen bringing in a boatload of green turtles for the commercial market. The men usually kept a few turtles for their families. *BM, El Desemboque, summer 1953; ASM.*

exploitation of sea turtles. By the 1980s sea turtles were scarce, primarily due to overharvesting at the nesting beaches 1,500 km to the south at Maruata and Colola in Michoacán (Cliffton, Cornejo, and Felger 1982).

In summertime the men commonly hunted green turtles at night, using a variety of techniques and strategies that involved differences in tides, moonlight, and phosphorescence. There were numerous places where the Seri believed sea turtles lived, each of which was named as a different turtle *iime* 'home.' The hunters were able to locate these sites by aligning themselves with specific mountain peaks.

During cooler months dormant overwintering turtles were hunted in the Infiernillo Channel (Felger, Cliffton, and Regal 1976). These partially buried turtles were called *moosni hant coit* 'green-turtle land touch,' referring to their "touching down" or "landing" on the sea floor. These turtles also were located at specific turtle

"homes" or *iime* (singular). Some *iime* served as both summer retreats and wintering places.

Buried turtles were said to be situated 3 to 5 m apart along muddy or sandy edges of eelgrass (*Zostera*) beds. From approximately November through March, the hunters usually sought either buried turtles or those grazing on eelgrass. At this time of year, if the wind was not blowing, the water was often exceptionally clear. In the winter, sea turtles were seldom, if ever, seen to surface during windy weather, and they surfaced only rarely on sunny, calm days. However, they frequently surfaced in the warm months, whether or not it was calm or windy.

Dormant turtles were harpooned from boats during low tide, while buried at water depths of about 4 to 8 m. The harpoons for these turtles had mainshafts 7 to 10 m long. Buried turtles were located by looking for the exposed portion of the carapace, which often appeared as a circle or strip above the sand, sometimes only 10 to

15 cm wide. Turtles could be seen only during daytime, at low tide, when the sea was clear and calm, and there was no cloud cover. The Infiernillo is often murky but tends to clear after several days of cool, calm weather. The best conditions were at neap tides; January neap tides were considered best of all. Buried turtles were harpooned by young men with keen vision, although in summer even older men harpooned the active turtles.

The *moosni hant coit* was described as often having seaweeds and a muddy, sandy film on the exposed portion of the carapace, but the marginal scales were clean because they had been buried beneath the mud or sand. Turtles with much carapace seaweed were thin because they were said to have been buried one to three months without eating. Turtles with little or no carapace algae had more fat since they had been moving about feeding. We have seen turtles brought in during winter and early spring with the carapaces clean at the edges but otherwise with extensive epizoophytic growths of marine algae. We found more than twenty species of these epizoophytic marine algae on winter *Chelonia* captured in the Infiernillo Channel by the Seri.

When harpooned, a buried turtle had to be worked out of the substrate. In the process a great deal of mud and sand was raised and the turtle could not be seen until it moved or was pulled away. With a large turtle a second harpoon was usually set in order to pull it free. In contrast to summer turtles, these turtles were torpid and easily boated. By the 1970s dormant turtles had become scarce and were seldom hunted. Since the first publication of the discovery of buried, dormant sea turtles (Felger, Cliffton, and Regal 1976), there have been other reports of sea turtles overwintering on the sea floor elsewhere in the world (e.g., Ogren and McVea 1982).

The name *cooyam* referred to the young, migratory phase of the green turtle. These were the young turtles entering the region from the south for the first time. Flotillas of perhaps forty to seventy of these migrants swam close to shore,

the Seri said, to avoid predators. There were several places in particular, such as camps on the crests of high dunes, where the people watched for the schools of *cooyam* to arrive in spring and caught them in quantity. So numerous were these turtles that men would run at them in shallow water and either harpoon or just grab them.

The *cooyam* was distinguished in part by its white plastron and dark carapace (often mottled with red or amber, like that of a hawksbill); it also had substantial amounts of fat when it arrived, but the gut was empty and the intestines thin like a pencil. *Cooyam* means to travel for a long time in the sea or go through a mountain pass. These turtles were noted as coming from afar and swimming fast just beneath the surface of the water.

As an adolescent sea turtle growing up in the region, the *cooyam* was reclassified as *cooyam caacöl* 'cooyam large,' and matured into the *ipxom haquíma* 'its-fat the-most,' or "the fattest one." This turtle was found off the north and west coasts of Tiburón Island and between San Esteban and Tiburón islands—regions of deep, cool water. It was said that it was easily harpooned when found feeding on *copsíij*, Portuguese man-of-war (*Physalia* sp.), because it closed its eyes to avoid the stinging pneumatocysts. An older name for *Physalia* translated as "what the fattest one eats."

Green turtle meat is delicious and was a preferred food. After a turtle had been killed, the women cleaned it at the shore. With the turtle lying belly up, the throat was cut open transversely, and a woman reached in and pulled out the crop, stomach, and intestines. The stomach was washed in sea water and saved. A fire of *Frankenia* brush was burned on top of the plastron (Figure 3.11). When the fire had burned down, the plastron was cut off and the succulent white meat adhering to it was passed around like hors d'oeuvres. The remaining meat was then cut into large chunks and distributed according to a well-defined, although somewhat complicated, system of sharing. The recipients cooked the meat at their own hearths. The meat was roasted on a skewer or boiled (Sheldon 1979:

110; Whiting 1951:7). The fat or oil was extensively employed for cooking and blending with other foods, and was used whenever available. Sometimes the shell was broken up and boiled to obtain additional oil (Sheldon 1979:110).

Sea turtles do not nest in the region except on very rare, isolated occasions. The eggs, therefore, were not part of the Seri diet. Egg sacs of female turtles were sometimes found filled with unshelled, immature eggs. The very few times hatchling turtles were found were well-remembered, noted events. Sometimes the baby turtles were scooped up and kept as pets, but they soon died.

In addition to food, sea turtles were used for many other purposes. For example, turtle stomachs were made into water bags and all-purpose containers. The leather was used to make sandals. The carapaces were stacked against the base of brush houses and used as multi-purpose containers (see *Agave cerulata*, *Stenocereus thurberi*, and *Zostera*). They were not, however, used as cradles, as claimed by McGee (1898:187).

The olive Ridley, hawksbill, and loggerhead sea turtles were only occasionally eaten because the flavor was considered inferior to that of the green turtle. Tortoiseshell, from the hawksbill, was an item of trade since Spanish colonial times. In addition, the Seri made tortoiseshell finger rings for their own use (Griffen 1959:8).

Other common reptiles utilized for food include the desert tortoise (*Gopherus agassizi*; Felger, Moser, and Moser 1983), common chuckwalla (*Sauromalus obesus*), San Esteban chuckwalla (*S. varius*), spiny-tailed iguana (*Ctenosaurus*), and rattlesnakes (e.g., *Crotalus atrox*). Other snakes were apparently not eaten. The desert tortoise, chuckwallas, and iguana were particularly important, since they could easily be transported and kept alive for days or weeks. The Seri ate the bones of chuckwallas and iguana but did not eat the bones of larger animals, as claimed by McGee (1898:197).

Poisonous reptiles in the region include the Gila monster (*Heloderma suspectum*), rattlesnakes (*Crotalus* spp.), Arizona coral snake (*Micruroides euryxanthus*), and Pacific sea snake

Figure 3.11. Initial cooking of a green turtle. María Félix burning *Frankenia palmeri* on a freshly butchered turtle. *RSF, Campo Víboras, April 1973.*

(*Palemis platurus*), which is occasionally seen in the region during the summer. The Seri did not consider the coral snake to be dangerous and often had no compunction about handling it. They also regarded the sea snake as being nonpoisonous. However, a number of nonpoisonous lizards were thought to be dangerous, e.g., banded gecko (*Coleonyx variegatus*), desert night lizard (*Xantusia vigilis*; Felger 1965), and desert iguana (*Dipsosaurus dosalis*; also see Malkin 1962).

Birds

The general, or distributive, term for bird is *ziic*. Nearly every local species of land and sea bird was distinguished by name. At dawn pelicans, gulls, and other sea birds fly in long lines across the sky, heading for their daily fishing grounds. In the heat of the day frigate-birds soar in thermals above the sea, as do hawks and vultures over the land. Ospreys build huge stick nests in *cardón* branches, and egrets and Great Blue Herons are a common sight in the mangroves. The tops of scattered *cardones*, white with guano, serve as favorite roosts for Turkey Vultures, Ospreys, ravens, and caracaras. Hummingbirds feed from bright-colored tubular flowers of such plants as *Galvezia*, *Justicia*, and *Penstemon*. This group of plants was called *noj-oopis*, which implies a flower sucked by a hummingbird.

Most kinds of larger birds which could be captured were eaten, except for vultures, ravens, and hawks. Prominent among those hunted for food were the Brown Pelican, gulls, cormorants, booby, Royal Tern, killdeer, Greater Yellowlegs, Great Blue Heron, Great Horned Owl, Osprey, and various waterfowl. Most birds were hunted on a seasonal basis. The Great Blue Heron was considered the best-tasting bird (see Sheldon 1979:139).

The Seri apparently did not make traps nor use nets for capturing birds or any other kind of animal. For this reason quail and doves were not a regular part of their diet. Men hunted birds at night with torches and clubs, and during the day with harpoons, bows and arrows, or rocks. Women occasionally killed birds with sticks fashioned into makeshift clubs. Sling shots, perhaps a modern introduction, were occasionally used to kill smaller birds.

Brown Pelicans, cormorants, boobies, Greater Yellowlegs, and terns were taken in organized nighttime hunts at rookeries on the small islands of Alcatraz, Turners, Patos, and sometimes San Esteban. Predawn hours on dark, moonless nights were preferred (McGee 1898:190). It was said that on moonlit nights pelicans often slept on the water. The men would sneak up on the birds and then, often at a distance of about 50 m, light torches made from the wood of organ pipe (*Stenocereus thurberi*), charge and club the birds while they were still groggy and confused (Sheldon 1979:139). A hunting party of only a few men could kill several hundred birds during one of these hunts. The clubs were often made from ocotillo (*Fouquieria splendens*). Bunches of pelicans, held by their bills, were slung over the shoulder and carried back to boats on the shore or to camp.

Brown Pelicans were usually hunted during winter and spring, because they were fatter and had "more meat" than they did during hot weather. However, children often killed young pelicans at any time of year, probably for amusement as much as for food, as they did with any animals they could catch. Young gulls were hunted in March.

Certain large birds were harpooned when they had overeaten and were unable to fly away quickly. For example, White Pelicans, seasonal migrants through the region, were harpooned at Estero de la Cruz after they had gorged on large mullet. Certain larger terrestrial birds, such as the Great Horned Owl (see *Jacquinia*) and Osprey, were hunted by day with bow and arrow. Ospreys were frequently hunted when there were eggs in the nest. It was said that at such times the females did not leave the nests, or did not leave for very long. The hunters were thus able to approach close enough to kill them with bow and arrow. Quail and doves, only occasionally hunted, were killed with bow and arrow or slingshot.

When walking through the desert, people took eggs from the nests of almost any kind of bird and ate them raw. Egg-collecting trips were made to the larger sea bird rookeries, such as those at Patos and Alcatraz islands, and the mangrove estero at Punta Perla on the northeast side of Tiburón Island.

Pelican pelts were used extensively by men and women for clothing and sleeping mats and blankets. They were fashioned into kilts, or robes (see Chapter 12). During the late nine-teenth century Seri men made excursions by foot or balsa to Guaymas to trade pelican-skin robes or pelts for such items as liquor and cloth (see Figure 1.5). Usually six pelican pelts were incorporated into a robe or kilt, although four or eight might also be used. They were sewn together with sinew. Pelican bill pouches were made into breechclouts and baby clothes. Several pouches sewn together served as all-purpose personal carrying or storage bags.

Various species of young birds were sometimes kept as pets. A quail's head, stuffed with cloth, served as a child's toy, and bird bones were made into dolls. Birds also featured prominently in oral tradition (see *Atamisquea, Bursera hindsiana, Suaeda* spp., and *Zostera*).

Hawk and other raptor feathers were used for fletching arrows (McGee 1898:198). Raven feathers were occasionally made into a relatively elaborate headdress styled after Apache feathered headdresses (Bowen and Moser 1970a: 171–173). In the seventeenth century Adamo Gilg showed Seri men wearing feathers in their hair and as decorations on their arms (Di Peso and Matson 1965:51). Feathers were incorporated into fetishes (Hardy 1829:294) and used to decorate crowns worn by men (see *Jatropha cinerea*). They were also used in a game called *ziic ina xpaléemelc ano cöcazám* 'bird feather cone-shell(*Conus princeps*) in that-which-is-put,' or "the feather in the cone shell."

Mammals

Larger land mammals in the region included the mountain lion, bobcat, mule deer, white-tailed deer, pronghorn "antelope," desert bighorn sheep, javelina or collared peccary, coyote, badger, raccoon, black-tailed jackrabbit, antelope jackrabbit, and desert cottontail. The mountain lion, bighorn sheep, and pronghorn were rare or virtually extinct in the region by the 1960s or earlier. The Seri had knowledge of and names for both the spotted and black phases of jaguar.

The terrestrial mammalian fauna, relatively diverse on the mainland, is reduced in diversity on Tiburón Island. On San Esteban Island the only terrestrial mammals are a few species of rodents. Mule deer, coyote, jackrabbit, and several genera of rodents, including wood rat (pack rat) and kangaroo rat, are particularly numerous on Tiburón Island (Sheldon 1979:111, 112, 113). Sheldon (1979:118) thought that on Tiburón Island the abundant mule deer ranged into the niches usually filled by white-tailed deer and bighorn sheep on the mainland. On the island mule deer are most often found in the mountains, particularly near the water holes (Sheldon 1979:118). Native cats, white-tailed deer, pronghorn, javelina, and badger are absent from the island. Bighorn sheep were also not present but were introduced there by the Mexican federal wildlife department in 1975.

The mule deer (*Odocoileus hemionus*) was the most important terrestrial food animal. Terminology and knowledge associated with it was extensive. There were two terms for mule deer, *hap* and *ziix heecot quiih* 'thing desert that-which-is(on),' or "thing that lives in the desert." Four terms existed for kinds or variants of mule deer found only on Tiburón Island, and six terms were used to classify bucks by number and size of antler points.

There were terms for eight different methods which the men employed in hunting mule deer, involving the following techniques:

1. tracking and heading off a fleeing deer
2. locating deer in summertime as they rested during the heat of the day
3. luring a doe by holding a bleating fawn or by blowing through a leaf of ashy limberbush (*Jatropha cinerea*) to imitate a fawn's cry
4. using a deer-head decoy headdress (Bowen and Moser 1970a:174)
5. hiding behind a blind made from branches of elephant tree (*Bursera microphylla*) at a water hole on moonlit nights
6. hunting with dogs by both men and women

7. clubbing deer from a boat in summer while they were swimming near the shore in the Infiernillo Channel (the Seri said the deer try to cool off in the water)
8. chasing deer into the sea at a narrow, coastal peninsula and clubbing them from a boat

The first six methods involved the use of bow and arrow. These strategies allowed the hunter to get close enough to shoot the deer with a high degree of success.

The stuffed deer head was used as a decoy during the mating season by lone hunters (see *Jatropha cinerea*, *Larrea*, and *Prosopis*; Sheldon 1979:110, Figure 5–13). When the hunter encountered a buck he removed his clothing (because of the scent), hid his bow and arrow or rifle behind him, and pretended to browse on a bush while wearing the stuffed deer head. The buck was usually an easy target as it approached very close, lowering its head for a fight (Bowen and Moser 1970a:174). The stuffed deer head was also used in an abbreviated Yaqui-style deer dance (Bowen and Moser 1970a:173; Griffen 1959:17).

Sometimes a deer being stalked would suddenly see the hunter. The hunter might then distract or lull it into returning to feed by twisting or rotating the bow back and forth in front of him.

A mule deer usually headed toward the mountains when pursued, so the hunter could predict where to find it again. Sometimes, after pursuing a deer, the hunter would encounter it out of breath, leaning up against a sahuaro or *cardón*. A pregnant doe was sometimes tracked until she tired and could be approached at close range and killed. Spread hoof tracks told the hunter the doe was pregnant.

Meat of mule deer and other large animals was commonly roasted on a spit (Sheldon 1979:115, 126, 127). The fat of the mule deer, like that of most other food animals, was particularly relished and carefully conserved after the animal was killed (Sheldon 1979:118). The bones were cracked and the marrow eaten (Quinn and Quinn 1965:204; Sheldon 1979:139). Wild tepary beans (*Phaseolus acutifolius*) were particularly savored when boiled with deer meat and bones. Deerskins served many purposes, such as for clothing, sandals, and a surface for threshing and winnowing. The hides were expertly tanned (see Chapter 11). The awl for basketmaking was made from the metatarsal bone (see Chapter 15).

Pronghorn and white-tailed deer were occasionally seen but infrequently hunted because of the abundance of mule deer. Pronghorn were stalked during the mating season, when they could be approached and dispatched by bow and arrow. Men stalked desert bighorn sheep in the mountains and killed them with bow and arrow.

Javelinas were hunted primarily by the people of the Libertad Region. Men used dogs to locate them and then killed them with clubs. Sometimes a large stick was thrust down an angry javelina's throat, and other men clubbed or speared it with a metal-pointed harpoon. The Libertad Region people also hunted badgers, but elsewhere badgers were only occasionally eaten.

Jackrabbits, especially the antelope jackrabbit, were important food animals. Men hunted them with arrows tipped with solid knobs of heavy wood rather than risk breakage of the usual projectile points. A hit disabled the animal so that the hunter could approach and kill it. Jackrabbits were stalked with stuffed rabbit-head decoys and sometimes distracted by the hunter's rotating a stick (about a meter in length) or bow back and forth. One woman attracted jackrabbits with a split-cane musical instrument and then clubbed them (see *Phragmites*).

The cottontail was generally not regarded as a food animal. However, it was hunted by the people of the Libertad Region. Women opened wood rat (*Neotoma albigula*) nests, and the rodents as well as their caches of fruit and seeds were collected for food. The rock squirrel, round-tailed ground squirrel, and Harris antelope squirrel were also commonly eaten. A certain burrowing rodent, *hemeja*, perhaps a pocket gopher, was an important food resource on the coastal plain south of Kino Bay. These rodents were hunted with trained dogs.

Rabbits and rodents were commonly roasted on a spit or baked in a pit. Ground squirrels were prepared by removing the intestines and singeing off the hair; they were then baked in a hole dug in the ground. Wood rats were skinned and roasted in the ground under the coals of a hearth or campfire.

Many of the small, nocturnal rodents are conspicuous, particularly on Tiburón Island, because of their runways spreading across the desert floor. These include the white-throated wood rat, Merriam's kangaroo rat, pocket mice, and white-footed mice.

Coyotes and dogs were believed to have once been people. Dogs were not killed and coyotes were seldom killed. However, coyote fat was occasionally used for medicine, and the pelts were made into breechclouts prior to the twentieth century. In common with most Indians of western North America, the Seri regarded the coyote, *oot*, as a trickster and fool; as such, *oot* was often featured in stories (see *Phragmites* and *Prosopis*). False items and things of little use or value were commonly associated with the coyote (see *Bursera microphylla*, *Jatropha cinerea*, and *Passiflora arida*). One coyote story tells of the *pinacate* beetle (*Eleodes*) and the coyote.

One day a pinacate beetle was sitting in the road.

Along came a coyote passing that way. He saw the pinacate sitting there, and said to him, "What is my friend doing there, what is he doing sitting here? I am going to chew my friend," thus said the coyote.

Then the pinacate said, "Hmmmm! I am really doing something here."

Then the coyote said, "Whatever could he be doing?"

Then the pinacate said, "I'm listening to what they are saying down in the earth, that's why I am here."

Then the coyote said, "My friend, what are they saying? Don't hide their talk from me."

He said that, and the pinacate said thus: "All right, how will it come out? What will be the meaning of it? This is what they are saying. They are going to kill everything that is out on the road," said the pinacate.

Then the coyote said, "Well then, I'm going to run away," and, saying it, he ran. He ran from the thing sitting there.

He ran away, and the pinacate ran into a hole in the ground. He had lied to the coyote. He had lied and gotten away.

When disturbed, and also when running from possible danger, the *pinacate* raises its abdomen so that its head seems to touch the ground. This position seems to account for the beetle's "listening to what they are saying down in the earth."

Numerous dogs have been a characteristic feature of Seri camps and villages. Several people said that they acquired dogs from nearby ranchers in the nineteenth century. Dogs were trained to hunt mule deer, certain rodents, and desert tortoises (Felger, Moser, and Moser 1983). The Seri said that in the past there were feral dogs on Tiburón Island.

Among marine mammals, the California sea lion (*Zalophus californicus*)—seasonally abundant on San Esteban Island—was a locally important food resource. The San Esteban men stalked them during the day, and killed sleeping sea lions by throwing rocks at their heads. These people were known to be expert marksmen with rocks, and were able to mortally wound a sea lion even after it was in the surf heading for deep water. The sea lion was known as *xapóo* 'sea javelina.' The hide was used as protection from rain, as a ground cloth to sit or lie on, and to make sandals.

Porpoises and whales were well known to the Seri but were not hunted; they had no means or need to capture them. Porpoises were often accidentally drowned in turtle and fish nets, but neither the Seri nor the Mexican fishermen made use of them. Whales occasionally became stranded on the beaches and died, but the meat was apparently not utilized. The Seri believed that killer whales tear out the tongues of other whales. They said they had heard the moaning or crying of these dying whales across the water.

Horses, donkeys, and cows, introduced by the Spaniards, were extensively preyed upon or rustled for food. This became the major source of friction leading to bloodshed between the Seri and their Spanish-Mexican neighbors (Sheldon 1979:136). In fact, it was the basic cause of the Encinas Wars, which decimated the Seri in the late nineteenth century (McGee 1898:113).

PART II

Biological Ethnography

The Desert and the Sea in Seri Culture

4. The Seri Calendar

The Seri year consisted of twelve moons. Ecological information indicates that the beginning of each Seri moon corresponds approximately to the middle of the Gregorian calendar month. The name and cultural significance of each moon are briefly given below:

July: *icóozlajc iizax*
    'when-one-sprinkles moon'
    "moon to sprinkle"

This name derived from the sprinkling of hot sand on mesquite pods. This process was done to aid in toasting the pods before they were pounded in a mortar. This moon signaled the beginning of the new year to the Seri, as well as to other Sonoran Desert peoples. It is significant that the new year began when the common desert trees and shrubs dropped their seeds and the brief Sonoran Desert monsoon began, for this was the time of greatest renewal of life in this part of the world.

August: *hant yail iháat iizax*
    'land its-greenness time-of-ripening
      moon'
    "moon when the land is green"

Greenness refers to new vegetation or to the greenness of new foliage and summer ephemerals.

September: *azoj imaal icózim quih ano caap*
    'star what-does-not-accompany hot-
      season the in what-stands'
    "moon of no star during hot weather"

The Seri said that the moon brought a star with it except during two moons (months). This was one of the two months when the moon was not accompanied by a star. The word *azoj* can mean either a star or planet.

October: *queeto yaao*
    'Aldebaran its-path'

The star named *queeto* (Aldebaran, a red star in the constellation Taurus) appeared at the end of this moon. *Queeto* controlled other stars, bringing them in turn from below the horizon into the sky.

November: *hee yaao*
    'jackrabbit its-path'

This was the moon when the star named jackrabbit passed by.

December: *naapxa yaao*
    'turkey-vulture its-path'

This was the moon when the star called turkey vulture passed by.

January: *azoj imaal cmizj*
  'star what-does-not-accompany clear'
  "moon of no star"

This was the second month of the year (along with September) in which the moon was not accompanied by a star.

February: *cayaj zaac*
  'hunters small'
  "moon of few turtle hunters"

This was the moon when the *cooyam* first arrived and there were few hunters pursuing them. These were young green turtles (*Chelonia*), entering the Gulf of California for the first time.

March: *cayaj-aacoj*
  'hunters large'
  "moon of many turtle hunters"

During this moon many hunters pursued the *cooyam*, which were migrating northward into the region in large numbers. These turtles arrived in schools close to the shore and were so numerous that the people of the Sargento Region stood in the water and harpooned them with ease.

April: *xnois iháat iizax*
  'eelgrass-seed time-of-ripening moon'
  "moon of the eelgrass seed"

This was the season or time of the eelgrass harvest.

May: *iquéetmoj iizax*
  'shade-for-sitting moon'
  "moon for sitting in the shade"

This was the time to sit in the shade of a shelter—such as a ramada—out in the open, where there was a breeze, and enjoy the warm spring weather.

June: *imám imám iizax*
  'its-fruit when-it-is-ripe moon'
  "moon of the ripening cactus fruit"

This was the month when the giant cactus fruit began to ripen. The first term *imám* above refers to the fruit of the sahuaro, *cardón*, and organ pipe. The second instance of *imám* derives from the verb meaning 'to be ripe,' or 'cooked.'

5. Seri Classification of the Biological World

The Seri had names for most of the plants and animals in their territory. We have recorded 427 Seri names for plants (see Appendix A). Thirty-seven of these are for exotic plants which do not grow naturally in the region, such as sugarcane, cotton, and vegetables. Depending upon the region in which they lived, each person or group of people probably knew the names for about 350 to 400 plants, and at least half again as many animals.

There was no clear-cut system which the Seri were aware of for classifying plants and animals. There were some general patterns, but they had numerous exceptions. In discussing Seri classification and naming of plants, we are necessarily presenting information from a non-Seri point of view, and there is the problem of inventing concepts not present in the culture. Although some quantified data are given, these should be regarded as relative rather than precise values.

The classification of plants and animals was generally similar, although there were a few more higher categories for animals (Figure 5.1). The overall term for any animal, *ziix ccam* 'thing that-is-alive,' would have been used until one knew what the specific animal was. For example, one would use it for a larva if he did not know what it might turn into (see *Lycium fremontii*).

The general term for fish was *zixcám*, which derives from the overall term for animal. Another general term for fish was *zixcamáa* 'fish true.' The name for a large sea bass, *zixcám caacoj* 'fish large,' was also used as a general term for large fish. The generic word for bird was *ziic*.

A human being is *ziix quiisax* 'thing with-soul (or spirit or breath).' Humans are not considered to be a kind of *ziix ccam*, even though a human has *icám* 'its-life' and a *ziix ccam* has *ihíisax* 'its-breath.'

Various small lizards could collectively be called *haquímet*, although each species also had a distinct name. Excluded from this category were chuckwallas (*Sauromalus*), spiny-tailed iguana (*Ctenosaurus*), Gila monster (*Heloderma*), horned lizard (*Phrynosoma*), *huico* or whip-tail lizards (*Cnemidophorus*), and probably one or two others. Non-poisonous snakes, particularly large ones such as the gopher snake (*Pituophis*) were sometimes referred to as *ziix coimaj hant cöquiih* 'thing non-rattlesnake land what-is-on.'

Xica ccam heecot cocoom 'things with-life desert what-is-lying-on' was a seldom used life-form category. It was said to include animals which are fairly large, with fur, four legged, and stand off the ground. It therefore included larger terrestrial mammals.

Most of the Seri plant names were verified

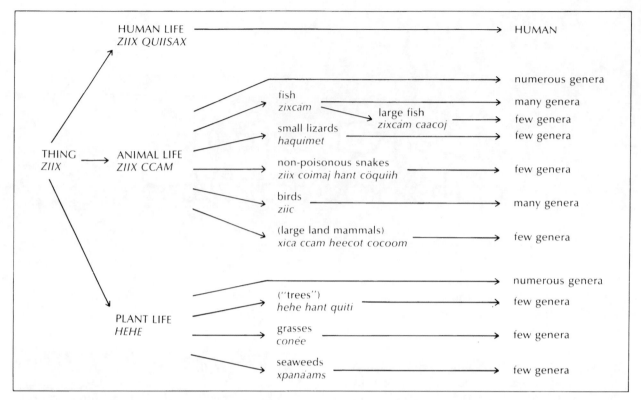

Figure 5.1. Seri classification of the biological world—the higher categories.

with living plants in the field at least several times during the span of our studies. Furthermore, after extensive study of the language, we did not find additional plant names. Like most folk taxonomies, the Seri classification put plants into numerous folk genera, fewer folk species, very few varietals, and few higher categories.

Hehe (*hehet*, plural) was the general term for any plant, regardless of size. This term could also signify wood, log, stick, branch, tree, pole, club, harpoon handle, meter stick, or mast. In addition, it was used to designate the very life of a plant (see Chapter 7). *Hehe* was incorporated into 34 descriptive plant names (see Appendix A). Most of these plants were not of major cultural significance. *Hehe* represented neither a folk genus nor a relationship among these plants.

Most Seri folk genera were not grouped into more inclusive or higher categories (see Figure 5.1). The only commonly used higher taxonomic categories for plants, other than *hehe*, were *xpanáams* and *conée*. *Xpanáams*, apparently derived from *xepe an ihíms* 'sea area its-fringe,' was used as an overall term for marine algae (seaweeds). It was also the folk genus (ethno-genus) for sargasso weed (*Sargassum*) and about one dozen other marine algae. In addition, *xpanéezj* 'sea membrane,' was sometimes used in a vague distributive manner for seaweeds not otherwise designated with a specific name. Eelgrass (*Zostera*) and ditch-grass (*Ruppia*) generally were not considered to be *xpanáams*.

The concept of *conée* covered at least a dozen species of ordinary-looking grasses. Other members of the grass family, such as saltgrasses and reedgrass, were not included. Examples of *conée* are:

Aristida spp., threeawn
Bouteloua spp., grama grass
Brachiaria arizonica, Arizona panic grass
Digitaria californica, cottontop
Eragrostis sp., lovegrass
Muhlenbergia microsperma, six-weeks muhly
Sporobolus cryptandrus, sand dropseed

Even though higher category plant terms were generally lacking, a number of relationships were understood and might be called covert categories or relationships (see Berlin, Breedlove, and Raven 1968). There was no single term for cactus or even for columnar cactus and also none for century plant (*Agave*). However, the Seri clearly knew that the various kinds of cacti were related, similar kinds of plants. There were similar concepts for the century plants and others.

Life-form terms for plants were few and seldom used. Plants which grow large and live for very many years could be considered as *hehe hant quiti*, but only when fully grown. *Hehe hant quiti* can be roughly equated to "tree," and glossed as 'plant land firmly-established,' or "permanent plant." *Hant quiti* implies something constant, long lasting, or even eternal. For example, a permanent water hole or spring is *hax hant quiti* "permanent water." Among the columnar cacti, only *cardón* (*Pachycereus*) was considered as an *hehe hant quiti*. Examples of other plants in this category included:

Acacia willardiana, *palo blanco*
Cercidium floridum, blue palo verde
C. microphyllum, foothill palo verde
Ficus petiolaris, cliff fig
Guaiacum coulteri, *guayacán*
Olneya tesota, ironwood
Prosopis glandulosa, mesquite

Vining plants were occasionally called *hehe iti scahjíit* "let's fall on the plant," the specific name of the vine *Brandegea*. However, to assign a well-defined, life-form meaning to this vague term would be exaggerating Seri usage.

The Seri were characteristically inconsistent in applying names to some of the more inconspicuous and culturally unimportant plants (Tables 5.1 and 5.2; Figure 5.2). They were well aware that certain names were synonyms for the same plant and that in some cases a single name covered more than one taxon or kind of plant. For example, although there are seven Seri names for the various barrel cacti or *siml* (*Ferocactus*), both the Seri and botanists realize there are only three kinds, or species, in the region. Some variation and synonymy might be the result of names originating from people of different geographical backgrounds and dialects.

Sometimes a newer name replaced an older one. When the name of a plant or animal was associated with someone young who died, the name was often changed. For example, the name *casol cacat* (*Hymenoclea monogyra*) became taboo because it was the favorite medicine of a young woman who died. However, the taboo did not seem to be widespread, or it was forgotten, because we found both the original and newer names in use (also see *Stenocereus gummosus* and *Zizyphus*). The newer names were descriptive, while the older names contained less descriptive terms.

The 427 Seri plant names or ethno-taxa (monotypic ethno-genera and ethno-species) represent approximately 411 botanical species. From this, one might conclude that overall the Seri tended to slightly over-differentiate plants (and probably animals, too). However, as mentioned above, this listing includes many names known to the Seri to be synonyms. Obvious over-differentiation was limited to a very few important animals and plants, such as sea turtles and mule deer (see Chapter 3), certain chollas (*Opuntia arbuscula* and *O. fulgida*), desert wolfberry (*Lycium andersonii*), and possibly blue palo verde (*Cercidium floridum*). However, it seems that in some cases the Seri recognized micro-races or populations unknown to biologists.

The total Seri flora of 427 plant names can be grouped into 310 ethno-genera (folk genera). Eighty-two percent (255) of the ethno-genera are monotypic; the remaining nearly 18 percent (55) are polytypic (two or more ethno-species) and contain 127 ethno-species (Table 5.3). Botanical classification places the 55 ethno-genera and 172 ethno-species into approximately 115 genera and 174 species. Ethno-varietal distinctions occur only among *Opuntia arbuscula* (*heem*) and *Pectis*.

Large, conspicuous, or culturally important plants tended to have unanalyzable names (Table 5.4). Approximately 26 percent (110) of the

plant names were unanalyzable terms. Many of the more conspicuous and culturally important ones were known by a name composed of a single word with only one or a few syllables (a monomorphemic lexeme). Such terms (unanalyzable monomial folk genera), referred to linguistically as primary names, characteristically suggest considerable cultural antiquity (Berlin 1973). Examples are *haas* (mesquite), *ool* (organ pipe), *siml* (barrel cactus), and *tis* (catclaw).

Nearly 75 percent (263) of the plant names

TABLE 5.1
Plants With Three or More Seri Names

Scientific Name	Seri Name	Scientific Name	Seri Name
Achyronychia cooperi	hant yapxöt tomítom hant cocpétij tomítom hant cocpétij caacöl	*Opuntia violacea*	heel saapom ziix istj captalca
Amphiroa beauvoisii *A. van-bosseae*	xepe oohit xpanáams ccapxl ziix hant cpatj oohit	*Pectis papposa*	cacátajc casol heecto casol ihasíi tiipe
Asparagopsis taxiformis	moosníil ihaquéepe moosn-oohit taca imas ziix hant cpatj oohit	*Phaseolus acutifolius* (domesticated)	teepar teepar cmasol teepar coopol teepar coospoj teepar cooxp
Digenia simplex	tacj oomas xepe oohit xpanáams coopol	*Rhodymenia divaricata*	moosni ipnáil ptcamn iha xpanáams cheel
Dithyrea californica	hant istj hehe imixáa moosni iha	*Rhodymenia hancockii*	moosni iha xpanáams coozlil xpanáams cöquihméel xpanáams ool
Ferocactus covillei	caail iti siml siml caacöl siml cöquicöt siml yapxöt cheel	*Tribulus terrestris*	cosi cahóota cözazni caacöl hee inóosj heen ilít hehe cosyat
Galaxaura fastigiata	hast iti coocp nojóo ixpanáams xepe oil xpanáams ccapxl	*Zinnia acerosa*	cmajíic ihásaquim mojépe ihásaquim cmaam saapom ipémt
Hymenoclea salsola	casol cacat casol coozlil casol ziix ic cöihíipe		
Machaeranthera parviflora	hee imcát hehe cacátajc zaah coocta		

*Plant not native to the Seri region.

TABLE 5.2
Seri Names Applied to Three or More Species

Seri Name	Scientific Name	Seri Name	Scientific Name
caatc ipápl	Abutilon incanum	hehe quiijam	Ipomoea sp.
	A. palmeri		Janusia californica
	Horsfordia alata		J. gracilis
cmajíic ihásaquim	Ambrosia deltoidea	isnáap ic is	Bouteloua aristidoides
	A. divaricata		B. barbata
	A. dumosa		B. repens
	A. magdalenae		Lepidium lasiocarpum
	Zinnia acerosa	moosni iha	Dithyrea californica
cocóol	Cleome tenuis		Palafoxia arida
	Descurainia pinnata		Rhodymenia hancockii
	*Sisymbrium irio	moosni ipnáil	Cryptonemia obovata
conée ccapxl	Brachiaria arizonica		Halymenia coccinea
	Eragrostis sp.		Padina durvillaei
	Erioneuron pulchellum		Rhodymenia divaricata
cpooj	Digitaria californica	noj-oopis caacöl	Galvezia juncea
	Sporobolus cryptandrus		*Nicotiana glauca
	unidentified grass		Penstemon parryi
hant iipzx iteja	Mammillaria microcarpa	tomítom hant cocpétij	Achyronychia cooperi
	M. sheldonii		Euphorbia petrina
	M. spp.		E. polycarpa
hant iipzx iteja caacöl	Mammillaria estebanensis		E. setiloba
	Echinocereus engelmannii	xcoctz	Ambrosia deltoidea
	E. fendleri		A. divaricata
	E. grandis		A. dumosa
	E. pectinatus		A. ilicifolia
hap oacajam	Caesalpinia palmeri		A. magdalenae
	Echinopterys eglandulosa	xepe an ihíms	Hypnea valentiae
	Thryallis angustifolia		Laurencia johnstonii
hasoj an hehe	Andrachne ciliato-		Spyridia filamentosa
	glandulosa	xepe an impós	Gelidiopsis variabilis
	Teucrium glandulosum		Gymnogongrus johnstonii
	unidentified Cruciferae		Hypnea valentiae
hast ipénim	Camissonia californica		Laurencia johnstonii
	Kallstroemia grandiflora		Spyridia filamentosa
	Eschscholzia parishii	xepe oohit	Amphiroa beauvoisii
haxz iiztim	Calliandra eriophylla		A. van-bosseae
	Hoffmanseggia intricata		Digenia simplex
	Krameria parvifolia	xpanáams ccapxl	Amphiroa beauvoisii
hee inóosj	Lotus salsuginosus		A. van-bosseae
	L. tomentellus		Galaxaura fastigiata
	*Tribulus terrestris	xtamóosn-oohit	Chaenactis carphoclinia
hehe cotópl	Cryptantha angustifolia		Chorizanthe brevicornu
	C. maritima		Fagonia californica
	C. spp.		F. pachyacantha
	Mentzelia adhaerans	ziix hant cpatj oohit	Amphiroa beauvoisii
	M. involucrata		A. van-bosseae
	Perityle emoryi		Asparagopsis taxiformis
hehe czatx	Cryptantha angustifolia		Dictyota flabellata
	C. maritima		Galaxaura arborea
	C. spp.		
	Mentzelia adhaerans		

*Plant not native to the Seri region.

TABLE 5.3
Polytypic Ethnogenera

Seri Name	Scientific Name	Seri Name	Scientific Name
MORE THAN TEN ETHNOSPECIES		caail iti siml	Ferocactus covillei
		mojépe siml	F. acanthodes
xpanáams	Sargassum herporhizum	siml	F. wislizenii
	S. sinicola	simláa	F. wislizenii
xpanáams caacöl	Sargassum sinicola	siml caacöl	F. covillei
xpanáams caitic	Dasya baillouviana	siml cöquicöt	F. covillei
xpanáams ccapxl	Amphiroa beauvoisii	siml yapxöt cheel	F. covillei
	A. van-bosseae		
	Galaxaura fastigiata	**FIVE ETHNOSPECIES**	
xpanáams cheel	Rhodymenia divaricata		
xpanáams coil	Enteromorpha	hahöj an quinelca	Lycium brevipes
	acanthophora	hahöj cacat	L. fremontii
xpanáams coopol	Digenia simplex	hahöj-enej	L. andersonii
xpanáams coozlil	Rhodymenia hancockii	hahöj ináil coopol	L. andersonii
xpanáams cöquihméel	Rhodymenia hancockii	hahöj-izij	L. californicum
xpanáams cquihöj	Lomentaria catenata		
	Plocamium cartilagineum	heem	Opuntia arbuscula
xpanáams hasít	unidentified marine alga	heemáa	O. arbuscula
xpanáams isoj	Sargassum herporhizum	heem icös cmasl	O. arbuscula
	S. sinicola		O. cf. burrageana
xpanáams itojípz	Lomentaria catenata	heem icös cmaxlilca	O. arbuscula
xpanáams mojépe	unidentified marine alga		O. versicolor
xpanáams oaf	Sargassum sinicola	hepem ihéem	O. versicolor
xpanáams ool	Rhodymenia divaricata		
xpanáams xaasj	unidentified	teepar	*Phaseolus acutifolius
			(domesticated)
		teepar cmasol	*P. acutifolius (domesticated)
		teepar coopol	*P. acutifolius (domesticated)
		teepar coospoj	*P. acutifolius (domesticated)
SIX OR MORE ETHNOSPECIES		teepar cooxp	*P. acutifolius (domesticated)
casol caacöl	Baccharis sarothroides	**FOUR ETHNOSPECIES**	
	Dodonaea viscosa		
casol cacat	Haplopappus sonorensis	caay ixám	*Cucurbita pepo
	Hymenoclea monogyra	coxi ixám	*Citrullus lanatus
casol coozlil	H. monogyra	xam	*Cucurbita mixta/moschata
	H. salsola	xam coozalc	*C. pepo
casol heecto	Pectis papposa		
casol heecto caacöl	P. palmeri	conée caacöl	unidentified grass
casol ihasíi tiipe	P. papposa	conée ccapxl	Brachiaria arizonica
casol itac coosotoj	Hymenoclea salsola		Eragrostis sp.
casol ziic ic cöihíipe	Haplopappus sonorensis		Erioneuron pulchellum
	Hymenoclea monogyra	conée cosyat	E. pulchellum
		conée csai	Aristida californica
coquée	Lepidium lasiocarpum		
coquée caacöl	*Capsicum annuum	hataj-en	Plantago insularis
coquée coil	*C. annuum	hataj-ipol	Suaeda moquinii
coquée coopol	*Piper nigrum	hataj-isijc	Atriplex canescens
coquée quitajij	*Capsicum annuum		A. linearis
	annuum	hataj-ixp	A. polycarpa
coquée quizil	C. annuum aviculare		A. linearis

TABLE 5.3
(continued)

Seri Name	Scientific Name	Seri Name	Scientific Name

FOUR ETHNOSPECIES (cont.)

Seri Name	Scientific Name
heel	*Opuntia violacea*
heel cooxp	**O. ficus-indica*
heel hayéen ipáii	*O. phaeacantha*
heel cocsar yaa	**O. ficus-indica*
hehe pnaacoj	*Sideroxylon leucophyllum*
pnaacoj hacáaiz	*Laguncularia racemosa*
pnaacoj-iscl	*Avicennia germinans*
pnaacoj-xnazolcam	*Rhizophora mangle*
spitj	*Atriplex barclayana*
spitj caacöl	*Sesuvium verrucosum*
spitj cmajíic	*Abronia maritima*
spitj ctamcö	*Sesuvium verrucosum*
xapij	**Arundo donax*
	Phragmites australis
xapij-aacöl	**Arundo donax?*
xapij-aas	**bamboo?*
xapij coatöj	**Saccharum officinarum*
ziim caacöl	**Salsola kali*
ziim caitic	*Amaranthus fimbriatus*
ziim quicös	*A. watsonii*
ziim xat	*Chenopodium murale*

THREE ENTHOSPECIES

Seri Name	Scientific Name
haamxö	*Agave subsimplex*
haamxö caacöl	*A. colorata*
	A. fortiflora
haamxöíi	**Ananas comosus*
	Hechtia montana
hamíp	*Boerhaavia coulteri*
hamíp caacöl	*B. erecta*
hamíp cmaam	*Allionia incarnata*
hant-oosinaj	*Abronia villosa*
hant-oosinaj cooxp	*Oenothera californica*
hant-oosinaj ctam	*Camissonia claviformis*
sahmées	**Citrus sinensis*
sahmées ccapxl	**C. limon*
sahmées hamt caháacöl	**C. paradisi*
sea	*Opuntia bigelovii*
sea cotópl	*O. fulgida*
sea icös cooxp	*O. fulgida*

Seri Name	Scientific Name
tee	*Chorizanthe corrugata*
	Eriogonum inflatum
tee caacöl	*Senecio douglasii*
tee cmaam	*Eriogonum trichopes*
xoop	*Bursera microphylla*
xoop caacöl	*B. laxiflora*
xoop inl	*B. hindsiana*

TWO ETHNOSPECIES

Seri Name	Scientific Name
caal oohit	*Triteleiopsis palmeri*
caal oohit caacöl	*Brodiaea pulchella*
caháahazxot	*Baileya multiradiata*
	Dyssodia concinna
caháahazxot ctam	*Phacelia ambigua*
cocazn-ootizx	*Trixis californica*
cocazn-ootizx caacöl	*Verbesina palmeri*
cocóol	*Cleome tenuis*
	Descurainia pinnata
	*Sisymbrium irio
cocóol cmaam	*Draba cuneifolia*
coote	*Opuntia bigelovii*
coteexet	*O. fulgida*
cozi	*Condalia globosa*
cozi hax ihapóin	*Tephrosia palmeri*
cözazni	*Cenchrus palmeri*
cözazni caacöl	**Tribulus terrestris*
	*Xanthium strumarium
haap	*Phaseolus acutifolius*
haamoja iháap	*P. filiformis*
hanaj iit ixac	*Marina parryi*
hanaj itámt	*Dalea mollis*
hant iipzx iteja	*Mammillaria microcarpa*
	M. spp.
hant iipzx iteja caacöl	*M. estebanensis*
	Echinocereus spp.
hapats imóon	**Phaseolus lunatus*
yori imóon	**Vigna unguiculata*

consist of analyzable terms, or contain an analyzable or translatable portion. The names often indicate contrast with another well-known plant. The majority of the analyzable names reflect descriptive characteristics of the plant, cultural function or use, physiological function or effect on humans or animals, or relationships with animals. Some plant names consist of combinatons of these attributes.

Approximately 22 percent (93) of the Seri plant names reflect descriptive characteristics of color or morphological features, such as size, shape, spines, and resemblance to another plant (Table 5.5). Thirty-three of these names indicate

TABLE 5.3
(continued)

Seri Name	Scientific Name	Seri Name	Scientific Name
TWO ETHNOSPECIES (cont.)		tomítom hant cocpétij	Euphorbia polycarpa
hapis casa	Nicotiana trigonophylla		E. spp.
hapis coil	*Cannabis sativa		Achyronychia cooperi
		tomítom hant cocpétij caacöl	Euphorbia tomentulosa
hast ipénim	Camissonia californica		
	Eschscholzia parishii	xazácöz	*Argemone mexicana
	Kallstroemia grandiflora		A. pleiacantha
hast ipénim ctam	Senecio douglasii	xazácöz caacöl	A. subintegrifolia
hee imcát	Machaeranthera parviflora	xoját	Amoreuxia palmatifida
	Perityle emoryi	xoját hapéc	*Solanum tuberosum
hee imcát caacöl	P. leptoglossa		
		xomxéziz	Fouquieria splendens
hehe czatx	Cryptantha angustifolia	xomxéziz caacöl	Chorizanthe rigida
	C. maritima		
	C. spp.	xonj	Proboscidea altheifolia
hehe czatx caacöl	Argythamnia neomexicana	xonj caacöl	P. parviflora
	A. lanceolata		
		xneejam is hayáa	Viscainoa geniculata
hehe quina	Notholaena standleyi	xneejam-siictoj	Stegnosperma halimifolium
hehe quina caacöl	Selaginella arizonica		
		xpanéezj	Enteromorpha
jcoa	Sphaeralcea ambigua		acanthophora
	S. coulteri		unidentified marine alga
hepem ijcóa	Hibiscus denudatus	xpanéezj cheel	unidentified marine alga
jöene	Passiflora palmeri	xtoozp	Physalis crassifolia
oot ijöéne	P. arida	xtoozp hapéc	*Lycopersicon esculentum
noj-oopis	Justicia californica	zaaj iti cocái	Eucnide rupestris
noj-oopis caacöl	Galvezia juncea	zaaj iti cocái cooxp	Mentzelia involucrata
	*Nicotiana glauca		
	Penstemon parryi	zamij cmaam	Brahea aculeata
		zamij ctam	Sabal uresana and/or
ool	Stenocereus thurberi		Washingtonia robusta
ool-axö	S. gummosus		
		zazjc	Castela polyandra
taca oomas	Codium simulans	zazjc caacöl	C. emoryi
	Galaxaura arborea		
taca oomas cooxp	Codium amplivesticulatum		

*Plant not native to the Seri region.

TABLE 5.4
Unanalyzable Plant Names

Seri Name	Scientific Name	Seri Name	Scientific Name	Seri Name	Scientific Name
aaxt	Phorandendron californicum	heecoj	Roccella babingtonia	snaazx	Castela polyandra
	P. diguetianum	heejac	Pithecellobium confine	snapxöl	*Parkinsonia aculeata
caaöj	Baccharis salicifolia	heel	Opuntia violacea	spitx	Atriplex barclayana
cap	Acacia willardiana	heem	O. arbuscula	tacs	Allenrolfea occidentalis
coap	Cnidoscolus palmeri	heepol	Krameria grayi	tincl	Ambrosia ambrosioides
cocásjc	Jouvea pilosa	heetes	unidentified	tis	Acacia greggii
cocóol	Cleome tenuis	heme	Agave cerulata	tomáasa	Oligomeris linifolia
	Descurainia pinnata	hexe	Sarcostemma cynanchoides	tonóopa	Vallesia glabra
	*Sisymbrium irio	iipxö	Opuntia leptocaulis	xaasj	Pachycereus pringlei
cof	Jacquinia pungens	iiz	Cercidium floridum	xam	*Cucurbita mixta/moschata
comáanal	Anemopsis californica	impós	Aristida adscensionis	xapij	*Arundo donax
comíma	Brickellia coulteri		Muhlenbergia microsperma		Phragmites australis
	Eupatorium sagittatum	insáacaj	Physalis cf. pubescens	xasáacoj	Stenocereus alamosensis
comítin	Olneya tesota	jcoa	Sphaeralcea ambigua	xazácöz	*Argemone mexicana
comot	Matelea cordifolia		S. coulteri		A. pleiacantha
coote	Opuntia bigelovii	jöene	Passiflora palmeri	xcocoj	Struthanthus haenkeanus
cos	Maytenus phyllanthoides	maas	Cercidium praecox	xeescl	Hyptis emoryi
coptoj	Agave angustifolia	mahyan	unidentified	xjii	*Lagenaria siceraria
coquée	Lepidium lasiocarpum	mocni	Guaiacum coulteri	xloolcö	Erythrina flabelliformis
cototax	Fouquieria columnaris	mojépe	Carnegiea gigantea	xojásjc	Sporobolus virginicus
cozi	Condalia globosa	mooj	*Gossypium spp.	xoját	Amoreuxia palmatifida
cöset	Atamisquea emarginata	najcáazjc	Asclepias albicans	xomcahíift	Lippia palmeri
cötep	Monanthochloe littoralis		A. subulata	xomcahóij	Opuntia marenae
cpooj	Digitaria californica	najmís	Phacelia ambigua		O. reflexispina
	Sporobolus cryptandrus	nas	Matelea pringlei	xomée	Marsdenia edulis
	unidentified grass	ool	Stenocereus thurberi	xométe	Psorothamnus emoryi
eaz	Zostera marina	paaza	Bumelia occidentalis	xomxéziz	Fouquieria splendens
haaca	Zizyphus obtusifolia	paij	Salix gooddingii	xonj	Proboscidea altheifolia
haacoz	*Melilotus indica	pat	Typha domingensis	xooml	Koeberlinia spinosa
haalp	Randia thurberi	paxáaza	Ambrosia cf. confertifolia	xoop	Bursera microphylla
haamxö	Agave subsimplex	pnaacöl	Simmondsia chinensis	xpaasni	Ficus petiolaris
haap	Phaseolus acutifolius	ptaacal	Celtis pallida	xtisil	Porophyllum gracile
haas	Prosopis glandulosa	ptaact	Colubrina viridis	xtooxt	Neoevansia striata
haat	Jatropha cuneata	pteept	Euphorbia eriantha	xtoozp	Physalis crassifolia
hamat	Yucca arizonica	saapom	Opuntia violacea	zai	unidentified grass
hamíp	Boerhaavia coulteri	sahmées	*Citrus sinensis	zazjc	Castela polyandra
hamísj	Jatropha cinerea	sapátx	Bebbia juncea	ziij	Cercidium floridum
hamoc	Agave angustifolia	satóoml	Ruellia californica	ziipxöl	Cercidium microphyllum
hapxöl	*Zea mays	sea	Opuntia bigelovii	ziizil	Muhlenbergia microsperma
hasac	Setaria macrostachya	seepol	Frankenia palmeri		Setaria liebmannii
heecl	Jatropha cardiophylla	siml	Ferocactus wislizenii	znaazj	Castela polyandra

*Plant not native to the Seri region.

67

size contrast of the entire plant or part of it (such as the fruit): 29 of these are distinguished as large (*caacöl*) but only four as small (*heecto*). These size-related terms are formal names used in the culture, and they generally reflect botanical relationships. The diminutive *Chorizanthe rigida* is called "large ocotillo," and *Pectis pal-*

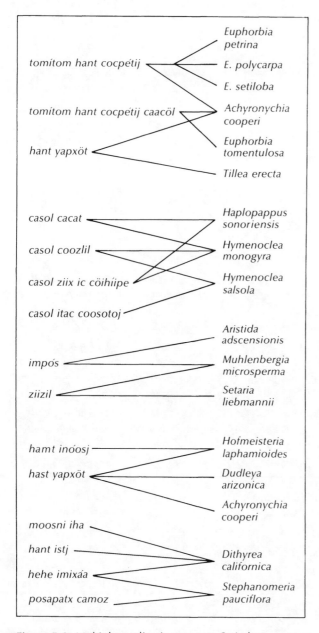

Figure 5.2. Multiple applications among Seri plant names.

meri, one of the very few varietals, is "large small-*casol*." Low, spreading plants have female connotations, while tall, slender ones are male: six plants are named as female, five as male.

Color, or even the lack of it (drab), is a modifying term in 32 plant names. The name for *Lantana* "flames of fire," refers to the color change from orange to red as the flowers age. Spinose condition is apparent in a number of plant names with terms such as prickly, spiny, tangled, pinches, clings, and stickery. Glabrous plants, or ones lacking objectionable pubescence and spines, were termed *caitij* 'soft (plural)' plants and were culturally important for roofing, food preparation, and other uses. For example, *Amaranthus fimbriatus* was named "soft *ziim*" in contrast to other kinds of *ziim*—such as *A. watsonii*, called "prickly *ziim*."

Other names descriptive of morphological or anatomical features reflect characters of the fruit (e.g., *Lycium brevipes*), wood (e.g., *Coursetia*), root (e.g., *Stephanomeria*), and shape (e.g., *Opuntia ficus-indica*). *Teucrium* was called "ribbed stemmed plant," in reference to its square stems, a characteristic of the mint family. *Pitaya agria* (*Stenocereus gummosus*) was known as "organ pipe excrement," perhaps because of its untidy appearance as compared with organ pipe (*S. thurberi*). Fifteen plant names indicated resemblance to other, more common and culturally important ones. These were poetically named as plants which "think" that they are another plant (e.g., *Acacia farnesiana*).

Seven plant names indicate physiological or growth functions of the plant (Table 5.6). For example, the common desert lupine (*Lupinus*) and desert aster (*Machaeranthera*) were named "sun watcher," referring to their tracking the sun—the phototropic response of the leaves and flower heads turning through the day to face the sun. The legume shrub, *Caesalpinia palmeri*, was called "burst open." When mature, the pods explosively burst open, flinging the seeds away from the parent plant.

Some names combine a descriptive morphological character and another feature such as habitat, function, or cultural analogy. For exam-

TABLE 5.5
Names Reflecting Morphological, Anatomical, or Descriptive Features of the Plant

Seri Name	Scientific Name	Seri Name	Scientific Name
LARGE		xica caacöl	*Zea mays*
caal oohit caacöl	Brodiaea pulchella	"large things"	
"large blue sand lily"		xomxéziz caacöl	Chorizanthe rigida
casol caacöl	Baccharis sarothroides	"large ocotillo"	
"large casol"		xonj caacöl	Proboscidea parviflora
casol heecto caacöl	Pectis palmeri	"large devil's claw"	
"large cinchweed"		xoop caacöl	Bursera laxiflora
cocazn-ootizx caacöl	Verbesina palmeri	"large elephant tree"	
"large *Trixis*"		xpanáams caacöl	Sargassum sinicola
com-aacöl	Antirrhinum kingii	"large seaweed"	
"com-large"	Trianthema portulacastrum	zazjc caacöl	Castela emoryi
conée caacöl	unidentified grass	"large *Castela*	
"large grass"		polyandra"	
coquée caacöl	*Capsicum annuum	ziim caacöl	*Salsola kali
"large chiles"		"large *ziim*"	
cözazni caacöl	*Tribulus terrestris		
"large sandbur"	*Xanthium strumarium	**SMALL:**	
haamxö caacöl	Agave colorata	casol heecto	Pectis papposa
"large agave"	A. fortiflora	"small *casol*"	
hamíp caacöl	Boerhaavia erecta	coquée quizil	Capsicum annuum
"large spiderling"		"small chiles"	aviculare
hant iipzx iteja caacöl	Mammillaria estebanensis	hahöj-izij	Lycium californicum
"large fishhook cactus"	Echinocereus spp.	"little *Lycium*"	
hee imcát caacöl	Perityle leptoglossa	hehe is quisil	Lophocereus schottii
"large rock daisy"		"small-fruited plant"	
hehe caacoj	*Ricinus communis		
"large plant"		**FEMALE**	
hehe czatx caacöl	Argythamnia neomexicana	cocóol cmaam	Draba cuneifolia
"large stickery plant"	A. lanceolata	"female tansy mustard"	
hehe quina caacöl	Selaginella arizonica	hamíp cmaam	Allionia incarnata
"large rock fern"		"female spiderling"	
noj-oopis caacöl	Galvezia juncea	mojépe ihásaquim cmaam	Zinnia acerosa
"large chuparosa"	*Nicotiana glauca	"female sahuaro	
	Penstemon parryi	hairbrush"	
siml caacöl	Ferocactus covillei	spitj cmajíic	Abronia maritima
"large barrel cactus"		"female coastal saltbush"	
spitj caacöl	Sesuvium verrucosum	tee cmaam	Eriogonum trichopes
"large coastal saltbush"		"female desert trumpet"	
tee caacöl	Senecio douglasii	zamij cmaam	Brahea aculeata
"large desert trumpet"		"female palm"	
tomítom hant cocpétij	Euphorbia tomentulosa		
caacöl	Achyronychia cooperi	**MALE**	
"large spurge (*Euphorbia*		caháahazxot ctam	Phacelia ambigua
polycarpa/spp.)"		"male desert marigold"	
xapij-aacöl	*Arundo donax	hast ipénim ctam	Senecio douglasii
"large reedgrass"		'rock splattered-with	
xazácöz caacöl	Argemone subintegrifolia	male'	
"large prickly poppy"		hant-oosinaj ctam	Camissonia claviformis
		"male sand verbena"	

69

TABLE 5.5
(continued)

Seri Name	Scientific Name	Seri Name	Scientific Name
MALE (cont.)		teepar coospoj	*Phaseolus acutifolius
spitj ctamcö	Sesuvium verrucosum	"spotted tepary"	(domesticated)
"male coastal saltbush"		teepar cooxp	*Phaseolus acutifolius
zamij ctam	Sabal uresana and/or	"white tepary"	(domesticated)
"male palm"	Washingtonia robusta	xica imám coopol	Zizyphus obtusifolia
		"black-fruited things"	
COLOR		xica is cheel	*Phaseolus vulgaris
coquée coil	*Capsicum annuum	"red-seeded things"	
"green chile"		xica ihíijim coopl	*Vigna unguiculata
coquée coopol	*Piper nigrum	"black-eyed things"	
"black chile"		xneejam-siictoj	Stegnosperma halimifolium
hacáiin cooscl	Croton californicus	"xneejam red"	
"drab windbreak"		xpanáams cheel	Rhodymenia divaricata
hahöj ináil coopol	Lycium andersonii	"red seaweed"	
"black-barked Lycium"		xpanáams coil	Enteromorpha
hamácj inoloj	Lantana horrida	"green seaweed"	acanthophora
"flames of fire"		xpanáams coopol	Digenia simplex
hant-oosinaj cooxp	Oenothera californica	"black seaweed"	
"white sand verbena"		xpanáams cquihöj	Lomentaria catenata
hapis coil	*Cannabis sativa	"red seaweed"	Plocamium cartilagineum
"green tobacco"		xpanáams cöquihméel	Rhodymenia hancockii
heel cooxp	*Opuntia ficus-indica	"purple seaweed"	
"white prickly-pear"		zaaj iti cocái cooxp	Mentzelia involucrata
heem icös cmasl	Opuntia arbuscula	"white Eucnide	
"yellow-spined pencil	O. cf. burrageana	rupestris"	
cholla"		ziix is cmasol	Cucurbita digitata
hehe imoz coopol	Viguiera deltoidea	"yellow-fruited thing"	
"black-barked plant"		ziix is cquihöj	*Phaseolus vulgaris
hehe iix coil	Lycium macrodon	"red-seeded thing"	
"blue/green-sapped			
plant"			
hehe iix cooxp	Euphorbia xanti	**SURFACE, SPINE CONDITION, TEXTURE**	
"white-sapped plant"		an icös	Phaulothamnus spinescens
hehet ináil coopl	Cordia parvifolia	"thorny inside"	
"black-barked plant"		casol coozlil	Hymenoclea monogyra
pnaacoj-iscl	Avicennia germinans	"sticky casol"	H. salsola
"drab mangrove"		conée cosyat	Erioneuron pulchellum
sea icös cooxp	Opuntia fulgida	"spiny grass"	
"white-spined teddybear		cözazni	Cenchrus palmeri
cholla"		'tangled'	
siml yapxöt cheel	Ferocactus covillei	heem icös cmaxlilca	Opuntia versicolor
"red-flowered barrel		"stiff-spined pencil	
cactus"		cholla"	
taca oomas cooxp	Codium amplivesticulatum	hehe coozlil	Horsfordia alata
"white Codium"		"sticky plant"	
teepar cmasol	*Phaseolus acutifolius	hehe cosyat	*Tribulus terrestris
"yellow tepary"	(domesticated)	"spiny plant"	
teepar coopol	*Phaseolus acutifolius	hehe cotázita	Mimosa laxiflora
"black tepary"	(domesticated)	"plant that pinches"	

TABLE 5.5
(continued)

Seri Name	Scientific Name	Seri Name	Scientific Name

SURFACE, SPINE CONDITION, TEXTURE (cont.)

hehe cotópl	*Cryptantha angustifolia*	hehe is quiixlc	*Punica granatum
"plant that clings"	*C. maritima*	"round-fruited plant"	
	C. spp.	hehe itac coozalc	*Teucrium cubense*
	Mentzelia adhaerans	"ribbed-stem plant"	*T. glandulosum*
	M. involucrata	hehe-monlc	*Antirrhinum cyathiferum*
	Perityle emoryi	"curly plant"	
hehe czatx	*Cryptantha angustifolia*	inyéeno	*Agave pelona*
"stickery plant"	*C. maritima*	"faceless"	
	C. spp.	isnáap ic is	*Bouteloua aristidoides*
	Mentzelia adhaerans	"whose fruit is on one	*B. barbata*
hehe quina	*Notholaena standleyi*	side"	*B. repens*
"hairy plant"			*Lepidium lasiocarpum*
sea cotópl	*Opuntia fulgida*	ool-axö	*Stenocereus gummosus*
"clinging teddybear		"organ pipe's excrement"	
cholla"		pnaacoj-xnazolcam	*Rhizophora mangle*
taca-noosc	*Eucheuma uncinatum*	"crisscrossed mangrove"	
"its jaw's papillae"		tomítom hant cocpétij	*Euphorbia polycarpa*
tacj-anóosc	*Gigartina johnstonii*	"prostrate *tomítom*"	*E. spp.*
"porpoise's papillae"	*G. pectinata*		*Achyronychia cooperi*
xepe zatx	*Dictyota flabellata*	xam coozalc	*Cucurbita pepo
'sea glochid'		"ribbed squash"	
xpanáams caitic	*Dasya baillouviana*	xica coosotoj	*Oryza sativa
"soft seaweed"		"thin things"	
xpanáams coozlil	*Rhodymenia hancockii*	xica quiix	*Setaria macrostachya*
"sticky seaweed"		"globular things"	
ziim caitic	*Amaranthus fimbriatus*	xoop inl	*Bursera hindsiana*
"soft ziim"		"elephant tree's fingers"	
ziim quicös	*Amaranthus watsonii*	ziim xat	*Chenopodium murale*
"prickly ziim"		'ziim hail'	
ziix is quicös	*Matelea pringlei*	ziix istj captalca	*Opuntia violacea*
"prickly-fruited thing"		"wide-leaved thing"	

SHAPE, GROSS MORPHOLOGY

NAMES SUGGESTING A COMPARISON TO ANOTHER PLANT

casol itac coosotoj	*Hymenoclea salsola*	hehe pnaacoj	*Sideroxylon leucophyllum*
"thin-stemmed *casol*"		"mangrove plant"	
com-ixaz	*Cocculus diversifolius*	mojépe siml	*Ferocactus acanthodes*
"com-rattle"		'sahuaro barrel-cactus'	
coquée quitajij	*Capsicum annuum	paar icomíhlc	*Pisum sativum
"pointed chiles"		"padre's mesquite seed"	
hahöj an quinelca	*Lycium brevipes*	paar icomítin	*Cicer arietinum
"empty *Lycium*"		"padre's ironwood seed"	
hahöj-enej	*Lycium andersonii*	poháas camoz	*Acacia farnesiana*
"empty *Lycium*"		"what thinks it's a	*Desmanthus covillei*
hasot	*Agave chrysoglossa*	mesquite"	
'narrow'		ponás camoz	*Lyrocarpa coulteri*
hehe ctoozi	*Coursetia glandulosa*	"what thinks it's a	
"resilient plant"		*Matelea pringlei*"	
hehe imixáa	*Stephanomeria pauciflora*	posapátx camoz	*Stephanomeria pauciflora*
"rootless plant"	*Dithyrea californica*	"what thinks it's a sweet-	
		bush"	

ple, fishhook cactus (*Mammillaria microcarpa*) is "arroyo's bladder," dodder vine (*Cuscuta*) is "soil's intestines," and the large common mushrooms (*Battarrea* and *Podaxis*) are "land's foreskin."

Indications of habitat or of relationship with the land can be found in 34 plant names (Table 5.7). For example, "cliff hanger" is the name for *Eucnide*, a plant characteristically found on cliffs or rock faces. The name for the marine alga *Cutleria* "what sways on rock" alludes to both wave action and the holdfast attached to rock.

Thirty plant names indicate a cultural function or relationship and subsistence (Table 5.8). These names allude to such uses as water and food gathering, food preparation, cosmetic use, facepainting, and gambling. Several cultivated exotic plants are named as analogs of native plants, e.g., the common potato (*Solanum tuberosum*) and tomato (*Lycopersicum*). Only three names provide indication of use as food, and only one refers directly to medicinal usage, although medicinal application is implied for *Aristolochia*. Only five ethno-species are named as parts of the body:

hataj-ipol "black vulva"
  Atriplex canescens
  A. linearis

hataj-isijc "immature vulva"
  Atriplex canescens
  A. linearis

hataj-ixp "white vulva"
  A. linearis
  A. polycarpa

hataj-en "inside vulva"
  Plantago insularis

xpeetc 'sea scrotum'
  Colpomenia tuberculata

Twenty-one plant names reflect human physiological effects (Table 5.9). Seven names are indicative of taste, five of toxicity, five of smell,

TABLE 5.5
(continued)

Seri Name	Scientific Name
NAMES SUGGESTING A COMPARISON TO ANOTHER PLANT	
potács camoz "what thinks it's an iodine bush"	*Heliotropium curassavicum*
xepe an impós "small grass in the sea"	*Gelidiopsis variabilis* *Gymnogongrus johnstonii* *Hypnea valentiae* *Laurencia johnstonii*
xpanáams mojépe "sahuaro seaweed"	unidentified marine alga
xpanáams ool "organ pipe seaweed"	*Rhodymenia hancockii*
xpanáams xaasj "cardón seaweed"	unidentified marine alga
xonj itáast cmis "like devil's claw's tooth"	*Musa sapientum*
xoxát hapéc "cultivated *saiya*"	*Solanum tuberosum*
xtoozp hapéc "cultivated desert ground cherry"	*Lycopersicon esculentum*

*Plant not native to the Seri region

TABLE 5.6
Names Reflecting Physiological or Growth Functions of the Plant

Seri Name	Scientific Name
cmapöjquij "bursts open"	*Caesalpinia palmeri*
hacx cahóit "what wastes it"	*Tulostoma* sp.
hehe iti scahjíit "let's fall on it"	*Brandegea bigelovii*
hehe quiijam 'plant that-curls-around-it'	*Ipomoea* sp. *Janusia californica* *J. gracilis*
hehe quiinla 'plant that-rings'	*Cassia covesii*
hehe yapxöt imóxi "plant whose flower doesn't die"	*Salvia columbariae*
zaah coocta 'sun watcher'	*Lupinus arizonicus* *Machaeranthera parvifolia*

two of psychotropic effect, and two of allergy. *Caháahazxot* is an onomatopoeic name for several spring wildflowers (particularly *Phacelia*) known for their allergic properties.

Animals feature in more than 38 plant names (Table 5.10). Fourteen are descriptive of a mor-phological or anatomical feature, four suggest an animal's action, eleven are possesions of an animal, eight are named as food of certain animals, and one as the opposite (not eaten).

Six plants have the same name as that used for an animal (the first four names are unanalyzable,

TABLE 5.7
Names Reflecting Habitat and Relationship
With the Land and Sea

Seri Name	Scientific Name	Seri Name	Scientific Name
caail iti siml "dry lake barrel cactus"	*Ferocactus covillei*	hant yapxöt "land's flower"	*Achyronychia cooperi* *Tillaea erecta*
caail oocmoj "dry lake's waist cord"	*Juncus acutus*	hast iti coocp "what grows on rock"	*Galaxaura fastigiata*
hamt ináil "soil's skin"	unidentified soil algae	hast iti coteja "what sways on rock"	*Colpomenia tuberculata* *Cutleria hancockii*
hamt inóosj "soil's claw"	*Hofmeisteria laphamioides*	hast yamása 'rock lichen'	foliose and crustose lichens
hamt itóozj "soil's intestines"	*Cuscuta corymbosa* *C. leptantha*	hast yapxöt "rock's flower"	*Dudleya arizonica* *Hofmeisteria laphamioides*
hant caitoj 'land creeper'	*Vaseyanthus insularis*	haxoj ano ihímz "fringes of the shoreline"	*Spyridia filamentosa*
hant iit "land's lice"	*Mollugo cerviana*	hoinalca 'low hills'	*Holographis virgata* *Croton sonorae*
hant ipépj 'land *ipépj*'	*Lotus salsuginosus* *L. tomentellus*	xepe an ihíms "fringes of the sea"	*Hypnea valentiae* *Laurencia johnstonii*
hant iipzx iteja "arroyo's bladder"	*Mammillaria microcarpa* *M. sheldonii* & spp.		*Spyridia filamentosa*
hant istj "land's leaf"	*Dithyrea californica*	xepe oohit "what the sea eats"	*Amphiroa beauvoisii* *A. van-bosseae*
hant iteja "land's bladder"	*Colpomenia tuberculata*		*Digenia simplex*
hant-oosinaj 'land-*oosinaj*'	*Abronia villosa*	xepe xpeetc "sea *Colpomenia*"	*Galaxaura arborea*
hant ootizx "land's foreskin"	*Battarrea diguetii* *Podaxis pistillaris*	xepe yazj 'sea its-membrane'	*Cutleria hancockii*
hant otópl "what the land sticks to"	*Heliotropium curassavicum*	xepe zatx 'sea glochid'	*Dictyota flabellata*
hasoj an hehe 'river area plant'	*Andrachne ciliato-glandulosa*	xnaa-caaa "what calls for the south wind"	*Salicornia bigelovii*
	Teucrium glandulosum unidentified Cruciferae	xpanáams "fringe of the sea"	*Sargassum herporhizum* *S. sinicola*
hast ipénim "splattered against rock"	*Camissonia californica* *Eschscholzia parishii*	xpanéezj 'sea membrane'	*Enteromorpha acanthophora* unidentified marine alga
	Kallstroemia grandiflora	xpeetc 'sea scrotum'	*Colpomenia tuberculata*
hant yax "land's belly"	*Maximowiczia sonorae*	zaaj iti cocái "cliff-hanger"	*Eucnide rupestris*

TABLE 5.8
Names Reflecting Cultural Function

Seri Name	Scientific Name	Seri Name	Scientific Name
ADORNMENT AND HAIR CARE		caatc ipápl	Abutilon palmeri
cmajíic ihásaquim	Ambrosia deltoidea	"what grasshoppers	Horsfordia alata
"what women brush	A. divaricata	are strung with"	
their hair with"	A. dumosa	cozi hax ihapóin	Tephrosia palmeri
	A. magdalenae	'Condalia water what-	
	Zinnia acerosa	it-is-closed-with'	
conée csai	Aristida californica	haamxöii	Hechtia montana
'grass hairbrush'		"trimmed agave"	*Ananas comosus
icapánim	Agave schottii	hasahcápöj	Lophocereus schottii
'what one washes hair		'has-what is chewed'	
with'		hatáam	Zostera marina
heel hayeen ipáai	Opuntia phaeacantha	'what is harvested'	
"prickly-pear used for		hax quipóin	Cardiospermum corindum
face painting"		'water what-is-closed'	
xpanáams hasít	unidentified marine alga	iix casa insíi	Astragalus magdalenae
'seaweed earring'		"who doesn't smell his	
		putrid water"	
CULTIVATED PLANTS		moosni iti hatépx	Croton californica
hocö hapéc	*Tamarix aphylla	"what sea turtle meat	
"cultivated tree"		rests on"	
xoját hapéc	*Solanum tuberosum	queejam iti hacníix	Acalypha californica
"cultivated saiya"		"what out-of-season	
xtoozp hapéc	*Lycopersicon esculentum	fruit is dumped on"	
"cultivated desert		saapom ipémt	Zinnia acerosa
ground cherry"		"what purple prickly-	
		pear is rubbed with"	
		xpaxóocsim	Batis maritima
EQUIPMENT		'xpax-chew and spit	
hant ipásaquim	Abutilon californica	out'	
"broom"			
pnaacoj hacáaiz	Laguncularia racemosa		
'mangrove spear'		**MISCELLANEOUS**	
		coxi ihéet	Errazurizia megacarpa
MEDICINE		"dead man's gambling	
casol ziix ic cöihíipe	Haplopappus sonorensis	sticks"	
"medicinal casol"	Hymenoclea monogyra	halít an caascl	Tidestromia lanuginosa
hasla an ihoom	Abutilon incanum	"causes dandruff"	
"ear is its place"		hatáaij	Ipomoea sp.
hatáast an ihíih	Aristolochia watsonii	'what is spun' (like a	
"what gets between		top)	
the teeth"		sahmées hamt caháacöl	*Citrus paradisi
		"orange that enlarges	
		the breast"	
SUBSISTENCE		xneejam is hayáa	Viscainoa geniculata
caal oohit	Triteleiopsis palmeri	"xneejam whose seeds	
"what the companion-		are owned"	
child eats"			

*Plant not native to the Seri region.

74

while the last two seem to be derived from descriptive terms):

iiz
   blue palo verde (*Cercidium floridum*)
   pampano

ool
   organ pipe (*Stenocereus thurberi*)
   whale shark

paaza
   bebelama (*Bumelia occidentalis*)
   Gila monster

xaasj
   cardón (*Pachycereus pringlei*)
   a certain fish

xasáacoj
   sina (*Stenocereus alamosensis*)
   boa constrictor
xcocoj
   Struthanthus haenkeanus
   housefly

Fifteen species have names that consist of or contain borrowed, foreign terms (Table 5.11). Eleven of these are non-native plants. The names for black-eyed pea (*Vigna*) and lima bean (*Phaseolus lunatus*) contain words of Yaqui origin, while the name for broom-rape (*Orobanche*) is based on the Papago name for the plant. *Tee* may be derived from *té*, the Spanish word for tea. The names for the pea (*Pisum*) and garbanzo (*Cicer*) suggest introduction by Catholic

TABLE 5.9
Names Reflecting Physiological Effects on Humans

Seri Name	Scientific Name	Seri Name	Scientific Name
ALLERGY AND SNEEZING		xpanáams ccapxl	*Amphiroa beauvoisii*
caháahazxot	*Baileya multiradiata*	"bitter seaweed"	*A. van-bosseae*
'what causes sneezing'	*Dyssodia concinna*		*Galaxaura fastigiata*
hehe cocóozxlim	*Tragia amblyodonta*	ziix is ccapxl	*Stenocereus gummosus*
'plant that-causes-rash'		"sour-fruited thing"	
SMELL		**TOXICITY**	
casol ihasíi tiipe	*Pectis papposa*	cacátajc	*Pectis papposa*
"fragrant casol"		'what causes vomiting'	
cotx	*Encelia farinosa*	hehe cacátajc	*Machaeranthera parviflora*
'acrid smell'		'plant that-causes-vomiting'	
hapis casa	*Nicotiana trigonophylla*	hehe coanj	*Sapium biloculare*
"putrid tobacco"		"poisonous plant"	
hehe casa	*Desmanthus fruticosus*	hehe hatéen captax	*Bumelia occidentalis*
"putrid plant"		'plant mouth punctured(plural)'	
hehe ccon	*Allium cepa		
'plant that-reeks'		siml cöquicöt	*Ferocactus covillei*
		"killer barrel cactus"	
TASTE		**PSYCHOTROPIC**	
casol cacat	*Haplopappus sonorensis*	hehe camóstim	*Datura discolor*
"bitter casol"	*Hymenoclea salsola*	'plant that-causes-grimacing (from being crazy)'	
conée ccapxl	*Brachiaria arizonica*		
"sour grass"	*Eragrostis* sp.	hehe carócot	*Datura discolor*
	Erioneuron pulchellum	'plant that-makes-one-crazy'	
hahöj cacat	*Lycium tremontii*		
"bitter *Lycium*"			
sahmées ccapxl	*Citrus limon		
"sour orange"			
xapij coatöj	*Saccharum officinarum		
"sweet reedgrass"			

*Plant not native to the Seri region.

TABLE 5.10
Animal-related Plant Names

Seri Name	Scientific Name	Seri Name	Scientific Name
ANIMAL PART		haamoja iháap	Phaseolus filiformis
cocazn-ootizx	Trixis californica	"pronghorn's tepary"	
"rattlesnake's foreskin"		hee xoját	Tiquilia palmeri
hanaj iit ixac	Marina parryi	'jackrabbit saiya'	
"nits of the raven's		hehe coyóco	Melochia tomentosa
lice"		"dove plant"	
hanaj itáamt	Dalea mollis	hepem ihéem	Opuntia versicolor
"raven's sandals"		"white-tailed deer's	
hap itapxén	Solanum hindsianum	pencil cholla"	
"inner corner of mule		hepem ijcóa	Hibiscus denudatus
deer's eye"		"white-tailed deer's	
haxz iiztim	Calliandra eriophylla	globe-mallow"	
"dog's hipbone"	Hoffmanseggia intricata	moosni iha	Palafoxia arida
	Krameria parvifolia	"sea turtle's	Dithyrea californica
haxz oocmoj	Mascagnia macroptera	possessions"	
"dog's waist cord"		oot ijöéne	Passiflora arida
hee inóosj	Lotus salsuginosus	"coyote's passion	
"jackrabbit's claw"	L. tomentellus	vine"	
	*Tribulus terrestris	nojóo ixpanáams	Galaxaura fastigiata
heen ilít	*Tribulus terrestris	"spotted sand bass's	
"cow's head"		seaweed"	
hepem isla	Mirabilis bigelovii	ptcamn iha	Rhodymenia divaricata
"white-tailed deer's		"lobster's possessions"	
ear"		tacj iha	Gelidiopsis variabilis
moosni ipnáil	Cryptonemia obovata	"porpoise's	
"sea turtle's skirt"	Halymenia coccinea	possessions"	
	Padina durvillaei		
	Rhodymenia divaricata	**ANIMAL FOOD**	
taca imas	Asparagopsis taxiformis	hee imcát	Machaeranthera parviflora
"triggerfish's body	Hofmeisteria fasciculata	"what the jackrabbit	Perityle emoryi
hair"		doesn't bite off"	
taca-noosc	Eucheuma uncinatum	hohr-oohit	Nama hispidum
"triggerfish's papillae"		"what donkeys eat"	
tacj-anóosc	Gigartina johnstonii	mojet oohit	Allium haematochiton
"porpoise's papillae"	G. pectinata	"what mountain sheep	
xeezej islítx	Nicotiana clevelandii	eat"	
"badger's inner ear"		moosníil ihaqué e	Asparagopsis taxiformis
		"what blue turtle	
ANIMAL ACTIONS		likes"	
hap oacajam	Caesalpinia palmeri	moosn-oohit	Asparagopsis taxiformis
"what mule deer flay	Echinopterys eglandulosa	"what sea turtles eat"	Palafoxia arida
antlers on"	Thryallis angustifolia	noj-oopis	Justicia californica
sipöj yanéaax	Suaeda esteroa	"what hummingbirds	
"what the cardinal		suck out"	
washes his hands		xtamáaij-oohit	Nemocladus glanduliferus
with"		"what mud turtles eat"	
taca oomas	Codium simulans	xtamóosn-oohit	Chaenactis carphoclinia
"the cord that the	Galaxaura arborea	"what desert tortoises	Fagonia californica
triggerfish twined"		eat"	F. pachyacantha
tacj oomas	Codium simulans	ziix hant cpatj oohit	Amphiroa beauvoisii
"the cord that the	Digenia simplex	"thing that the	A. van-bosseae
porpoise twined"		flounder eats"	Asparagopsis taxiformis
POSSESSED BY AN ANIMAL			Dictyota flabellata
caay ixám	*Cucurbita pepo		Galaxaura arborea
"horse's gourd"			

*Plant not native to the Seri region.

priests, probably the Jesuits, and indicate comparison with native analogs also in the legume family. The name for wheat was probably derived from the Seri terms for Castilian eelgrass.

Terms for vegetation, morphological characteristics, and plant parts are listed below. In addition, there is specialized terminology for specific parts of certain plants, such as eelgrass (*Zostera*), *mala mujer* (*Cnidoscolus*), and mesquite (*Prosopis*).

VEGETATION

haas an 'mesquite area,' *mezquital*, or mesquite scrub

hant quinej 'land empty,' a desert area with vegetation only, devoid of signs of habitation, people, or civilization

hant yail 'land its-greenness,' ephemeral vegetation which appears on the desert after a rainy period; the month of August, when the effects of the summer rains usually appear, is commonly a time when ephemeral vegetation is present, and is called *hant yail iháat iizax* 'land its-greenness time-of-ripening moon'

heecot 'desert,' an uninhabited terrestrial area

hehe an 'plant area,' a place of terrestrial vegetation outside or away from camp; also refers to desert vegetation in general

hehe cpoin 'plant closed' refers to riparian vegetation, or riparian or semi-riparian desert scrub found along washes or small arroyos; it was described as a long and narrow stretch of dense desert growth or vegetation

pnaacoj an 'mangrove area,' mangrove scrub, or mangrove vegetation

MORPHOLOGICAL FEATURES AND PARTS OF PLANTS

In the following list of terms, *hehe* refers to any plant. To indicate the root, trunk, bark, leaf, etc., of a specific plant, such terms must follow the name of the plant. For example, *hesen it* 'ironwood its-trunk,' or "ironwood trunk."

Roots

hehe ixái 'plant its-root,' the root of any plant

Stems and Related Parts

hehe it 'tree(or shrub) its-base,' the trunk of a shrub or tree

hehe iseja 'plant its-branch'

iyat 'twigs' or 'upper branches,' the tip or tender new growth of a branch, or a twig; may include leaves (= herbage)

TABLE 5.11
Plant Names With Borrowed Terms

Seri Name	Scientific and Common Names
caay ixám "horse's gourd"	*Cucurbita pepo squash
camótzila 'guamúchil'	*Pithecellobium dulce guamúchil
caztaz "Castilian eelgrass"	*Triticum aestivum wheat
haaxo 'ajo'	*Allium sativum garlic
hapats imóon "Apache's beans"	*Phaseolus lunatus lima bean
hohr-oohit "what burros (donkeys) eat"	Nama hispidum
limóon 'limón'	*Citrus aurantifolia lemon
matar (Papago name for broom-rape)	Orobanche cooperi broom-rape
meróon 'melón'	*Cucumis melo cantaloupe
oeno-raama 'buena rama'	Acacia constricta whitethorn
paar icomíhlc "padre's mesquite seed"	*Pisum sativum garden pea
paar icomítin "padre's ironwood seed"	*Cicer arietinum chickpea, garbanzo
tee 'té'	Chorizanthe corrugata Eriogonum inflatum
teepar 'tepary'	*Phaseolus acutifolius tepary (domesticated)
yori imóon "Mexican's beans"	*Vigna ungiculata black-eyed pea

*Plant not native to the Seri region.

hehe ihasnáilc 'plant its-roughness,' the outer, rough bark, thick or woody; not all trees and shrubs have this type of bark; it can be pulled off in big chunks

hehe ináil 'plant its-bark' or 'plant its-skin,' green or smooth bark or persistent epidermis (e.g., the palo verdes), or the smooth "inner" bark which may be found underneath the *hehe ihasnáilc*

zatx 'glochid' or 'small spines,' small, obnoxious hairs or spines such as on *Argythamnia*, *Cryptantha*, and *Opuntia*; may be on stems, leaves, or fruit; the large fireworm *Eurythöe complanata* (a polychaete) was called *xepeno zatx* 'sea-from glochids'

imoz an 'its-heart place,' the core, center, or pith of a trunk

ipxási 'its meat,' the cortex area of a fleshy stem plant, specifically the cortex of a cactus

itac 'its bone,' the woody rib of a columnar cactus; also the petiole of a leaf

hehe iti ipáxoz 'stick on its-raising up,' a splinter

hehe iix 'plant its-liquid,' sap

hehe ooxö 'plant/tree its-excrement,' resin or gum; for example *coote ooxö* is a gum from *coote*, or jumping cholla (*Opuntia fulgida*)

csipx is resin or lac, such as from brittlebush (*Encelia*), creosotebush (*Larrea*), or *sámota* (*Coursetia*)

Leaves
hehe istj 'plant its-leaf'

ziix quih caxt 'thing the that-which-is-tender,' or "tender thing," greens, or any leafy green vegetable

itac 'its bone,' the petiole

coipj 'what is elliptical,' a long, narrow leaf; includes lanceolate leaves and a long, narrow compound leaf, such as that of *Bursera microphylla*

coosotoj 'what is narrow,' a very long and narrow leaf, such as an *Acacia willardiana*, *Agave*, *Jacquinia*, or *Yucca* leaf

cpetij 'what is circular,' a leaf which is "round" in outline, such as one of ashy limberbush (*Jatropha cinerea*)

Flowers and Inflorescences
itöj, the young emerging inflorescence of an *Agave* or *Yucca*, usually less than about 30 cm tall

icáp, the tall, mature inflorescence of an *Agave* or *Yucca*

ihíyoz 'bud,' a flower bud

hehe yapxöt 'plant its-bursting-open,' a flower

ipaöjam 'its tail fin,' the dry perianth remaining attached to the fruit of a columnar cactus

Fruit and Seeds
ihiyáxa, the young, developing ovary of a flower, described as the small new fruit "ball" appearing at the base of the flower

ihic 'seed' or 'pit (of a fruit)'

is, unripened fruit

imám, ripe fruit

queejam, out-of-season fruit

6. Water and Food Quest

Food was generally abundant; water was not. The Seri obtained their food from a broad spectrum of animals and plants. The traditional diet was rich and highly varied. Scarcity of drinking water was more important than food in limiting the distribution and size of Seri population. Food and water resources were quite different from place to place and not evenly distributed. This fact must have had a strong influence on settlement patterns and on the way people conducted their lives.

However, with the cultural disruptions and hardships during the Spanish and Mexican hostilities, there were times of food shortages. Some of the older people remembered harsh times and hunger. Also, although food resources were abundant, obtaining food was not necessarily easy work. When some Seri were asked how there could have sometimes been hunger earlier in the twentieth century in view of abundant food resources, such as the green turtle, the answer was that "there were some Seri who were very lazy."

Drinking Water

Fresh water was precious. Men had the arduous and critical task of supplying drinking water. Although sometimes situated close to water holes, the camps often were located by the sea, and water had to be brought from water holes in nearby mountains sometimes five to ten or more kilometers away. It was said that any trail would lead to water. According to the Seri, the location of water and access to it was the single most important factor in determining the placement and duration of the camps and the movements and activities of the people.

Men carried two large, water-filled, pottery vessels (ollas), each suspended in a mesquite cord net from the end of a carrying yoke (Figure 6.1). In May, 1721, the Jesuit missionary Juan de Ugarte landed at the coast near Kino Bay, and reported that "early the next day the Indians appeared in troops, and all with water vessels; the men each with two in nets hanging from a pole across their shoulders, and the women with one (Venegas 1759: vol. 2, 48)." Where water was locally available, such as at Tecomate, on Tiburón Island, or at Kino Bay, women brought it home in ollas or cans carried on their heads (Figure 6.2). As soon as twenty-liter (five-gallon) metal cans became available in the twentieth century, the use of pottery vessels for water transport was phased out.

Trails leading from shoreline camp sites to major water holes were strewn with potsherds (Quinn and Quinn 1965: 153). There were tragic tales of men who walked long distances only to stumble and fall, breaking the water jugs. It was also lamented that sometimes they walked all day only to find that the water hole was dry.

The men usually went for water every two

Figure 6.1. Jesús Ibarra with ollas suspended in nets from a carrying yoke. The photograph was taken at a camp on the east side of Tiburón Island in December 1921. In 1983 the Seri identified the net on the right as mesquite root fiber and the one on the left as human or horse hair. The gourd ladle tied to the yoke was identified as *xjii* (*Lagenaria siceraria*). CS, courtesy Neil Carmony and the Desert Bighorn Sheep Society.

to three days. They generally went in the early morning, particularly if it was summertime, and it was common for four or five men to go together. It was also commonplace for a man to borrow half a jug of water from another family, especially when he was going hunting, with the promise that he would pay it back.

Edward H. Davis described two men going from a camp on the shore of Tiburón Island to bring water from the mountains:

> Each took a *palanca*, or short shoulder pole with knotched [sic] ends, fitted empty five-gallon cans in nets of twisted fiber, and hooked the nets in the end of the notched stick and started off nine or ten miles to the water hole in the heart of the mountains. . . . The two men were gone six hours and when they returned they were almost staggering under the heavy loads of water . . . as the cans were full (Quinn and Quinn 1965:157–158).

It was April 28, 1922 (Davis 1922:304). They were at a camp on the east shore of Tiburón Island, probably several kilometers north of Palo Fierro. The men apparently walked to the southwest across the desert and into the rugged base of Sierra Kunkaak, where there are several water holes.

From the shore, it took men about an hour to reach another water hole, called *Pazj Hax* "graphite water" (*pazj* is derived from the term *xpazj* 'graphite'), and about two hours to return with full loads of water. This water hole, located in a canyon at the base of the southeast side of Sierra Kunkaak (Map 6.1), was named Tinaja Anita by McGee (1898:16, 19).

The large pottery vessels used to transport water from the water holes to camp were characteristically made of the Tiburón Plain ("eggshell pottery") ceramic tradition (Bowen and Moser

Figure 6.2. Women carried water in containers balanced on headrings. Note the plants stuffed into the tops of the cans and bucket to prevent water loss by sloshing. The men on the left carry water in cans suspended from yokes. *EHD, Kino Bay, 1929; HF-24015.*

1968, Bowen 1976:53–72). These large vessels, called *hamázaj* 'earth globular,' were of a hard and unusually thin plain ware, varying in thickness from 2 to 5 mm, averaging about 3 mm (Bowen 1976:53–54). McGee (1898:183) was impressed by the thinness, noting that economy of weight would be highly advantageous for long-distance transport. María Luisa Chilión ". . . said that she saw thin pottery being made when she was a young girl. She said the purpose of making vessels so thin was to make them lighter (and therefore easier to transport) (Bowen and Moser 1968:128)." Seri water-carrying vessels were among the largest and thinnest in the world.

Large pottery ollas were occasionally buried and used for water storage, sometimes near a water hole that was about to dry up. The vessel was sealed with a lid and covered with brush (Bowen and Moser 1968:120).

Pottery vessels and cans used to carry water did not have lids. Various *caitij* 'soft(plural),' sweet-smelling or odorless plants were stuffed into the top of ollas or water cans to prevent sloshing and water loss along the trail (see Figure 6.2). Plants used for this very important purpose included *Bebbia, Boerhaavia coulteri, Cardiospermum,* and *Tephrosia.*

Water was also carried in certain kinds of animal stomachs and bladders used as canteens. The stomach or bladder was cleaned, inflated and dried, and tied with tendon or gut. It was very common for a man to carry water in a green turtle stomach, called *moosni icsám* "sea turtle's stomach." The mule deer stomach, *hap anoyáhit* 'mule-deer stomach,' which generally has a larger capacity, was also much used. Water and other liquids were carried also in the bladder, *iteja*, of such animals as the green turtle, horse, and mule deer.

Permanent Water Holes

The Seri knew of many more temporary water places than permanent ones. For example, they had names for approximately forty-three water

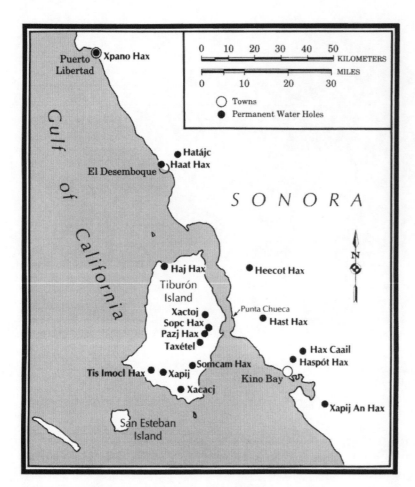

Map 6.1. Approximate locations of permanent water holes on Tiburón Island and the Sonoran coastal region from near Kino Bay to Puerto Libertad.

places on Tiburón Island. Only twelve or thirteen of these were considered permanent. However, permanent water holes were unevenly distributed on the island as well as on the mainland (see Map 6.1).

Permanent water places were particularly sparse along the mainland coast. From Puerto Libertad to Kino Bay, a distance of approximately 200 km, there were only five places along the coast with permanent water. From north to south these were:

1. *Xpano Hax* 'sea-in water' at Puerto Libertad, where fresh water flows out of the sand below the high-tide line
2. *Haat Hax* 'limberbush(*Jatropha cuneata*) water' at the mouth of the Río San Ignacio and others, such as *Hatájc* (Pozo Coyote), inland along the same dry riverbed

3. *Heecot Hax* 'desert water' (Pozo Posado) on Pico Johnson in the Sierra Seri
4. *Hast Hax* 'mountain water' (Pozo Peña) at the south end of the Sierra Seri
5. Kino Bay vicinity: *Hax Caail* 'water wide' (Pozo Carrizo), *Haspót Hax* 'haspót water,' and *Xapij An Hax* 'reedgrass inside water'; *Hax Caail* and *Haspót Hax* are north of Estero de la Cruz and *Xapij An Hax* is south of the estero

Permanent water holes were either springs or shallow wells, and probably all had at least some recharge from subsurface ground water. The largest permanent water places in the entire Seri region were at two places with small, spring-fed streams—on the order of one kilometer in length—trickling intermittently over rock and gravel: *Xapij* 'reedgrass' on the south side of

Figure 6.3. One of the bedrock pools at *Xapij (Sauzal)*. These pools constitute the largest permanent water source on Tiburón Island. Luis Guicho is on the left, José Torres on the right. *EHD, 1936; HF-23927*.

Tiburón Island (Figure 6.3) and Nacapule Canyon, north of San Carlos Bay. All of the other permanent springs were very small and flowed into bedrock depressions, or tinajas. Each of these springs usually supplied a pool of water no more than one to several meters across. A number of these water holes, such as *Pazj Hax* "graphite water hole," were in deep canyons and supported giant cane or reedgrass (Figure 6.4).

Most of the wells, usually located along floodplains and drainageways, were dug, enlarged, and maintained by the people. Examples of improved wells include *Haj Hax* 'any water' (Tecomate), at the north end of Tiburón Island, and

Figure 6.4. *Pazj Hax* (Tinaja Anita), a typical small permanent water hole at the base of Sierra Kunkaak on Tiburón Island. A) The water hole is located at the head of a deep canyon. B) A dense, localized stand of *carrizo* (*Arundo donax*) marks the site of the water hole. C) A small pool free of *carrizo*. *RSF, April 1983*.

Haat Hax, at the mouth of Río San Ignacio drainage near El Desemboque.

According to the Seri, each family or group of families considered a certain *ihízitim*, or territory, to be their home. They also identified with the nearest permanent water place or places. However, there was not a permanent water hole in each *ihízitim*, or even close to each one. The Seri pointed out that during favorable times, when both food and water were plentiful, each family would return to its *ihízitim*. However, during the frequent times when the temporary water places dried up, they had to coalesce in order to be within carrying distance of permanent water (see Moser 1963a and Sheridan 1982).

Temporary Water Holes

Following times of sufficient rainfall, temporary water holes filled and allowed the people more freedom of movement. A long walk to another camp by a group of people, especially if they had children with them, was usually made just after a rain, so that they could secure water along the trail at temporary water holes. A small

Figure 6.5. A playa, or dry lake bed, temporarily filled by rainwater. *MBM, about 3.5 km east of Cerro Tepopa, December 1978.*

group of adults could go longer with less water. Thus, the lack of rain could keep a whole camp from moving to a new location, even though the men had to bring water to camp from considerable distance. Dry lakes or playas—scattered along the coastal lowlands on the mainland—sometimes held large quantities of water (Figure 6.5).

When drought set in and recharge ceased, some of the small temporary water holes could be quickly depleted. Many of the small bedrock tinajas held only a thousand liters or less. For example, *Hast Eemla Hax* 'mountain jagged water' consisted of an upper and lower pool (Figure 6.6). The upper, smaller one, exposed to the full sun, dried up long before the lower one. The lower and larger pool, shaded in the winter and in the afternoon, had a capacity of not more than a thousand liters. Recharge from the gravel bed above the pools probably ceased no later than several weeks after the last major rainfall. Several men taking ten to twenty liters each would have removed substantial amounts of the water in a month or so.

One or two of three nearby coastal camps were usually occupied when *Hast Eemla Hax* had water. These camps were *Quipcö Quih An Icahéme* 'dune the on to-have-a-camp,' *Zaaj Cooxp* 'cliff white,' and *Comítija* 'ironwoods.' After a big rain the people knew there would be water, so they set up one or two of these coastal camps. Marine resources along this region were particularly rich, and it was a pleasant place to camp, but fresh water was particularly scarce, since there were no absolutely permanent water places between the vicinity of El Desemboque and Pico Johnson in the Sierra Seri. The men even went to *Hast Eemla Hax* from camps as far away as 10 or 20 km.

Modification of Water Holes

Shallow wells or depressions where water accumulated were often excavated and enlarged (Figure 6.7). The soil was often dug out with sea turtle bones and large shells of the giant egg cockle clam (*Laevicardium elatum*) and the dirt was carried away in a sea turtle carapace. In anticipation of rain, the people dug holes in dry lake beds, such as at the south end of Playa San Bartolo, so that they would fill with water. A channel was often cut through reedgrass, or *carrizo* (*Phragmites* or *Arundo*), that was growing at a small spring both to allow the water to flow through and to give access to it.

At *Hax Hapáfc* 'water pounded,' between El Dátil and Puerto Libertad, water came out of a small hole in the rock. As indicated by its name, the people pounded, or chipped away, the rock to enlarge the water hole. Trails leading to water holes had to be maintained. For example, in the canyon near the *Pazj Hax* water hole on Tiburón Island the trail was lined with rocks cleared out of the path.

Emergency Liquid

A substitute for drinking and cooking water was obtained from barrel cactus (*Ferocactus wislizenii*). Use of this liquid was apparently a fairly common practice, particularly during the long dry season in late spring and early summer. It was said that an entire camp sometimes depended on barrel cactus juice for days or weeks at a time. Possible deleterious effects were avoided if barrel cactus juice was taken only on a full stomach; when taken on an empty stomach it was said to cause diarrhea. Also, some people were more sensitive to it than others, but undesirable side effects were avoided by resting after drinking the juice. It was apparently harder on children than adults. According to Seri oral history, several men marooned on San Esteban Island survived the summer heat for a number of days by drinking juice extracted from cooked century plants (see *Agave cerulata*). People in the Central Desert of Baja California also used agave as an emergency source of liquid, and apparently they resorted to it rather frequently (Aschmann 1959:60). In one case a newborn infant was kept alive with cactus fruit juice (see *Stenocereus thurberi*).

Another source of emergency liquid was obtained from sea turtle blood. After a large turtle

Figure 6.6. *Hast Eemla Hax*, a temporary water hole at the north end of the Sierra Seri. A) The upper pool. B) The lower pool. The small trees are cliff fig, *Ficus petiolaris*. RSF, April 1983.

(*Chelonia*) was butchered, the blood and body fluids were collected and allowed to stand for a few hours. It was said that the redness (blood sediment) settled to the bottom, leaving a clear, potable liquid—like water—on the top. This clear liquid, the serum, was drunk as a substitute for water. It was said that this water substitute could be obtained only from a large turtle, because the liquid from a small one was not potable. If the blood and fluid of a small turtle was

Figure 6.7. A temporary source of water was obtained by digging small wells following times of sufficient rain. This emergency well was dug in the desert 0.5 km northwest of El Desemboque. The water slowly seeped in after the hole was dug slightly more than one meter in depth. Daniel Romero is on the left. *MBM, 1958.*

allowed to stand, it was said to remain red, even though some of the blood did settle. The serum would be lower in salt content than sea water, but close to that of human blood. Under dire circumstances one could use this serum as a water substitute, but probably not over an extended period of time (Clayton J. May, personal communication, 1983).

Animals As Food

During the middle of the twentieth century there was a sharp decline in dependence on hunting, fishing and gathering. This decline coincided with an increase in a cash and store-bought food economy. By the late 1970s various food animals—such as the *totoaba*, the green turtle, bighorn sheep, and pronghorn—had become scarce or locally extinct. However, a wide range of native land and sea animals and plants were still eaten.

Larger animals, sea turtles, and most fish were obtained by men, since the use of boats, harpoons, and bows and arrows were strictly in their domain. The hunting of smaller animals, which did not require such equipment, might involve men or women, or both. Women gathered plants and mollusks and hunted various smaller terrestrial animals with clubs and trained dogs. Men often assisted the women with the gathering of mollusks and with the larger harvests, such as for mesquite, eelgrass, and century plants. In historic times the San Esteban century plant (*Agave cerulata*) was harvested by men who travelled to the island for that purpose.

Meat of marine and terrestrial animals, invertebrate and vertebrate, was prepared by a variety of methods, including roasting—either in coals or on a skewer, boiling, broiling on a grill (a twentieth-century method), steaming, and pit baking. Hardwoods, particularly mesquite (*Prosopis*), creosotebush (*Larrea*), and jojoba (*Simmondsia*), were fashioned into a single-pronged meat skewer or double-pronged one for larger pieces of meat. The skewers were about ⅔ m long and were stuck into the ground slanting over hot coals (see *Prosopis*). A stick made from a slender hardwood branch, such as creosotebush, was used to remove meat from a cooking pot.

Although most meat was cooked, some of the choicer kinds were enjoyed rare and sometimes uncooked (Sheldon 1979:98, 119, 120). Green turtle and desert tortoise livers were often eaten raw, as were bird eggs encountered in the desert. However, the numerous earlier accounts of the Seri eating raw or even spoiled meat may be somewhat exaggerated or secondhand information, and concerned defeated people often at the edge of starvation due to war, disease, and chaotic social conditions.

Different kinds of meat were dried and stored, particularly that of the Gulf grunion, various larger fish, the green turtle, mule deer, and horse. Dried meat was called *hatíin* (e.g., *moosni hatíin* 'sea-turtle dried-meat'). Grunion were roasted, but larger fish and other kinds of meat were sun dried and not salted. Dried meat was kept in sealed pottery vessels concealed in small caves. The entrances were covered with rocks. Dried meat was most used in cool weather, when there was not much other food (meat) because wind and rain curtailed fishing and turtle hunting, as well as terrestrial hunting.

Various places were used to dry meat and fish out of the reach of dogs, such as on a house roof, brush house frame, meat rack, tree or shrub (see *Fouquieria splendens*, *Maytenus*, and *Olneya*). A meat rack, called *hapát*, was often made of forked posts with a branch or pole laid horizontally in the forks, or it might consist of a pole between a roof and a nearby branch (Figure 6.8). It was of varying length but high enough to keep the meat away from dogs. A tree or shrub with the limbs bent or broken to make a flat place on which to put meat or other objects was also called *hapát*. Sometimes deer meat or other food was tied in a bundle and hung from a stick driven into a *cardón* (*Pachycereus*).

In general animals constituted a larger portion of the diet than did plants, particularly in terms of protein. The Seri did not consider that they had eaten a proper or real meal unless it included a hearty portion of meat. In 1921 Sheldon (1979:139) observed that meat and fish were their staple foods, "and even when they have plenty of other food they still crave meat." The large stature of the Seri has been consistently noted since the sixteenth-century Spanish chronicles. The common height (1.8 m [6 ft.]) of the Seri must have been indeed striking to the Europeans of the time. Their size undoubtedly was largely due to the high protein component of their diet.

Plants as Food

The Seri did not practice agriculture within the confines of their territory: the climate was too arid. During Spanish colonial times, some Seri moved inland and from time to time accepted an agricultural life (Spicer 1962:105–107). Most of the Seri, however, remained hunter-gatherers.

Figure 6.8. An ocotillo pole served as a rack to dry meat at the *Saps* camp in March 1981. Angelita Torres is on the right. *MBM.*

Gathering and Preparation

A gathering party often consisted of several women and children of an extended family. From an established camp they went to specific areas where certain desired plants were known to occur. The party often spent several hours, or even an entire day, walking and gathering. This was still a common practice in the 1980s for gathering the fruits of jumping cholla, *pitaya agria*, *cardón*, organ pipe, and desert wolfberry. Although they generally did not gather every day, the women made frequent trips during the various major harvest times.

When windy weather prevented the men from going to sea to hunt turtles and fish, they usually went hunting in the desert. At such times a group of women and children might go to the mountains to gather specific plants, such as *mala mujer* (*Cnidoscolus*), teddybear cholla (*Opuntia bigelovii*), and desert mistletoe (*Phoradendron californicum*).

Most fruits and seeds were picked by hand (Figure 6.9). Roots and century plants were harvested with two kinds of hardwood tools. The *hahéel*, flattened at one end, was used as a chisel and pry bar for severing and dislodging century plants (see *Agave subsimplex*). It was made from the wood of trees and shrubs such as:

Acacia greggii, catclaw
A. willardiana, *palo blanco*
Cercidium floridum, blue palo verde
C. microphyllum, foothill palo verde
Colubrina viridis, *palo colorado*
Olneya tesota, ironwood

The *hapoj* digging stick had a pointed end for digging roots such as *saiya* (*Amoreuxia*), *mala mujer* (*Cnidoscolus*), and devil's claw (*Proboscidea altheifolia*). It was made from hardwoods such as catclaw, *palo blanco*, or *sámota*

(*Coursetia*). Elongated stone implements found in the region were not digging tools. The Seri were adamant that these objects were weapons of combat used by the Giants.

There were three kinds of fruit-gathering poles for the four major columnar cacti—one for *cardón* and sahuaro, and one each for organ pipe and *pitaya agria*. These poles were made from the flower stalk of a large century plant, the dry woody rib of sahuaro or *cardón*, ocotillo, or the stem of reedgrass. Several pieces were tied together to make the pole long enough. The end pieces and the spikes for dislodging the fruit were made from hardwoods such as *Lippia* (*orégano*), *Larrea* (creosotebush), and *Prosopis* (mesquite).

Women carried harvested plants back to camp in baskets (Figure 6.10) or 20-liter cans balanced on headrings. By the 1960s metal and plastic buckets were commonplace. Pottery vessels were not used for most food gathering. However, small fruit or berries, such as those of

Figure 6.9. Harvesting the fruit of a desert wolfberry, *Lycium fremontii*. A) Angelita Torres picking the berries beside a large *cardón* in the dense riparian desertscrub of Arroyo San Ignacio. B) The berries were picked one by one from among the tangled, spiny branches. *RSF, March 1983.*

A

B

Figure 6.10. Two young women bringing home *cardón* fruit; sticks have been inserted half way up the load to increase the amount that can be carried. Ernestina Morales holds the cactus fruit gathering pole. *MBM, El Desemboque, July 1963.*

desert wolfberry (*Lycium fremontii*) were put into a special pottery bowl when picked (Bowen and Moser 1968 : 122). When large quantities of columnar cactus fruit were harvested, probably for wine making, the men carried the fruit back to camp in large-mouthed ollas or 20-liter cans. Women generally carried the load on their heads, while men used the carrying yoke (*peen*) or carrying pole. A container for carrying honey was made from the barrel cactus (*Ferocactus wislizenii*). The "boot" from the sahuaro or *cardón* was occasionally used to carry fruit. Men sometimes carried columnar cactus fruit strung on strips of limberbush (*Jatropha cuneata*) or sticks of *Cordia*.

There were specific terms indicating one who gathered certain plants. For example:

caasax 'one who gathers from *haasax* (a pack rat nest)'
coomxö 'one who gathers *haamxö* (*Agave subsimplex*)'
cotáam 'one who gathers *hatáam* (*Zostera*)'

Pack rat (*Neotoma albigula*) nests were opened and robbed of food stored by the animals for winter. Considerable amounts of food were obtained, and in earlier times the pack rats were also eaten. In this manner the time of harvest was extended for several months or more. The nests were most often robbed in fall, probably September through November. The Seri said the pack rat picks fruit with its paws and stuffs it into pockets at the sides of its jaws and then runs to its nest and caches it. The Seri found several kinds of food cached in a single nest, and pointed out that each kind was segregated from the others. They said pack rats do not gather very small seeds or fruit. Fruits and pods of the following trees and shrubs were gathered from the nests:

Bumelia occidentalis, bebelama
Cercidium floridum, blue palo verde
Cercidium microphyllum, foothill palo verde
Lycium fremontii, desert wolfberry
Maytenus phyllanthoides, mangle dulce
Olneya tesota, ironwood
Prosopis glandulosa, mesquite
Zizyphus obtusifolia, white crucillo

Tongs (Figure 6.11) for handling cholla (*Opuntia bigelovii*) stems were made from meter-long pieces of the dry woody ribs of organ pipe (*Stenocereus thurberi*). The spines and glochids (*zatx*) of cholla and prickly-pear (*Opuntia* spp.) fruit were removed by sweeping them on the ground with leafy branches of such plants as white bur-sage (*Ambrosia dumosa*), creosotebush (*Larrea*), and desert zinnia (*Zinnia*).

A stout ocotillo (*Fouquieria splendens*) stick was used to thresh eelgrass (*Zostera*). Certain plants which are spineless and without objectionable hairs were called *caitij* 'soft (plural)' plants. These were used to make a bed on the

Figure 6.11. Wooden tongs were used to handle cholla. Ramona Casanova placing the stems of teddy-bear cholla (*Opuntia bigelovii*) on brush firewood. She made these tongs from organ pipe wood picked up in the desert at the time of harvesting. *MBM, 2 km north of El Desemboque, February 1964.*

ground or to line a sea turtle carapace or basket where meat was placed in order to keep it clean. Examples are sand croton (*Croton californicus*), *mangle dulce* (*Maytenus*), pickleweed (*Salicornia*), sea purslane (*Sesuvium*), and eelgrass.

The work basket served as a platform for preparation of many kinds of fruit, seed, and other foods (see Chapter 15). Edward Moser (1973:133–134) described four techniques of winnowing with baskets (by the 1960s a plastic or metal pan was used instead):

1. *coospx*. Grain, seed, or small fruit were tossed into the air repeatedly so that the chaff would blow away.
2. *coomen*. The basket was rolled and rotated so that larger sticks, leaves, and other impurities came to the surface, where they were removed by hand.
3. *cooctz*. Tapping the basket on a hard surface brought lighter impurities and pieces of husk to the surface. They were then scraped away by hand over the edge of the basket.
4. *iqui quispx*. This method was used to refine powder and flour. The basket was tapped on a

hard object to bring the finer powder to the surface. The basket was then tilted and the powder allowed to spill over the edge into another receptacle. This method was used to obtain *heepol* (*Krameria grayi*) powder; it was also used for mesquite and eelgrass flour.

Hai cöquisj 'wind *cöquisj*,' or "wind winnowing," was another common method to clean seeds and grain. Inflorescences or small plants with ripe seeds or grain were commonly rolled between the hands and wind winnowed to separate the chaff from the seeds or grain (Figure 6.12).

Plant-derived foods were prepared as gruel (*atole*), boiled, roasted in hot ashes or coals, grilled, pit roasted, sun dried, or eaten fresh. Many foods were cooked or mixed with sea turtle oil, which was the favorite cooking oil. Other kinds of animal fat, such as that of brown pelican, mule deer, sea lion, horse, and cow, were also used in cooking. Many kinds of food were sweetened with honey or sugar, both of which became available only at about the turn of the century.

Most seeds were parched or toasted, then

ground into flour on a metate, cooked in water, and consumed in the form of gruel. Various other plant-derived foods, such as flour from mesquite pods, were similarly prepared as gruels. By increasing the surface area of food particles, this method of preparation, common throughout southwestern North America, effectively conserved water, fuel, and time required for cooking. Fruits of the cacti, desert wolfberries, rock fig, and other plants were eaten fresh or boiled with or without honey or sugar. Fruit of the giant cacti was also dried. Only two plants were used as greens, *Amaranthus fimbriatus* and *Boerhaavia coulteri*, and these, too, were cooked in water.

Most Seri medicines were also prepared with water, generally as teas. Some of these teas undoubtedly had nutritional value. It seems that the predilection for watery foods and medicines was in part a response to the very arid environment. Ironwood (*Olneya*) seed was the only food cooked with a change of water. In view of the short supplies of fresh water, the general absence of foods requiring leaching or a change of water is both striking and fortunate.

Storage

Different kinds of food were stored. Seeds (both whole or ground into flour), dried fruit, mesquite and century plant cakes, and dried fish, sea turtle, and deer meat were kept in large pottery vessels, or ollas (Bowen and Moser 1968 : 118–120). These vessels had pottery, rock, or clamshell (e.g., *Laevicardium elatum*) lids sealed with creosotebush lac (*csipx*). Storage vessels were often cached in small caves (Figure 6.13). Parching or cooking food prior to storage and storing freshly harvested seeds in tightly sealed pottery vessels helped prevent spoilage and losses from rodents. Plant-derived foods from the following species were commonly stored:

Agave spp., century plant
Amaranthus watsonii, *bledo*
Carnegiea gigantea, sahuaro
Cercidium microphyllum, foothill palo verde

Figure 6.12. Seeds of small desert plants were often prepared in the desert before returning to camp. A) Aurora Colosio rolling Indian wheat plants between her hands to free the seed. B) Wind winnowing the seeds. *RSF, 8 km east of Pozo Coyote, April 1983.*

Chenopodium murale, goosefoot
Pachycereus pringlei, *cardón*
Plantago insularis, wooly plantain
Prosopis glandulosa, mesquite
Stenocereus thurberi, organ pipe
Zostera marina, eelgrass

Food Plants

We have recorded Seri information for 374 botanical species of seed plants (flowering plants) occurring naturally in the region, of which 25

Figure 6.13. Storage ollas as they were found in a small cave on Pico Johnson. The ollas, on the order of 60 cm high, contained *cardón* seeds. The rock lids were sealed in place with lac. *EHD, February 11, 1929; HF-24026.*

percent (94 species) were food plants. The Seri used 62 species (nearly 22 percent) of the plants on Tiburón Island and 25 species (23 percent) of the flora on San Esteban Island for food. Non-seed, or lower, plants were not eaten. Table 6.1 lists the species and various parts of plants used by the Seri for food (also see Felger and Moser 1976).

Basic staples were obtained from the four large-fruited columnar cacti (sahuaro, organ pipe, *pitaya agria*, and *cardón*), eelgrass (*Zostera*), mesquite (*Prosopis*), century plants (*Agave*), *mala mujer* (*Cnidoscolus*), and seeds of various ephemerals. Among the more important ephemerals were *Amaranthus* spp., goosefoot (*Chenopodium*), wooly plantain (*Plantago*), and wild tepary (*Phaseolus acutifolius*). Other im-

portant food plants included jumping cholla (*Opuntia fulgida*), desert wolfberry (*Lycium fremontii*), palo verde (*Cercidium floridum* and *C. microphyllum*), and *saiya* (*Amoreuxia*).

Certain major perennials—such as mesquite, with its deep roots, and the columnar cacti, with their succulent water-storage tissues—generally produce fruit even during a year of drought or other unfavorable conditions, although the yields may be substantially reduced. However, during a year of extremely severe conditions, such as 1979, following an exceptionally cold winter and local drought, mesquite groves near Playa San Bartolo north of Kino Bay may fail to produce fruit. In contrast, eelgrass (a marine plant), which is not affected by rainfall or local weather conditions, produces seed each year.

TABLE 6.1
Plants Used by the Seri for Food

Scientific Name	Part Used for Food					
	Seed (s)	Fruit (f)	Root, Bulb (r)	Stem, Caudex (st)	Flower, Bud (fl)	Leaves (herbage) (l)
Agave angustifolia				st		
A. cerulata				st		
A. chrysoglossa				st		
A. colorata/fortiflora				st		
A. pelona?				st		
A. subsimplex				st		
Allenrolfea occidentalis	s					
Allium haematochiton			r			l
Amaranthus fimbriatus	s					l
A. watsonii	s					
Amoreuxia palmatifida		f	r		fl	
Batis maritima			r			
Boerhaavia coulteri						l
Bouteloua barbata	s					
Bumelia occidentalis		f				
Capsicum annuum	s	f				
Carnegiea gigantea	s	f				
Celtis pallida		f				
Cercidium floridum	s	f			fl	
C. microphyllum	s	f			fl	
Chenopodium murale	s					
Cnidoscolus palmeri			r			
Echinocereus engelmannii		f				
E. fendleri		f				
E. grandis		f				
E. pectinatus		f				
Ferocactus acanthodes	s	?			fl	
F. covillei	s	f			fl	
F. wislizenii	s	f			fl	
Ficus padifolia		?				
F. petiolaris		f				
F. radulina		?				
Fouquieria splendens					fl	
Jacquinia pungens		f				
Lantana horrida		f				
Lippia palmeri						l
Lophocereus schottii		f				
Lycium andersonii		f				
L. brevipes		f				
L. fremontii		f				
Mammillaria estebanensis		f				
M. microcarpa		f				
M. sheldonii & spp.		f				
Marsdenia edulis		f				
Matelea cordifolia		f				
M. pringlei		f				
Maytenus phyllanthoides		f				
Muhlenbergia microsperma	s					
Neoevansia striata		f				

TABLE 6.1

(continued)

Scientific Name	Part Used for Food					
	Seed (s)	Fruit (f)	Root Bulb (r)	Stem Caudex (st)	Flower Bud (fl)	Leaves (herbage) (l)
Oligomeris linifolia	s					
Olneya tesota	s					
Opuntia arbuscula		f				
O. bigelovii		f		st		
O. cf. burrageana		f				
O. fulgida		f		st*		
O. leptocaulis		f				
O. phaeacantha		f				
O. versicolor		f				
O. violacea		f				
Orobanche cooperi				st		
Pachycereus pringlei	s	f			fl	
Passiflora arida		f				
P. palmeri		f				
Phaseolus acutifolius	s					
Phoradendron californicum		f				
Physalis crassifolia		f				
P. cf. pubescens		f				
Plantago insularis	s					
Proboscidea altheifolia			r			
Prosopis glandulosa	s	f				
Randia thurberi		f				
Rhizophora mangle		f**				
Ruellia californica					fl	
Sabal uresana and/or *Washingtonia robusta*	s	f				
Sarcostemma cynanchoides					fl	
Setaria liebmannii	s					
S. macrostachya	s					
Simmondsia chinensis	s					
Sporobolus virginicus				st		
Stenocereus gummosus		f				
S. thurberi		f			fl	
Trianthema portulacastrum	s					
Triteleiopsis palmeri			r			
Vallesia glabra		f				
Yucca arizonica		f				
Zizyphus obtusifolia		f				
Zostera marina	s					
hatoj caihöj (unidentified)		f				
mahyan (unidentified)		f				
TOTALS	25	53	6	10	11	3

* Refers to the gum.
**Edible part is the sprouting embryo.
? Seri use is hypothetical.

94

During extended drought ephemerals fail to appear, substantially fewer century plants are edible because they do not become reproductive, and many other major perennials have reduced fruit production. Nevertheless, it appeared that most of the Seri were able to locate significant quantities of edible plants at any time of the year (Table 6.2). During extended drought on San Esteban Island, however, the only important food plant available was *Agave cerulata*, which is rich in carbohydrate but low in protein and lipids. When conditions were severe the people of the San Esteban region must have depended upon animals rather than plants for food.

In addition to animal protein, substantial quantities of protein were obtained from seeds of columnar cacti, eelgrass, foothill palo verde, and various desert ephemerals. Prepared seeds of columnar cacti, especially those of *cardón*, provided high levels of vegetable oil. Significant amounts of carbohydrate were obtained from mesquite pods, ironwood seeds, century plant "hearts," columnar cactus fruits, desert wolfberry fruits, eelgrass seeds, *saiya* roots, and the roots of *mala mujer*. Sugar-rich foods, available only at certain times of the year, were eagerly anticipated, particularly by the children. Major sugary or sweet foods were obtained from columnar cacti, mesquite, century plants, and honey.

Foods were often mixed in nutritionally balanced or complementary manners. For example, *cardón* seeds, which are low in carbohydrate but high in oil content, were often prepared with eelgrass seeds, which are virtually lacking in oil content.

Very few available edible plants were overlooked, and the exceptions do not seem to be dietetically significant or are species uncommon or poorly developed in the region. Several marine algae and the herbage of certain halophytes such as *Batis*, *Salicornia*, and *Suaeda* are edible (Kirk 1970), but were apparently not eaten by the Seri. These plants are salty and their preparation would probably require generous use of fresh water, which would be rendered non-potable. The Seri also did not use reedgrass (*Phragmites*) for food. Chia (*Salvia columbariae*), at least one small grain (*Brachiaria*), and several other plants

TABLE 6.2
Harvest Times of the Major Food Plants

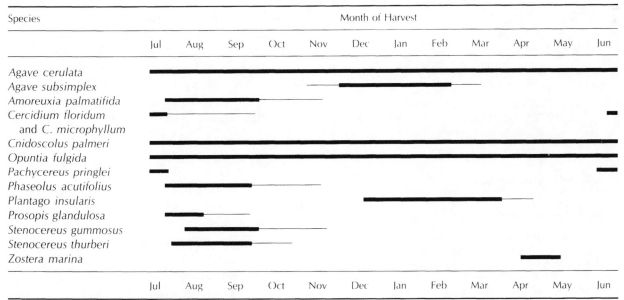

Species	Month of Harvest											
	Jul	Aug	Sep	Oct	Nov	Dec	Jan	Feb	Mar	Apr	May	Jun
Agave cerulata												
Agave subsimplex												
Amoreuxia palmatifida												
Cercidium floridum and *C. microphyllum*												
Cnidoscolus palmeri												
Opuntia fulgida												
Pachycereus pringlei												
Phaseolus acutifolius												
Plantago insularis												
Prosopis glandulosa												
Stenocereus gummosus												
Stenocereus thurberi												
Zostera marina												

━━━ Usual times of major harvest.
───── Times of reduced or possible harvest.

used elsewhere for food were probably too sparse or poorly developed in the region to warrant exploitation.

The Seri ate seeds and fruit of 70 species of plants while the vegetative parts of only 28 species were exploited for food. Furthermore, of the dozen or so plants of major nutritional or dietary significance, only century plant and *mala*

mujer are not seed or fruit-derived foods. Thus, there seems to be a tendency for this desert flora to yield edible fruit and seeds but non-edible herbage. This pattern is quite understandable in the case of an arid region where predation pressure of plants by herbivores is great.

Beverages

Nonalcoholic beverages were made from the pods and seeds of mesquite, the gum of jumping cholla (*Opuntia fulgida*), and the roots of saltwort (*Batis*). Alcoholic beverages were made with the fruit of the giant cacti (*cardón*, organ pipe, *pitaya agria*, and sahuaro), mesquite pods, and century plants (*Agave* spp.).

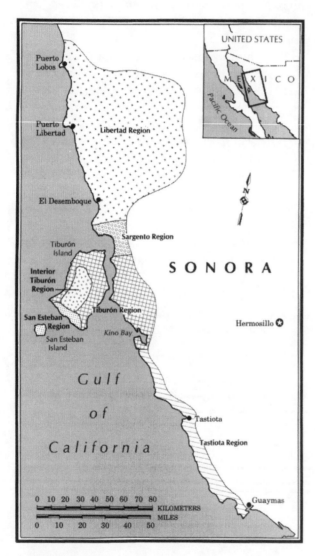

Map 6.2. The major Seri areas in the nineteenth century. The modern Seri language is derived from the dialect spoken by the people of the Libertad, Tastiota, Tiburón, and Interior Tiburón regions. Two other dialects, spoken by the people of the Sargento Region and the San Esteban Region, became extinct during the nineteenth century. According to Seri oral history the Sargento people originally came from the area around Guaymas.

Regional Variations in Food and Water Resources

The diets of the people of the major Seri regions (Map 6.2) varied because of floristic, vegetational, faunal, and environmental differences. Each season and place yielded a different array and quantity of wild crops. The major food resources, particularly the plants, found in the six regions are discussed below. Interpretation is based on information given us by the Seri, our understanding of the biogeography, and the description by E. Moser (1963a) of the regions. It may be assumed that the boundaries and distinctions between these regions were not precise and that movement of the people was somewhat fluid (Sheridan 1982; also see Bahre 1967).

Libertad Region

The *Xica Hai Ic Coii* 'things true-wind side live' or "they who live toward the true wind," were also known as the Tepocas or Salineros. They inhabited the region from about Puerto Lobos southward to Cerro Tepopa, and they "had campsites as far north as the mouth of the Colorado River (E. Moser 1963a:15)." In Seri oral history these people were known for their terrestrial hunting and gathering. They often mixed with the Papago.

Their territory was in the northern part of the overall Seri region, which generally receives more winter rainfall and less summer rainfall than the regions to the south. Thus, in most years there is a noticeably rich growth of winter-spring ephemerals, such as wooly plantain (*Plantago*) and goosefoot (*Chenopodium*). Conversely, summer ephemerals, such as amaranth, are generally not as abundant as they are farther to the south, and winter-spring populations of *Amaranthus watsonii* seldom develop north of the Infiernillo region.

The four large-fruited columnar cacti and the edible chollas (*Opuntia*, subgenus *Cylindropuntia*) are common, although *pitaya agria* (*Stenocereus gummosus*) occurs only as far north as El Desemboque. Other major food plants include a small century plant (*Agave subsimplex*), mesquite (*Prosopis*), ironwood (*Olneya*), palo verdes (*Cercidium* spp.), and desert wolfberry (*Lycium fremontii*). Some of the people reported that eelgrass (*Zostera*) was once abundant in the vicinity of El Desemboque, although by the 1960s it was rarely found there.

Tiburón Region

The *Tahéöjc Comcáac* 'Tiburón-Island people' have been referred to in the literature as the Seri or Tiburones. They occupied the northern and eastern coasts of Tiburón Island and the opposite mainland coast of the Infiernillo Channel. The climate is locally buffered by the Infiernillo Channel and the surrounding mountains. This region is rich in sea life and supports a diverse terrestrial flora.

There is a more extensive development of eelgrass here than anywhere else in the Gulf of California, and the green turtle (*Chelonia*) was extremely abundant. Mullet, crabs (*Callinectes* spp.), oysters, and clams abound in the esteros. Mule deer, desert tortoises, and jackrabbits are abundant on the island.

All of the large, edible fruit species of columnar cacti and chollas (*Opuntia* spp.) are common. Mesquite, ironwood, and foothill palo verde (*Cercidium microphyllum*) are likewise

abundant. Century plants (*Agave subsimplex* and *A. chrysoglossa*), saiya (*Amoreuxia*), and *mala mujer* (*Cnidoscolus*) are found in quantity in the mountains on both sides of the Infiernillo Channel. There are extensive mangroves at various places along both coasts and *mangle dulce* (*Maytenus*) occurs along much of the Infiernillo coast. Summer-fall ephemerals are usually well developed.

Interior Tiburón Region

The *Heno Comcáac* 'desert-of people,' or the "desert people," occupied the interior of Tiburón Island. They were said to live in the desert away from the coast, although they visited the shore from time to time. They lived largely on mule deer, jackrabbits, pack rats, and an array of food plants. They also ate desert tortoises and consumed much honey (honey was not available until about 1900).

These people were said to have eaten considerable quantities of the wild tepary (*Phaseolus acutifolius*). One part of their territory has dense stands of sahuaro and another part has large numbers of *senita* (*Lophocereus*). The fruit of both was eaten in quantity. Other plants which probably provided important staples include:

Agave chrysoglossa
A. subsimplex, century plant
Amoreuxia palmatifida, *saiya*
Cnidoscolus palmeri, *mala mujer*
Ferocactus wislizenii, barrel cactus
Lycium andersonii, desert wolfberry
L. fremontii, desert wolfberry
Olneya tesota, ironwood
Opuntia bigelovii, teddybear cholla
O. fulgida, jumping cholla
Phoradendron californicum, desert mistletoe
Proboscidia altheifolia, devil's claw
Randia thurberi

Apparently they did not use very much mesquite in their diet. Mesquite is far less common in this area than in neighboring regions.

Sargento Region

The *Xnaa Motat* 'south-wind they-came-from,' or "they who came from the direction of the south wind," occupied the coastal territory surrounding the extensive mangrove esteros in the vicinity of Punta Sargento. The natural resources are essentially similar to those of the Tiburón Region. Eelgrass is abundant at the southern margin of the region. Century plants (*Agave subsimplex* and *A. chrysoglossa*), *saiya*, *mala mujer*, and cliff fig can be obtained in the nearby mountains. Other potentially important food-producing plants which are notably common in this area include all the edible-fruited columnar cacti, teddybear cholla, jumping cholla, ironwood, and the palo verdes.

San Esteban Region

The *Xica Hast Ano Coii* 'things mountain in-who-are,' or "they who live in the mountains," inhabited one of the hottest and most arid environments ever to be permanently occupied by humans. During the final period of their existence they resided all year on San Esteban Island except when their meager supply of water gave out and drove them to the opposite south coast of Tiburón Island. There is no doubt that fresh water was the dominant limiting factor of these remarkable people, and their population certainly must have been modest. The climate is classified as Extremely Arid (sensu Meigs 1953), and the average annual rainfall is probably on the order of only 100 mm (Felger 1976a:25). These peaceful people were exterminated by the military in a tragic incident, possibly in the 1860s.

The San Esteban people obtained a rich supply of food from the land and sea. The men hunted sea lions, killing them with skillfully thrown rocks. They also hunted sea birds and harpooned fish and sea turtles (*Chelonia*). They ate the large insular chuckwalla (*Sauromalus varius*). There are no land mammals on San Esteban except two species of small rodents which, if eaten, would probably have been of minor importance.

Sahuaro is absent and organ pipe is not common on the island. Mesquite is likewise present but not common enough to be dietetically significant. Eelgrass does not occur near the island nor along the opposite south shore of Tiburón Island.

The San Esteban century plant (*Agave cerulata dentiens*) was unquestionably their most important food plant. It is abundant on the island, and unlike other species of century plant, edible plants can be obtained at any time of the year. Other important food plants on the island are *cardón* (*Pachycereus*), ironwood (*Olneya*), and *pitaya agria* (*Stenocereus gummosus*). *Pitaya agria* is particularly abundant. Dense stands of *Amaranthus watsonii* may occur at any season of the year. During a period of fifteen years we observed several years in which there were no protein-yielding plant crops available on San Esteban Island or the adjacent south shore of Tiburón Island. We concluded that these people sustained themselves primarily with animal-derived foods, and that during the frequent extended droughts virtually their entire source of protein and lipid was derived from animals.

Tastiota Region

The people who lived in this region were called the *Xica Xnai Ic Coii* 'things south-wind side live,' or "they who live toward the south wind." This group, known as the *Tastioteño*, "roamed the coastal area from Guaymas to Bahia Kino (E. Moser 1963a:16)." Most of this region has not been occupied by the Seri since at least the early part of the twentieth century. In the memory of the elderly Seri in the mid–twentieth century, these people had lived almost completely off the land rather than the sea. The Guaymas, the southernmost Serian-speaking people, were missionized by the early seventeenth century (Spicer 1962:106). It is not known if these people would have been considered as *Xica Xnai Ic Coii*.

The area from near Tastiota south to Guaymas, a major portion of the territory, has a flora with more species and supports richer vegetation than the northern areas. This area undoubtedly contains plants for which Seri information is no longer available. A notably large century plant (*Agave colorata*), several species of *Randia*, three genera of palms, and several species of native fig, in addition to a number of other edible plants, provide potential food resources not found in the more northern areas. The rock fig (*Ficus petiolaris*) is far more abundant and commonly develops into a larger tree in this region than it does farther north. As in the regions to the north, there is the usual array of columnar cacti (although *pitaya agria* is absent), cholla, *mala mujer* (*Cnidoscolus*), *saiya* (*Amoreuxia*), and leguminous shrubs and small trees with edible seeds and pods.

The northern part of this territory, from Tastiota northward to the south end of Kino Bay, encompasses the broad alluvial plain known as the Llano de San Juan Bautista. Until about the 1950s it supported vast *mezquital*, or patchy forests of mesquite-dominated vegetation. Mesquite pods from this region were said to have superior flavor. In addition to the great quantities of mesquite, there are distinctive zones with dense stands of *cardón* (*Pachycereus*) and sahuaro (*Carnegiea*) paralleling the dune-fringed shore. Other important local food plants include:

Bumelia occidentalis, bebelama
Jacquinia pungens, San Juanico
Lycium spp., desert wolfberry
Maytenus phyllanthoides, mangle dulce
Olneya tesota, ironwood
Opuntia fulgida, jumping cholla
Phoradendron californicum, desert mistletoe
Vallesia glabra

By the 1970s the few remnants of the llano which had not been plowed for irrigation farming were rapidly turning into dusty wasteland flats. It is difficult to visualize the rich vegetation which previously persisted there. It was probably the greatest terrestrial food-producing part of the Seri region. However, there was a scarcity of permanent fresh water. To overcome this problem the people dug shallow wells along the coast.

Agave subsimplex occurs on Cerro San Nicolas near the south end of Kino Bay. There are mangrove esteros at Kino Bay, Punta Baja, and Tastiota. At San Carlos Bay and elsewhere in the Guaymas region, including Guaymas Harbor, there were impenetrable mangrove thickets.

7. The Supernatural

Religious experience was highly individual and based largely on the vision quest and shamanism. Although elements of traditional beliefs still persisted in the 1980s, the original religion generally ceased to function in the rapidly acculturating Seri society of the mid–twentieth century. Some traditional Seri beliefs are briefly discussed below.

Origin Stories

As with so many aspects of Seri culture, there were many different versions of the Origin Myth. The following interpretations, based in part on excerpts from the unpublished field notes of Edward Moser, were common elements of Seri origin mythology.

In the beginning there was no land and there was no life. Then *Hant Caai* 'land maker,' or "he who made the land," created a number of land and sea animals and placed them on a huge balsa (reedgrass boat) on the sea. *Hant Caai* caused a male green turtle (*Chelonia*) to assist in the forming of the land. When this was accomplished, *Hant Caai* caused the personage *Hant Quizim* "he who hardens the land" to appear and make the ground firm.

As yet, there were neither people nor vegetation on the land. Then *Hant Caai* caused a tree to grow. This first tree was a red elephant tree (*Bursera hindsiana*; Figure 7.1). Next he created a man, a woman, and a horse and placed them under the red elephant tree. This first man and woman were Giants.

Then *Hant Caai* tested the man and woman. He told the man to ride the horse but the man failed (he fell off). So *Hant Caai* told the man to paddle a balsa. This the man was able to do. He paddled out to sea and harpooned a sea turtle. Upon returning to land, the man found that he had no knife, so he split a stalk of reedgrass and used one of the sharp edges to butcher the turtle. These two tests showed the man was a capable fisherman, although incapable of working the land. The woman's task was to prepare food and do household chores. She attempted to do the work, but did a poor job.

The first man and woman had children and eventually a number of Giants inhabited the land. The land was flat, without mountains or even sand dunes, so it was natural that floods should occur. Floods were accompanied by fire, smoke, and earthquakes. After one such disaster, *Hant Caai* saw that because the land was flat, the people had little chance to escape the destruction. So he sang a song, causing mountains, hills, and dunes to form. These were to provide protection for the people during floods.

In one such flood a group of Giants from the south fled northward to the mountains south of Puerto Libertad. There the flood overtook them and changed them into boojum trees (*Fouquieria columnaris*), which still occur there (Figure 7.2).

Figure 7.1. Red elephant tree, *Bursera hindsiana*, the first plant created by *Hant Caai*. *RSF, 8 km east of Estero Sargento, April 1983.*

Figure 7.2. According to oral tradition, a number of Giants were changed into boojum trees during a great flood. A stand of boojum trees at Las Cuevitas. *DLB, 1978.*

The last flood was the Great Flood. The water covered even the tallest mountains and killed all the Giants, changing them into various other living things, such as the barrel cactus (*Ferocactus wislizenii*).

With the passing of the Giants, a different personage, *Hant Hasóoma* 'land ramada,' appeared on Tiburón Island. *Hant Hasóoma* was male, short, fat, and dirty, and wore a breechcloth and a hat with an exceptionally wide brim. He was the owner of all wild animals and some say he always carried fruit from jumping cholla (*Opuntia fulgida*) with him for food. In another version his principal food was the black gum also found on jumping cholla. He was associated with the chief spirit of the desert. *Hant Hasóoma* caused the first Seri to be formed, and it happened on Tiburón Island. Some say that *Hant Hasóoma* is the real God of the Seri, that he has always existed, and is closely related to the sun, the eye of God.

Then *Hant Caai* caused another personage to appear who was called *Hant Iha Quimx* "he who tells what there is on the land." He told the people the names of the animals and plants on the land and in the sea.

Somewhat contradictory to this concept is the tradition of *Cmaacoj Cmasol* 'old-man yellow,' or "old yellow man," a person who taught unmarried youths the names of all living things. As each animal appeared before him, he taught the animal's name to the youths and then the animals disappeared. He taught the Seri how to have fiestas and the taboos that should be observed. One day he told them he must leave, and then left. His skin was yellow, and he spoke good Seri. Some people feared him. According to one version, he taught the people at a camp near Tecomate on Tiburón Island. Some said he lived in historical times, and that the father of Loreto Marcos knew him.

Plants were created by *Hant Caai*. The first plant was a red elephant tree (*Bursera hindsiana*), the second was *Frankenia*, and the third was *senita* (*Lophocereus*). In other versions iodine bush (*Allenrolfea*) and *senita* were among the first plants formed. All maintained that *senita* was powerful and that, together with red elephant tree, was among the first to be formed. The local jimson weed (*Datura discolor*) and desert lavender (*Hyptis*) were also often named as being among the first ones formed. These plants were generally associated with supernatural powers but were not important food plants.

According to traditional Seri belief, all peoples live in the same world. Each people has its own deity, and each was created separately. The Seri deity is male. To the Seri the following three names represent the same being:

Hant Caai "he who made the land"
Ziix Hamíime Com Ano Quiij "the one who is seated in the sky"
Yooz 'God,' from the Spanish word *Dios*

Spirit Power of Plants

In addition to being the word for "plant," the term *hehe* was used to designate the very life of a plant. This life, or spirit, of a plant was formed by *Icor*, an invisible power said to resemble a hairy bug. It had fine hairs like cobwebs growing from its back, which waved and floated around in the air. *Icor* controlled the spirit of each plant. When the plant died, its spirit left.

With the aid of *Icor* the shaman prepared special powder which he put in small reedgrass tubes. These were rented from the shaman by the people to bring good luck; one could also use them to cure himself of illness or place a curse on someone. The shaman cautioned the people that if they opened the reedgrass container to look in it, the special powder, made from the hairs of *Icor*, would disappear.

Power was in the leaves and green branches. Every kind of plant had its *Icor*, but several had very special power. However, some said that *Icor* was associated only with *senita* and barrel cactus (*Ferocactus wislizenii*). The boojum tree was considered one of the most powerful and potentially dangerous plants. These plants were particularly potent when in flower. Nevertheless, *Icor* was not dangerous nor was it feared.

The *hehe* of the plant manifested itself as the "dust" of the plant and could be black, white, red, or green-blue. Actually it was the hairs on *Icor* that caused the dust. Sometimes this dust drifted into the air and formed white clouds. All clouds were formed in this manner. Rain and storms were also caused by *Icor* (see *Ferocactus wislizenii*).

Vision Quest

The term for shaman, *haaco cama*, derives from *heecot cama* "he who lives in the desert." Both men and women could become shamans (see Griffen 1959:49–51). Spirit power to become a shaman was sought through the vision quest. Each person sought his vision in his or her own chosen way, but followed a generalized pattern. The quest was almost always conducted alone, in isolation and fasting, and away from camp or habitation. Any unmarried young man or woman could choose to go on a vision quest. Not everyone cared to do so, and not every supplicant succeeded in obtaining a vision or power. The Seri found it very amusing that one man went to seek his vision but took his lunch with him.

Ideally the vision quest was done in spring, and was supposed to last four days, although some said that three days would suffice. Griffen (1959:51) reported that "Usually only one person at a time goes to a cave, but it has been known that as many as two or three have gone at once, and generally all or none are successful."

During the approximately four days of fasting, the vision seeker often drank tea from such plants as *Vaseyanthus*, *Brandegea*, elephant tree (*Bursera microphylla*), or ironwood (*Olneya*), or might even have sipped sea water (Griffen 1959:16). A vision might be sought in a vision circle, cave, brush hut built by the sea or in the desert, or even by walking along the shore.

Vision circles, located in lofty places—such as the tops of mountains and hills (Figure 7.3)—are apparently quite old, and not much was known about them by the modern Seri. Beneath each vi-

sion circle there was an underground camp inhabited by spirits. Certain caves were known for "miraculous power" (Griffen 1959:16–17), and were involved with the underground spiritual world, such as a cave on Tiburón Island with doors to an underground spiritual place with brilliant light (Xavier 1941:35). (The concept of doors was probably the result of outside influences, as they were not a part of original Seri culture.) There were also concepts of spiritual animals, beings, and houses (camps) beneath the sea which more or less mirrored ordinary terrestrial life (Xavier 1941).

Figure 7.3. Vision circles were often located in lofty places. Richard E. Schultes sits in a circle on the northeast side of Tiburón Island. *RSF, April 1974.*

During the initial phases of a successful vision quest the supplicant often saw patterns of jagged light and light spots (Bowen and Moser 1968: 110). These patterns were incorporated into most Seri decorations, such as designs on a leather-back turtle (Figure 7.4), fetishes, pottery, face-painting, and tattoos. Although these designs had religious significance, they were also used for purely decorative purposes.

The common elephant tree (*Bursera microphylla*) was strongly involved in the vision quest and religious concepts and might be considered as the Seri "holy bush" (Figure 7.5). *Icóocmolca* 'what (blame) is put on(plural)' fetishes carved from the wood of the red elephant tree served as messengers to the spiritual world (see *Bursera hindsiana*). Also involved in the quest for spiritual power were desert lavender (see *Hyptis emoryi*) and the several above-mentioned teas.

Griffen (1959:50–51) stated that the suppli-cant built a temporary hut for the quest and on the fourth day went to a cave to seek power. He also mentioned fasting at home for four days, and then going to a cave. Towards the end of the vision quest supernatural powers were sum-moned with the bullroarer made of ironwood (see *Olneya*). Upon returning to camp, a newly enlightened shaman might also use a bullroarer to announce his success (Griffen 1959:51).

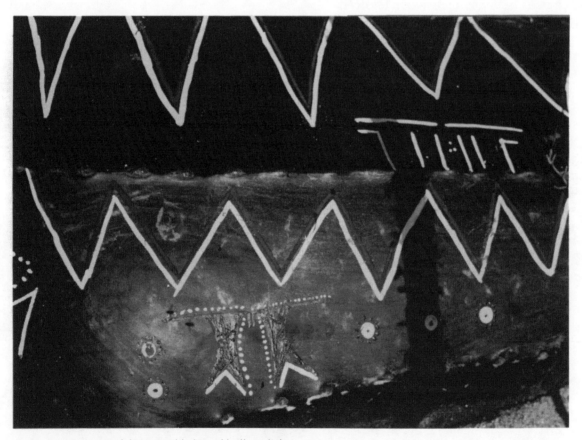

Figure 7.4. Patterns of the jagged light and brilliant light spots seen during a successful vision quest were often rep-resented in Seri decorations. These designs were painted on a leatherback turtle during a fiesta. *RJH, Saps, March 1981.*

Shamanism

Shamanistic abilities could be used for good or evil. Curses or malevolence could result in the death of the one being cursed. Some shamans had more power than others, probably due mostly to acquired reputation and experience.

Figure 7.5. The common elephant tree, *Bursera microphylla*, was involved in supernatural concepts and religious practices. *RSF, vicinity of Campo Víboras, April 1983.*

Because of their knowledge and the frequent need to resort to it, shamans exerted economic and socio-political power over others (Griffen 1959:50). Yet many of the adults were shamans. Griffen (1959:50) indicated that 14 percent of the people at El Desemboque practiced shamanism, and that the custom served as a means of economic distribution outside of the kinship and *hamác* systems of sharing.

The treatment of sickness often involved a shaman. However, the shaman's curing practices did not include the administering of medicines. He or she customarily held a handful of branches or a wand made from elephant tree (*Bursera microphylla*) or branches of desert lavender (*Hyptis*) or chewed the leaves while practicing curing powers (Figure 7.6). These plants, particularly the elephant tree, were used because of their powerful spirits.

The Seri "santo" or *icóocmolca* (Figure 7.7) was carved by the shaman from wood of red elephant tree (*Bursera hindsiana*) and was rented from the shaman as a household or personal fetish. Reedgrass containers with magical properties were also rented to the people by the shaman. The rented reedgrass containers and *icóocmolca* fetishes brought good luck and could also be used for curing. Crosses made from twigs of the shrub *Desmanthus fruticosus* were worn around the neck as an amulet to give protection from sickness. Also involved in shamanistic curing of illness were crosses painted on the patient with Seri Blue (see *Guaiacum*).

Other Beliefs

Various plants and animals were solicited for supernatural powers or treated with caution because of their strong powers. For example, shamans often sought power from an animal, such as the coyote. A person with coyote power was called *oot quiho* 'coyote that-sees' (see Bowen and Moser 1970a:169–170; also see *Phragmites* and *Jatropha cinerea*). The boojum tree (*Fouquieria columnaris*) and *senita* (*Lophocereus*) were treated with particular circum-

Figure 7.6. Shaman curing a child. Porfirio Díaz chanting "*zop, zop, zop . . .*" through a bundle of desert lavender. His wife, Juana, holds their grandchild. *MBM, El Desemboque, 1957.*

spection. Cutting boojum tree or *Sideroxylon* branches caused the wind to blow. Because of its power the *senita* was employed in placing a curse against an enemy.

Clamshells or occasionally other objects, such as arrows, were wedged in columnar cactus stems, ironwood (*Olneya*) trees, and rock crevices as good luck offerings or good luck caches (Figure 7.8). When a rock crevice was used it was called *hast heeyolca* 'rock what-are-given-to' or "what are given to the rock" (see *Pachycereus*). A prayer was often repeated when this practice was performed and also when the women went to gather the first cactus fruit of the season. However, some people carried out these practices and did not say anything. Other plants involved with seeking good luck included *Castela polyandra*, coral bean (*Erythrina*), rock fern (*Notholaena*), and *Sideroxylon.*

Several plants were employed in attempts to

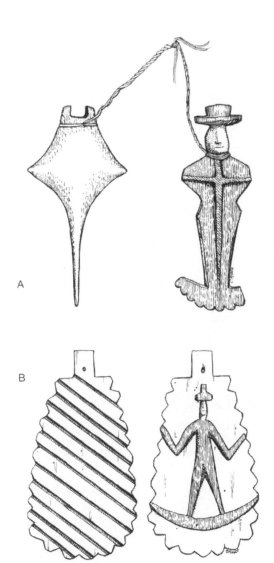

modify weather conditions. To bring rain one resorted to barrel cactus (*Ferocactus wislizenii*), *cardón*, or rock fern. To stop rain one used sahuaro (*Carnegiea*) and desert saltbush (*Atriplex polycarpa*). To bring a desired wind, *Frankenia* was used; to calm the sea, ironwood (*Olneya*); to calm or stop the wind, organ pipe (*Stenocereus thurberi*) or a rock covered with marine algae (*xpanéezx*). The gathering of *Selaginella* was thought to cause cloud formation.

There were concepts of former epochs—worlds or times in which present-day animals and plants were people. These people had been changed into their present non-human forms during great floods or, as some said, by The Flood. Although there were many different stories surrounding these changes and how they occurred, most of the animals and plants associated with such changes were large or unusual,

Figure 7.7. These Seri fetishes, carved from wood of *Bursera hindsiana*, were collected by Gywneth Harrington Xavier in July 1941. A) Fetish made by Jesús Ibarra, who told Xavier (1941:42) that the male figure was for the husband and the manta ray figure for the wife, and that both figures should be hung in the doorway to protect the couple against death and disaster. He said that the male figure, which is janus-faced, stands on the new moon and can look both ways with his two faces—below the earth and above the sky—and comes to the shaman at night. (Male figure 8.4 cm long, hat reddish maroon, face natural wood, body reddish maroon with blue cross, other side blue with reddish maroon cross. Manta ray 8.6 cm long, blue on dorsal side, natural wood on ventral side; *ASM E-957 and E-958.*) B) Fetish made by Santo Blanco (Xavier 1941:38, 40). This was a popular type of fetish for men. The figure stands on the new moon. The notched edge background represents the earth, and the scalloped margin the edges of the world. "On the back are the roads of the gods, the roads from heaven to earth." The object on the head was said to be a cross and not a hat. (Fetish is 8.9 cm long; body painted blue, "hat" and crescent blue, diagonal grooves on back alternately red and blue, rest of figure natural wood. *ASM E-819.*) CMM.

Figure 7.8. Clamshell (*Chione californiensis*) wedged into a *cardón* stem as a good luck cache at Estero Sargento. *RSF, April 1983.*

conspicuous, or important in Seri culture. Bowen and Moser (1970a : 172) stated that "During one of the floods of Seri mythology people were transformed into animals of their choice by one of the spirits, which the Seri are unable to identify." The usual taboo or reluctance to kill the raven, coyote, or dog involved the concept that these animals were once people.

Many customs and objects were associated with the Giants. Bowen (1976 : 103–107) and Moser and White (1968) have provided convincing evidence that these Giants were the ancestors of the Seri and that they generally predated oral history, which we found to extend back in time more than a century to the middle of the 1800s.

We have recorded eight or nine species of plants involving taboos. Girls were to observe a first-fruit-gathering taboo involving *cardón* (*Pachycereus*) in order to prevent them from becoming lazy throughout their lives. First-born children were supposed to avoid touching columnar cactus buds lest their lives be shortened. Girls were cautioned not to touch the flowers of *Melochia* or they would become promiscuous. Many people would not touch the native Jimson weed (*Datura*) because it was one of the first plants formed. However there were no such taboos involving other first-formed plants. Extensive beliefs and taboos also surrounded baskets and basketmaking (see Chapter 15).

There were several taboos surrounding the *csai* hairbrush (see *Aristida californica*). It was dangerous to use a dead person's hairbrush. If one used the *csai* brush at night his spirit would wander off toward *coaxyat*, the place of the dead.

When a person died young, and that person was associated with or known by a name which was the same, or even very similar, to the name of a plant or animal, that name often became taboo. The older name for the *pitaya agria* cactus was *ool-axö*. A child who was called *oláöj*, the term for a kind of fish, died, and the name of the cactus was changed because both names sounded so similar (see *Stenocereus gummosus*). The name *casol cacat*, apparently the original term for *Hymenoclea monogyra*, became taboo because it was the favorite medicine of a woman who died. It seems to have been commonplace that these name taboos were either not widespread, or else they were eventually forgotten or no longer followed, because the original name often came back into use.

There were apparently few strong taboos surrounding animals as food. Some were generally not eaten because it was believed they had once been people (e.g., coyotes and dogs). There was also somewhat of a taboo against eating leatherback turtle meat (see the section on reptiles in Chapter 3).

The wood of *palo zorrillo* (*Atamisquea*) had a powerful spirit and was used in several smoking ceremonies. One such ceremony was performed to cure a fussy, crying baby. The ceremony called up the spirit of a certain bird by burning the bird's nest with twigs of elephant tree (*Bursera microphylla*), *Frankenia*, and *palo zorrillo*. Another smoking ceremony was performed for curing the sea turtle hunter's harpoons and to bring him success in hunting. It involved burning the wood of elephant tree, *Frankenia*, iodine bush (*Allenrolfea*), *palo zorrillo*, and pieces of sea turtle shell.

8. Medicinal Uses of Plants and Animals

Most Seri medicines were prepared from plants. More than 106 species of plants in 88 genera, 17 species of animals, and one mineral featured in Seri pharmacopoeia (Felger and Moser 1974b). By the 1960s commercial pharmaceuticals had largely replaced native medicines, and the people often sought treatment from medical doctors at clinics and hospitals. However, it would be difficult to state or predict which Seri medicines were no longer resorted to, at least on occasion. For example, creosotebush continued to be much used for medicinal purposes in the 1980s (Figure 8.1).

The use of traditional remedies was highly pragmatic and the Seri generally did not have specific explanations for the efficacy of their medicines. When asked why a particular remedy was used, the usual answer was "because it works." A common manner for a remedy to enter the culture was for someone to try or experiment with it, be satisfied with the results, and then tell others about the experience. This explanation was given when we inquired why black vulture meat was utilized for heart trouble.

The administering of medicines was generally individual and personal, and there was variation and overlap in usage. Different remedies were variously esteemed by different people. In some cases it was known that neighboring people had used a certain remedy. The Seri said that the Papago made medicinal use of desert zinnia (*Zinnia acerosa*) to cure diarrhea. In other cases remedies were learned from non-Seri people residing in their territory. For example, the use of a poultice of *Datura* leaves to treat boils was learned by a Seri from a local Mexican.

Plants with aromatic foliage, such as those containing terpenes and other volatile oils and resins, were particularly valued, e.g., *Bursera*, *Hymenoclea*, *Larrea*, and *Porophyllum*. The aromatic nature was generally brought out by heating, and medicinal teas were prepared from no less than 64 different species. The most common methods of preparing medicinal plants were (1) cooking in water or (2) heating in hot coals or ashes. The predilection for medicinal teas and aqueous preparations may have been an adaptation to the extremely hot, arid environment, and an effective method of preparing heat-labile compounds. Seri pharmacopoeia did not include the use of plants for psychoactive effects.

Preparations were simple, and there were relatively few mixtures. The absence of complicated formulas facilitated non-professional and individual preparation and administration. It also encouraged the rapid diffusion of information

This chapter is revised and expanded from R. S. Felger and M. B. Moser, "Seri Indian Pharmacopoeia," Economic Botany (1974) 28:414–436. Used with permission.

Figure 8.1. Elvira Valenzuela brewing a medicinal tea of creosotebush in her home at El Desemboque. *RSF, March 1983.*

through the culture. Native remedies were used most commonly for diarrhea or dysentery (three kinds of dysentery were distinguished by name), headache, and problems experienced by women around the time of parturition.

The concept of "hot" and "cold" plants or medicines was nonexistent in Seri culture. Furthermore we know of no remedies for broken bones or fractures, although there was a remedy for bruises and sprains (see *Cercidium praecox*). The Seri seldom suffered from accidental fractures. When injuries of this nature were sustained in fights they apparently did not treat them.

The vast majority of Seri medicines were prepared from plants. Table 8.1 lists the plants that were used to treat various physical problems, from headache and colds to heart pain and remedies related to childbirth. Detailed information on the preparation and specific application of each plant can be found in Part III under the species name listed in the table.

In addition to the plants listed in Table 8.1, various animals and a type of clay were also used for medicinal purposes. The clay, called *casíime*, was mixed with water and drunk as a remedy for diarrhea. It was also used with a lichen, *Roccella*, to alleviate diarrhea and to reduce fever. It was described by the Seri as being chalky, pinkish, and soft, and as occurring between two layers of rock. They knew of it only from the walls of an arroyo about 0.5 km from the shore in a mountain about 10 km east of Tecomate on Tiburón Island. A hand-sized specimen was pinkish, bounded by slickened surfaces, compressed, cohesive, but not indurated, and not gritty. It showed no sedimentary features.

The fat of the coyote (*oot, Canis latrans*) was cooked and drunk to relieve a chest cold or tired muscles. Two birds were used for medicine: the black vulture (*Coragyps atratus*) and an unidentified species of hummingbird. The Seri name for the black vulture was *col quiimet* 'high who-have-dwellings,' or "who dwell up high." The meat, which was very smelly or rank, was boiled, and the meat and broth were consumed to cure epileptic attacks. To cure an earache, heated fat and excrement from this bird were placed in the ear. As a remedy for heart trouble, the meat was dried, crushed, mixed with cinnamon and eaten. This remedy was said to have entered the culture because one person tried it and was successful, and then told others about it.

The meat of a certain hummingbird called *xeenoj* was eaten to cure fainting and epileptic attacks. According to the Seri, this meat also was very rank.

Reptiles, both terrestrial and marine, were also used in preparing medicines. Although any rattlesnake would probably have been used in the

TABLE 8.1
Pharmacological Uses of Plants

BIRTH
  After Parturition
    Baccharis salicifolia
    Larrea divaricata
    Mascagnia macroptera
  Childbirth
    Ambrosia ambrosioides
    Phoradendron diguetianum
    Simmondsia chinensis
    Trixis californica
  Control Loss of Blood After Parturition (Hemostatic)
    Argemone spp.
    Baccharis salicifolia
  Difficult Delivery
    Porophyllum gracile
  Expulsion of Torn Placenta
    Abronia maritima
    Argemone spp.
    Cercidium praecox

BITES, STINGS, AND LICE
  Black Widow Spider Bite
    Bursera laxiflora
    Sarcostemma cynanchoides
  Head Lice
    Bursera microphylla
    Lippia palmeri
  Rattlesnake Bite
    Stegnosperma halimifolium
  Scorpion Sting
    Bursera laxiflora
  Stingray Wound
    Atriplex barclayana
    Bursera microphylla
    Larrea divaricata

BODY ACHES AND PAINS
  Aching Feet
    Larrea divaricata
    Tidestromia lanuginosa
  Bruises and Sprains
    Cercidium praecox
  Body Aches, Sore Muscles, and Rheumatism
    Anemopsis californica
    Baccharis salicifolia
    B. sarothroides
    Carnegiea gigantea
    Dodonaea viscosa
    Ferocactus covillei
    Hymenoclea monogyra
    Koeberlinia spinosa
    Larrea divaricata
    Pachycereus pringlei
    Stenocereus thurberi

  Swellings
    Dalea mollis
    Euphorbia polycarpa/spp.
    Larrea divaricata
    Neoevansia striata
DERMATOLOGICAL
  Boils
    Datura discolor
  Cuts, Sores, and Burns
    Anemopsis californica
    Battarrea diguetii
    Encelia farinosa
    Krameria grayi
    Podaxis pistillaris
    Roccella babingtonia
    Tulostoma sp.
  Falling and Split Hair
    Camissonia californica
  Headsores
    Bursera microphylla
    *Ricinus communis
    Simmondsia chinensis
    Zizyphus obtusifolia
  Scratches (Prevent Scarring)
    Bursera microphylla
  Skin Rash
    Hymenoclea monogyra
    Vallesia glabra
  Thorn Removal
    Tidestromia lanuginosa
DISEASES, COMMUNICABLE
  Epidemics (Measles, etc.)
    Koeberlinia spinosa
  Chicken Pox
    Cassia covesii
  Measles
    Boerhaavia erecta
    Cassia covesii
    Oligomeris linifolia
    Vallesia glabra

FATIGUE
  Ruellia californica

GASTRO-INTESTINAL
  "Clean Out" Stomach
    Cassia covesii
  Diarrhea/Dysentery
    Acacia constricta
    Allionia incarnata
    Euphorbia misera
    Erythrina flabelliformis
    Guaiacum coulteri
    Jatropha cinerea
    Koeberlinia spinosa

    Krameria grayi
    Mascagnia macroptera
    Maytenus phyllanthoides
    Opuntia fulgida
    O. marenae
    O. reflexispina
    Phoradendron diguetianum
    Porophyllum gracile
    Rhizophora mangle
    Roccella babingtonia
    Solanum hindsianum
    Sphaeralcea spp.
    Struthanthus haenkeanus
    Zinnia acerosa
    Zostera marina
  Emetic
    Atriplex canescens/linearis
    Larrea divaricata
    Olneya tesota
    Prosopis glandulosa
    Simmondsia chinensis
  Intestinal Disorders
    Passiflora arida
  Laxative
    Prosopis glandulosa
  Stomachache
    Cleome tenuis
    Euphorbia misera
    *Melilotus indica
    *Parkinsonia aculeata
    Plantago insularis
    Tiquilia palmeri
  Upset Stomach
    Acacia constricta
    Krameria grayi

HEAD AND THROAT
  Eardrops
    Mammillaria spp.
    Jacquinia pungens
  Eyedrops
    Descurainia pinnata
    Horsfordia alata
    Phaulothamnus spinescens
    Prosopis glandulosa
    Ruellia californica
    Sarcostemma cynanchoides
    Simmondsia chinensis
    *Sisymbrium irio
    Sphaeralcea spp.
  Dizziness
    Jacquinia pungens
    Koeberlinia spinosa
    Lantana horrida
    Larrea divaricata
    Ruellia californica

following situations, among the most common species in the region are the diamondback (*Crotalus atrox*), called *cocázni*, and the sidewinder (*Crotalus cerastes*), called *ctamjij*. Cooked rattlesnake fat was put on open, festering sores. This continued to be a common practice in 1983.

The skin of the snake was wrapped around the area of internal pain, such as the arm or chest. The oil was used to soothe a baby's mouth that was sore from nursing. Rattlesnake meat was roasted; then the meat and bones were finely ground and added to the food of one suffering

TABLE 8.1
(continued)

HEAD AND THROAT (cont.)
  Headache
    Andrachne ciliato-glandulosa
    Asclepias spp.
    Baccharis salicifolia
    Bursera microphylla
    Larrea divaricata
    Ruellia californica
    Sarcostemma cynanchoides
    Stegnosperma halimifolium
    Tidestromia lanuginosa
  Hearing Improvement
    Hyptis emoryi
  Loose Tooth
    Encelia farinosa
  Mouth Sores
    Desmanthus fruticosus
    Horsfordia alata
    Jatropha cinerea
  Sore Throat
    Abutilon palmeri
    Bursera laxiflora
    Datura discolor
    Maytenus phyllanthoides
    Pithecellobium confine
    Simmondsia chinensis
    Sphaeralcea spp.
  Swollen Throat
    Datura discolor
  Toothache
    Anemopsis californica
    Asclepias spp.
    Aristolochia watsonii
    Encelia farinosa
    *Euphorbia polycarpa/*spp.
    Hyptis emoryi
    Opuntia fulgida
    Porophyllum gracile

HEART AND INTERNAL
  Illness "Inside Body"
    Phoradendron spp.

Liver Ailments
  Cassia covesii
Kidney Pain
  Argemone spp.
  Cassia covesii
  Cenchrus palmeri
  Xanthium strumarium
Heart Pain
  Asclepias spp.
  Encelia farinosa
  *Euphorbia polycarpa/*spp.
  Opuntia fulgida

INFANTS
  Harden Baby's Fontanel
    Neoevansia striata
  Heal Umbilicus
    Bursera microphylla
  Help Child to Walk
    Marina parryi
  Sore Mouth
    Jatropha cinerea
  Tonic for Pregnant Woman's Child
    Marina parryi

RESPIRATORY AND FEVER
  Colds and Flu
    Aloysia lycioides
    Baccharis sarothroides
    Bursera laxiflora
    Eriogonum inflatum
    Frankenia palmeri
    Hyptis emoryi
    Mascagnia macroptera
    Pithecellobium confine
    Porophyllum gracile
    Ruellia californica
    Senecio monoensis
    Simmondsia chinensis
    Suaeda moquinii
    Tiquilia palmeri
  Cough
    Bursera laxiflora

    Pithecellobium confine
  Febrifuge
    Roccella babingtonia
  Dyspnea (Asthma, etc.)
    Bursera hindsiana
    Hyptis emoryi
    Larrea divaricata
    Olneya tesota
    Opuntia fulgida
    Roccella babingtonia
  Pain in Trachea and Lungs
    Hymenoclea salsola

UROGENITAL AND
GYNECOLOGICAL
  Contraceptive
    Baccharis salicifolia
    Larrea divaricata
    Viguiera deltoidea
  Conception
    Anemopsis californica
    Cassia covesii
    Notholaena standleyi
  Diuretic
    Argemone spp.
    Cenchrus palmeri
    Opuntia bigelovii
  Stop Menstrual Flow
    Cenchrus palmeri
  Urinary Problems
    Argemone spp.
  Venereal Disease
    Bursera microphylla
    Euphorbia misera

WEIGHT AND APPETITE
  Appetite Stimulant
    Cassia covesii
  Weight Gain
    Struthanthus haenkeanus
  Weight Loss
    Baccharis salicifolia

*Plant not native to the Seri region.

from tuberculosis ("vomiting blood"). The Gila monster (*Heloderma suspectum*) was called *paaza*; the skin of this large, poisonous lizard was heated and placed on the forehead to cure a headache.

The green sea turtle (*Chelonia mydas—moosni*) was used for several medicinal purposes. The penis was dried and made into a tea which was drunk to induce conception. This method of inducing fertility was still in practice in the 1980s. A piece of dried turtle intestine was boiled in a small amount of water and the liquid drunk as a remedy for sea sickness and in modern times also for car sickness.

The *cooyam* was a young green turtle noted for its relatively large amount of fat. To cure a "crazy" person, the fat was burned and the patient sat in the smoke: "Make him stay in it, he will get well." Also, when one went crazy during his vision quest, the others cured him by rubbing *cooyam* grease into his head and face. The grease was obtained by frying the fat.

Only one fish—mullet—was used medicinally. To aid in the expulsion of "bad blood" after parturition, a woman sometimes was given a broth made from boiled mullet. The striped mullet (*Mugil cephalus*) was called *ziix coafp* 'thing that-which-jumps,' or "jumping thing."

A number of invertebrates, most of them marine species, had medicinal properties valued by the Seri. A small octopus (about 2.5 cm long) called *hapaj cosni* 'octopus small' was commonly found in empty seashells in tidepools. It was dried, crushed, and cooked with *hierba de manso* (*Anemopsis*) and the resulting tea was given to a person or horse to make him run fast. Sara Villalobos said that this tea was taken by someone fleeing from his enemies. Either the dried octopus or the prepared, dried mixture was kept until needed.

The shells of either of two mollusks were ground, mixed with water, and applied to an infant's umbilicus if it was not healing. These mollusks are *cotópis ináil* 'cotópis its-shell' (*Turbo fluctuosus*) and *satoj ináil* 'satoj its-shell' (*Modiolus capax*).

A large polychaete worm, *xepe ano zatx* 'sea

Figure 8.2. The case of the wingless, female bagworm moth (*Oiketicus* sp.). Tea made from this species was taken "to make one thin." *RSF, Punta Santa Rosa, April 1983.*

in glochids' (*Eurythöe complanata*), was used with bur grass (*Cenchrus*) to cause cessation of menstrual flow. It is the most common intertidal fireworm in the Gulf of California.

Pyoque was a term for three species of starfish which, although not distinguished by name, were recognized as being different. Two kinds were utilized for medicinal purposes: *Luidia phragma* was boiled and the liquid drunk by a woman to check postpartum hemorrhaging. This starfish was said to be rather hard to find. *Phataria unifascialis*, a slender-rayed starfish, was toasted in hot coals and then ground into a paste with salt and some beef kidney fat. The paste was rubbed onto a swollen area on the body, such as a swelling from a blow.

The case containing a female bagworm moth (Psychidae; Figure 8.2) was boiled and the tea drunk to help one lose weight ("make one thin"). The bagworm was called *cacáöjc* 'that which causes to be dry' (*caöjc* is an archaic word for 'dry'). The nest of a paper wasp, *saaij* (*Polistes*, in the family Vespidae), was crushed and brewed as a tea taken by women to prevent conception.

9. Shelter and Fuel

Housebuilding was casual and the structures light and easy to construct. Living as semi-nomads in a mild climate, the people did not need more substantial housing. Housebuilding materials and fuel were almost always nearby and readily available.

Shelter

The Seri built three major kinds of traditional shelters: the brush house, the ramada, and the roofless windbreak. The term *haaco* was used for any dwelling. The brush house was specifically called *haaco hahéemza* 'house what-were-made-curved-down.' *Hahéemza* refers to the house poles that were bent down into the ground. In certain places, particularly in the inland mountainous regions and San Esteban Island, the people often occupied caves, especially during rainy weather.

Brush House

The brush house was the usual dwelling place (e.g., McGee 1898:221–224; Bowen 1976: 43–46; Sheldon 1979:139). Generally shaped like a Quonset hut, it was of light construction and was temporary (Figure 9.1). Abandoned brush houses usually disintegrated rapidly and evidence of their existence could disappear in a few years. The brush house was commonly

about 1.5 m tall and 3 to 4 m long (e.g., Bowen 1976:44; McGee 1898:221). Sometimes two houses were joined, or one was extra long, and in some cases part of the roof was ramada-like and flat, or the house was box shaped (essentially an enclosed ramada). The brush house was usually built by women.

The entrance was on the side away from the prevailing wind. Often the entire end was left open or uncovered to serve as the entrance. In other cases the entrance was on the curved-wall side (Figure 9.2). Occasionally the side entrance was a small vestibule, constructed with one or two arches perpendicular to the main axis of the house (Bowen 1976:43). The position of the entrance probably varied according to season and local weather conditions.

The framework for the brush house was made from slender, flexible poles set in the ground in two parallel rows. Two, three, or even four poles were often intertwined or twisted together for added strength (Figure 9.3). The poles were bent over, overlapped and intertwined, and lashed together with convenient materials—such as strips of limberbush (*Jatropha cuneata*), cord, cloth, or, when they became available, with wire or fishing line (plastic or nylon). Each brush house commonly had three to six arches. The framework was most often ocotillo (*Fouquieria splendens*), although *palo blanco* (*Acacia willardiana*) and other woods were also used. The

Figure 9.1. Brush houses and wattle and daub houses at Campo Dólar. Note large drum at center of photo, where fresh water was stored. *ETN, March 1951.*

Figure 9.2. A brush house with the entrance on the curved wall side. The leafy branches of many different plants were used to cover the sides and roof. *CMM.*

115

Figure 9.3. The framework of a brush house was often made from the flexible poles of ocotillo lashed together. Each house commonly had from three to four arches. *Drawn from a photo by LMH, Kino Bay, 1935. CMM.*

cross members or tie sticks were ocotillo, ribs of organ pipe (*Stenocereus thurberi*), commercial wood, or any other readily available slender poles. *Carrizo*, or *xapij* (*Arundo* or *Phragmites*), was extensively used in the framework at camps located close to where it grew. The entire structure could be erected in a few hours, with several women working together to gather the poles (McGee 1898:221–224; Bowen 1976:43–46).

For protection from the hot summer sun and the damp chill of night during the cooler months, the walls and roof were covered with spineless, or "soft," plants. Also incorporated into the roofing and walls were various animal pelts and skins (including those of the brown pelican and sea lion), blankets, cloth, dried flat sponges, and driftwood. Sea turtle carapaces were stacked around the base or put on the roof for possible use at a later time. Protection from rain was provided by adding such plants as saltgrass (*Monanthochloe*), *mangle dulce* (*Maytenus*), or sargasso weed (*Sargassum*). However, the brush house generally afforded poor protection from rain. Plants used for covering the walls and roofing included:

Acalypha californica
Allenrolfea occidentalis, iodine bush
Atriplex barclayana, coastal saltbush
A. canescens, four-wing saltbush
A. linearis, narrow-leaf saltbush
A. polycarpa, desert saltbush
Avicennia germinans, black mangrove
Bebbia juncea, sweet-bush
Frankenia palmeri, coast shrub
Hymenoclea salsola, burro brush
Laguncularia racemosa, white mangrove
Maytenus phyllanthoides, *mangle dulce*
Monanthochloe littoralis, saltgrass
Sargassum spp., sargasso weed
Stephanomeria pauciflora
Suaeda spp., sea blite
Zostera marina, eelgrass

Supplies, food, basketry splints, headrings, and a wide range of other materials were kept on top of brush houses to keep them away from dogs, as well as for convenience. Personal items, such as hairbrushes and dried plants for medicinal purposes, were stored in nooks behind the crosspieces. Fetishes made from elephant tree

(*Bursera hindsiana* and *B. microphylla*) were often hung from the roof. Girls decorated the walls with freshly picked bouquets of wild-flowers, such as sand verbena (*Abronia villosa*) and evening primrose (*Oenothera*). The floor was sand or dirt. Hearths were located either inside or in front of the house, depending on weather. When the hearth was outside, there was usually a short brush wall which served as a windbreak. In December 1921, at a camp on Tiburón Island, Sheldon (1979:121) observed that "Little fires burned in all the jacales." At that time the Seri did not have blankets and slept next to their fires (Sheldon 1979:123, 131).

In the 1980s people continued to build the typical brush house at temporary camps set up for fishing or other activities away from the permanent villages. Canvas was often used as roofing since about the 1950s. In the 1970s and 1980s a sheet of polyethylene plastic was sometimes stretched over the frame before the brush roofing was applied, or the roofing was made from corrugated tar paper (*lamina de cartón*).

When an occupant died, the family of the deceased moved into the brush house of the burial sponsor (*hamác*), and the *hamác* family built a new house. Family members of the deceased left their belongings, which the *hamác* took to his new house as his property. The *hamác* gave the bereaved family equivalent items and then burned the brush house. In this way the spirit power of the dead was eliminated.

According to the Seri, the people of San Esteban Island "lived in huts made of driftwood and in caves" (Moser 1963a:24). Flower stalks of century plants (*Agave cerulata*), wood ribs of *cardón* (*Pachycereus*), and sea lion hides were available and could have been incorporated into their shelters. However, ocotillo, *palo blanco*, and other plants with long, flexible, but durable stems or wood are absent from the island (Felger and Lowe 1976); these people, therefore, probably did not build the usual brush houses. Furthermore, the need for such shelters was not great, since the winters are milder and there is substantially less rainfall on San Esteban than elsewhere in the region.

Twentieth-Century Housing

In the 1920s the Seri began to build wattle and daub Mexican-style houses at their more settled sites. Men took over the job of building these more permanent houses. However, women still did the roofing when brush was used and they also worked on the stick and mud-plastered walls. Later, frame and tar-paper shacks replaced the wattle and daub houses, although house construction remained relatively casual. After the death of one of the occupants, these houses were often dismantled instead of being burned, and most of the building material was used to construct a new house (Griffen 1959:7). However, this practice was discontinued by about the 1960s.

The house posts and beams of these more permanent houses were often made from trees and shrubs such as the following:

Acacia willardiana, palo blanco
Bumelia occidentalis, bebelama
Laguncularia racemosa, white mangrove
Olneya tesota, ironwood
Prosopis glandulosa, mesquite

Commercial wood, when available, was also used. The wattle and daub walls were generally made with ocotillo stems and the dry, woody ribs of columnar cacti, and the spaces were filled with mud or adobe (Figures 9.4, 9.5). By the late 1970s most Seri lived in modern cement block houses provided by the Mexican government.

Other Shelters

The ramada, *hasóoma*, was an open-sided structure designed primarily to provide shade during hot weather. The ramada traditionally consisted of four forked corner posts and roofing of brush (Figure 9.6). The posts were made of mesquite or other sturdy woods, such as *palo blanco*. In the 1980s the ramada continued to be popular during warm weather (Figure 9.7).

Another means of coping with hot weather was to remove or leave out the lower portions of roofing and walls of a brush house so that it

Figure 9.4. Wattle and daub houses at Kino Bay. *JDH,
1940; ASM-21043.*

Figure 9.5. Details of a wattle and daub house. The forked
corner posts and main cross members are mesquite and
the horizontal sticks are ocotillo with mud plaster. The
roofing is ocotillo, columnar cactus ribs, brush, and dirt.
Drawn from a photo by LMH, Kino Bay, 1935. CMM.

Figure 9.6. Details of a traditional ramada. The posts and
main roof supports are mesquite; the smaller roofing poles
are ocotillo. The roof is covered with brush, pieces of corrugated tar paper, and cloth. *Drawn from a photo by
EWM, Campo Víboras, 1967. CMM.*

118

Figure 9.7. The framework of this modern ramada is commercial pine, and the roofing consists of leafy branches of white mangrove (*Laguncularia racemosa*). RSF, *Punta Chueca, April 1983.*

functioned as a ramada (Figure 9.8). Hardy described a ramada he saw at Tecomate on Tiburón Island on August 9, 1826:

> . . . at a distance of about twenty-five yards from the shore, I found a hut, constructed of four perpendicular stumps, which supported a quantity of bushes, and bones of the tortoise [sea turtle], serving as a shade to the inhabitants beneath. Suspended to the posts, were skins of the eagle [probably pelican], bows and arrows; the four sides being completely open to the winds. The height of this hut was about four feet, and beneath it were lying women and children, and one man (Hardy 1829:282).

The semi-circular windbreak, *hacáiin hant hapáhtolca* 'windbreak down what-is-put(plural),' was made for overnight use during a journey (Figure 9.9). The size depended upon the number of people using it. The walls were made

with spineless, or "soft," brush, such as brittlebush (*Encelia*), sand croton (*Croton californicus*), indigo bush (*Psorothamnus emoryi*), or sweet bush (*Bebbia*). Entire bushes were pulled up, roots and all, and stacked in together upside down. This shelter consisted only of brush; no rocks were used at the base. A windbreak was not always constructed, particularly when the men were hunting only overnight. In December 1921 Sheldon observed that the Seri "pick out splendid places to camp and always sleep in a spot protected from the wind (Sheldon 1979: 126–127)."

When people were traveling in the desert, they sometimes made a temporary shelter from the sun by bending together the branches of a living ocotillo and covering them with brush. A special shelter was built to protect a leatherback turtle during the four-day fiesta held in its honor.

Figure 9.8. During warm weather the walls of a brush house were sometimes omitted so that the structure served as a ramada. The large poles were undoubtedly collected from the beach drift. *WCS, southeastern corner of Tiburón Island, April 1936; AHF.*

Figure 9.9. This windbreak for overnight use would accommodate one man; machete and clamshell (*Laevicardium elatum*) in the foreground. *CMM.*

These resemble small brush houses with two or three ocotillo arches, but open at each end and roofed only on about the upper two thirds (see Figure 3.7A).

Fuel

Good firewood and materials to start fires were abundant and almost always readily available near camp. During earlier, traditional times firewood was not stockpiled or transported over long distances.

Firewood

Most firewood was gathered by women (Sheldon 1979:122). They carried it back to camp on their heads, tied in bundles set directly on a headring or in a basket resting on a headring (Figure 9.10). Men occasionally gathered firewood and usually transported it in bundles suspended from a carrying yoke or carried over one shoulder (see *Salix*).

Dry, dead wood was gathered because it burned better than green wood, was lighter, and was easier to break. Firewood was usually broken up by hand. Mesquite and ironwood were the most important firewoods, and mesquite was the preferred cooking fuel. There were other desirable hardwoods, such as *palo colorado* (*Colubrina*) and *Condalia*, but these were not as widespread or common as mesquite and ironwood. Many others, both hardwoods and softwoods, were used when available or needed.

Mangle dulce (*Maytenus*) and mangrove driftwood were often the only desirable hardwood fuels along the shore. When children were asked to gather mangrove driftwood, they were told to bring *xepe itoj iipz* "eyelashes of the sea."

Softwoods such as palo verde and ocotillo were commonly used because they were readily available. The fast-burning, soft, resinous wood of the elephant trees (*Bursera* spp.) was used as kindling, as were the twigs of most woody plants, such as desert saltbush (*Atriplex polycarpa*). Grass (*Aristida californica*) mixed with rabbit dung served as tinder.

Figure 9.10. Bringing firewood home. Cathy Moser, on the left, balances the load with two hands—her Seri friends use only one hand for balancing. *MBM, El Desemboque, July 1963.*

As shown in Figure 9.11, pottery was fired with wood from shrubs such as desert saltbush, elephant tree (*Bursera microphylla*), or occasionally burro brush (*Hymenoclea salsola*). The fire was sometimes covered with turtle carapaces to help contain the heat (Figure 9.12; Bowen and Moser 1968:112–114). Brush, such as desert saltbush, was also used to burn off the ends of basketry splints. *Frankenia* was the preferred fuel for cooking plastron meat of a sea turtle. With its many small twigs and fairly hard wood it burned quickly and there was no need to use kindling.

The Seri had a remarkable system of smoke signals, which were much used for communicating across the Infiernillo Channel and to arrange channel crossings (Quinn and Quinn 1965:178; Sheldon 1979:107, 140). Smoke signals were usually made from a smoldering fire of a saltgrass (*Monanthochloe*) or organ pipe (*Stenocereus thurberi*), although other woods or brush were also used. Elephant tree wood (*Bursera microphylla*) was considered the best fuel for smoking out bees, although organ pipe was similarly used.

Plants known to have been used as firewood included the following:

> *Atamisquea emarginata*, *palo zorrillo*
> *Atriplex polycarpa*, desert saltbush
> *Avicennia germinans*, black mangrove
> *Bursera* spp., elephant tree
> *Cercidium floridum*, blue palo verde
> *C. microphyllum*, foothill palo verde
> *Colubrina viridis*, *palo colorado*
> *Condalia globosa*
> *Fouquieria splendens*, ocotillo
> *Frankenia palmeri*, coast shrub
> *Hymenoclea salsola*, *jécota*
> *Laguncularia racemosa*, white mangrove
> *Maytenus phyllanthoides*, *mangle dulce*
> *Olneya tesota*, ironwood
> *Pithecellobium confine*
> *Prosopis glandulosa*, mesquite
> *Rhizophora mangle*, red mangrove
> *Salix gooddingii*, willow

The Seri sometimes sold firewood from the late 1920s and into the 1940s, primarily at Kino Bay. The wood was brought into Kino Bay by the boatload (Figure 9.13). Fausto Topete, governor of Sonora from 1927 to 1929, gave the Seri a sailboat so they could take firewood to

Figure 9.11. María Antonio Colosio firing pottery with wood from desert shrubs. *EWM, El Desemboque, 1967*

Figure 9.12. Turtle carapaces helped contain the heat when pottery was fired. *EWM, El Desemboque, 1967.*

Guaymas to sell. Different men made a number of such trips in this boat, probably taking ironwood (*Olneya*), from the Tiburón Island and Kino Bay region. In the 1970s and 1980s the Seri often purchased ironwood for sculpture-making from Mexican woodcutters. The scrap pieces and chips were used in the hearth.

Firemaking

According to legend the people learned to make fire from the fly (Kroeber 1931:13). The way a fly rubs its forelegs together to groom looks like it is using a firedrill.

> This story happened almost at the beginning of the world. The fly, by rubbing his front legs together, makes fire by friction-making-motion, just like a person. Thus when the fly is with a dead animal and makes a fire by friction and makes smoke signals, the vulture flying along finds the carcass on the desert. That is why it happens like that. Even today when any carcass is out of sight on the desert, the vulture, due to the fly making fire by friction and making smoke signals, as it were, sees it and finally gets to the carcass. That is how it happens (modified from E. Moser 1968:364–365).

Firedrills were made and used by men. The firedrill was operated by hand; the Seri did not use a bow to drive the drill. Firedrills were fashioned from one of at least five kinds of softwood: cliff fig (*Ficus petiolaris*), elephant tree (*Bursera microphylla*), red elephant tree (*Bursera hindsiana*), ashy jatropha (*Jatropha cinerea*), mesquite root (*Prosopis*; the root is a softwood), or sahuaro rib (*Carnegiea*). Cliff fig and *satómatox*, mesquite root driftwood, seem to have been the preferred materials. However, these were not always readily available. The cliff fig firedrill was carved from green wood and dried before it was used. Firedrills made from the other materials were carved only from dry, dead wood. It was said that cliff fig caught fire right away and that sahuaro rib also ignited quickly but took a little bit longer.

The firedrill, called *caaa*, consisted of a fireboard, *caaa cmaam* 'caaa female,' and the drill or drillstick, *caaa ctam* 'caaa male' (Figure

Figure 9.13. Men bringing firewood into Kino Bay to sell. Jesús Félix is in the foreground; Nacho Romero is behind the boat. *JWM, c. 1929; ASM-27983.*

9.14). The fireboard and drill or drill tip were made from the same kind of wood. The drill was simple or compound. The simple drillstick was made from a single piece of wood. The compound drill had a drill tip attached to a handle of another material; they were usually bound together with mesquite root twine. The handle was made from any convenient wood such as a sahuaro rib. Compound drills were made when one did not have a piece of wood sufficiently long and straight to make an entire drillstick from a single piece of wood. Suitable wood from the two *Bursera* species, the *Jatropha*, and the *satómatox* mesquite root were seldom long

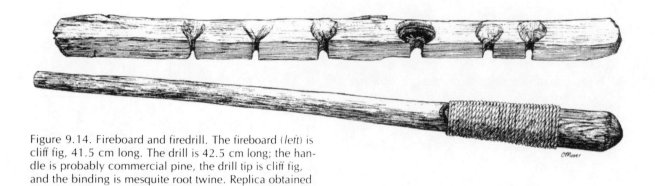

Figure 9.14. Fireboard and firedrill. The fireboard (*left*) is cliff fig, 41.5 cm long. The drill is 42.5 cm long; the handle is probably commercial pine, the drill tip is cliff fig, and the binding is mesquite root twine. Replica obtained in El Desemboque, 1957. *CMM.*

enough to make a simple drillstick, and in certain places cliff fig was relatively scarce or far from coastal camps. One compound drill collected in 1923 from a cache had the drill tip inserted into a handle made from the mainshaft of an arrow (Schindler 1981:239–240). This arrow was made from cane (*Phragmites* or *Arundo*).

It was said that there were three things men always carried in their arrow quivers: arrows, a supply of lac (*csipx*), and a firedrill. The firedrill carried in a quiver was usually made from one of the preferred woods (cliff fig or *satómatox*). The other materials, common and almost always readily available, served the purpose when the men did not already have a firedrill with them. *Bursera* and *Jatropha* firedrills were made only from a dry, dead branch attached to a tree or shrub; on ashy jatropha this special wood was called *hamísj yapáain*, and the term *yapáain* 'what caused it to fall' was sometimes also applied to a similar piece of wood on the two elephant trees.

Román Borbón, from the Yaqui community in Hermosillo, described firedrills which were occasionally used by Yaqui men working on Sonoran ranches (personal communication, Tucson, May 1983). They made firedrills from at least four species of native woods. These were *sámota* (*Coursetia*; stem), *álamo* (*Populus fremontii* and/or *P. dimorpha*; root or branch), *batamote* (*Baccharis salicifolia*; stem), and *guayavilla* (*Acacia coulteri*; stem). Two of these, *sámota* and *batamote*, occur in the Seri region, but the

Seri did not report them as firedrill materials. *Sámota* is a hardwood. Like the Seri the Yaqui made the fireboard and drill from the same wood. However, unlike the Seri they sometimes also combined different woods: *álamo* with *batamote*, or *sámota* with *guayavilla*. The drillstick was made from a single piece of wood.

Although a man working alone could start a fire with a firedrill, usually two men worked together (Figure 9.15). It was important that the fireboard did not move, so one man held it down while the other rotated the drill between his hands. Any thin, flat object, such as a knife or sandal, was placed below the fireboard to contain the igniting punk. Jackrabbit dung was often put around the fire hole as tinder. An experienced team could start a fire in several minutes or less. However, it was hard work. One's hands could become blistered and bloody if he had not used a firedrill for some time.

There was considerable variation in size and construction of the firedrill. The preferred size for the fireboard would be wide enough to allow room for four or five fire holes and for a man to hold the board down with one hand at each end while giving the other man enough room to rotate the drill. Although fireboards were made as small as the one shown in Figure 9.16, they were usually larger. The drill was usually about 50 cm long. Schindler (1981:237) pointed out that some museum specimens, made for sale, were outside the usual functional size range and were improperly constructed.

Because of the difficulty and skill involved in

Figure 9.15. Demonstrating the use of a firedrill. A) Alberto "Vaquero" Molino prepares the tinder. B) Vaquero holds the fireboard while Guadalupe Astorga rotates the drill. Roberto Thomson times them with a stopwatch. C) Vaquero holds the igniting punk while Guadalupe blows on it. *EHD, 1929; HF-23886, HF-24029, HF-23887.*

Figure 9.16. A small makeshift firedrill, made by Antonio López in about five minutes from *xoop yapáain*—a dead branch on a live elephant tree (*Bursera microphylla*). To be functional, the drill tip would be tied to a handle. *RSF, 5 km southeast of El Desemboque, April 1983.*

using a firedrill, it was commonplace to transport hot coals, even over considerable distance, rather than use a firedrill in camp. Hot coals were carried between several sticks held together, or a burning stick was taken to start a fire. Sometimes live coals were carried to the coast from as far away as Hermosillo. Firedrills were phased out in the twentieth century as soon as commercial matches became available. However, as late as the 1950s, when they ran out of matches—especially on Tiburón Island—the men resorted to using a firedrill.

10. Equipment

In most cases there were no special craftsmen. Each person usually made his own equipment (or made it for a member of his family) or was assisted by relatives or close associates. Although Seri equipment tended to be functional and spare, it was often decorated.

Bow and Arrow

Men used the bow and arrow (Figures 10.1, 10.2) for terrestrial hunting, combat, and as a musical instrument. They made a simple bow, called *haacni*, from a single piece of flexible hardwood carved to shape (McGee 1898:197; Hayden 1942:28). *Mauto* (*Lysiloma*) was often the preferred wood for the bow, although desert hackberry (*Celtis*) was also highly regarded for this purpose. Excellent bows were also made from catclaw (*Acacia greggii*), *Caesalpinia*, and *sámota* (*Coursetia*).

Most Seri bows were usually 1.5 to 2 m long and rather narrow, although longer, shorter, and wider ones were also made. Seri bows were notably long compared with most other North American bows (see Mason 1894). On occasion the men continued to use bows and arrows for hunting until the 1950s, particularly when camped on Tiburón Island for several months at a time and they ran out of bullets. (Although they began to acquire rifles in the late nineteenth century, bullets were usually very difficult to obtain).

The bow was often decorated, such as with alternating stripes or zigzag designs of red ochre and Seri Blue. The bow was strung with twisted sinew—including that from mule deer, horse, or cow (Bowen and Moser 1970b:182)—or with mesquite root cord (McGee 1898:200).

In 1692 Gilg illustrated two kinds of bows (see Figure 1.4). One kind had a deep double curve, and the other had a simple single curve. Although most Seri bows from the late nineteenth and early twentieth centuries had a single arc, one collected in 1922 or earlier from Tiburón Island (ASM-1640), had a shallow double curve. In addition, double-curved bows are known from elsewhere in southwestern North America. For example, Frank Russell (1908: plate XIII) illustrated a Pima bow with a shallow double curve and called it a "war bow." The Apache also made double-curved bows (Mails 1974:265).

The arrow was called *haxáaza* 'what was oozed onto.' Seri arrows commonly consisted of a projectile point, foreshaft, mainshaft, and feathers (see Figure 10.2; McGee 1898:197, facing 200). The point, or arrowhead, was stone or occasionally ironwood (*Olneya*), but metal (particularly hoop iron) was quickly adopted when it became available. When made from stone, it was called *hast hax* 'stone arrowhead';

Figure 10.1. Demonstrating the use of bow and arrow. *EHD, Kino Bay, 1929; HF-24011.*

otherwise the arrowhead was called *hax*. The point was hafted with lac from creosotebush (*Larrea*) or resin from brittlebush (*Encelia*), and also wrapped with sinew.

The foreshaft, *iti icozám* 'on to-attack-with,' was made from hardwood, particularly desert wolfberry (*Lycium fremontii*). McGee (1898: 197) stated that

> The foreshaft is 8½ inches long, of hard wood carefully ground by rubbing with quartzite or pumice into cylindrical form, about three-eighths of an inch in diameter at the larger end and tapering slightly toward the point; the larger end is extended by careful grinding into a tang [point] which is fitted into the main shaft, the joint being neatly wrapped with sinew.

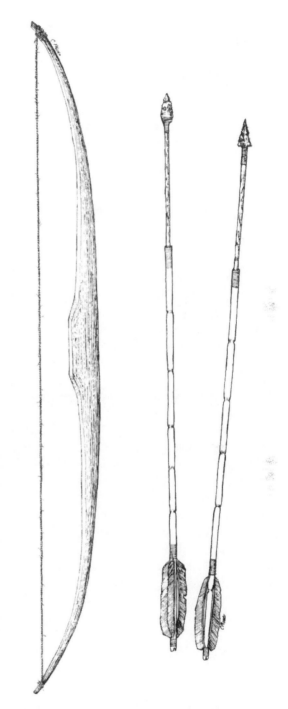

Figure 10.2. Details of bow and arrows. The bow, 110 cm long, is made from *palo blanco* (*Acacia willardiana*) and strung with mesquite root twine. The arrows, 87.5 and 89 cm long, have *carrizo* (*Arundo* or *Phragmites*) mainshafts, desert wolfberry (*Lycium fremontii*) foreshafts, stone points, and sinew bindings. These replicas were made by Nacho Morales in 1961. *CMM.*

A groove, called *ihatáamj* 'where it is forked,' was cut into the wood all the way around the foreshaft about three-fourths of the distance away from the point. This was done to facilitate its breaking at this point and making the arrow more deadly because it would be more difficult for an animal or enemy to pull it out. Seri arrows were often decorated; McGee (1898:198) mentioned that the foreshaft was commonly painted a bright color, particularly red.

The mainshaft was made from a selected piece of reedgrass, or cane (see *Phragmites*). The people of the Libertad Region in the northern part of the Seri territory used the flowering stalk of *amole* (*Agave schottii*) because of scarcity of reedgrass in their area. There were three feathers, *ihásim* (singular), often from a hawk or osprey, attached with sinew wrapping. Beneath this transparent sinew the shaft was marked with thin black rings made with *csipx* from creosotebush (*Larrea*). This was done so that one could tell whether or not an arrow had been made by a Seri. The arrow nock was also called *ihatáamj*. Quivers were made from animal hide (McGee 1898:facing 216).

Arrows for boys and for hunting small game were made of hardwood, such as desert wolfberry, with the tips sharpened and fire hardened (McGee 1898:197). Hardwood arrows with blunt, knobbed ends, rather than sharp points, were used for hunting rabbits and birds.

Arrow poison for hunting and warfare usually was made from the sap of *hierba de la flecha* or Mexican jumping bean (*Sapium*). Sometimes it was used in combination with sap of *Marsdenia*. Sap from limberbush (*Jatropha cinerea* and *J. cuneata*) was also put on arrow points. These poisoned arrows were greatly feared by enemies of the Seri, and are mentioned in many of the Spanish colonial and mission documents.

Harpoons

Harpoons were extensively employed for hunting and fishing. The generic term for a harpoon or the point is *tis*, which is also the name for catclaw (*Acacia greggii*). There were specific harpoons for sea turtles, fish, and crabs.

The turtle harpoon had a mainshaft made of several pieces of strong but flexible wood, such as white mangrove (*Laguncularia*), mesquite root (*Prosopis*), or driftwood. They were joined with mesquite root cord. Edward H. Davis reported a harpoon pole "comprised of five pieces of driftwood (Quinn and Quinn 1965:171)." The modern harpoon shaft was made from several pieces of commercial pine bound together with heavy-duty fishing line.

The harpoon shaft for winter turtle hunting was about 6 to 10 m long (three poles for winter use measured in 1975 were 6.3, 8.9, and 9.3 m in length). The mainshaft for warm weather use was about half as long; these shorter harpoons were also used for hunting white pelicans by boat. Harpoons for winter use were longer because the turtles were not at the surface. Harpoons were also occasionally used for hunting terrestrial animals, such as javelina (peccary).

Young men often decorated their harpoons. After the hunter cut a suitable branch, such as from white mangrove, he carved designs into the bark. The bark not within the outline of the design was excised. The shaft was then scorched, and then the remaining bark removed. This process produced a natural, light-colored ("white") design. Each man created his own design.

The harpooner stood at the prow of a wooden boat, searching for signs of a turtle, holding the shaft of the harpoon horizontally. After sighting a turtle, the hunter signaled to the boatman how to pursue it by complex movements of the harpoon, such as raising or lowering one end or the other while describing circles of varying size and orientation with the harpoon ends.

The sea turtle harpoon point was embedded in a short, detachable wooden foreshaft (Figure 10.3). The foreshaft was split into halves, and a groove or channel was cut in each half to receive the metal harpoon point. The point and the wooden halves were hafted with lac from creosotebush (*Larrea*), *sámota* (*Coursetia*), or brittlebush (*Encelia*). There was a wide notch in the foreshaft to accommodate the string or cord

Figure 10.3. Sea turtle harpoon head. The metal point is cold-hammered and filed; the harpoon head is probably mesquite wood wrapped with commercial cord; and the rubber gasket is a piece of tire. This harpoon head (11.5 cm long, including the point) was made by Pedro Comito around 1965. *CMM.*

which helped hold the two halves together and to which the toggle line was attached. The toggle line was originally mesquite cord, but this was replaced by commercial fishing cord and later by heavy-duty monofilament plastic line when these items became available. The modern foreshaft was still made from the native hardwoods: mesquite, ironwood (*Olneya*), *palo colorado* (*Colubrina*), or *palo blanco* (*Acacia willardiana*). Ascher (1962) provides a detailed description of Seri turtle harpoons.

Although harpoon points have long been made of metal, they were originally made of *tis*, catclaw (*Acacia greggii*), and the generic term for a harpoon or a point remained *tis*. Wooden turtle harpoon points probably went out of use sometime during the nineteenth century. The original turtle harpoon point was called *tis comíhj* "smooth harpoon point." It was smooth, straight, round in cross section, and made from catclaw wood. The turtle was harpooned in the flipper, and considerable skill must have been needed to bring it in. Apparently, the first metal harpoon points were also of this design.

The *tis coozalc* "ribbed harpoon point" later replaced the *tis comíhj*. The *tis coozalc* was made of metal and had a shaft which was square

in cross section. This point was also used to harpoon the turtle in its flipper.

Even as early as the 1820s the Seri had obtained metal for harpoon points. Hardy's description of a turtle harpoon point matches that of the *tis coozalc*, described to us 150 years later. Hardy (1829:296–297) described the turtle harpoon as follows:

> An Indian paddles himself from the shore on one of these [balsas] by means of a long elastic pole of about twelve or fourteen feet in length, the wood of which is the root of a thorn called Mesquite, growing near the coast; and although the branches of this tree are extremely brittle, the underground roots are as pliable as whalebone, and nearly as dark in color. At one end of this pole there is a hole an inch deep, into which is inserted another bit of wood, in shape like an acorn, having a square bit of iron four inches long fastened to it: the other end of the iron being pointed. Both the *ball* and *cup* are first moistened, and then tightly inserted one within the other. Fastened to the iron is a cord of very considerable length, which is brought up along the pole, and both are held in the left hand of the Indian. So securely is the nail thus fixed in the pole, that although the latter is used as a paddle, it does not fall out.

Modern turtle harpoons had barbed metal points, and were known as *enim tis* 'metal harpoon-point.' McGee (1898:193) illustrated a barbed turtle harpoon point which he mistakenly labeled as a "fish-spearhead."

The double-pronged fish harpoon, *tis hatéemla* "serrated harpoon point," was used for fishing from a balsa and for harpooning mullet while wading in esteros (also see *Olneya*). The shaft was similar to that of the turtle harpoon. The points were made from the heartwood of catclaw, or from creosotebush (*Larrea*) if catclaw was not available. The serrated edge faced inward to hold the fish (Figure 10.4). Hardy (1829:290) described it as follows:

> They have a curious weapon which they employ for catching fish. It is a spear with a double point, forming an angle of about 5 degrees. The insides of these two points, which are 6 inches long, are jagged; so that when the body of the fish is forced between them it cannot get away on account of the teeth.

Figure 10.4. Replica of a wooden double-pronged fish harpoon, made by José Torres, 1961. The points are iron-wood (each 19 cm long) and are tied with mesquite root twine to a shaft made of commercial pine. (The points were traditionally made from catclaw or creosotebush.) *CMM.*

The modern fish harpoon, *tist hahóocj* "twin harpoon points," was similar in design except that it had a pair of metal points (usually pieces of steel reinforcing rods) in place of the paired wooden ones (Figure 10.5).

The single-pronged fish harpoon, *hacaaizáa* '*hacáaiz* true,' was used from the shore. It has not been in general use since about the 1930s. It was essentially similar to the double-pronged fish harpoon, except that it was single-pronged. The serrated point was made from the same materials. The shaft, about 2.5 to 3 m long, was commonly made from desert lavender (*Hyptis*), *sámota* (*Coursetia*), or *Caesalpinia*. These woods were used because they are strong but resilient. *Hyp-*

tis would seem to be less desirable than the other two preferred woods. However, it is readily available on San Esteban Island, the south shore of Tiburón Island, and elsewhere in the arid lowlands where the others do not occur.

The *hacaaizáa* was used for harpooning fish, mostly larger ones, among rocks and in tidepools. It had to be thrust with considerable force. Techniques often used in conjunction with this harpoon included luring fish out of their hiding places by using reedgrass (*Phragmites*) to stir up the water or throwing pieces of giant rainbow cactus (*Echinocereus grandis*) into the water as bait. The San Esteban men were well known among the Seri for their expertise with the *hacaaizáa* harpoon; the men from Tiburón Island were not so skillful with it. Porfirio Díaz, said by the Seri to be the last San Esteban man, was respected for his prowess with this harpoon.

Smaller harpoons, called *hacáaiz*, were used to spear swimming crabs (*Callinectes bellicosus*).

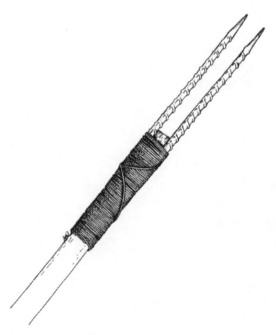

Figure 10.5. Modern double-pronged fish harpoon. The prongs are steel reinforcing rods filed into sharp points. They are bound with commercial cord to a shaft made from commercial pine. *Drawn from a photo by RSF, Punta Chueca, April 1983. CMM.*

The shaft was often made from reedgrass, but any convenient wood probably sufficed. McGee's (1898:187) illustration of a cane-shafted "turtle harpoon" may have been used for crabs, because reedgrass would not be strong enough to contain a turtle.

Traditionally the Seri did not use fishhooks (McGee 1898:194; Kroeber 1931:19), traps, or nets for hunting or fishing. However, their neighbors had all of these devices. We can only surmise that the Seri did not use these items because of the abundance of game, particularly sea turtles, and also because traps would have been cumbersome to carry from camp to camp. Nets were certainly not necessary in earlier times when sea turtles were abundant, and cordage was undoubtedly too valuable to be used so lavishly. However, since the advent of commercial fishing, the men have made extensive use of gill nets and turtle nets. The Seri also did not engage in communal drives for hunting rabbits or other game, although several hunters sometimes chased or harassed a deer until they could get close enough to shoot it (see Mammals in Chapter 3).

Boats

The reed boat, or balsa, was a vital part of the pre–twentieth century culture of the seafaring Seri. Soon after the turn of the century the balsa was replaced by the wooden boat. Early wooden boats were paddled and powered by sails; later, outboard motors replaced sails and paddles. These transitions brought profound socio-economic changes. During previous centuries there had often been imposed cultural pressures, but the Seri themselves chose to make these newer cultural adaptations. In the 1970s wooden boats were replaced by fiberglass boats. The building or purchase, maintenance, and operation of boats were responsibilities of the men.

Balsas

The *hascám*, or balsa, was made from the stems of reedgrass (see *Phragmites*) or giant cane (*Arundo*) fastened into bundles and bound with mesquite (*Prosopis*) root cordage (Figure 10.6). It was a flexible, double-ended craft, and was paddled with the knees resting on pads of grass (*Aristida californica*) or seaweed (*Sargassum*). The paddle was double-bladed. The blade was usually made from ironwood (*Olneya*) because it entered the water easily, due to its own weight. The handle was made from any convenient, strong wood, such as white mangrove (*Laguncularia*). In addition the balsa was often poled with the ever-present harpoon (Figure 10.7). When not in use, the harpoon might have been held in place with pegs set into the balsa (Quinn and Quinn 1965:166). Cargo was secured to hardwood spikes, called *icócaöj*, usually made from creosotebush (*Larrea*) or catclaw (*Acacia greggii*), driven into the sides of the balsa.

Neither passengers nor cargo remained dry. The flexible balsa could ride the waves, perhaps even allowing the men to go to sea during moderately windy weather. The San Esteban men were noted for their prowess at sea and being undaunted by rough water. It was said that the men of San Esteban Island were first-class men of the sea; it didn't matter if it was windy—they went to sea anyway.

A balsa, being relatively light-weight, could be carried well above the high-tide zone, even to the top of beach dunes (see *Fouquieria splendens*). Thus, it was generally not necessary to camp at a site with protected water. Among the seafaring Seri, probably every able-bodied man owned a balsa. When the owner died, his balsa was burned.

María Antonia Colosio described a childhood trip from Tiburón Island to San Esteban Island in the first decade of the twentieth century. This account was recorded in El Desemboque in 1978.

We were at *Cyajoj*. We were on two balsas that were tied together, side by side. We were going to . . . what's it called, the mountain out in the sea? I didn't want to go. We were going to *Coftécöl*. They had put blankets and water on, jugs as big as this [gesturing, indicating a large water vessel]. Since the jugs were full of water, they were tied in place. Plants were stuck in the mouth of the

Figure 10.6 This large balsa, taken from Tiburón Island by William J. McGee in 1895, was photographed at the Smithsonian Institution. *WD, NAA-4282.*

jugs, and the jugs were in carrying nets. I didn't want to get on, but my father caught me and put me on. After he caught me and put me on, he tied me behind a blind man who went along to paddle. Then I cried a lot, but he didn't pay any attention to me. That's how we went to *Coftécöl*. It was so dangerous when we almost entered the area called *Ixötáacoj* [Big Whirlpool]. The sea just swirled and churned. The wind wasn't blowing but the water was choppy. It just churned, it was dangerous. The sea was going around. Everything just roared. The children and old women all cried. The old man Pozoli just said, "We'll land really soon." As we were going to land, he sang to the shore. And it seemed we landed right away. The men paddled with all their strength, and we landed near the rocks.

The open water distance from *Cyajoj*, at the south shore of Tiburón Island, to San Esteban is about 15 km. The trip was made in spring, because they gathered eggs of sea birds. They also gathered the San Esteban century plant, *Agave cerulata* subsp. *dentiens*. It is interesting to note that the two balsas were tied together. María said that she ran and tried to hide from her father, because she was so frightened of the perilous trip. *Coftécöl*, the name for San Esteban Island, is derived from *coof*, the term for the large endemic chuckwalla (*Sauromalus varius*), and *caacöl* 'large (plural).' "Soft" plants were stuffed into the tops of water vessels to prevent water loss due to sloshing (see Chapter 6). María was secured to the blind man to prevent her from being swept overboard. It was said the blind men were exceptional paddlers. *Ixötáacoj* 'whirlpool large' is an area of the sea between Tiburón and San Esteban islands known for treacherous, swirling currents, or giant whirlpools.

On occasion temporary rafts for hunting sea turtles were made from driftwood logs lashed

Figure 10.7. Ramón Blanco poling a one-man balsa with a turtle harpoon at Kino Bay, 1922. This was one of the last Seri balsas. *EHD; HF-24086.*

together with mesquite root cord. When there was a need for food and no boat was available, a few men might take a turtle harpoon and some rope, and walk along the shore to pick up several driftwood logs. A raft would be made by tying the logs together. After they got a few turtles (there were so many that it did not take long) and brought them ashore, they would dismantle the raft and discard it but carefully save the rope.

An emergency raft was once made from the dry flower stalks of the San Esteban century plant (see *Agave cerulata*). This agave-stalk raft was used to cross the channel from San Esteban Island to Tiburón Island.

Wooden Boats

The first wooden boat was made around 1900 by Juan Mata, who copied boats used by Mexicans from Guaymas (Griffen 1959:9). He went

on foot "all around Tiburón Island" getting little pieces of driftwood (on the order of 30 to 60 cm long) to make the first wooden boat. It was made at a camp on the north shore of Tiburón Island known as *Canóaa Quih An Ipáii* 'boat the on its-making.'

Subsequent early wooden boats were also made of pieces of driftwood and flotsam. Men walked the shores of Tiburón Island to collect enough for one boat. The thicker boards were cut lengthwise into two thinner pieces with saws filed from hoop iron, also found along the shore. (Although we were told that hoop iron, from pieces of barrels found in the beach drift, was used to make these saws, probably any flat piece of iron would have sufficed. At that time the Seri could have obtained discarded scrap iron at ranches and from Hermosillo.) Files for making the saws were obtained on infrequent trading excursions to Hermosillo. The ribs and vertical stem at the bow and stern of the boat were made of elephant tree (*Bursera microphylla*). It was used because Seri nails, fashioned from barbed wire cut from fences of nearby ranches, would not penetrate a hardwood, such as mesquite (Griffen 1959:10). These nails were about 3.5 to 4 cm long.

Edward H. Davis described one of these plank boats that he saw in the early 1920s.

A seagoing boat which I saw in Guaymas, captured by Mexican fishermen after killing the Seri crew, was composed of fully one hundred pieces of driftwood fastened with nails made of wire, and tied with fiber, and all the seams were filled with a black composition of native origin resembling pitch (Quinn and Quinn 1965:171).

The undersides of these early plank boats were extensively coated with tar-like pitch or caulking compound (Figure 10.8) prepared from organ pipe (*Stenocereus thurberi*), *pitaya agria* (*S. gummosus*), or elephant tree (*Bursera microphylla*) mixed with oil from any one of several animals, such as sea turtle, pelican, sea lion, horse, or cow. Davis stated that even the insides of these boats were painted with pitch (Quinn and Quinn 1965:201). The pitch-covered boats

Figure 10.8. Covering the bottom and sides of a wooden boat with caulking pitch. The pot in the foreground contains the pitch. Note that Seri boats were made without a keel. *EHD, 1924; HF-23805.*

were said by the Seri to be ugly but seaworthy—to a degree (see *Stenocereus thurberi*). Kroeber (1931:24) reported that "Their modern plank boats are weak in plan, badly joined, and always leaky; but they serve." They leaked so much that a boy accompanied the men on turtle hunting expeditions to bail with the carapace of a *cooyam*, a small green turtle (*Chelonia*). Balancing himself in the center of the boat and bailing constantly, he won the men's sympathy and was given first choice of the turtle meat.

These early plank boats were outfitted with paddles and a single sail. The paddle was often used to steer the boat (Quinn and Quinn 1965: 202). In 1921 Sheldon (1979:109) crossed the Infiernillo Channel with the Seri in a boat he described as a "heavy, old rotten affair like a long dory" which was powered by men using "five clumsy paddles." The mast was made from commercial wood, white mangrove (*Laguncularia*),

or driftwood—including *xapij-aas* (possibly bamboo). The paddle had a single blade and also could be made from white mangrove. The mast and gunwales were often decorated with red and blue designs or solid colors. The sail was usually made of muslin, although the fisherman's blanket sometimes served as a sail.

Unlike in a balsa, people and cargo were not constantly wet in a plank boat, more cargo could be carried, and cargo and people could more easily be accommodated with much less danger of being lost. Plank boats were more maneuverable and faster than balsas. Crossing the Infiernillo Channel and moving camp by boat was therefore much easier. When a turtle was brought aboard a balsa, it generally had to be killed and carefully secured in a balanced position. However, it could be carried live in a wooden boat without any particular regard to balancing it. Live turtles could be kept much longer, espe-

Figure 10.9. A wooden boat in March 1937. A group of Seri men paddling out to greet the Velero III research vessel off the southeast shore of Tiburón Island. Chico Romero has official papers for the visitors. *WCS; AH-2314.*

cially in hot weather, and had commercial value because they were brought to market alive. After acquiring wooden boats, the pragmatic Seri discontinued the custom of burning a boat upon the death of the owner (Griffen 1959:59).

Wooden boats also influenced choice of camp sites. For example, before they had wooden boats, the people used to camp at the mouth of the Río San Ignacio, where there is a plentiful supply of fresh water. They moved to the present site of El Desemboque, about 4 km to the southeast, when they obtained wooden boats, because of the better protection of the bay. When they camped at the mouth of the arroyo, water was right at hand at the *Haat Hax* "limberbush water hole." However, after moving to the new site, it took the men about an hour to go to *Haat Hax* and carry water back to El Desemboque.

On occasion, during the early part of the twen-

tieth century, the Seri came into possession of dugouts, called *canóaa hehe tazo* 'boat wood one' or "boat of a single wood." María Luisa Chilión related that at the time of the birth of one of her sons her people were fleeing from the military, who burned several dugouts belonging to the Seri. This event took place in the first decade of the twentieth century. From about the mid–twentieth century, wood from discarded boats was used to build coffins for adults.

Later wooden boats, such as those made from the 1930s onward, were much better built (Figure 10.9). The ribs were made from sturdy hardwoods, such as mesquite (*Prosopis*), black mangrove (*Avicennia*), or *bebelama* (*Bumelia*). Outboard motors, introduced in the 1940s, were used in conjunction with sails until about the mid-1950s (Figure 10.10). After that time, the use of sails was discontinued as the men began

Figure 10.10. Beaching a sailboat at the end of the day in El Desemboque. *BM, summer of 1953; ASM.*

using larger outboard motors. These boats were strikingly painted with bold combinations of red, blue, green, or yellow. Several Seri men became well known for their boat-building skills. Both Seri and Mexican fishermen cleared rocks away from intertidal areas to provide sandy-bottom slips or boat landing ramps (Figure 10.11; Bowen 1976:36).

Panga is the local Spanish term applied to a small fishing boat. The Seri term is *canóaa,* derived from the Spanish term *canoa* 'canoe.' Seri wooden boats, powered by sails and later by outboard motors, vastly increased mobility and hunting and fishing capabilities. However, the cost of outboard fuel was a major factor in bringing the Seri into a cash economy. Without

wooden boats the Seri could not have entered into commercial fishing ventures. By adopting wooden boats the Seri progressed from a hunting and gathering economy into a cash economy.

By the mid-1970s commercial fiberglass *pangas* had almost entirely replaced the older wooden boats. However, the fiberglass boats were not as maneuverable as the wooden ones. Because of the lack of a keel on the boat and the intricate system of signaling between harpooner and boatman, Seri fishermen could turn their wooden boats in tight circles in rapid pursuit of a sea turtle. As sea turtles edged towards extinction in the late 1970s, the Seri ceased to use their wooden boats and by 1983 these wooden boats were virtually a thing of the past.

Figure 10.11. A boat slip cleared in the rocks at Campo Dólar. Inside this relatively new boat are a single-blade paddle and a harpoon. *ETN, March 1951.*

Carrying Equipment, Cordage, and Containers

When it came time to move camp the people went by balsa or boat, or overland on foot. Heavy items, such as stone metates, pestles, and some pottery, were stashed nearby to be used when the camp was re-occupied. Other items were transported to the new camp. Men usually carried a cargo suspended from each end of a carrying yoke (*peen*) by a rope or a carrying net (*cool*) (Figure 10.12). The net was made from mesquite root fiber (see *Prosopis*), human hair, or horse hair cordage. The yoke, called a *palanca* in Spanish, was usually made from legume hardwoods, such as:

Acacia greggii, catclaw
Acacia willardiana, palo blanco
Caesalpinia palmeri
Olneya tesota, ironwood
Prosopis glandulosa, mesquite

The men often decorated their carrying yokes with red and blue bands and other typical Seri patterns and designs. Makeshift carrying yokes were poles or large sticks cut to a suitable length and notched near each end to hold the net or rope. A particularly large carrying yoke, made from driftwood or willow (*Salix*) was used to carry firewood. When water was brought from distant water holes, men carried it in ollas suspended from *cool* nets (see Figure 6.1). During the 1920s 20-liter cans supplanted the breakable pottery ollas. These water cans were suspended by a carrying net or rope handles tied either directly onto the top of the can or to a stick placed through it.

Men also carried cargo, such as an olla, 20-liter can, or a single bundle of goods, suspended from one end of a carrying pole. Items such as a block of wood might be carried on a rope slung over the back. Desert tortoises were lashed to carrying poles with strips of limberbush (*Jatropha cuneata*). When men carried ripe cactus fruit back to camp they often strung them on a stick, such as *Cordia*, or on a strip of limberbush.

Women almost always carried cargo on their heads, usually on a headring (Figure 10.13) or in a basket balanced on a headring. The headring, *hatxíin*, was almost always made with limberbush stems (see *Jatropha cuneata*), although

Figure 10.12. Antonio Herrera using a carrying yoke to transport his personal belongings and a green turtle carapace. *EWM, Kino Bay, 1951.*

mesquite (*Prosopis*) was sometimes substituted. All of the twentieth-century headrings we have seen were wrapped or bound with variously colored fabric, such as cloth, scraps of blankets or clothing, canvas, or yarn. There was considerable variation in quality of construction and aesthetic appearance; some were bound with two or three selected colors of cloth or yarn incorporated into bold patterns. A piece of cloth deftly looped into a ring on the head served as a makeshift headring (Burckhalter 1976:30, 36).

The *hatxíin* was in daily use by women to balance and cushion a wide array of objects carried on the head. Even metates, loads of rocks, and babies on cradleboards were carried on top of their heads (Figure 10.14; also see Figure 1.4). Children too old to carry on a cradleboard but too young to walk were carried on the hip by their mothers or female relatives. Women car-

Figure 10.13. Lupe Comito carrying home a freshly butchered sea turtle balanced directly on a headring. *EWM, El Desemboque, 1963.*

Figure 10.14. Josefita Torres carrying her child on a cradleboard balanced on top of a basket on a headring. *EWD, 1936; HF-23914.*

ried water when the distance was short, balancing a water olla or a 20-liter can on their heads as shown in Figure 6.2.

Infants were kept on a cradleboard, called *hasítj*, fashioned from any one of a number of different woods (Figures 10.15 and 10.16). The curved outer frame was made from a flexible pole from such shrubs as *sámota* (*Coursetia*), *hierba de la flecha* (*Sapium*), *mauto* (*Lysiloma*), *Caesalpinia*, and ashy jatropha (*Jatropha cinerea*). When *hierba de la flecha* was used, it was said to give the infant a long life. The cross pieces were made from slats from commercial wooden crates, but in earlier times almost any wood sufficed. The child was swathed in rabbit fur and cloth.

The most frequently used cordage was made from mesquite roots (see *Prosopis*). It was important and highly valued. Rope or twine was also spun from human hair by means of a wooden

Figure 10.16. Details of a cradleboard frame; 69 cm long. The curved frame is ashy jatropha (*Jatropha cinerea*); the slats are from a commercial crate and fastened with commercial nails. String and cloth were used to bind the baby onto the cradleboard. Obtained in 1956 from María Félix, who had used it for her daughter. *CMM.*

Figure 10.15. Infants up to two years old were kept on a cradleboard a few hours each day. Andrea Romero nursing her child in the summer of 1953. *BM, ASM.*

Figure 10.17. Rope twister with human hair rope. This replica, 25.5 cm long, was made from commercial wood by Jesús Morales in 1962. *CMM.*

hair spinner called *icóonzx* 'to spin (string)' or *halít hapámas* 'hair that-which-is-spun' (Figure 10.17). Hair spinners were made from almost any kind of wood, including ocotillo and commercial pine. Cordage made from mesquite roots or hair was used to make the *cool* carrying net, waist cords, and necklaces, and to tie objects and for many other purposes. The *cool* net was started on a stick held between the toes (Figure 10.18). Strips of *Jatropha cuneata* were used

Figure 10.18. Antonio Herrera starting a carrying net. *MBM, El Desemboque, 1964.*

Figure 10.19. Green turtle bladder, partially filled with turtle oil, hung in a creosotebush. *EWM, El Desemboque, 1970.*

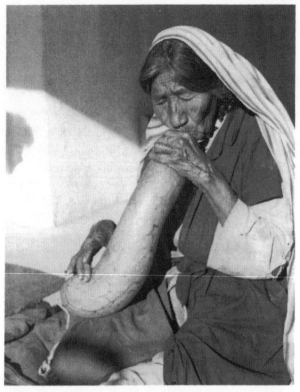

Figure 10.20. Sara Villalobos inflating a green turtle stomach bag. She has cleaned and washed the stomach and tied it at one end. Such containers were dried while inflated. *RSF, El Desemboque, March 1973.*

for binding a wide variety of objects, including the roofing and framework of a brush house.

In addition to pottery vessels and turtle carapaces, containers included small wooden boxes made from *Bursera hindsiana*, *cardón* and sahuaro "boots," and reedgrass cylinders. Bags were made from the bill pouches of brown pelicans, and the stomachs and bladders of various animals, notably the green turtle (Figures 10.19, 10.20). Women used stems of limberbush (*Jatropha cuneata*) to make baskets, which were very important containers in traditional Seri culture (see Chapter 15).

11. Adhesives, Tanning, and Pigments

Materials for adhesives and sealants were obtained from biological materials, and those for tanning from biological materials and salt. Pigments were prepared from both biological and mineral sources.

Adhesives and Sealants

Lac from certain scale insects (*Tachardiella* spp.) was obtained from stems of creosotebush (*Larrea*) or *sámota* (*Coursetia*; Figure 11.1). The lac on both shrubs was said to be the work of a certain black ant called *cootaj ccapxl* 'ant sour.' It is plastic when heated but hardens again on cooling, forming a strong bond closely akin to commercial sealing wax (Bohrer 1962, Euler and Jones 1956). Men carried a supply of lac in their arrow quivers.

Lac was used for hafting harpoon points, arrow points, and knife blades, for sealing lids on storage vessels, and as an all-purpose glue. Resin from brittlebush (*Encelia*) and *Frankenia* was similarly employed, although it did not form as strong an adhesive. Creosotebush lac was also extensively used in ironwood sculpture as a wood filler. The generic term for lac and resin is *csipx*.

Pottery vessels were sealed with resin from brittlebush or *Frankenia* or with a scum formed by boiling flour made from seeds of eelgrass (*Zostera*) or foothill palo verde (*Cercidium microphyllum*). Pottery was also sealed by burning pieces of the outer surface of a sea turtle carapace inside the vessel (Bowen and Moser 1968: 115). Dry flowers of the sahuaro (*Carnegiea*) and creosotebush herbage were used to seal pottery vessels (ollas) for cactus wine (Bowen and Moser 1968: 115). Small holes in pottery vessels were sometimes plugged with mashed leaves of *palo colorado* (*Colubrina*). More often such holes or cracks in pottery were repaired with creosotebush lac or pitch.

A black, tar-like pitch, or caulking compound, called *hocö ine* 'wood its-mucus,' was prepared from organ pipe (*Stenocereus thurberi*), *pitaya agria* (*S. gummosus*), or elephant tree (*Bursera microphylla*), and animal fat. It was essential for the wooden boats made during the earlier decades of the twentieth century (see Figure 10.8 and *Stenocereus thurberi*).

Tanning

The Seri traded tanned deer hides at least since early historic times. In the seventeenth century Adamo Gilg stated that "They tan the skins of wild beasts very nicely with the animals' own brains and tallow (Di Peso and Matson 1965:

141

Figure 11.1. Lac on *sámota* (*Coursetia glandulosa*). *RSF, Rancho Estrella, April 1983.*

55)." In 1921 Charles Sheldon (1979:138) also observed that deerskins were tanned with brains, and that the skins were used for trading with the Mexicans. Presumably, these were mostly the hides of mule deer.

To protect a deer skin being prepared and to keep it off the ground and clean, various "soft" plants, such as iodine bush (*Allenrolfea*) or coastal saltbush (*Atriplex barclayana*) were placed beneath it. A common method of tanning involved applying to the hide mashed seeds of ironwood (*Olneya*) or castor bean (*Ricinus*) mixed with salt. The hide was also soaked in sea water. In another process a grass (*Aristida californica*) was mixed with animal brains, dried into cakes, and rubbed into the hide. In still other methods the seeds of sahuaro (*Carnegiea*) or *cardón* (*Pachycereus*) or the green, leafy twigs of

ironwood were mashed and rubbed into the hide to soften it. *Cardón* or sahuaro seeds were used because of their high oil content.

The hide was draped over a stout pole or branch while it was being scraped. Red elephant tree (*Bursera hindsiana*) and *palo blanco* (*Acacia willardiana*) were among the trees which served this purpose. The hide was scraped with a beamer fashioned from a cow or horse rib. About five beamers were used, each wrapped with cloth at the end to make it easier to grip. In 1970 Jesús Morales and his wife María Antonia Colosio demonstrated the process of tanning a mule deer hide (Figure 11.2). When she handed him a sharpened beamer he gave her a dull one which she then sharpened. He kept rotating the five different beamers. The tanning process spanned a week, during which time the hide was repeatedly soaked, wrung out, dried, treated, and scraped.

Paints and Dyes

The most common pigments were reddish brown and blue. These were used to decorate pottery, boats, fetishes, musical instruments, beads, arrows, carrying yokes, digging sticks, gambling sticks, game sticks, headdresses, wooden boxes, deer hides, cloth, and a wide variety of other objects. Pigments were commonly stored in powdered form in reedgrass containers (see *Phragmites*). Paints were often mixed in clamshells (see Figure 3.7) and then stored in the same shells (McGee 1898:186).

Reddish or reddish brown paint was most often obtained from elephant tree (*Bursera microphylla*). Other sources included *Jatropha cuneata*, *Caesalpinia*, or, rarely, *Jatropha cardiophylla*. When bruised or cut, the thick succulent stems and roots of *Bursera* and *Jatropha* ooze copious reddish sap which looks like blood. Another important source of reddish paint was crushed red ochre, *xpaahöj*, dissolved in water.

Blue pigment, known as Seri Blue, played an important role in Seri aesthetic expression. It was one of the few items made by them which

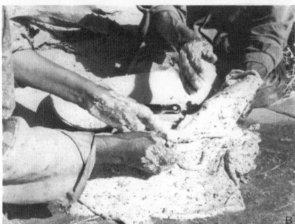

Figure 11.2. Jesús Morales and María Antonia Colosio demonstrating the tanning of a deer hide. A) Scraping the inside of the hide (after soaking it) with a horse rib beamer; second day. A bed of coastal saltbush (*Atriplex barclayana*) was placed beneath the hide to help keep it clean. B) Rubbing castor bean mash onto the hide to cure it; second day. C) Re-soaking and softening the hide with sea water; the fifth day. *MBM, El Desemboque, 1969.*

involved the mixing of more than two different materials (M. Moser 1964; Pierce 1964). The ingredients, none of which was blue, were obtained from *guayacán* (*Guaiacum*), white clay, and bur-sage (*Ambrosia* spp.). Canyon ragweed (*Ambrosia ambrosioides*) or *Stegnosperma* was sometimes substituted for bur-sage (see *Guaiacum*).

Patterns in Seri baskets have been made almost exclusively with basketry splints dyed reddish brown or black. Black was nearly always in combination with a reddish brown pattern. Reddish brown could be made from the roots of several shrubs. However, it was almost always made from *cósahui* (*Krameria grayi*). Alternative sources were obtained from *Hoffmanseggia*, a mixture of *cósahui* and *Hoffmanseggia*, or—on rare occasions—*Melochia* or *Jatropha cuneata*. *Cósahui* sometimes was used to dye objects besides baskets.

Black basketry dye was usually made from mesquite (*Prosopis*). It was also obtained from sea blite (*Suaeda moquinii*), which was sometimes used in combination with pomegranate (*Punica*), *Pithecellobium confine*, and desert lavender (*Hyptis*). Occasionally mesquite and sea blite dyes were mixed or roots of *Jatropha cuneata* were cooked with the sea blite dye. If mesquite and sea blite were not available, red mangrove (*Rhizophora*) could be used. By the 1960s commercial black dye had almost entirely replaced the native black dyes.

Undecorated plain baskets were occasionally stained orange with sand croton (*Croton californicus*) or yellow with indigo bush (*Psorothamnus*). However these colors soon faded or rubbed off.

12. Dress and Adornment

In early times the Seri made extensive use of bright, gaily colored, and imaginative ornaments and dress. In 1692 Adamo Gilg (Di Peso and Matson 1965:51–52) reported that teenage boys were the most indulgent. With personal possessions at a minimum, the Seri carried face-painting, and apparently also body painting, to a highly developed art form. By their own choice, they replaced animal and bird skins and pelts with textile clothing.

Clothing

Indigenous clothing included breechclouts, aprons, skirts, and robes or blankets made from the pelts and skins of the brown pelican and mule deer, although several other mammals were also used—notably, cottontail, jackrabbit, and even coyote (see Di Peso and Matson 1965:51, 52, 54; Hardy 1829:281, 289; McGee 1898:10, 224–232).

Robinson and Flavell (1894:May 25) reported that the Seri wore "clouts of rushes" (see *Typha*), and McGee (1898:10) claimed that "garments were formerly woven of coarse thread or cords made from native vegetal fibers." These reports seem incorrect. We found no evidence that the Seri used native plant materials for clothing, except for the waistcord.

The general pattern seems to have been that young children and old men did not wear clothing, and that the people usually wore very little

clothing—at least during the hotter times of the year and in the warmer regions, such as San Esteban Island and the south side of Tiburón Island.

Men often wore a breechclout of jackrabbit or coyote fur or the soft skin from the pouch of a brown pelican. The breechclout was tied to a waistcord made from mesquite root cord or hair cord. According to the Seri, the San Esteban men and women wore a pelican pouch breechclout. In 1692 Adamo Gilg wrote that "married men hang fox skins over the lower part of the abdomen, behind and before (Di Peso and Matson 1965:54)." Although the grey fox occurs in the Seri region, Gilg may have been referring to the coyote. Both men and women used pelican pelts for kilts, skirts, and robes. The Seri said that when pelican pelts were worn as clothing, the larger feathers were removed and the pelts were turned feather-side inward (Figure 12.1). However, when used as robes, they were worn with the feather side out.

Until about the middle of the twentieth century most men wore a cloth kilt over their trousers, and a few older men continued to wear it in the 1980s. This cloth kilt derived from the older animal pelt breechclout. At the end of the nineteenth century and into the twentieth century some men wore only a cloth kilt (McGee 1898: facing 121, 154, and 158).

Women wore aprons or skirts of varying length, depending on the geographic region and season. As mentioned above, the San Esteban women were said to have worn an apron of pelican pouch skin. Hardy (1829:298) stated that "Their dress is a sort of blanket, extending from hips to the knees. But most of the old women have this part of the body covered with skins of the eagle, having the feathers turned *towards the flesh*. The upper part of the body is entirely exposed . . . ." He was at Tecomate at the north end of Tiburón Island, and undoubtedly mistook pelican pelts for those of the eagle (which was too rare and difficult to obtain).

According to the Seri, women living on Tiburón Island and the mainland often wore a deerskin skirt. This skirt was called *hapáptim* 'mule deer *-aptim* (an archaic and untranslatable term).' The hide was tanned and softened (see Chapter 11), and dyed red with *torote* (*Jatropha cuneata*) root dye. Several pieces of deer hide were sewn together with the tendon of the deer, the holes being punched with a deer bone awl. Sometimes alternating sections of the skirt were dyed red and left white, and the lower margin of the skirt also was left white, with a fringe cut along the bottom of it. It was about knee length, although the Seri also told us about longer deer hide skirts.

As soon as cloth became available to them, the Seri began adopting it for their clothing. At different places this occurred at different times. In 1692 Gilg stated that ". . . from childhood on, the women cover the lower part of the body with a skirt of animal skins, if they can get nothing else (Di Peso and Matson 1965:54)." On Tiburón Island, Hardy observed a Seri man wearing "a red baize shirt, and a blanket around his legs (Hardy 1829:281)." The earliest known photograph of the Seri, taken in 1874, shows them wearing textile clothing.

Blankets were made of rabbit and pelican pelts. The Seri said the larger feathers (the flight feathers) were removed from pelican pelt blankets which were worn as clothing, but when they were used as a blanket for sleeping, all of the feathers were left on. Gilg described skillfully made rabbit blankets:

Against the cold they clothe themselves with rabbit skins, which they cut into four cornered oblong strips, which they sew together lengthwise, and the strips each separately are sewed together like a tube, in such a way that the fur is turned outside all around. These fur tubes they join together so skillfully with little thongs that both outside and inside only a continuous fur is seen. Formerly, canons in Europe wore the like (Di Peso and Matson 1965:54).

Figure 12.1. Franciso Aguilar wearing a pelican pelt kilt with the feathered side facing inward—tied with a waistcord. Note his necklace with long beads, which may be made from *Asclepias* or *Baccharis salicifolia*. William J. McGee, on the right, persuaded him to pose for this photograph. *WD, Costa Rica, November 1894; NAA-4263.*

Footwear

Sandals were made of leather from the green turtle, mule deer, sea lion, or cattle. Leather from the tail of a male green turtle yielded the best quality. A pair of sandals, called *moosni in-áil hatáamt* 'sea-turtle its-skin sandal,' could be made from the tail of one male turtle (male sea turtles are distinguished by their very large tails). A single layer of sea turtle leather sufficed, but deer and sea lion leather was doubled because the hides of these animals are thinner. Sea lion sandals were said to be inferior to the others because there was too much fat in the hide. A pair of cowhide sandals made for us by Jesús Morales had soles with a single layer and seemed as sturdy as sea turtle leather.

It was customary for each man to put his identifying mark on the bottom of his sandal so that people could identify the footprints. Makeshift sandals were occasionally fashioned from pads of creosotebush (*Larrea*) bound with balloon vine (*Cardiospermum*). In the late eighteenth century Gilg mentioned soles of deerskin wound together crosswise just like the old Roman shoes (Di Peso and Matson 1965:54) and illustrated them on his map (see Figure 1.4).

Headpieces

A leafy crown or wreath, called *hehe yail iti hapácatx* 'plant its-greenness on what-is-left-on,' was woven from leafy stems, often with the flowers included. It was made from vines and shrubs, such as balloon vine (*Cardiospermum*), elephant tree (*Bursera microphylla*), and *Mascagnia* (Figure 12.2). Men and women wore it in summertime for protection against the sun and to keep the hair in place. *Hehe yail iti hapácatx* also refers to a band of flowers worn on a girl's head as decoration, a practice which was still current in the early 1980s.

Two kinds of decorated crowns, made of ashy jatropha (see *Jatropha cinerea*), were worn by men in dances and combat and, later, around the camps. The *hehe hamásij* 'wood that-is-opened-up' was a type of crown with a framework consisting of hoops and strips of jatropha wood fastened with sinew (later often replaced by string or nails). This crown terminated in a peak to which was fastened a tuft of white breast feathers of a gull. The feather tuft was later commonly supplanted by a knob or bird carved from red elephant-tree (*Bursera hindsiana*). Designs on the wooden parts of the crown, typically blue and red, were said to be purely decorative (Figure 12.3; also see Hayden 1942).

Coyote men carried bows, wore *hehe hamásij* crowns and were ferocious (see *Phragmites*, Oral Tradition). These crowns, through their association with supernatural coyote men, brought bravery to a warrior. As a result, men began wearing them in combat, probably in the nineteenth century. Sometimes they were also worn around camp to indicate warrior status.

The crowns ultimately became common and many men wore them, although belief in their supernatural properties evidently diminished. During the 1920s and 1930s the crowns were still worn but their popularity was declining. By the late 1940s they had all but disappeared from use (Hayden 1942:24), and since about that time were made mostly for sale to outsiders. In the 1970s and 1980s, however, *hehe hamásij* crowns were often worn by dancers at fiestas. Schindler (1981:51) believed that these crowns were derived or inspired by the headdress of *Matachín* dancers, which have been common features of Indian Catholic ceremonialism in Mexico for centuries.

The other type of crown headdress, called an antenna headpiece by Bowen and Moser (1970a), was known as *ano cojozim quih itáamalca* 'inside who-flees the his-horns' (Figure 12.4). It usually consisted of a cloth headband into which were stuck two knobbed sticks carved from ashy jatropha wood, each stick about 30 cm tall. The antenna headpiece was originally worn by each of two men who were the leaders of a dance called *ziix coox cócoila* 'thing all who-dance,' or "they who dance about everything," which was performed at girls' puberty ceremonies. The Seri said that Juan Mata initiated the use of this headpiece (it would have been around the turn of the century). In a dream vision he was instructed by *cama*, the manta ray, to introduce it into the *ziix coox cöcoila* dance. The antenna headpiece was said to hold no special significance other than to identify the dance leaders.

In May, 1721, Padre Juan de Ugarte saw archers wearing "a kind of helmet of feathers (Venegas 1759, vol. 2:50)." In the last century men occasionally wore a raven-feather headpiece (Bowen and Moser 1970a). Obviously non-Seri in origin, it was probably introduced through friendly contact with the Apache during the nineteenth century. Bowen and Moser (1970a) have given a detailed discussion of Seri headpieces.

Figure 12.2. María Antonia Colosio wearing a leafy crown made from vines of *Mascagnia macroptera*. MBM, El Desemboque, 1969.

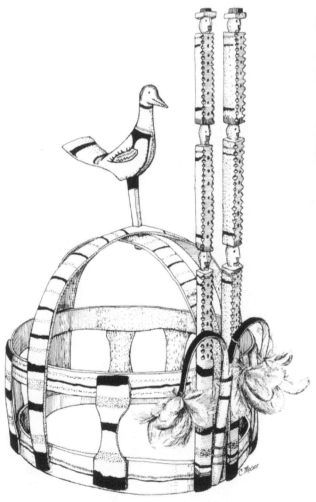

Figure 12.3. The bird and anthropomorphic columns of this wooden crown were carved from red elephant tree (*Bursera hindsiana*). The rest of the headpiece was made from ashy jatropha (*Jatropha cinerea*) tied with cotton string and decorated with white feathers. The colors are blue (represented here by solid black), red (hatching), orange (stippled), and natural wood (plain). This fine specimen (AF-2767), 35 cm in height, shows considerably more artistic skill than later examples. Made by Jesús Morales in September 1956, it was collected by William N. Smith at Punta Chueca in September 1956. *CMM*.

147

Figure 12.4. An antenna headpiece, made from wood of ashy jatropha and decorated with white chicken feathers. It has red (represented by hatching) and blue (solid) pigments, and natural wood and is 27 cm in height. This example is more elaborate than most headdresses of this type. Replica made by Jesús Morales at El Desemboque in 1968. *CMM.*

Ornaments

An imaginative array of native materials was used for fashioning personal ornaments—primarily necklaces, although pendants, rings, and other decorative objects were also worn. Necklaces were made from such diverse objects as snake and fish vertebrae, sea shells, lobster antennae, octopus suction cups, seeds, stems, roots, buds, flowers, fruits, seaweed, and clay. Certain of these materials were commonly stored until needed. More than 25 plant and 29 animal species were used in Seri necklaces and other personal ornaments (Table 12.1). This listing includes items used by the Seri themselves as well as those made strictly for sale. However, most of these materials were used as ornaments by the Seri themselves (see McGee 1898, 1971; Griffen 1959; Di Peso and Matson 1965; Bowen 1976). Figure 12.5 shows examples of Seri necklaces using a variety of native materials.

Necklaces were made with a single kind of color of bead or object, or different kinds of beads and colors were combined. The beads were often dyed or bleached; reddish and blue were favorite colors. The reddish color was most often obtained from *cósahui* (*Krameria grayi*), and the blue from Seri Blue (see *Guaiacum*); by the 1960s commercial dyes were generally used. Commercial thread, string, yarn, or cord have long been used for stringing (Figure 12.6). In earlier times gut, sinew (Sheldon 1979:137), or cordage made from mesquite root or human hair was used. Thread was often obtained by unraveling woven or knitted garments. Commercial beads were worn since early historic times.

There were no special artisans, and ornaments were usually worn by the person who made them or by a family member, although women often exchanged necklaces. Shamans made their own special objects. The majority of women and girls wore decorative necklaces. However, since the late 1950s they have almost entirely ceased to wear their own necklaces, except on special occasions (see Griffen 1959:8). At about that time the women and girls began making necklaces in great diversity and profusion which they sold to outsiders at relatively modest prices. Figure 12.7 shows a woman making a commercial necklace from a variety of beads.

Men wore several kinds of necklaces (Figure 12.8) but did not wear them as frequently as did the women, at least not since the late nineteenth century. Like facepainting, this custom was discontinued by the men long before it went out of fashion among the women. Two strands of clay beads—crisscrossed, one under each arm—were worn by a bachelor to show he was eligible for marriage. Beads made from stems of seep willow (*Baccharis salicifolia*) were pinned or tied to a shirt called a *coton*. During a fiesta a dancer usually wore a good-luck necklace, such as one made with rock fern (*Notholaena*).

Since ancient times olivella shells were one of the most common sources of beads for the Seri, as well as for many other peoples (Bowen 1976: 87). *Olivella dama* was the most common species in Seri necklaces. The tip of the spire of the

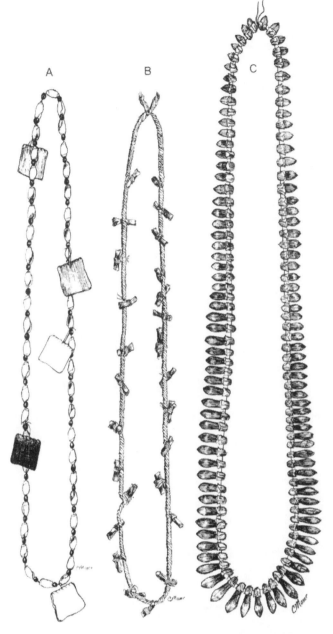

Figure 12.5. Necklaces. A) Bleached olivella shells, *Viscainoa* seeds, and variously colored cloth bags containing pieces of desert snapdragon (*Antirrhinum cyathiferum*); 80 cm circumference, purchased at El Desemboque, 1965. B) Pieces of cliff fig (*Ficus petiolaris*) bark tied with commercial thread onto a piece of red yarn; 74 cm circumference, obtained in El Desemboque, 1958. C) Fresh ocotillo (*Fouquieria splendens*) buds, 66 cm circumference, made by Elvira Valenzuela in El Desemboque, April 1963. *CMM.*

Figure 12.6. Carlota Herrera stringing a flower necklace made from corollas of *San Juanico* (*Jacquinia pungens*) at El Desemboque in 1963. The flowers were brought from the vicinity of Kino Bay. *MBM.*

shell was perforated for stringing. Although natural, brownish shells were often strung, more commonly they were bleached white before stringing by roasting them in hot sand. Until about the 1960s women made long necklaces of tiny, closely matched olivellas. The shells were often juvenile *Olivella dama* and *O. steveni*, called *xica cooxp* 'things white.' These necklaces were made for their own use, as gifts to other women, and occasionally to sell.

At low tide small groups of women and children combed exposed tide flats for live olivella and other mollusks. These were collected by picking them out of the wet sand or mud at the end of their small, raised trails. The mollusks were carried back to camp in buckets, cans, or improvised pockets in the women's skirts. Other shells, such as the jingle shell (*Anomia adamas*) and coffee bean cowrie (*Trivia solandri*), were collected from beach drift.

Beads were made from raw clay or ground pottery sherds. The clay was moistened, formed by hand into beads, pierced, and air dried or baked in ashes. Necklaces of clay beads were still commonplace in the early 1980s, but they were usually made for sale to outsiders.

The small black seeds of *Viscainoa* were much sought after for necklace making. Some tradi-

TABLE 12.1
Biological Materials Used for Personal Adornment

Species	Parts Used	Species	Parts Used
Necklaces			
		Anachis coronata crowned dove shell	shell
PLANTS		*Anomia adamas* jingle shell	upper shell or valve
Agave subsimplex century plant	seed, bud	*Cerithidea mazatlanica*	shell
A. schottii amole	flower	*Cerithium stercusmascarum*	shell
Amoreuxia palmatifida saiva	seed, root	*Conus perplexus* cone shell	shell
Antirrhinum cyathiferum desert snapdragon	seed	fish, e.g., *Rhizoprinodon longuria* Pacific sharpnose shark	vertebrae
Baccharis salicifolia seep willow	stem	*Heterodonax pacificus* a small clam	shell
Brickellia coulteri	herbage	*Jenneria pustulata*	shell
Cercidium floridum palo verde	seed	*Lyropecten subnodosus* scallop	lower valve or shell
C. microphyllum palo verde	seed	*Mitrella guttata*	shell
Desmanthus fruticosus	flower, leaflet	*Nassarius moestus* mud snail	shell
Errazurizia megacarpa	seed	*N. tiarula* mud snail	shell
Eucheuma uncinatum marine alga	branch (thallus)	*N. sp.* basket whelk	shell
Ficus petiolaris cliff fig	bark	*Octopus* sp. octopus	suction cup
Fouquieria splendens ocotillo	bud, flower, fruit	*Oliva incrassata* olive shell	shell
Gigartina johnstonii marine alga	branch (thallus)	*O. spicata* olive shell	shell
Jacquinia pungens San Juanico	flower	*Olivella dama* olivella	shell
Lycium brevipes desert wolfberry	fruit	*O. steveni* olive shell	shell
L. macrodon desert wolfberry	fruit	*Pandion haliaetus* osprey	talon
* *Malva parviflora*	fruit	*Panulirus inflatus* spiny lobster	antennae
* *Melilotus indica* sweet clover	herbage, flower	snakes, e.g., *Crotalus atrox*— diamondback rattlesnake	vertebrae
Notholaena californica desert rock fern	leaf	*Tetriclita stactalifer* barnacle	shell
Proboscidea altheifolia devil's claw	seed	*Trivia solandri* coffee bean cowrie	shell
Rhizophora mangle red mangrove	"fruit"		
Simmondsia chinensis jojoba	seed	Finger Rings	
Viscainoa geniculata	seed		
unidentified herb coquéeen	root	ANIMALS	
		Eretmochelys imbricata hawksbill turtle	carapace plates
ANIMALS		*Heloderma suspectum* Gila monster	hide from tail
Agaronia testacea olive shell	shell		

150

TABLE 12.1
(continued)

Species	Parts Used
Pendants	
PLANTS	
Jacquinia pungens	fruit
San Juanico	
ANIMALS	
Tachardiella larraea	lac from
lac insect on *Larrea*	creosotebush
	(*Larrea*)
Pinctada mazatlanica	shell
pearl oyster	
Pteria sterna	shell
pearl oyster	
Lip Plug	
Tachardiella larraea	
lac insect on *Larrea*	

*Plant not native to the Seri region.

Figure 12.8. Chico Romero wearing a traditional-style necklace. The alternating natural wood and darkened beads were probably made from reedstem milkweed (*Asclepias subulata*) or seep willow (*Baccharis salicifolia*). EHD, 1924; HF-23817.

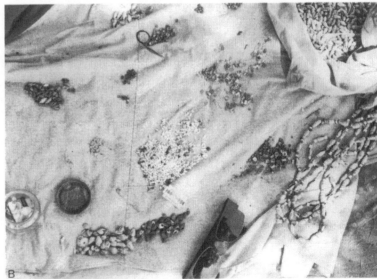

Figure 12.7. Necklace making became an important local industry during the 1970s and 1980s. A) Lupe Comito making a commercial necklace. B) She chose beads from a variety of shells, fish vertebrae, and seeds spread out on a blanket. *RSF, El Desemboque, April 1983.*

tional necklaces consisted of plant materials in small cloth bags tied on a string of beads, yarn, cord or other material, or pieces of the plants were tied onto a decorative cord or yarn. Such plants were often associated with good luck or supernatural properties or with pleasant fragrance, e.g., cliff fig (*Ficus petiolaris*), rock fern (*Notholaena*), desert snapdragon (*Antirrhinum*), and *Brickellia*.

Both men and women frequently wore simple bracelets of cord or cloth. Sometimes the cord was obtained from the shaman for curing purposes. Women also wore bracelets of commercial beads. Women with pierced ears kept the holes open by inserting a twig, such as one of *Abutilon incanum*, but earrings were not common. Finger rings were made from tortoiseshell and thin cross sections of the tail of the Gila monster.

Early records indicate the Seri wore a wide variety of ornaments made from shells, seeds, fruits, feathers, bone, fur, stones, and other objects. Men perforated the nasal septum to hold bone ornaments or brightly colored stones, often described as blue-green pebbles, on a cord (Och in McGee 1898:78; Hardy 1829:170; Di Peso and Matson 1965:51). McGee (1898:170) reported that this custom was obsolete by the 1890s.

Ear and lip plugs, clamshell bracelets, shell beads, and other possible ornaments have been recovered from archaeological sites (Bowen 1976). The Seri claim that these objects were worn by the Giants. The clamshell bracelets were fashioned from *Glycymeris gigantea*. However, Bowen (1976:87) has shown that these bracelets were made by the Trincheras people, and not the indigenous inhabitants. The ancient shell beads were often made from *Olivella* shells.

Facepainting

Facepainting was developed into an elegant and skillful art. Facepaint was applied as protection from the sun, for purely aesthetic or decorative purposes, for curative and supernatural

protective purposes, and to influence nature. Kroeber (1931:27) stated that "It is a true art of high order within its narrow compass" and pointed out its similarity to the river Yuman facepainting. Our findings essentially agree with Xavier's detailed notes (1941) and published work (1946).

People of all ages wore facepaintings, often with a fresh design put on each day (Sheldon 1979:123; Whiting 1951:9, 22). The custom persisted as a common practice among the women and some men (particularly when camped on Tiburón Island) until the mid-twentieth century (Whiting 1951). The men generally discontinued the practice earlier than did the women. Each person commonly applied his own facepainting (Figure 12.9), although women often painted members of their family (Figure 12.10), and mothers their children (Figure 12.11). Outsiders were occasionally painted in the spirit of fun (Hardy 1829:286; Xavier 1946). In 1921 Charles Sheldon (1979:109, 123, 141) found most of the Seri, including all of the women, engaged in facepainting, using blue, white, and red colors.

Facepaint was usually applied with brushes made from a twig, such as indigo bush (*Psorothamnus emoryi*), or human hair (McGee 1971:384). The latter brush consisted of a few strands of hair, pulled over to the mouth, bitten off about 15 cm long, and moistened with water or saliva to hold them together. Before commercial mirrors were available, a small bowl or shell filled with water was placed in the shade and used as a mirror (McGee 1898:166). The pigments were mixed in a clamshell (e.g., Hardy 1829:286) or ground on a miniature metate.

Individual designs commonly consisted of several colors applied in a variety of intricate patterns extending across the nose and cheeks (Figures 12.12–12.14). Xavier (1946:16–17) reported that blue, red, and white were by far the most usual colors. The pigments were used fresh, or stored in reedgrass (*Phragmites*) containers or various clam or other bivalve shells (see McGee 1898:165, facing 261). A particular color might be derived from a single material or from several different ones ground or mixed to-

Figure 12.9. Matilde Blanco applying her own facepainting. Her materials include white gypsum on a small flat rock (behind the coffee can), other pigments in clamshells (one on each side of the rock), and water in an enamel cup. Evangelina López is on the left and Anita López on the right. *ETN, El Desemboque, March 1951.*

Figure 12.10. Lolita Astorga applying a facepainting to Miguel Barnet. *EHD, Kino Bay, 1934; HF-24044.*

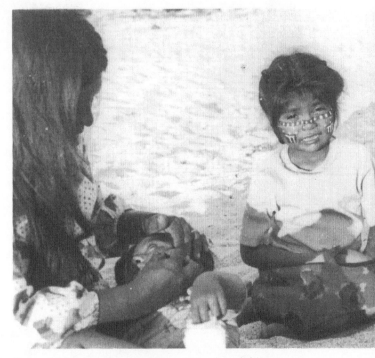

Figure 12.11. Panchita Moreno painting her children's faces during a leatherback turtle fiesta at the *Saps* camp in March 1981. The child on the right, María de la Luz Díaz, wears a completed design. *MBM.*

Figure 12.12. María Burgos beginning a delicate facepainting pattern. She used a slender brush made from her own hair. *JWM, 1959; ASM-25049.*

gether. More than sixteen species of plants and at least six other kinds of materials were used for making facepaints (Table 12.2).

Facepaintings for protection from the sun were large, simple designs (Figures 12.15, 12.16; see also McGee 1971:139, 154, 159, 357, facing 181; Whiting 1951:21). Women painted their faces for this purpose particularly when traveling at sea (Xavier 1946:17) and in summer when out in the desert.

Among the decorative and curative or protective patterns, certain elements were feminine, others masculine. Apparently both unmarried and married men wore similar patterns, but unmarried and married women generally used different classes of designs. There were, however, certain designs worn by any woman. Some chil-

dren's designs were distinct, while others were the same as some of their mothers' designs.

A highly stylized manta ray was a married woman's design (the manta ray was also used as a woman's household fetish—see *Bursera hindsiana*). Patterns representing lightning, rain, and clouds were powerful or strong designs for married women, while others were "just pretty, just decorative." Girls and unmarried women put on designs which included flowers, leaves, waves, and "things to bring good luck." There were patterns worn by unmarried women indicating that they were eligible for marriage. Men wore designs representing such things as knives, snakes, shark's teeth, and the moon as seen in dreams.

Special curative or protective designs included those which were purely Seri and those which

Figure 12.13. Ramona Díaz wearing a simple facepainting. *FFD, Tiburón Island, May 1937; ASM-1327.*

Figure 12.15. María Antonia Colosio with the style of large facepainting which gave protection from the sun. This particular design also includes decorative elements. She used a slender twig as a brush. *GHX, El Desemboque, 1940; ASM-903.*

Figure 12.14. Graciela Sesma with an elaborate facepainting. *JWM, March 1957; ASM-24906.*

were probably influenced by Spanish contacts. Crosses (Figure 12.17), which may have predated Spanish influence, were considered by the Seri to be "*católica*" and were said to bring good luck (Xavier 1941:34). To the Seri "*católica*" meant any strong religious power. Another design with supernatural connotation represented doors to a sacred cave on Tiburón Island that was involved with the vision quest. Such doors, drawn as rectangles, sometimes had flowers on them (the concept of rectangular doors was non-Seri in origin). Also included in this class of designs were mountains, lightning, rain clouds and rainbows, with some or all of these subjects incorporated into a single facepainting. Some facepaintings were done to cure a "crazy" person. One such design included a whirling top (toy), said to be the same as the "mind whirls" (Xavier 1941:50). The curing image of "mind whirls" refers to the spinning of a disturbed person's mind.

Other facepaintings included stylized representations of a wide array of subjects, such as fish tails (Figure 12.18), porpoise tails, certain flowers, snakes (including the coral snake), the sun or a flower with a face said to represent God in the sky, very abstract hands representing the hands of God, marks or bars, and numerous others (Xavier 1941, 1946).

In the seventeenth century Adamo Gilg reported that the Seri also practiced body painting (Di Peso and Matson 1965:53). According to oral tradition, unmarried youths among the San Esteban people painted stripes and other ornamental designs on their bodies and extremities (Moser and White 1968:146).

Tattooing

In preparation for tattooing, *Condalia* leaves were mashed and wrapped in a piece of cloth; water was added and the resulting liquid

TABLE 12.2
Facepaint Materials

Color	Species or Materials	Part or Substance Used
Red or reddish	*Bursera microphylla*	sap
	Ferocactus wislizenii	base of spine
	Opuntia phaeacantha	fruit
	ochre (an iron ore)	powdered rock
	ochre and	powdered rock
	honeybee and	honeycomb
	Acacia greggii or	flowers
	Cercidium floridum or	flowers
	C. microphyllum	flowers
Orange-red	honeybee and	honey
	Lycium fremontii	fruit
Rose or pink	*Ferocactus wislizenii*	base of spine
	Opuntia violacea	fruit
Yellow	*Fagonia laevis*	herbage
	Ferocactus wislizenii	flower
	Typha domingensis	pollen
Blue (Seri Blue)	*Guaiacum coulteri* and	resin
	Ambrosia spp. or	root
	Stegnosperma and	root
	white clay	powdered clay
Brown	*Agave subsimplex*	leaf base
Black	*Prosopis glandulosa*	sap or pitch
	Randia thurberi	seed or mesocarp of fruit
	charcoal and	powdered
	animal grease	fat
White	gypsum	powdered rock

squeezed through the cloth. Ironwood (*Olneya*) ashes were then mixed into the liquid, and the ashy mixture rubbed on the area to be tattooed. In another method juice of *cardón* (*Pachycereus*) fruit, mixed with charcoal, was used. The design was pricked into the skin with three or four *Condalia* thorns tied together. When the skin

bled, a strip of fresh limberbush (*Jatropha cuneata*) was pulled back and forth over the area to stop bleeding and to help it heal.

Edward Moser and Richard White (1968: 145–146) indicated that tattooing and other forms of body decoration were similar to the decorative styles found on clay figurines which

Figure 12.16. Sara Villalobos with a large, simple face-painting that provided protection against the sun. *JWM, April 1960; ASM-25096.*

Figure 12.17. A young woman on Tiburón Island in 1936 with a facepainting including crosses, which were said to bring good luck. *WCS; AH-1853.*

Figure 12.18. Andrea Romero with a "fishtail" pattern facepainting. *EHD, 1929; HF-23875.*

were apparently used as dolls. A common *ha-pázt* 'tattoo' for women was a line from the lower lip through the center of the chin. Moser and White (1968:145) reported that "Parents often tattooed a single line down the middle of their daughter's chin. This served to identify her as Seri in the event she was taken captive."

Other tattoos, including one made of dots along the inside of the lower lip, were done near the age of puberty. The mouth hurt so much that one could not talk, and so learned patience and silence. The older people counseled the young person, male or female. The young person then went into seclusion to contemplate the instruction. This practice has not been done since

the nineteenth century (Moser and White 1968:146).

One old man remembered that his grandfather had a zigzag line enclosed within two parallel lines down his chin, and several arrows tattooed on his forehead, with the points toward his hairline. A common area for a tattoo was on the arm. The arm tattoo was one's own design, and each adult had one which was his or her identifying mark. When out in the desert, one sometimes made the same design in the sand to show he had been there (Figure 12.19).

"As prescribed by a shaman, a woman often would have a cross topped by two horizontal lines tattooed on the upper portion of each

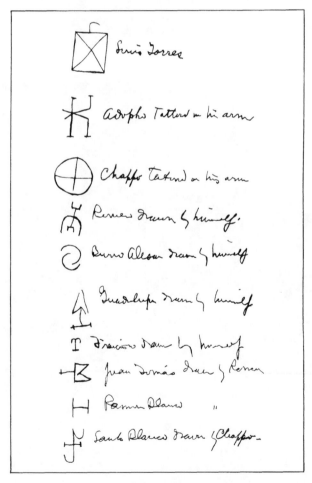

Figure 12.19. Examples of Seri men's personal identification symbols recorded by Charles Sheldon in his journal, December 29, 1921. *Courtesy UAL and the Desert Bighorn Sheep Council.*

breast. This was to assure good health for her future children (Moser and White 1968:145)." Shamans often tattooed crosses over an afflicted area of a patient. Occasionally other tattoos were added, but these were done for decorative or light-hearted reasons.

In the 1970s tattoos could still be seen on the arms of many of the adults. Sara Villalobos once had a tattoo line from her lower lip through the center of her chin, although it was no longer visible in 1972. However, several other women still had visible tattoo lines on their chins.

On at least one occasion the Seri were tattooed by an outsider. George Flavell, a San Francisco reporter in company with several other Americans, visited Tiburón Island in May, 1894 (McGee 1898:117–120). Flavell wrote in his log, "I had promised to tattoo some of them, so I went ashore again with the needles and ink and commenced work (Robinson and Flavell 1894: May 26)." According to Porfirio Díaz—who was present on that occasion and whose account of the visit was recorded by E. Moser in January, 1956—one of the Americans tattooed the arms of a number of women.

Hair Care

The hairbrush, *csai*, was fashioned from the wiry roots of a perennial grass, *Aristida californica*; when this plant was not available, *Jouvea* or *Sporobolus cryptandrus* was used. Several to many plants were severed close to the root-crown, tied into a bundle—with yarn, strips of colored cloth, or sinew (Quinn and Quinn 1965: 160)—and evened off by burning. The people of San Esteban Island, lacking these grasses, made hairbrushes from the roots of the local century plant (*Agave cerulata*).

Considerable taboo surrounded the hairbrush. If one used the *csai* brush at night his spirit would wander off toward *coaxyat*, the place of the dead. If one dared to use it a second time his hand would swell. Such a person was then known as *csai himo cöhaalajc* "the brush user who finishes last." A discarded hairbrush or a brush used by one who had died was considered dangerous.

Shampoo made from seeds of jojoba (*Simmondsia*) was highly esteemed and continued to be used in 1983. The aromatic twigs of elephant tree (*Bursera microphylla*) were often added to the jojoba shampoo. The pounded leaves of *amole* (*Agave schottii*) were said to make an excellent shampoo which softened the hair and made it grow long. Shampoo was also made from the roots of *Brandegea* or *guayacán* (*Guaiacum*), the leafy twigs of desert saltbush (*Atriplex polycarpa*), or the bark of white crucillo (*Zizyphus*). Shampoo made from the leaves of *Vaseyanthus* was said to promote luxurious hair growth. Mashed jojoba seeds were used by a man desiring to grow a beard.

13. Recreation, Music, and Oral Tradition

Fiestas, held to celebrate special events, were a major means of social exchange beyond the usual extended-family interactions. Many of the games, dances, and songs were performed almost exclusively at fiestas, while others were part of everyday recreation. Seri song-poetry was a highly developed art form. It included songs which were powerful and religious in context as well as others which were strictly secular. Many of the games, dances, and instruments show similarity to those which were once widespread throughout the southwestern part of North America, or perhaps over an even wider region. There was considerable influence in historic times from the Yaqui on Seri musical instruments, songs, and dances.

Games

About a dozen games were played, and these usually involved gambling. Men bet for larger stakes than did women. Games were played at fiestas and during everyday idle times (e.g., Sheldon 1979:110).

Women and children both played *camóiilcoj*, the circle game, at fiestas. The circle, called *hamóiij*, was laid out with sets of five cross sections of stems from organ pipe (*Stenocereus thurberi*;

Figure 13.1; also see Griffen 1959:14). If organ pipe stems were not available, rocks were used instead. The size of the circle and number of sets of cross sections or rocks depended upon the number of players. A circle 3 to 4 m across might have eight sets of cross sections. Each player had a counter, called *hapéxz*, usually made from sticks of ashy jatropha (see *Jatropha cinerea*). The cactus stem cross sections were discarded after the games, and the sticks were given away but not used again. The scoring was done with game sticks, called *hemot*, used as dice. They were made from decorated pieces of reedgrass (*Phragmites*). Bets for this game consisted of inexpensive items, such as pins, thread, thimbles, and buttons (Griffen 1959:14).

Various plants were rubbed or carried on the person or otherwise employed to bring good luck in gambling. These plants included *Castela polyandra*, tansy mustard (*Descurainia*), coral bean (*Erythrina*), *Lotus* spp., rock fern (*Notholaena*), and *Sideroxylon*. Gambling chips were made of such objects as pieces of rush (*Juncus*) "stems" or its fruit ("seeds"). Pieces of reedgrass (*Phragmites*) culms were made into large dice.

A game called *xtapácaj caahit* 'tower-shell he-who-causes-to-eat,' or "he who feeds the tower shell," was played with a section of reedgrass (see *Phragmites*) and a tower shell (e.g., *Turri-*

Figure 13.1. The *camóiilcoj* circle game at a girl's puberty fiesta. The circle is outlined with sets of pieces of organ pipe cactus, and the markers are sticks of ashy jatropha. (*Jatropha cinerea*) with strips of colored cloth tied to the ends. *MBM, El Desemboque, 1975.*

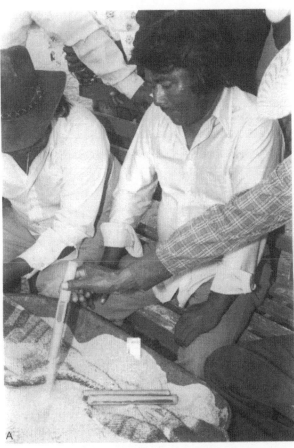

Figure 13.2. Playing *xapij caanlam* at a fiesta. A) Rogelio Romero watching a teammate empty the sand from a cane tube. B) Keeping score with the 100 small stick counters. *RSF, El Desemboque, April 1983.*

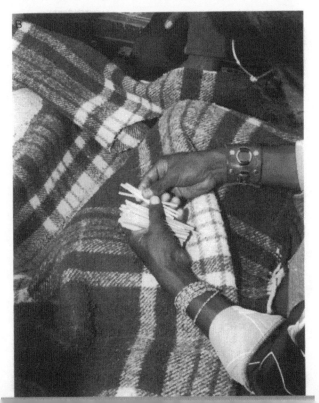

tella gonostoma). Two people at a time played, and the winner challenged the next contestant. Some said it was played only by men, while others said that both men and women played it. It was probably primarily a men's gambling game played at fiestas. Some men were expert at this game.

Xapij caanlam 'reedgrass enclosed(plural),' a men's gambling game, continued to be popular in 1983 (Figure 13.2; see *Phragmites*). It was played with two teams of four men on a side, using four differently decorated reedgrass tubes and one hundred small game sticks. The tubes were open at one end and a marker—such as a twig, coral bean seed, or match stick—was hidden in one of the tubes. All four were filled with sand beneath a blanket to disguise the one with the marker, and then passed to the opposite team. A member of this team emptied them one by one and tried to pick the one with the marker on the third attempt. The turns for emptying and filling the reed tubes were rotated with each team, and each man placed a bet in order to play. *Xapij caanlam* was played at fiestas as well as on any occasion. Similar games were once widespread in southwestern North America (Culin 1907:354–381).

A solid ironwood or mesquite ball, about the size of an ordinary softball, was used in the men's football race called *hehe hahójoz* 'wood what-is-made-to-flee.' The ball was kicked and the race run over a considerable distance, probably on the order of 4 to 10 km. The men ran informal relays along the beach, where cactus, thorns, and brush would not interfere. Both spectators and participants gambled on the outcome (Griffen 1959:15). Football relay races with a wooden ball were widespread in the region (Beals 1932:190).

Hehe hahójoz cmaam 'wood what-is-made-to-flee female,' the women's hoop rolling race, was said to be similar to the men's football race. The hoop, made from mesquite root (see *Prosopis*), was rolled or thrown with sticks, often made of *Melochia* or *Vallesia*. Kroeber (1931: 16) indicates it was played by the women prior to the men's football race. The modern Seri knew very little about this game.

Sahuaro fruit "skins" were used in the men's game called *cacómaloj* 'stick throwers.' The players sat opposite each other, each with three pointed throwing sticks and several fruit skins cut into narrow strips. They became skilled at piercing very narrow strips, called *hacóml* 'pierced things.' Men played for large stakes, such as knives and guns. Boys used a much wider strip called *haxz iipl* "dog's tongue." They played to win a bet of a certain number of cactus fruits which the losers had to gather in the intense heat of the desert.

Men raced boats, using the balsa (see *Phragmites*) or, later, wooden boats. According to Griffen (1959:15), there was usually only one man to a balsa, or occasionally two. Each wooden boat had a team of six or seven men, each one paddling. Men enjoyed playing cards (Sheldon 1979: 110)—the cave at *Hast Hax* (Pozo Peña) was a favorite retreat for card playing and gambling. According to the Seri, the San Esteban men played a dangerous game of "chicken" by sliding in a turtle shell down a steep slope above a sea cliff on the island (Figure 13.3). The one who went closest to the edge won, but sometimes a man went too far. It was said that as he was heading for doom, the first man to yell "your wife is mine" could take her as a second spouse (Bowen 1976:33–34).

Boys played a game called *hasahcápöj pte cjeaatim* 'senita(*Lophocereus*) together hitting,' or "hitting each other with *senita*." They cut the spines off slabs of the cactus and then, choosing sides, threw them at each other in playful fights.

Toys and Play

Girls played with dolls fashioned from sea plants, such as eelgrass (see *Zostera*) or sargasso weed (see *Sargassum*). Wooden dolls were carved from the wood of red elephant tree (see *Bursera hindsiana*). Dolls were also made from stuffed quail or jackrabbit heads, turtle bones (Figure 13.4), pelican bones, crab claws, (e.g., *Callinectes bellicosus*), clay figurines, and rocks (Moser and White 1968:148–153). Doll clothing was made from pieces of cloth or seaweed, such as *Gigartina*. Sometimes toy houses were made with turtle bones (Figure 13.5). Girls thought it fun to outline their hands with San Juanico (*Jacquinia*) leaves by sticking the spine-tipped leaves into their hands. They decorated their homes with wildflowers, such as sand verbena (*Abronia villosa*) and evening primrose (*Oenothera*), by sticking the flowers into crevices in the dry walls.

Children played with gourds (*Cucurbita digitata*) and balls made from a mule deer or bighorn sheep scrotum stuffed with eelgrass. Boys played with a wheel toy cut out of the stem of either of two columnar cacti—senita (*Lophocereus*) or organ pipe (*Stenocereus thurberi*).

Figure 13.3. The bare gravel slope (*center*) on San Esteban Island where men engaged in dangerous contests to see who dared slide nearest the edge of the cliff in an up-turned turtle carapace. *RSF, 1965.*

The cactus wheel toy was rolled along the ground by pushing it with a forked stick (Figure 13.6).

Children also played with the fluffy seed appendages of a milkweed (*Matelea pringlei*), and playfully threw spore-spewing *Tulostoma* mushrooms at each other. Along the beach they pretended that seaweeds, such as *Codium* and *Gigartina*, were pieces of turtle or fish meat, and also played with the hollow seaweed *Colpomenia*.

The dry shell from a desert tortoise found in the desert was made into a toy rattle. Boys played with toy flutes, whistles, and harpoons made from reedgrass (*Phragmites*). Also made from reedgrass were sliver guns, squirt guns, and peashooters using ammunition such as the fruit of elephant tree (*Bursera microphylla*) and coral beans (*Erythrina*). Boys also played with disk tops, made with a pottery or wooden disk. The pottery disk was made from a sherd rounded or trimmed down to the desired size and shape.

One of the favorite pastimes for boys was to slide down a hill in a sea turtle carapace, using it like a sled (Figure 13.7; Smith 1974:157). Boys also played with model boats carved by their fathers or other male relatives from red elephant tree (*Bursera hindsiana*) or occasionally from the common elephant tree (*B. microphylla*). In

Figure 13.4. A girl's turtle bone doll with cloth skirt, made in 1975 by Carlota Colosio. *CMM.*

163

Figure 13.5. Girls playing along the beach used sea turtle bones and shells as doll houses and dolls. A) Doll houses made from turtle bones. B) The bones served as dolls and the shells as dishes and other objects. *RJH, El Desemboque, December 1971.*

Figure 13.6. Boys pushing cactus wheel toys made from pieces of organ pipe cactus. *MBM, El Desemboque, 1978.*

Figure 13.7. A boy "sledding" in an upturned turtle carapace. *EHD, Tiburón Island, 1936; HF-23894.*

the 1950s and 1960s, and perhaps earlier, men made elaborate model boats, or *pangas*, for the boys, using commercial wood and including such details as miniature wooden outboard motors. Small toy boats were sometimes made from the "fruit" of red mangrove (*Rhizophora*). Boys also fashioned toy boats from the leaves of cattail (*Typha*).

Toy bows (Figure 13.8) were made from the flexible stems of shrubs, such as *Cordia*, and arrows from desert wolfberry (*Lycium fremontii*). Boys also used toy harpoons, made from reed-

Figure 13.8. José Astorga with a toy bow. *EHD, 1929; HF-23866.*

grass and various other woods. A large mesquite spine often served as the harpoon point. They speared crabs (*Callinectes*) or small fish, including small stingrays, in shallow water along the shore or in esteros (Sheldon 1979:125).

Smoking

Five native plants were smoked: canyon ragweed (*Ambrosia ambrosioides*), *Trixis*, creosotebush (*Larrea*), *rama parda* (*Ruellia*), and two wild tobaccos (*Nicotiana clevelandii* and *N. trigonophylla*). The parts smoked were a certain leaf gall of the creosotebush, the flowers and/or the leaves of *rama parda*, and the dried leaves of the others. The tobaccos were much preferred, and *N. trigonophylla* was preferred to *N. clevelandii*. There were certain more desirable popu-

lations or stands of *N. trigonophylla*. The men made special expeditions to gather it in quantity at these specific places and bring it back to camp.

The dried leaves, flowers, and leaf galls were smoked in pipes made of fired clay, stone, or a piece of reedgrass (*Phragmites*). These stone and clay pipes are common in archaeological sites in the Seri region (Bowen 1976:73–74, 79). The stone pipes were tubular, wider at one end, and often decorated. The clay pipes were tubular or stemmed. The tubular clay pipes were similar to the stone pipes and were more common than the stemmed ones, which seem to be newer—probably mostly nineteenth and early twentieth century (Bowen 1976:74). A tube worm shell, *xtozaöj* (*Tripsycha tripsycha*), was a popular pipe for smoking the native tobaccos or marijuana.

Smoking was for personal pleasure rather than for ceremonial or ritual purposes. However, a shaman sometimes blew smoke over a patient as part of a curing procedure. Commercial tobacco, when it became available, replaced the native smoking materials.

Some people indicated that *satóoml* (*Ruellia*) was hallucinogenic, although it is probably only mildly so. *Datura* was not used by the Seri for its psychotropic effects. However, they knew this property of the plant as evidenced by its two names: *hehe camóstim* 'plant that-causes-grimacing' and *hehe carócot* 'plant that-makes-one-crazy.'

Fiestas

Fiestas were the major social gatherings for celebration, ceremony, and social exchange. The more important fiestas spanned four consecutive days. They usually began in the afternoon or early evening and lasted until about midnight, although the girl's puberty celebration lasted through the night on the last day. There were games, dancing, singing, and feasting. Traditional Seri fiestas included religious aspects, and there was usually an element of potential danger or dangerous spirits which could be appeased by carrying out the fiesta which would "make the spirits happy and bring good luck." Fiestas were meant to be joyful and fun.

Fiestas were held to celebrate at least ten kinds of special events. These can be divided into two major categories: those held for special personal occasions and those given in honor of a special large object, animate or inanimate. The Seri attached special significance to unusually large objects, and the fiestas were held to appease their potentially dangerous spirits. Of the major kinds of traditional fiestas listed below, only the girl's puberty ceremony and leatherback fiesta were commonly held after the middle of the twentieth century.

1. A girl's puberty fiesta was the most vigorous and common fiesta during the twentieth century. In fact, this fiesta seems to have undergone a renaissance during the 1970s and 1980s, perhaps due in part to the improved economic situation of the people (see Hinton 1955). The fiesta was occasioned by a girl's first menses. The rationale for music and dancing during the fiesta was to make the girl happy and to insure that she remain awake. It was believed that should she fall asleep and dream of misfortune, the dream would come true.

2. The boy's puberty fiesta, last held about 1923, was similar to the girl's fiesta and continued for four nights. It was said to be much simpler than that held for a girl, but this information may refer to times when the boy's fiesta was already declining.

3. In former times a fiesta was sometimes given because of a loved one's recovery from a serious illness or survival of an accident.

4. A marriage celebration, lasting only a single evening, took place at the home of the groom upon the arrival of the bride. It was last held in about 1920 or 1921. However, this fiesta was not a traditional Seri custom and probably resulted from Mexican influence.

5. A victory fiesta was held when the men returned from a battle after killing their enemies.

The fiesta was conducted outside the camp, since the spirits of the dead enemies would be following the warriors to do them harm. The dances performed at this fiesta were a device to remove danger of retribution. The fiesta spanned four days. The festivities lasted through the early part of the night and kept the warriors happy during those especially dangerous hours when they must not sleep.

A scalp or a bloody piece of clothing of one of the slain enemy warriors was attached to the top of a pole. Men and women danced around the pole singing special songs (Kroeber 1931:14). No musical instruments were used. The last victory fiesta probably took place about 1930.

6. The capture of a leatherback sea turtle was the occasion for a four-day fiesta similar to the girl's puberty celebration. (For a description of the leatherback fiesta see the section on reptiles in Chapter 3).

7. Another occasion which prompted a fiesta was the capture of a giant sea bass called *lamz* or *mero* (*Epinephelus itajara*). It was a rare event, and probably this fiesta has not been held since at least the middle of the twentieth century. *Lamz* is the old name for a large *mero*; the newer name is *zixcám caacoj* "large fish."

8. The celebrations conducted to appease the spirit of a giant basket, called *sapim*, are described in Chapter 15.

9. Fiestas were also sometimes given when a very large pottery storage vessel, or olla, was found. Such vessels were found partially buried in the desert or in caves (Bowen and Moser 1968: 123–124). Prior to the early 1900s such vessels were usually reused, but after the early part of the century their utilitarian value decreased. At about that time, "a shaman declared that the very large old storage ollas were closely associated with the spirits of their previous owners and should be considered too dangerous to use. Anyone finding such an olla, he contended, must give a fiesta and subsequently sell the vessel (Bowen and Moser 1968:124)."

10. A fiesta for the completion of a boat lasted only one night. This fiesta has not been performed since the 1960s. The purpose was to make the people happy as they wished the craft good luck in its future voyages.

Dances

Dances were performed primarily at fiestas, but some, especially the solo dances, might have occurred on a number of special occasions. Some of the more prominent fiesta dances are described below.

The only traditional dance still commonly performed in the 1980s was the *icoitáa* 'dance true.' It was danced solo, primarily during fiestas, but also on a number of other special occasions (Kroeber 1931:14).

In this dance a singer accompanied himself with a tin can rattle or gourd rattle (see *Lageneria*), while the dancer performed a pascola-like syncopated rhythm on the foot drum (Figure 13.9). Formerly a musical bow served as the accompanying instrument when a rattle was unavailable. The foot drum was originally a sea turtle carapace, but during the first third of the twentieth century it was replaced by wooden planks. Since about the middle of the twentieth century the dancer has been a man or a boy, but formerly women and girls also performed this dance. Similarly, women also once sang and played instruments at fiestas, but during the twentieth century singing and dancing at fiestas became an activity for men.

Songs for this dance, called *icoosáa* 'song(s) true,' were said to have been taught by the Yaqui. The repertory of these songs is extensive. Although the Seri said that the songs were taught in the Yaqui language, most of the words have been reduced to nonsense syllables. Only occasional Yaqui words could still be identified by the 1960s. Since these songs were not spirit songs, they were sung by anyone who cared to do so.

The deer dance, *hepem cöicóit* 'white-tailed-deer who-dance-about(concerning),' was said to

Figure 13.9. Chapo Barnet dancing on a foot drum and accompanying himself with a can rattle during a fiesta. *RSF, El Desemboque, April 1983.*

have been learned from the Yaqui. Only certain songs, called *hepem cöicóos* 'white-tailed-deer song(s)-about' were sung for this dance. There seem to be from ten to fifteen songs known, each containing mostly nonsense syllables intermixed with infrequent Yaqui words. The dance was accompanied by cocoon (later, tin can) rattles and either the musical rasp or musical bow (Figure 13.10).

In another dance, called *hant cötitóij cöcoila* 'land return-to(bend forward) who-dance-about,' girls danced in a circle around a man in the center. It was performed at the girls' puberty fiestas.

Figure 13.10. Antonio Herrera performing the deer dance and playing can rattles, with Jesús Morales playing the rasp. *EWM, El Desemboque, 1954.*

Songs for this dance, called *xepe an cöicóos* 'sea song(s)-about,' concerned things of the sea, such as fish, other sea creatures, sea birds, and boats. These were spirit songs with Seri words, and were sung without accompanying instruments.

Younger people of both sexes participated in *cmaam cöcoila* 'female who-dance-about.' The songs for this dance, sung in Seri and without accompaniment, were called *cmaam cöicóos* 'female song(s)-about.' The dancers formed a line standing side by side and moved forward as a line—with short, rapid steps—then backward; this movement was repeated throughout each song. If a dancer giggled or laughed, it was said that he or she would marry an old person. Consequently, the dancers sometimes poked each other and cracked jokes to cause somebody to laugh.

A men's circle dance was called *ziix coox cö-coila* 'thing all who-dance,' or "they who dance about everything." It was so named because the songs unique to it, called *ziix coox cöicóos* 'thing all song(s)-about,' contain a variety of references. These songs were sung in Seri. The men carried a split reed instrument in each hand which was shaken to make a rattling sound while dancing. This dance has apparently not been performed since the early part of the twentieth century (Griffen 1959:17).

The coyote dance, called *oot cöicóit* 'coyote who-dances-about,' was performed for fun. Despite the name of the dance, the songs were about both the coyote and the fox. The dance was performed by a man, usually when inebriated, to entertain the children at fiestas. The dancer fastened a piece of cloth to the back of his pants to represent a fox tail. The dancer sang as he danced rapidly, turning in a small circle pretending to try to see his tail. He accompanied his singing and dancing with the split cane instrument.

Musical Instruments*

Indigenous musical instruments were easily fabricated from only one or a few kinds of readily available materials; some instruments were already on hand (e.g., the hunting bow could also be used to make music). Many instruments seem to have been adapted from those used by the Yaqui. It may be that during earlier times the more limited range of native materials and tools, plus the Seri's semi-nomadic lifestyle, precluded the making and accumulation of many objects, such as the newer, borrowed musical instruments. For the most part Seri musical instruments were the same or similar to many that were widespread in southwestern North America, and some, such as the bullroarer, were ancient and found worldwide. The principal musical instruments used by the Seri are listed below:

*Much of the information in this section is based on T. Bowen and E. Moser, "Material and Functional Aspects of Seri Instrumental Music," The Kiva (1970) 35(4):178–200. Used with permission.

violin
mouth bow
musical bow

musical rasp
metal disk rattle
split reed instrument
gourd or metal can rattle

bullroarer
shaman's flute
shaman's whistle

floating gourd drum
foot drum

The violin, of European origin, may have come to the Seri via neighboring Indians, such as the Yaqui. The musical rasp, metal disk rattle, gourd rattle, floating gourd drum, split reed instrument, and foot drum were similar in structure and use to instruments played by the Yaqui. Furthermore, the Seri said that they had learned of most of these from the Yaqui.

Four instruments were used primarily or exclusively at fiestas and celebrations: musical rasp, split reed instrument, gourd or can rattle, and musical bow. The metal disk rattle was played at fiestas and at home to accompany lullabies, which were sung only by grandparents and other relatives of their generation. The violin and mouth bow were played at any time, primarily at home in the family setting. The bullroarer and the shaman's flute and whistle were used in vision quests and to gain or call supernatural powers.

The violin (Figures 13.11, 13.12), the most complex instrument made by the Seri, was made of several different materials. The violin body or box was usually made from red elephant tree (*Bursera hindsiana*) but occasionally from cliff fig (*Ficus petiolaris*), mesquite root driftwood (*satómatox*, *Prosopis*), or commercial produce crates. The cover of the violin box was driftwood, a piece of metal, or a piece of produce crate nailed onto the box. Hardwood, such as *Vallesia* or *Lycium fremontii*, served as the tuning peg, and the bridge was made from any of several woods. The string was usually twisted

Figure 13.11. Lauro Bustamante playing the one-string Seri violin. *EWM, El Desemboque, 1956.*

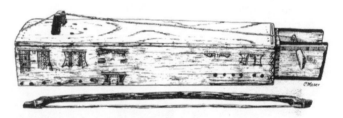

Figure 13.12. Details of a violin and bow made for sale by Jesús Morales. The entire violin is 48 cm long. The violin box is red elephant tree (*Bursera hindsiana*) wood and the cover is a piece of produce crate. The tuning peg and bridge are hardwood (probably *Lycium fremontii*) and the string is sinew. The side of the violin box is decorated with blue and red facepainting designs. The bow, 44 cm long, is made of wood and human hair. *CMM.*

mule deer sinew, although horse, cow, and pelican sinew, twisted badger intestine, and twisted sinew-like fiber from the intestines of the green turtle were sometimes used. Sap or gum from brittlebush (*Encelia*) served as rosin, or green ocotillo (*Fouquieria splendens*) bark was used as a substitute.

The violin bow was usually made from hardwood, such as catclaw (*Acacia greggii*), ironwood (*Olneya*), mesquite root, or desert wolfberry

(*Lycium fremontii*). However, even ocotillo was occasionally used for the bow. Horsehair or a girl's long hair served for bow string.

The violin was normally constructed by men and was played mostly by young men and women. Older men occasionally played it, but older women only rarely did. The Seri said it was not used as a courting instrument, as claimed by the Coolidges (1939). It was used primarily to play shamans' songs relating to the power of the spirits. It was usually played around the home in the morning and evening, but not after dark because the damp string did not produce a good sound and was likely to snap.

The musical rasp, *hexöp*, was usually made from ironwood (Figure 13.13), but occasionally from catclaw. The framework of the metal disk rattle was carved from red elephant tree (see *Bursera hindsiana*). Three instruments were made entirely or mostly from reedgrass (see *Phragmites*): split reed instrument, mouth bow (Figure 13.14) and shaman's flute (Figure 13.15). The shaman's whistle was ceramic or made from stone and was blown to call the spirits.

The gourd rattle was made with *Cucurbita digitata* or bottle gourd (*Lageneria*). The gourd was later replaced with a tin can. Bullroarers were of ironwood (*Olneya*). A man dancing the deer dance, as well as other dances, often wore

Figure 13.14. Jesús Morales demonstrating the mouth bow, which was played lying down. *EWM, El Desemboque, 1969.*

Figure 13.13. Jesús Morales playing a musical rasp, made of ironwood (*Olneya tesota*). The demonstration was at El Desemboque in 1969. *EWM.*

Figure 13.15. The shaman's flute demonstrated by Jesús Morales at El Desemboque in 1969. *EWM.*

Yaqui-style cocoon rattles around his legs (Shei-don 1979:87). *Xica quiinla* 'things that-rattle' is the Seri term for the string of rattles or a single cocoon; *teneboim* is the Yaqui-Mayo term for these rattles. The Seri put pieces of broken sea shells inside each cocoon to make it rattle, and the cocoons were strung on sinew. The cocoon was from a saturnid moth (*Eupachardia* sp.), a wild desert relative of the silkworm moth. The cocoons were usually picked off the upper branches of ashy limberbush (*Jatropha cinerea*) and other shrubs at a site on Tiburón Island near the coast opposite Campo Víboras. Cocoon rattles were probably of Yaqui origin (Kroeber 1931:14).

No distinction was made between hunting bows and musical bows, and bows were not made specifically for music. The special term *haacni coccáxz* 'bow who-hits' was used to designate one who played a musical bow. Two work baskets inverted over a small depression scooped out of the sand served as resonators. Each end of the bow was placed on a basket with the string up (Figure 13.16; also see Coolidge and Coolidge 1939:facing 207, and Quinn and Quinn 1965:180). Sound was generated by beating against the string with a hardwood stick or arrow. It provided a steady, accentless rhythm. It was played primarily at fiestas and accompanied by singing or chanting and the foot drum. It was sometimes used to accompany the deer dance.

The foot drum consisted of planks placed over scooped-out sand; a man or boy danced a pascola-like dance on it. In earlier times a sea turtle carapace served the same purpose, and women and girls also danced on it. The dancer steadied himself with a stick, often made from the woody rib of a giant cactus, and sometimes wore cocoon rattles around his ankles (Kroeber 1931:14; Quinn and Quinn 1965:180; Sheldon 1979:87).

Figure 13.16. Jesús Morales demonstrating the bow being played as a musical instrument. *EWM, El Desemboque, 1969.*

Oral Tradition

Seri culture has a rich tradition of oral history, story-telling, and song-poetry. Plants and animals featured prominently in many of these oral traditions. For example, see the accounts under *Atamisquea*, *Bursersa microphylla*, Cactaceae—columnar cacti, *Fouquieria columnaris*, *Jatropha cuneata* (Chapter 15), *Olneya*, *Opuntia bigelovii*, *Pachycereus*, *Phragmites*, *Prosopis*, *Stenocereus thurberi*, and *Stegnosperma*.

Accounts of oral history among the Seri generally seemed to be accurate for events which occurred within one hundred to one hundred twenty years, or back to about the mid- to late nineteenth century. For example, Lowell (1970:158), in her analysis of the legend of Lola Casanova, found that "Seri oral tradition appears to reflect historical events more accurately and completely in this instance than does the better known Mexican version which has been influenced by romantic literature." For examples of plants featured in oral history see desert mistletoe (*Phoradendron*), elephant tree (*Bursera microphylla*), and reedgrass (*Phragmites*).

There are many narratives featuring plants. For example, there is an interesting tradition involving the boojum tree (*Fouquieria columnaris*). In order to escape a great flood, a group of Giants from the south fled northward to the mountains between El Desemboque and Puerto Libertad. There the flood overtook them, and they were changed into boojum trees. The tall ones were men and the short fat ones were pregnant women. In another version, those who were turned into boojum trees were people from the Tastiota Region. The flood overtook them just after they had passed the crests of the mountains (heading northward) and were descending, and that is why the boojum trees are on the north sides of the mountains.

Song-poetry

Seri song-poetry was a rich and elegant art form. Spirit power song-poems were the most numerous, but purely secular ones were also very common. Seri poetry, like that of most cultures worldwide, was communicated by singing or chanting. Much of this tradition had been lost by the 1960s and 1970s, and when Jesús Morales died in 1975, the people lamented that there would be nobody to sing many of the old songs at fiestas.

Many songs were sung four times or in multiples of four (four being the sacred number). However, this trend was often not followed; the set might be repeated once or many times. The songs often consisted of a quatrain (four phrases), sometimes with an added couplet or phrase. This, too, was merely a common trend and by no means a hard and fast rule. It was commonplace to lengthen or add vowels, drop consonants, substitute sonorant consonants for non-sonorant consonants, and add syllabics to enhance the euphonic quality.

Although individual song-poems tended to be short, they often conveyed considerable information and emotion. Detailed and subtle concepts and observations embodied in some of these short song-poems required extensive background in the subject matter, and often involved complex ecological and cultural relationships.

Song-poems were sung on many occasions, but particularly during fiestas. Both men and women knew many of the songs, and formerly both men and women sang them at fiestas. When it was required, the shamans usually were the ones who sang or chanted through the night. Songs were sung either with or without accompaniment of a musical instrument. Other song-poems, sometimes comprised of only two phrases, were sung while working, such as when collecting cactus fruit. Songs were also sung by a shaman for curing.

The majority of Seri songs were probably "spirit songs" called *hacátol cöicóos* 'danger song(s)-about.' Each of these songs belonged to a particular shaman who learned the song from the spirits. Many of these were included in the category *xepe an cöicóos* "songs about the sea." These imparted special power and abilities to their shaman owner, as do all spirit songs of the Seri (see Griffen 1959:16–17). Spirit songs entered public domain after the death of their shaman owners. The songs were no longer considered to have their spirit power, and they could be sung by anyone at any time. Many such song-poems were known, and the names of the long-deceased owners were often remembered.

Songs were often learned in a dream or while one was on a vision quest. These songs might represent spirit power from a plant, an animal, the desert, or the sea. Others were learned by people overhearing a particular plant or animal singing its own song. Thus, a song-poem might seem as if the plant or animal were expressing its feelings, and the singer was merely the medium for conveying the song. Nevertheless, a shaman who learned such songs of spirit power made them his property. Many song-poems conveyed aesthetic feeling without supernatural connotation. Sometimes song-poems were made up following an important or notable event (see *Ferocactus covillei*).

Examples of two very different songs are given below. The pack rat's song consists of a single phrase repeated four times:

com-i-tin i-sii   com-i-tin i-sii   com-i-tin i-sii   com-i-tin i-sii

Music transcribed by Edward Garza.

The words *comitin isii* have been distorted from *comitin tiis* "Is the ironwood (*Olneya*) with seeds?" This song was explained by Roberto Herrera:

Hant com hantx ac ano cömiha cmaa cöipaxi taax zo ano cöitai itcmisiha xica ccam tacom xo comcaaciha thaax toc cömoii coi ziix haasax ano quiij quih cmique quih thaa ziix hehet cöquiij quih cmique quih thaa hap quih cmique quih thaa xica ccam quih hmaa coi comcaac coi thaax toc cöcoiiha. Cmoziimtoj hant toiima ziix haasax ano quiij quij cmique quih thaax toc cöquiij ox itaaam yoque ziix haasax ano quiij quij:
    —*Ctam aihi. Inyas pac haa poohca pameepit pacatol haa iihca isax haai—itai yoque ziix haasax ano quiij, ihmaa coi ox itaaam.*
    Ox tpactama:
    —*Iha, ihsaaiaha—ox tee ziix haasax ano quiij.*
    —*Cacatol z ihmaaha. Ihsaaihi.*
    Hipi yas teete yoque. Icoosaa cocaaati mos z ihmaaha xo ta yas teete yoque ziix haasax ano quiij quih cmique quih cöihaa ac itaai isaai ha teete yas. Itaai hant tiij hizaax oo cöimapacta. Icoos isojaa z ihmaa hax tahii hax ma ta yas teete ziix haasax ano quiij. Comitin xah iqueepe cah haa imiih hax tahii hax ma cöipacta ac.
    comitin isii
    comitin isii
    comitin isii
    comitin isii
Taax yas teete yoque ziix haasax ano quiij quih cmique quih thaa toc cömiij.

In the beginning when the earth was made, at that time the animals, but they were people, were there. Pack rat was a person, tortoise was a person, mule deer was a person. Other animals who were people were there. They were drunk. The others said to the pack rat: "Hey man. Sing one of your songs, even if it is a powerful, dangerous song," it is said that the others said to him. Then, "OK, I'll do it," the pack rat said thus. "It isn't dangerous. I'll do it." It is said that he said it was his song. It isn't a messed-up (complicated) song but he sang it, it is said, the pack rat who was a person, he did it, he

would do it, he said, his song. He did it, he sat, he did it like this. It wasn't a real song, it seemed, he sang it, the pack rat. His liking of ironwood (seeds) it seemed, its being like that. "Does the ironwood have seeds?," his song was that, it was said, the pack rat. He was a person, he was there.

A powerful song-poem involves the barrel cactus, *siml* (*Ferocactus wislizenii*). Clouds come from all of the barrel cacti. The spirit power *Icor* causes them to form fog (*xeele*), which makes the clouds from which rain comes and gives life to all plants. "Fog and clouds have life; they are alive. You don't see clouds coming out of *siml* but there is a relationship." Once a *siml* sang a song. A shaman heard the cactus sing and learned the songs from it. Sara Villalobos sang us a song of the *siml*:

Hant hipcom siml iti coocapii
Xeele iti mocaaya
Coox imcaamo
Hant ino quiyaaaya
Hant ino quiyaaaya

All of the barrel cactus that grow on this land
Fog coming from them
They all have life
It is the sound of the land
It is the sound of the land

The third line refers to the fog clouds. The sound of the land, called *hant iinoj* 'land its-roar/hum,' was a low, steady roar or hum caused by the *siml*. This sound was important for the spirits, because they employed it in their use of power. The Seri words are shown as they were sung, with lengthened vowels and extra syllables: *coocapii = coocp, mocaaya = moca, imcaamo = imcam, ino = iinoj,* and *quiyaaaya = quih yaa.*

14. Ironwood Sculpture

By the early 1970s the Seri were famous for their distinctive ironwood sculpture. This craft industry developed in the early 1960s, and is neither traditional nor introduced. It developed on the heels of drastic acculturation and decreases in sea turtles and certain commercial fish. Although the Seri had traditionally carved wood figures, many of which were abstract and stylized, they had necessarily been made of soft wood, generally red elephant tree (*Bursera hindsiana*). With greater access to modern files, hatchets, saws, and sandpaper, working hardwood became a possibility. Some earlier fetishes, such as porpoises and manta rays, were essentially identical to later ironwood sculptures, although the fetishes were much smaller than most ironwood pieces (see *Bursera hindsiana*). The sale of ironwood sculpture provided a means of cash income while allowing the Seri to remain in their territory. By the early 1970s more than half of the adults were engaged in ironwood carving, which provided a major source of income (Hills 1977; B. Johnston 1968, 1969; Ryerson 1976).

Sculpting was done by both men and women, and might involve several members of the family, with one person doing most of the carving and others finishing and polishing. An average piece was completed in one day. Subjects were taken mostly from nature, primarily sea and desert animals. Common subjects included desert bighorn, manta ray, osprey, owl, pelican, porpoise, quail, roadrunner, gulls, sea lion, sea turtles, sharks, stylized male and female human figures, and whales (Figure 14.1). There was much variation in quality and individual skill. The better artists had individual styles and consistently produced high-quality work, readily commanding higher prices. In 1972, Aurora Astorga began placing her initials on her work, and several others followed suit.

Seri ironwood sculpture was initiated by José Astorga (Figure 14.2; B. Johnston 1968). He was encouraged to carve ironwood by Helen Derwin, a resident of Oracle, Arizona, who was a frequent visitor to Puerto Libertad from the 1940s to the early 1960s. In 1961 Astorga made an ironwood carving in the form of a bar, as a gift-exchange for an American friend, Alexander "Ike" Russell. Astorga was pleased and amused to give it to Ike (Jean Russell, personal communication). Astorga then made a few hair barrettes, bowls, and spoons. These were not entirely successful, but he soon turned to making animal figures with obvious success.

The early ironwood sculpture was characterized by its simplicity, density, massiveness, rigid and solid form, and flowing outline. It was neither monumental nor intricate. The subject matter was mostly realistic but stylized, and intimacy with nature was often revealed in the work. Subtle details sometimes showed specific characteristics of the subject, or an essence which could be obtained only by an artist sensitive to the subject: e.g., a sea turtle with over-

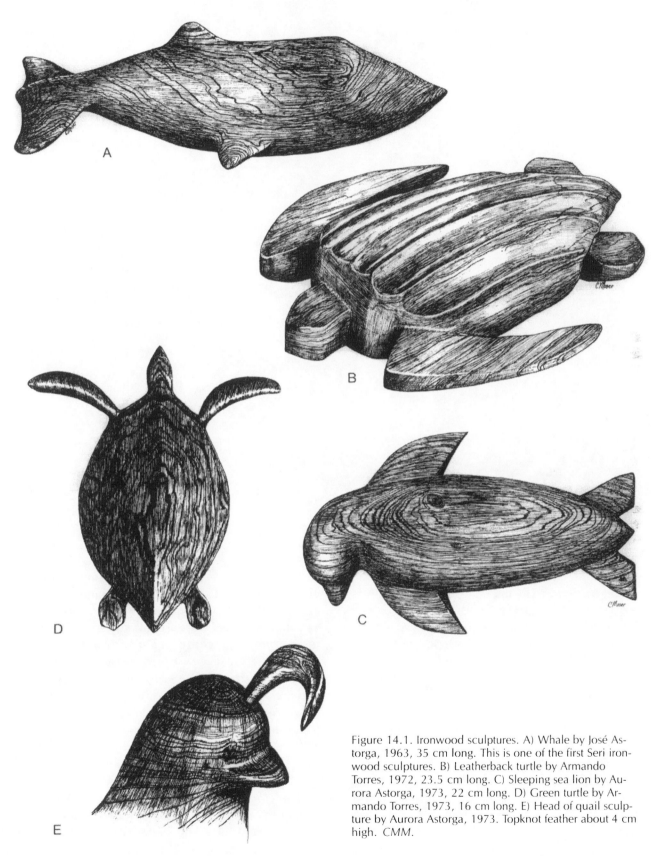

Figure 14.1. Ironwood sculptures. A) Whale by José Astorga, 1963, 35 cm long. This is one of the first Seri ironwood sculptures. B) Leatherback turtle by Armando Torres, 1972, 23.5 cm long. C) Sleeping sea lion by Aurora Astorga, 1973, 22 cm long. D) Green turtle by Armando Torres, 1973, 16 cm long. E) Head of quail sculpture by Aurora Astorga, 1973. Topknot feather about 4 cm high. *CMM*.

Figure 14.2. José Astorga, founder of the Seri sculpture industry, working at his home in El Desemboque. *HT, March 1969; ASM-21904.*

sized front flippers, representing a baby turtle, or a large toad crouching in the manner characteristic of a Colorado river toad (*Bufo alvarius*) when threatened by a potential enemy.

Usable ironwood soon became scarce near the villages, and the Seri purchased it at ever increasing cost from local merchants and Mexican woodcutters. During the first decade or so of the industry it was usually cut with hand saws and axes, and often came from the great mesquite-ironwood *bosques* along the floodplains of the Arroyo San Ignacio and its tributaries. In 1982 the Seri began to harvest the great ironwood stands on Tiburón Island. Crossing the channel by boat, they harvested wood only near the shore.

A good-quality ironwood log, about one-half meter long, yielded only one medium-sized figure. The dark heartwood, or *hesen*, was used for carving. The logs were irregular, and the sculptures commonly contained spaces and fissures. These were filled with *csipx*, the lac from creosotebush (*Larrea*). The Seri polished their earlier carvings with vegetable shortening, sea turtle oil, or motor oil. However, they readily accepted technological suggestions from outsiders and soon turned to high-quality sandpaper, shoe polish, or wax to produce a high-gloss finish (Figure 14.3).

While both men and women carved figures, generally only the women met tourist vehicles to sell (Figure 14.4). The better artists, however,

Figure 14.3. Nacho Molino polishing an ironwood sculpture to achieve a high gloss. *EWM, El Desemboque, 1976.*

Figure 14.5. This asymmetrical sculpture of a desert bighorn is 20 cm high. It was made by Panchita Moreno in February 1983. *MM, 1983.*

Figure 14.4. María Burgos meeting a tourist to sell a sculpture and necklaces. *HT, El Desemboque, March 1969; ASM-21930.*

Figure 14.6. Composite sculpture of seals on a rock was made by Herminia Astorga in El Desemboque in February 1983. It is 27 cm long. *MM, 1983.*

177

were sought after in their homes by knowledgeable buyers, and supply seldom met demand. During the early years of the industry, sales to tourists dropped sharply during the hot summer months, and the slack was taken up mostly by commercial buyers. Quality and appearance of the sculpture quickly evolved to meet market conditions and the purchasers' preferences. For example, the Seri quickly abandoned representing eyes with nail heads, carving a mouth on most animals, and making non-native and domestic animals. For the most part commercial buyers were discriminating in their purchases. They influenced the direction and development of the craft by selecting asymmetrical figures (such as animals with heads turned—Figure 14.5), more delicate and often intricate works, composite sculptures, and works with multiple figures (Figure 14.6).

As knowledge and appreciation of Seri ironwood sculpture spread, it was imitated locally and elsewhere in Sonora by non-Seris, but these pieces often lacked the vitality or quality of Seri work. Even though there were outside influences, the sculpture remained distinctly Seri, and became artistically and economically viable in the world art market.

15. Basketry

Baskets were important utensils in the Seri household. For the most part the making and use of baskets were the domain of women. All Seri baskets have been made from a species of limberbush called *haat* (*Jatropha cuneata*; Figure 15.1). This desert shrub is common throughout the Seri region. The basketry splints were prepared from the woody core of the flexible stems.

Commercial development of basketry began at least in the nineteenth century, when the Seri occasionally traded their baskets in Sonoran towns and ranches. Beginning in the 1930s baskets were bought primarily by tourists, and later also by art dealers. Simultaneously, the domestic uses of baskets declined—as metal and plastic containers came into general use—and basket sales increased. Some earlier forms disappeared and new shapes and designs emerged in response to the commercial market. By the 1960s baskets were made exclusively for sale and often displayed considerable skill and artistry. Some women achieved such excellence that their baskets ranked among the most expensive in the world, sometimes selling for thousands of dollars in fashionable art galleries. For a comparative view on Seri basketry, see Bowen (1973).

Other major works on Seri basketry are Burckhalter (1982), B. Johnson (1959), and Smith (1959).

Collection

Women of all ages searched the desert for suitable plants. Usually several of them, often accompanied by children, made the excursion together. They wore old clothes because the sap stained. Straight branches, about 1 m in length and 1.5 cm in diameter at the base, were selected after the plant was tested to see if the branches were sufficiently pliable. The test was made by splitting the tip of a branch and bending the split ends away from each other and downward (Figure 15.2). If they did not break, the plant was deemed suitable and its straightest branches were cut off at the base. Thin branches were preferred to thick ones, since their white "inner bark" and core were more pliant and had not begun to yellow (become woody). Larger twigs on a branch were removed immediately. Some women removed all of the "twigs" (shortshoots) and leaves before returning to the village.

Branches from several or more plants were arranged in a bundle and tied with rope or strips of bark taken from some of the same stems. Women carried the load in a basket on a headring or directly on a headring (Figure 15.3). Because the demand for baskets increased around

This discussion of Seri basketry is revised from the detailed report by Edward Moser, "Seri basketry," The Kiva (1973) 38 (3–4): 105–140. Used with permission.

Figure 15.1. A medium-sized shrub of *haat, Jatropha cuneata. EWM, 1972.*

Figure 15.2. Testing an *haat* stem for flexibility. *EWM, vicinity El Desemboque, 1968.*

Figure 15.3. Ramona Casanova carrying a bundle of freshly cut *haat* stems home. *EWM, vicinity of El Desemboque, 1968.*

the 1960s, men occasionally collected it with and for their wives. They carried it in their usual manner: two bundles suspended from each end of a *peen* carrying yoke or one large bundle slung over the shoulder on a rope. However, since the 1970s the men and women often went by pickup truck or automobile to harvest *haat* stems.

Preparation

Back in the village, if she had not already done so out in the desert, the basketmaker began preparation of the branches by removing any remaining twigs and leaves. Then she held them, a bunch at a time, in a fire of brushwood, turning them frequently until the bark was scorched (Figure 15.4A). This loosened the bark and made it peel more easily. After the branches cooled the woman bit into the middle of the stem to break open the bark. The flexible woody stem was pulled out from the bark with the fin-

gers like drawing back a bowstring. After accumulating a pile of peeled stems, each one was split lengthwise into halves by biting into the base of the stem and pulling half of it away.

The basketmaker then took each half stem and, using her teeth to split the base, separated it

Figure 15.4. Ramona Casanova carrying out some of the initial stages of preparation of the *haat* stems for basketmaking. A) Scorching the stems in the fire. B) Splitting the half stems with her teeth. C) Forming a roll of the outer strips of the stems. *EWM, El Desemboque, 1968.*

into two pieces; the second split was parallel to the first (Figure 15.4B). She then gathered the rounded outer strips and formed them into a roll called *hasáamij* 'thing looped.' She fastened the roll with strips of the roasted bark (Figure 15.4C) and set it aside. She then took the flat inner strips and, using her teeth and fingers, split each one into splints. The two strands peeled from the edges of this inner strip were called *itéeloj* 'edges' and served as stitching splints (Figure 15.5). These she formed into a roll, also called *hasáamij*, which she usually fastened with strips of the bark. Although *itéeloj* splints were soft and pliable, they lacked uniformity of width and normally were not used for the finer baskets. The remaining inner portion of the inner strip, composed of the coarse, less flexible core of the stem, was made into as many as seven or eight splints called *zeee* which were used only to make up the foundation bundle. The *zeee* splints were gathered into straight packets which were tied either with strips of the bark or several strips of *zeee*.

It was usually sometime later that the woman unfastened the roll of rounded outer strips which she previously had set aside and softened them in water for at least fifteen to twenty minutes. Using the point of an awl to split the base of each strip, she separated the strip into two pieces in the same manner that she had previously split the half stem. Of the resulting two strips, the thin curved outer layer was called *ipócj itac* 'back its-bone.' This soft and pliable strip was the source of the top grade stitching splints.

The remaining strip was called *izc ipót* 'front its-moving.' It was also made into stitching splints. However, since most of this section came from the relatively woody center of the stem, the splints into which it was split were somewhat stiffer than the *ipócj itac* splints and therefore considered inferior in quality and less desirable.

Using teeth and fingers, the woman further split both the *ipócj itac* and the *izc ipót* strips into a number of splints. The *ipócj itac* usually yielded two or three stitching splints; the *izc ipót* strip, from three to five.

Since the 1960s a basketmaker often gathered and prepared enough materials for the fabrication of several baskets. A woman occasionally helped another prepare basketmaking materials. She was likely to be paid with a roll of *itéeloj* stitching splints.

Awls

The tool used in the construction of coiled baskets was the bone awl (Sheldon 1979:137; Smith 1959). Formerly it was called *pac*, but in modern times it has been called *ziix icóop* 'thing what-one-sews-basket-with.' It was made from the proximal end of the metapodial bone of the mule deer (Figure 15.6).

Originally the awl was usually roughed out by a man, then finished by the basketmaker, who sharpened the point and shaft on a piece of pumice. However, some women made their own awls. A new awl was about 25 cm long. It was resharpened as it wore, and discarded after it was reduced to about 10 cm. A woman often incised the shaft with her identification mark (Moser and White 1968) or a decorative design (Figure 15.7). Since about the 1960s awls sometimes were marked with the owner's name or initials.

The tool was normally stored by inserting the point into the base of a packet of foundation splints. Sometimes it was stuck into the flat end of the *csai* hairbrush. Since about the 1960s steel awls (ice picks) have occasionally been used.

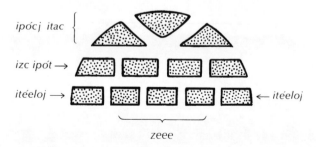

Figure 15.5. Diagram of a half stem of *Jatropha cuneata* showing the division into stitching splints (*ipócj itac, izc ipót*, and *itéeloj*) and foundation splints (*zeee*). The finest stitching splints are made from the *ipócj itac* layer. CMM.

Setting

Basketmaking took place outdoors. In hot weather the basketmaker sat in the shade of a ramada. Often two related women worked within easy talking distance of each other. Little girls stayed close to their relatives, observing and often participating in the work. Girls as young as seven or eight years old were encouraged to begin making small baskets.

In the typical working position, the basketmaker sat on a blanket or a large piece of cloth spread out on the sand. She folded and tucked her legs in close to one side. In that position her knees provided good support for the basket as she worked. Sometimes, however, she sat with her legs extended straight in front of her. At her side, within easy reach, she kept a bundle of dry foundation material (*zeee*) and one or more pans of water in which the stitching splints were soaking (see Sheldon 1979: Figure 5-14).

Figure 15.6. A new bone awl. *EWM, El Desemboque, 1968.*

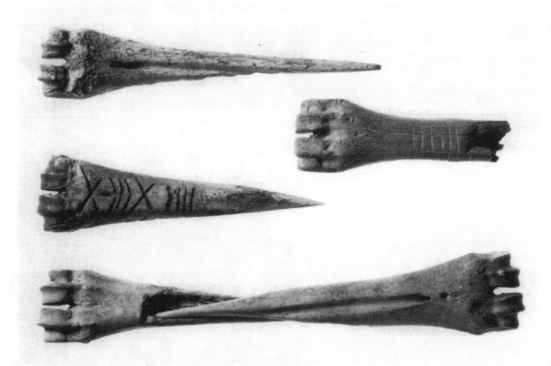

Figure 15.7. Used bone awls. The two in the center bear the owners' identification marks. The two at the bottom were found nested together in an olla in 1969; the longest one is 20.1 cm in length. *HT, 1972; ASM-32003.*

Salt water was the preferred soaking medium for the stitching splints (B. Johnson 1959). It was believed that the dyes used to color splints would remain fast if salt water rather than fresh water was used. However, women sometimes used fresh water when making undecorated work baskets for themselves or when living at an inland camp. Both salt and fresh water, when used for basketmaking, were called *icatáaxa*. Such water was not to be used for any other purpose (see Baskets and the Supernatural, below).

Fabrication

Except for an occasional plaited piece, all known Seri baskets were made by close coiling on a foundation bundle with non-interlocking simple stitches. Most baskets included split stitches and double stitching, although the most carefully made ones exhibited neither. Intricate or pattern stitching was unknown. Open stitching characterized one type of basket; in all other coiled baskets the foundation could not normally be seen between the stitches. The rim coil was usually tapered and attached to the previous bundle in the same manner as the body coils.

There were two kinds of basketry starts. The *cöhacápnij* 'made into a circle' was formed by tying a small bundle of foundation material or stitching splints into a single overhand knot (Figure 15.8A).

The Seri said that the interlocking start—a knot called *cöhafíz*, characteristic of modern baskets—was introduced to them by the Papago. It was incorporated into the earliest known entire Seri basket, a specimen collected by McGee (1898: Figure 24). The *cöhafíz* knot consisted of folding and then interlocking four pieces of stitching splint material into a square from which four double splints projected at 90 degrees from each other (Figure 15.8B). The projections were split into narrow strands which formed the beginning of the foundation bundle. For a medium or large basket, the interlocking start was constructed of *izc ipót* splints. For a small basket, this start was made of *itéeloj* splints.

Figure 15.8. Ramona Casanova demonstrating the two types of basket starts. A) Single overhand knot. B) Interlocking knot. *EWM, El Desemboque, 1968.*

A woman sometimes made several starts for future baskets at one sitting. A prolific basket-maker often fabricated the starts and the first two or three coils for several baskets while the materials were fresh. A start with two or three coils was called *hasaj hahíti* 'basket what-was-begun.'

The coiling and stitching process, using either of the basket starts, began by dampening the projecting strands of foundation material and folding them to the basketmaker's left. In the case of the interlocking start, the first of the four bundles combined with the second projection, which was also folded into the growing foundation. The process was continued until all four groups of projecting strands were incorporated into a single foundation bundle. The coil built up quickly and usually reached the desired thickness within the first few centimeters of a well-made basket. As the stitching proceeded, damp-ened foundation material was added to maintain uniform thickness of the coil. Usually a strand of foundation material was added by inserting the butt end into the bundle. Some women folded the end of the strand before adding it.

Baskets with the overhand knot start were be-gun by dampening the two opposite bunches of projecting foundation strands, folding the right-hand bunch 180 degrees to the left, and stitch-ing them into the coil. Additional material was added to the foundation bundle in the same manner as for baskets begun with the interlock-ing start.

The woman normally used an awl to split and trim the stitching splints to exact widths. Some-times she trimmed away the excess using only her teeth and fingers (Figure 15.9). If carefully done, this process resulted in a reasonably uni-form width of splint. Uniform splints made it easier to place each stitch flush against the previ-ous one and thus avoid gaps that might allow the foundation to be visible. This in turn enabled the careful basketmaker, on the next circuit, to place the hole precisely between the stitches of the previous coil and avoid splitting the earlier stitch or double stitching (to cover exposed foundation material). This resulted in a basket

Figure 15.9. Ramona Casanova trimming basketry splints. *RSF, El Desemboque, March 1983.*

of high technical quality according to Seri stan-dards. A few basketmakers achieved such excel-lence by stitching only with the pliant *ipócj itac* splints, but most women were unwilling to in-vest the extra time and effort which this entailed. Consequently, some split and double stitches were characteristic of the average basket.

The women worked facing the convex or underside of the basket with the strands of the foundation bundle extending to the left. This permitted a right-handed basketmaker to com-press and twist the bundle with her left hand, giving the basket added strength while manip-ulating the awl and stitching splint with her right hand (Figure 15.10). As a result the coils,

Figure 15.10. María Burgos manipulating an awl. *EWM, El Desemboque, 1968.*

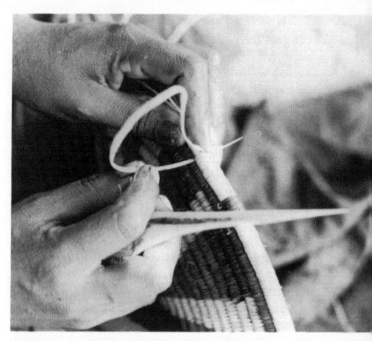

Figure 15.11. Stitching through a hole made with the awl. *EWM, El Desemboque, 1968.*

as seen from the interior or concave side, spiraled from the center to rim in a clockwise direction.

Coil width, typically one coil per centimeter, was generally uniform throughout a basket. A poorly made basket, however, often exhibited considerable variation in coil width. In an exceptionally well made basket, the central coils were very narrow (almost two per centimeter) and

gradually increased in width away from the center. Some miniature baskets of the 1970s and 1980s were made using a very short awl with a needle-sharp point.

The awl was held between the index and middle fingers with the base seated in the palm of the hand. This left the thumb and index finger free to insert and pull the stitching splint through the perforation in the foundation bundle without having to shift or lay aside the awl (Figure 15.11). After pulling the splint tight, the woman tapped the stitch twice with the shaft of the awl. The tapping was said to tighten the stitch.

The butt end of a stitching splint was normally folded into the foundation bundle. When a splint was nearly used up, the moving end was either folded into the foundation bundle and enclosed in the coil as the stitching continued, or it was left exposed. Some basketmakers left the first moving end exposed when each day's work was begun to mark their progress. When the basket was completed, exposed ends were removed by rubbing the basket, often with the *csai* hairbrush. Those which remained were pulled off with the fingers or cut off close to the coil with a razor blade.

The degree of slope of a basket's walls was determined by the angle at which the awl pierced the top of the previous coil. By slightly raising the base of the awl relative to the plane of the wall of the basket at each piercing, each succeeding coil was slightly offset from the preceding one; this process generated a moderate curve of the basket wall.

The number of stitches per centimeter varied according to the habit and care of the maker. Some poorly made baskets had as few as 3 per cm. Most of the better ones had 4 to 5 per cm. A rare few had as many as 7 per cm. Some miniature baskets contained 8 to 20 or more per cm.

The rim coil was stitched in the same manner as the body coils and was terminated by gradually tapering the foundation bundle. Usually the final taper was simply stitched to the top of the preceding coil. Once the woman had finished the basket and removed the protruding ends, she scrubbed it clean and set it in the sun to dry. It was then ready for use or sale. Baskets were usually kept wrapped in a cloth for protection until they were finished and sold.

Basket Repair

The longevity of a coiled work basket was said to be about two years. Broken stitches were easily replaced by stitching new splints, usually around the section that rested on the headring. A woman sometimes stitched a new section of coils into the bottom of a basket that had failed to resist wear. McGee (1898:208) reported that ordinarily when the bottom broke out the basket was discarded. However, he referred to one obtained on Tiburón Island in which the missing bottom was replaced with a piece of sealskin attached by means of sinew. One basket in the Heye Foundation collection has a thin piece of metal fitted into the open bottom, fastened to the adjacent coils with short lengths of wire.

Basket Types

The Seri classified baskets into fifteen named types, based on form and function. Seven of these were considered to be traditional; the other eight types were later innovations developed for the commercial market.

Traditional Baskets

Traditional baskets, listed below, had either utilitarian or ceremonial functions. Work baskets were usually not decorated.

1. The *hasajáa* "true basket" was the utilitarian or work basket shaped like a shallow bowl (Figure 15.12). When it was the most common type produced it was called *hasaj*, the term for basket. Later, however, the term *hasajáa* was applied to distinguish it from the similarly shaped decorated basket produced for sale which came to bear the name *hasaj*. A woman often marked her basket for identification by inserting a strand of yarn or cloth between the coils for two revolutions during the stitching process.

Since the *hasajáa* was designed for use rather than for sale, its construction was more sturdy than that of most of the newer, commercial *hasaj*. The *hasaj* was strengthened by tightly twisting the coil foundation splints during the stitching process. The walls were generally built up in a uniform curve, making this basket somewhat deeper in proportion to its diameter than the commercial *hasaj*. The effect of this kind of engineering was much the same as that of an inverted arch, making possible the support of a greater load.

2. An older type of tray basket, similar in form to, but slightly smaller than, the utilitarian *hasajáa*, was the *hasj itoj* 'basket its-eyes.' (Figure 15.13). The name refers to the basket's distinguishing technique—open stitching. In this technique the stitches were widely spaced, revealing the foundation bundle between stitches. The spaces between stitches were spoken of as "eyes." Conservation of time and materials was the reason given for making this style of utilitarian basket. It was often decorated with two

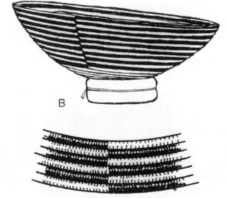

Figure 15.12. The work basket was shaped like a shallow bowl. A) These women, two with work baskets on their heads, were photographed while being detained by the military on Tiburón Island in 1904. The woman on the left was identified by the Seri as Manuela. *Photographer unknown; ASM-53991.* B) This detailed drawing of Manuela's basket shows that it was decorated with alternating rows of natural and dyed coils, a pattern called *cöhapái-jam* "wrapped." *CMM.*

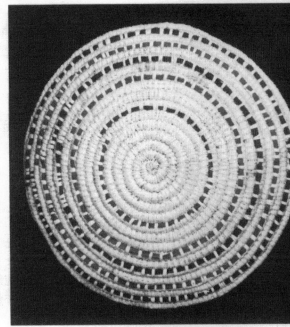

A   B

Figure 15.13. An open-stitched basket, known as "basket with eyes" because of the spaces between the stitches. A) The inside of the basket. B) Alternating red and blue bands of yarn are woven into the outer, convex side. This basket, made by María Antonia Colosio in 1972, is 37.5 cm in diameter. *MM, 1983.*

concentric reddish brown coils which were placed closer to the rim than the center. The number of stitches throughout the basket averaged about 2 per cm.

María Antonia Colosio, Sara Villalobos, and a few other women occasionally made this type of basket for sale. María Antonia inserted brightly colored ribbon into the convex or bottom side of the baskets (see Figure 15.13B). She said this was done so the decoration would not get dirty when the baskets were filled.

3. A shallow basket called *icáhaal* 'to accompany' was probably not over 30 cm in diameter and undecorated. It was essentially a small version of the *hasajáa* and used as an auxiliary vessel in which cargo was carried. The name derived from the fact that it was carried on the hand at shoulder height, thereby "accompanying" the *hasajáa* being carried on the head. The *icáhaal* apparently has not been used since about 1940.

4. The *hasaj caacoj* 'basket large' was a large version of the utilitarian basket *hasajáa*. It was said to have been at least 60 cm in diameter. Like most work baskets, it lacked decoration.

5. The *hasaj-ispoj* 'basket decorated(large)' was a giant-sized, decorated version of the work basket, *hasajáa*. The *hasaj-ispoj*, which has probably not been made since the nineteenth century, was said to be one of the oldest Seri basket types. Loreto Marcos recalled seeing one that was more than a meter in diameter. The concepts concerning this basket and its ceremonial function are described below in the section on baskets and the supernatural.

6. The *sapim* or *saptim* was a giant-sized, olla-shaped, decorated basket. Oral tradition surrounding a *sapim* which was made on Tiburón Island is discussed below in the section dealing with the supernatural and basketry. Although there may have been others, this is the

only *sapim* which we know was made by the Seri for their own use. It was probably essentially similar in shape to the large, olla-shaped, commercial baskets called *haat hanóohcö* or *haat hanóohcö caacoj* (see Commercial Basket Types, below). It was likely that the *sapim* was not as well made as its modern commercial counterparts and, therefore, did not take as long to make.

7. The only Seri basketry technique besides coiling was plaiting, which was used to produce a single kind of specialized basket called *icóasc*. This term is perhaps etymologically related to the verb *quiisc* 'to rustle.' This plaited basket was made with an open plain weave (over one–under one interval) and formed into a square or rectangular shallow box (Figure 15.14). It was used as a strainer to separate seeds from the pulp of *cardón* fruit (see *Pachycereus*). Punctured metal from a 20-liter can replaced the plaited basket strainer early in the twentieth century.

Commercial Baskets

Coiled baskets made specifically for sale are listed below. These were departures from traditional shapes, although there was little difference in the technique of manufacture. Baskets made for sale were almost always decorated.

1. The *hasaj*, the general term for basket, was the commercial version of the traditional work basket, *hasajáa*. It was generally more shallow, having gently sloping walls, and was seldom as closely stitched. Occasionally it was made in the form of a circular plaque. This has been one of the most popular shapes for commercial baskets. Like other baskets made for sale, it was usually decorated with a design in reddish brown, black, or both.

2. Several types of deep baskets were made, the most common taking its shape from the pottery olla. With the exception of the *sapim* (described above), olla-shaped baskets were not made by the Seri for their own use. Variations of this basket have been popular with buyers since the early part of the twentieth century.

The *haat hanóohcö* 'limberbush deep-concave-thing" ranged in height from approximately 10 to 45 cm (Figure 15.15). The height usually exceeded the diameter. The mouth, usually set on a short neck, varied greatly in width. Prior to about the middle of the twentieth century, a lid was made for each of these baskets, but since then lids were only occasionally produced. The *haat hanóohcö* was usually decorated. During the early part of the twentieth century these baskets were produced in considerable numbers for Sonoran ranchers and residents of Hermosillo. During the 1970s and 1980s the shape of this basket type was often modified: the lid was eliminated, the neck was shortened, and the basket was as wide as or wider than tall (Figure 15.16).

A modification of the earlier olla-shaped basket was the *haat hanóohcö it* 'limberbush deep-concaved-thing its-base.' It was the equivalent of the lower half of the olla-shaped *haat hanóohcö* and resulted in a bowl-shaped basket. This form was sometimes made intentionally; however, it usually resulted when the basketmaker closed off a half-finished olla-shaped basket, either because she was tired of working on it or because a visitor wanted to buy it at that time.

The *haat hanóohcö caacoj* 'limberbush deep-concave-thing large' was merely an oversized *haat hanóohcö* basket. Many were around 60 cm in height and 50 cm in diameter, and some were much larger (Figure 15.17). A basketmaker often required several hours a day over a period of two months or more to manufacture one of these large baskets.

Some enormous baskets of this type (Figures 15.18 and 15.19), made in the 1970s and 1980s, were sold for relatively large sums of money. Several families used the proceeds to purchase pickup trucks or automobiles. Apparently when this type of basket was first made commercially (probably about the middle of the twentieth century), it had ceremonial significance due to its large size. Later, as these huge baskets became more commonplace and religious beliefs changed, most of the supernatural connotations surrounding such baskets diminished.

Figure 15.14. Plaited basket for straining *cardón* seeds. This replica, made by María Antonia Colosio in 1968, is approximately 25 cm wide. *RSF, El Desemboque, 1968.*

Figure 15.16. A modified olla-shaped basket of exceptional quality. This commercial basket, made by Aurelia Molina in August 1982, is 25.5 cm in diameter and 17.7 cm high. It has 25 stitches and 7 bundles per inch. The design on the bottom of the basket is shown on the right. *RJH, 1982.*

Figure 15.15. Sara Villalobos with one of her olla-shaped baskets. This commercial style was called *haat hanóohcö. JWM, August 1958; ASM-25173.*

Figure 15.17. Two large olla-shaped baskets collected by Alexander Russell, c. 1970. The basket on the right, made by Ramona Díaz, is 36 cm in diameter; the one on the left is 45 cm in diameter. *HT, 1972; ASM-32005.*

Figure 15.18. A giant olla-shaped basket made by Elvira Valenzuela (*center*). Her daughter, Angelita Torres, and M. B. Moser admire the newly completed basket. *EWM, El Desemboque, 1973.*

There were about half a dozen other named commercial basket types. Most of these were experiments aimed at the developing commercial market around the middle of the twentieth century. Since most of these shapes were not popular with buyers, they were generally discontinued. These basket types are listed below:

a. The *hasaj quitxíin* 'basket have-head-ring' was an elaboration of the common *hasaj* and was distinguished by a shallow base built of coils (Figure 15.20), purportedly added to enable the buyer to balance it without the aid of a head-ring. This convenience was not appreciated by potential buyers.

b. The *haat haipj* 'limberbush what-was-made-oblong' was an elliptical basket which also was rarely made.

c. A variation of the above was the *haat haipj quislítxcoj* 'limberbush oblong-thing with-corners.' It had a flat, rectangular base and vertical walls about 5 cm high. A flat lid was sometimes attached with wire hinges.

d. The *haat hanóohcö quihízj* 'limberbush deep-concave-thing what-has-handle' was also

Figure 15.19. Elvira Valenzuela with a partially finished giant basket in her home at El Desemboque, March 1983. Note basketry materials. The plastic dishpan contains desert wolfberries. *RSF*.

rarely made. It was equivalent to the *haat hanóohcö it* basket but included a handle made of one or more extra coils arched over the mouth of the basket. This was a copy of the common woven Mexican hand basket which the Seri called *hatépen* 'limberbush carrying yoke.'

e. The *liitro*, named after the one-liter metal can, had a round, flat base. The vertical or slightly flaring walls extended from approximately 8 to 45 cm above the base and were usually decorated. Before prices escalated in the 1960s, these baskets were often purchased as waste baskets.

f. *Hamcanóiin quih iti ihíij* 'pan the on where-it-sits,' was a coiled circular plaque made to serve as a hot pad in the modern kitchen. It ranged from 13 to 26 cm in diameter and was plain or decorated. The finishing taper of the rim coil was sometimes made into a half-loop for hanging on a peg. These pads were not used by the Seri.

Figure 15.20. Basket with an attached headring made by Lolita Astorga. *EWM, El Desemboque, 1968*.

Decoration and Dyes

At the time of his brief visit, McGee (1898: 209) claimed that Seri baskets were "absolutely without decorative devices in weave, paint, or form." However, a photograph (NAA-4276) taken by his expedition photographer, William Dinwiddie, shows a work basket which seems to have a star-shaped design in it. This is the earliest known photograph of a Seri basket. The next known photographs, taken in 1904, clearly show decorated baskets (see Figure 15.12B).

María Luisa Chilión, born in the late 1880s, did not recall seeing decorated baskets as a young girl. She said most of the baskets at that time were of the simple open-stitched variety, the *hasj itoj*. She first saw a decorated basket about the time of her puberty fiesta (around 1900). Only the rim of the basket was made with dyed splints. She said solid reddish brown coils were next to be introduced and used in the development of new patterns.

On the other hand, oral tradition indicates that ceremonial baskets, the *hasaj-ispoj* and the *sapim*, were decorated. It may be that during the Seri-Mexican hostilities the Seri generally abandoned their practice of decorating baskets. The modern florescence of Seri basket decoration may have begun with the gradual decline of hostilities around the turn of the century. To meet the mounting interest in their baskets, women developed a number of new shapes, dyes, and decorations.

Patterns in Seri baskets were made almost exclusively with basketry splints dyed reddish brown and/or black. The dye ingredients were boiled and the basketry splints soaked in the dye pot.

The most common color and, according to Seri oral tradition, the only color used in the early part of the twentieth century was a reddish brown described by Barbara Johnson (1959:12) as a "rich burnt sienna color." This dye was usually prepared from the roots of *cósahui* (see *Krameria grayi*). Other sources of reddish brown dye were the roots of *Hoffmanseggia*, *Melochia*, and *Jatropha cuneata*.

Apparently the second color to be adopted was black. According to one Seri, sometime around 1930 a Mexican buyer ordered a basket decorated with a black cross to be placed on a grave in Hermosillo. Since that time, particularly since the 1950s, black became an increasingly popular color. It most frequently was used in combinations with reddish brown. Although black seems to be a modern innovation, the Seri had several sources of it. Since the 1960s commercial black dye generally replaced the native ones.

The favorite native black dye was derived from mesquite (*Prosopis*). A second, but seldom-used, black dye was prepared from red mangrove (*Rhizophora*). The third black dye, *haat an ihahóopol* 'limberbush in what-one-blackens-with,' was rarely used because it involved several ingredients: *Suaeda moquinii*, *Hyptis*, *Pithecellobium confine*, and pomegranate rind (*Punica*). Black dye was also obtained by boiling the roots of sea blite (*Suaeda moquinii*) with mesquite (*Prosopis*) bark or limberbush (*Jatropha cuneata*) roots.

Natural-colored stitching splints were occasionally dyed yellow with *Psorothamnus* or yellow-orange with *Croton californicus*. Some baskets were made using stitching splints stained with inexpensive commercial dyes as well as laundry bluing, but these had little appeal to potential buyers.

Design Technique

Decorated baskets were constructed in essentially the same manner as plain baskets. Designs were created solely by the use of dyed stitching splints. Foundation material was dyed only in the rare instances in which an interlocking start was included in the colored design, since the extensions of this start formed the beginning of the foundation bundle. Although a few baskets had dyed centers, normally the start and at least the first three coils (occasionally as many as the first fourteen coils) were left the natural color of the splints. This undyed center area was then usually enclosed by one or two dyed coils which served as the base of the design.

The rim coil was often entirely stitched with dyed splints or left the natural color. On some baskets the rim coil was decorated with alternating dyed and natural sections. In still others the main design extended through the rim coil.

Design Styles

The more elaborate designs of the twentieth century probably resulted from response to the commercial market. Several basket designs were similar to those on Papago baskets, and the Seri said that some of these were learned from the Papago. Traditional Papago and Pima designs

were rendered in black by using splints taken from the seed capsule of the devil's claw (see *Proboscidea parviflora*). Although devil's claw is common on the Seri coast, the Seri produced their designs with *Jatropha cuneata* splints dyed either reddish brown or black.

In the 1960s the Seri recognized more than a dozen design styles. Others, although distinct, were unnamed. As newer designs were developed, some of them also were named. Although several designs were sometimes incorporated into a single basket, the dominant one determined the design name of that basket. Some of the more prominent designs are shown in Figures 15.21 to 15. 23.

Pima and Apache designs were sometimes

Figure 15.21. Ramona Casanova holding a modern commercial basket. The inner design element, called *azoj canoj it ihíij* "star flame," was the most popular design pattern. The outer design element was called *ano hahéefcoj* "points on it." *EWM, El Desemboque, 1968.*

Figure 15.22. Carmelita Burgos holding one of her baskets with a design called *colequi cocmóonjc* 'zig-zagged.' This design was also associated with ceremonial baskets. *JWM, June 1957; ASM-24930.*

Figure 15.23. Large tray baskets. The basket on the left, 70 cm across, has an Apache design copied from a photograph. The other basket, 71 cm across, has a design called *seenel iti coxalca* "butterfly on it," which was one of the few zoomorphic figures on modern baskets (*ASM-E6049*). This design style was also attributed to the ceremonial basket called *hasaj-ispoj*. The butterfly style basket was made by María Elena Romero. Both baskets were collected by Alexander Russell c. 1970. *HT, 1972; ASM-32009*.

copied from photographs or drawings, and a variety of other designs were incorporated into baskets on request. Befitting the general pattern of Seri individuality, a woman felt free to choose whatever basket designs appealed to her. There was almost no duplication of precisely the same decoration.

Designs were not considered the property of families or individuals, and it appeared that no resentment was generated by copying. A woman often favored designs taught her by her mother, or she might prefer others or create her own. Thus, patterns were carried on without stifling the development of new and original ones, some of which, if successful enough, might be given names of their own.

Domestic Use: Work Baskets

Every Seri woman, young or old, owned a work basket which was her most important household utensil. Balanced on the head with the aid of a headring, the work basket was ex-

tensively used as a burden carrier to transport many items, such as food, firewood, personal possessions, trash, soil or sand, and even infants.

Shellfish, fish, meat, cactus fruit, mesquite pods, barrel cactus buds, *Cnidoscolus* roots, seeds, and other plant foods were carried in baskets. New baskets were sealed to prevent fruit juice and other liquids from draining through by coating the inside with an atole, or gruel, made of ground seeds. A large load of mesquite pods could be accommodated by building up successive layers with the ends of the outside pods intertwined to bind them together. This oversized load, called *hazái*, was also held together by tying it with strips made from long mesquite twigs. Another method to increase the load capacity of a basket was to extend the sides with sticks held in place by the pressure of the load itself (Figure 15.24).

Women often carried firewood in baskets for considerable distances. Dirt for making the mud which plastered the outside of ocotillo house walls and clean sand to cover the floor of beach camp houses were brought in baskets. Household

Figure 15.24. Basket load built up with sticks. Mariana Camou has just returned from the desert with a load of barrel cactus flowers. *MBM, El Desemboque, 1956.*

goods were transported from camp to camp in them, and on occasion trash was carted out in them to be dumped away from camp. A baby or small child was sometimes carried from one camp to another lying or seated in a basket placed inside a carrying net and suspended from one end of a man's carrying yoke. It was common for a woman to carry an infant on a cradleboard on top of a basket balanced on her head. This practice was illustrated by Gilg in 1692 (see Figure 1.4).

The work basket served as a platform in the preparation of various foods, which were kept clean by lining the bottom with certain "soft" plants. Winnowing was also done with baskets (see the section in Chapter 6 dealing with the gathering and preparation of food plants).

A basket was frequently used as a communal serving and eating dish for the nuclear family. *Hasj itoj* baskets were often used as individual eating dishes and bowls (clamshells—e.g., *Mac-*

tra dolabriformis or *Laevicardium elatum*— were used as spoons). A tortilla griddle, called *iti icatáscar* 'on what-one-makes-tortillas-with,' was made by pressing clay into the bottom of a flat basket. After firing, it was used as a trade item with neighboring ranchers.

In addition to its uses in transport and for food gathering and preparation, the work basket had a number of other domestic uses. For example, Seri babies, always born directly into the hands of a midwife, were often bathed in a basket. A basket filled with cloth-covered sand was often a temporary crib for the infant after it had been bathed, dressed, and fed sweetened water. A basket also often served as a basin for washing one's hair or clothing.

Baskets were used to winnow the clay powder and the ground dung temper used in the manufacture of pottery (Bowen and Moser 1968), and mud for plastering the walls of the wattle and daub, flat-roofed ocotillo house was prepared in a work basket. Baskets also served as work platforms for pottery making (Figure 15.25).

Although men seldom used baskets except to eat from, they did sometimes use a work basket during a game to hold the stakes (often bullets). Men also used baskets in conjunction with making music involving the musical bow, rasp, and gourd drum (see the section on music in Chapter 13).

The importance of baskets in traditional Seri culture was reflected in the fact that both decorated and plain baskets, as well as bundles of freshly cut *haat* stems and bundles of prepared basketmaking materials, were used as partial payment of the bride price and as gifts from one woman to another.

Baskets and the Supernatural

Fiestas, inspired by a shaman's dream or vision, were held to bring good luck upon the completion of three types of baskets: *sapim*, *hasaj-ispoj*, and *haat hanóohcö caacoj*. Most of these fiestas took place during early summer—a

Figure 15.25. Baskets served as work platforms for pottery making. A) María Luisa Chilión uses two work baskets to hold her pottery-making materials. The basket on the right holds rabbit dung, and the one on the left contains clay. B) The nearly completed pot rests on a bed of ashes and rabbit dung in a basket. *MBM, El Desemboque, 1967.*

time of year when food was plentiful and easy to obtain and there was much drinking of cactus fruit wine and partying (see Columnar Cacti). In the following account of a giant basket (*sapim*), the preparation for the fiesta was begun on Tiburón Island.

First, a huge tray basket (*hasaj-ispoj*) was made; then the *sapim* was begun. As the woman worked on the *sapim*, the awl ocasionally made a screeching sound as it pierced the coil. The sound was ignored until the base of the basket was completed. When the basketmaker began to form the walls, however, a dangerous spirit was said to have entered the basket. Thereafter the screeching of the awl was interpreted as the spirit wailing. This spirit, called *coen*, was feared because it could bring death to the basketmaker or one of her family.

To avoid this danger, the woman had to pacify the spirit. The *hasaj-ispoj* was filled with patties

of mesquite flour and little balls of cactus fruit. Whenever the awl screeched, some of the food was tossed onto the ground near the *sapim*; it was then retrieved and eaten by people who happened to be nearby. The purpose was to create a festive air and help pacify the spirit. In addition the basketmaker sang the following song four times:

> *Toii cöhahóteja toii cöhahóteja*
> *Toii cöhahóteja toii cöhahóteja*
> *Sapm iisax imíipox*
> *Toii cöhahóteja, toii cöhahóteja.*
>
> Rock it away, rock it away
> Rock it away, rock it away
> The *sapim* pulled out its soul (wailed)
> Rock it away, rock it away.

"Rock it [the basket] away" refers to the basketmaker rocking the big basket back and forth.

The song was directed to the spirit, who was told to pacify the basket. As soon as the song had been sung, the danger of death was averted.

The *sapim* was said to have been decorated. When the huge basket was finished, it was suspended from a pole and carried to the beach by two men. It was then placed on a large balsa and transported to the mainland. From there it was carried to a mountain pass just south of Johnson Peak, chosen because it was close to a supply of fresh water. There the people held a four-day fiesta. The pass was thereafter named *Sapmáyam* 'basket(*sapim*) where it crossed.'

Taking the basket, the group then proceeded farther inland to a camp later called *Sapmíscax* 'basket where it disintegrated.' There, joined by other Seri, they had a great celebration lasting eight days. The men hunted deer. There was much eating, drinking, and dancing. The people celebrated to make both themselves and the basket happy. At the end of the eight days, the spirit was said to have left the basket. The *sapim*, no longer important to the people, was abandoned.

Another narrative tells about the celebration of basketmaking with a fiesta. Long ago, one of the old coyote men went into the desert carrying water in a "boot" (woody callus tissue formed around a woodpecker nest-hole) from a *cardón* stem (see *Pachycereus*). There he made a basket, but he did not bring it back to camp. Later some coyote women were going into the desert to collect *comot* (*Matelea cordifolia*) and *nas* (*Matelea pringlei*) fruit. They carried with them pelican bill pouches which had been sewn partially shut to make bags. The coyote man told them to wait while he brought them something else in which to carry things. The coyote women asked him how that could be. He answered that the *haat* plant could be prepared and made into a basket. He added that someone had taught him how to make a basket and how to make a headring of the same plant. He warned them that the water used for making baskets must never be used for any other purpose. Then he brought his basket into camp and told the people that it was to be called *hasj itoj*. He said that it was to be cele-

brated with a fiesta at which many different songs were to be sung. He also told them he would make another basket of a different size which would be called *hasajáa*.

The old coyote man then made four *hasj itoj* baskets. A fiesta was held, and the coyote man brought out the four baskets to hold food prepared from mesquite pods. At the fiesta the man announced that he would teach the people to dance and that he would be the singer. He told them that all future fiestas would be patterned after this first one.

Another version of this story gives credit to a Seri shaman, rather than the coyote man, who went into the desert and made the first basket by means of spirit power obtained through dreams and visions. He returned with a large *hasaj*. The spirits also taught him that basketry materials were dangerous. This danger, they told him, was associated with *tear*, the chief of the evil spirits. The shaman supplied instructions regarding the fiesta songs and the special use of ceremonial objects. In both versions of this narrative, a man is associated with the origin of basketry.

In another story a male basketmaker is also involved. A Tiburón Island woman told the people that the Giants of Baja California had made and used baskets and that she knew how to make them. However, since she was old and crippled, she could not make one herself. She taught a man to make one. His first basket, an *hasj itoj*, was poorly made. From that time on the Seri made baskets.

In another account, the nighthawk, *coplim*, gave a basket to his daughter-in-law. With great anticipation he looked forward to sharing the cactus fruit she would collect. While his daughter-in-law went into the desert to gather fruit, nighthawk sang to his nephew about the delicious fruit that they would be eating. When the daughter-in-law returned with the basketful of fruit, she set it on the ground. Then she picked up a handful of dirt and threw it in night-hawk's face.

The references to male basketmakers in these accounts indicate that the Seri had some familiarity with the basketmaking of the Cochimí of Baja California (see Massey 1966). Perhaps there was a time when Seri men made baskets, as in the Cochimí tradition. Even in modern times there were occasional incidents in which men made or completed one or several baskets, although this may simply have reflected Seri individualism.

The Seri attached importance to objects and animals of exceptional size and generally treated them with a respect bordering on fear. Since the *hasaj-ispoj*, like the *sapim*, was very large, it, too, was said to possess a strong spirit. When the awl screeched during the fabrication of the *hasaj-ispoj*, the basketmaker was careful to stop work and sing the pacification song. The completion of the basket was also celebrated with a four-day fiesta.

The *hasaj-ispoj* was used primarily as a work basket. It was also the basket used to hold the gifts which were thrown into the air for those present at the end of a girl's puberty ceremony (Griffen 1959). When huge, olla-shaped baskets, called *haat hanóohcö caacoj*, were first made for sale, they were celebrated with a four-night fiesta upon completion.

The feared spirit of the basket, *coen*, was tiny, fat, old, and female, with a large face. She was a defiler and was especially active at night when she might enter a sleeping person. This contamination attributed to the *coen* was called *ya-cóene*. When a person had been contaminated, he became wan and weak. He was then said to be *coen oicö* 'spirit killed-by.' The *yacóene* was contagious and transmitted to objects that an afflicted person touched. As late as the 1920s most adults washed their hands each morning to rid themselves of this contamination.

Basketmaking materials were particularly susceptible to contamination by an afflicted basketmaker's hands. These items were, therefore, con-

sidered dangerous for others to touch or steal. Even the water used in basket construction was contaminated and was not to be used for any other purpose. The most frequently mentioned penalty for violation of *coen*-related taboos applied to women. Any woman who stole basketmaking materials would suffer a breech presentation on the birth of her next baby. A boy who happened to wash his hands in water used for basketmaking would pass the breech birth penalty on to his future wife.

While a finished basket held no apparent danger for the basketmaker, user, or buyer, there was somehow a tinge of evil associated with it. While no one expressed concern about this evil, it was admitted that baskets were somehow involved with spirit power. It was considered dangerous to mistreat or destroy a basket. The spirit *coen* caused the culprit to be harmed in some way. When a basket was worn out it was abandoned, never destroyed.

When a woman died, her baskets were buried with her. These, along with her other possessions, would be available for her use in *coaxyat*, the Seri hereafter.

Evil and danger were associated with baskets dyed solid or nearly-solid black or reddish brown. They were believed to be especially contaminated by *coen*, and the maker of such a basket was supposed to give it to her burial sponsor (*hamác*). However, by the 1960s or 1970s younger basketmakers would make one upon customer request.

A halo around either the sun or moon meant that the celestial object was making a basket. The name for the sand dollar is *xepe isx* "basket of the sea."

Baskets were sometimes involved as an intermediary in communication between the supernatural world and the shaman. The songs given below were communicated by a basket medium to Juan Mata (a shaman who died about 1943) and became his property. In the first song the cactus fruit (e.g., *cardón* or organ pipe) is chiding the woman who lacks the courage to go into the desert to gather the fruit in the oppressive summer heat.

Cmaam isoj ihíizat iqui yemetx
Hasajíil hatxiiníil
Hisoj haaco xoma
Icózim iinoj canoj ano hayoom.

The big strong woman inches her way toward the
shade
Her big basket and headring
I have spirit power
Where the sound of the summer heat is, I am.

In the next song the basket is speaking. The
fourth line is spoken by the woman who cannot
bring herself to go out into the heat.

Hant haa miifp
Icózim canoj heecot cöimáfija
Icózim canoj heecot coyoom
Hiisax cohyéepit hiisax yomátax
Icózim iinoj haaco xoma.

It is the new year
The sound of summer heat passes through the
desert
The sound of summer heat pervades the desert

I made the effort, I lost interest
The sound of the summer heat has spirit power.

The following song speaks of the communion
between a woman and her basket. The cactus
fruit is speaking. The woman was out in the
desert heat gathering fruit. While returning to
camp, she fell and lost the fruit, so she went back
into the desert and gathered more. The woman
was said to be Loreto Marcos. Juan Mata saw
this happen in a dream, and he taught Loreto the
song and made it hers.

Icózim iinoj himo cooyoj
Haa ano cooza ano hayoom
Hiyaal heecot ano cooza
Icózim iinoj quimáaxataj
Iti hayoom.

The sound of the summer heat is over there
They (cactus) talk in it, I am in it
My companions (the other desert plants) talk in
the desert,
The sound of the summer heat shimmering
I am in it.

PART III

Plants in Seri Culture

Species Accounts

Plants in Seri Culture: Species Accounts

The plants in this section are arranged alphabetically within several major categories. Nonflowering plants (Chapter 16) are divided into marine algae, non-vascular land plants, and ferns and fern relatives; within these categories the accounts are alphabetical by genus and species. The seed plants, or flowering plants, are treated alphabetically by family, genus, and species in Chapter 17. However, within the cactus family (Cactaceae) the columnar cacti are discussed as a unit apart from the rest of the cacti. The monocotyledons are not separated from the dicotyledons. Plants not native to the original region of Seri occupation are indicated by an asterisk (*). These include exotics, weeds, nonnative naturalized plants, plants cultivated in adjacent regions (even since ancient times), and store-bought fruits, vegetables, and grains.

Botanical nomenclature and taxonomy reflect our interpretation and understanding of the regional flora and existing literature. The nomenclature and taxonomy tend to be conservative. At the family level we usually use the more familiar name, such as Leguminosae for Fabaceae, Labiatae for Lamiaceae, and Graminae for Poaceae. Synonyms for the vascular plants are generally provided only for nomenclature which differs from that given by Wiggins (1964) in *The Flora of the Sonoran Desert*, and synonyms given by Wiggins are generally not repeated here. Synonyms are given in brackets [—].

For each species account the order is: scientific name, Seri name and gloss, free translation, and Spanish and English common names. When more than one Seri name is given for a single botanical entry or species account, we have attempted to list the more commonly used name first; otherwise the Seri names are alphabetical. The gloss—the word-for-word, or literal, translation—is indicated by single quotation marks ('. . .'). A gloss is given for each Seri name if we were able to translate it. As noted in the preface, we have somewhat simplified, as well as broadened, our interpretation of the gloss and the manner in which we translate plant names. For example, we often provide the scientific name in the gloss and the common name in the free translation.

Word order in the gloss is not altered from Seri word order. If a multi-word Seri name contains a term which requires more than one English word for translation, the English words for this term are hyphenated in order to preserve the word-for-word sense of translation. Literal translations from any language often seem awkward; for this reason, a free translation is usually given, to approximate more closely ordinary English usage and word order. The free translation is indicated by double quotation marks ("...").

In most cases we give only the singular form of most Seri terms (pluralizations in Seri are relatively complex). In some instances one term in a compound Seri name is a pluralized form. These are generally indicated as follows: no space between the Seri term and the word "plural" in parenthesis. For example, the gloss of *casol caacöl* is '*casol* large(plural),' indicating that the plural refers only to the term "large." In this example *casol* is singular and an untranslatable term.

16. Non-flowering Plants

Marine Algae*

The seaweed flora of the Gulf of California is rich and diverse (Dawson 1944a; Norris and Bucher 1976; Norris in prep.): there are about two hundred species in the Seri region. In general the marine algae are most abundant during the cooler months. Many of the common macro-algae tend to "disappear" during hot weather. Some of these species may persist through the summer as an alternate, possibly unrecognized form (see Lubchenco and Cubit 1980; West and Hommersand 1981).

Unlike their knowledge of flowering plants, the Seri's knowledge of seaweeds was uneven, with numerous synonyms. However, the larger and more conspicuous seaweeds generally had specific names, and information about them was relatively consistent.

The larger marine algae, or seaweeds, were collectively called *xpanáams*. This term is a contraction of *xepe ano ims* 'sea in fringe,' or "fringe of the sea." It was used as the specific name for the brown seaweed, *Sargassum*, or in a distributive, general manner for all seaweeds, or when the specific name of a seaweed was not known. The name *xepe an ihíms* 'sea in its-fringe' was probably only a variant of *xepe ano ims*. Sometimes *xpanáams* and *xepe an ihíms* were the only names applied to a given seaweed. In addition, the term *xpanéezj* 'sea membrane,' apparently the specific name of an unidentified seaweed, was also used in a vague, distributive manner.

All Seri names for seaweeds were descriptive—

there were no unanalyzable names. Seaweed names often contained descriptive information about the plant's habitat or, more often, ecological information concerning its association with major food resource animals, such as the green turtle and certain fish (Norris 1978). A major criterion for seaweed names and concepts was whether sea turtles, specifically the green turtle, ate a given kind of seaweed. Another criterion was the geographic distribution. The water in the shallow Infiernillo Channel and off the southern shore of Tiburón Island is warmer than the deeper water off the more exposed west and north coasts of the island and off the mainland coast north of the island. The water temperature strongly influences seaweed distributions and affects the feeding habits of sea turtles. The Seri were keenly aware of these distributional patterns and this knowledge was often incorporated into the concept of a given seaweed.

Some common Gulf of California seaweeds are edible (see Abbott and Williamson 1974; Uphof 1968). However, we have not found evidence that the Seri used any of them for food. Women became familiar with many kinds of seaweeds while cleaning sea turtles. In fact, a substantial portion of Seri information on seaweeds related to their extensive knowledge of sea turtles.

Various larger seaweeds found in the beach drift were used for roofing of the brush house and as knee pads while paddling a balsa (see *Sargassum*). At the beach children often used seaweeds in their play. To cause the wind to stop, a rock covered with seaweed (perhaps *Enteromorpha*) was taken from the sea, and

---

*The section on marine algae was contributed by James N. Norris in collaboration with the authors.

firewood was piled on top of it and then burned. When the fire had burned down the wind would stop.

In this section the green algae (Chlorophyta), brown algae (Phaeophyta), and red algae (Rhodophyta) are listed together alphabetically by genus and species. At the end is a separate list (alphabetical by Seri name) of unidentified seaweeds. Since the marine algae are arranged in alphabetical order according to scientific name, the division and family names are provided in brackets following the scientific name of the plant.

Amphiroa beauvoisii Lamour.
A. van-bosseae Lemoine
    [Rhodophyta; Corallinaceae]
xepe oohit 'sea what-it-eats'
    "what the sea eats"
xpanáams ccapxl 'seaweed bitter'
    "bitter seaweed"
ziix hant cpatj oohit 'thing land flat what-it-eats'
    "thing that the flounder eats"

These purple to purplish pink, articulated coralline algae are found in the beach drift and attached to rocks in the intertidal and shallow subtidal zones.

Asparagopsis taxiformis (Del.) Trev.   [Rhodophyta; Bonne maisoniaceae]
moosníil ihaquéepe 'blue-turtle what-it-likes'
    "what the blue turtle likes"
moosn-oohit 'sea-turtle what-it-eats'
    "what the sea turtle eats"
taca imas 'triggerfish its-body-hair'
    "triggerfish's body hair"
ziix hant cpatj oohit 'thing land flat what-it-eats'
    "what the flounder eats"

This red alga has feathery, rose-colored fronds. The erect axes are slender and cylindrical, and they arise from several entangled and stoloniferous basal portions (Figure 16.1). The plant grows subtidally on rocks and in protected intertidal areas, such as tide pools. The plant deteriorates rapidly upon exposure to the air.

The more commonly used name indicates that it was eaten by the blue turtle, *moosníil*. This was one of the more than eight kinds of green turtles (*Chelonia*)

Figure 16.1. *Asparagopsis taxiformis*, a red alga. *AHH, 1975.*

recognized by the Seri (see the section on reptiles in Chapter 3). The blue turtle was noted as being very large and strong, and occurring in open water north and west of Tiburón Island. It was said that when this animal was harpooned, the harpoon point, head, and line turned blue, as though they had been dyed. The blue turtle was rare, and by the mid-twentieth century had not been seen for many years. *Asparagopsis* contains unique halogenated compounds (bromines; McConnell and Fenical 1976), which may account for some of the bizarre information concerning the mysterious blue turtle.

Upon seeing a specimen preserved in 5 percent Formalin/sea water, María Antonia Colosio said, "What's wrong with this? It's usually much redder." The preserved specimen had faded. She then described it as "feathery" and said, "The blue turtle eats it—you know, the one that the blue fluid comes from."

Animal Food: Several people said the green turtle ate it, and one person said the triggerfish ate it.

Codium amplivesticulatum Setch. & Gardn.
    [Chlorophyta; Codiaceae]
taca oomas cooxp 'triggerfish what-it-twined white'
    "white *Codium*"

This seaweed was described as thinner, longer-stemmed, and lighter in color than *taca oomas* (*Codium simulans*), but with the same distribution. Based on the description given by the Seri, we believe that *taca oomas cooxp* is this species of *Codium*.

The branches are dark green, spongy, and up to 85 cm tall and 1.3 cm in diameter.

Animal Food: It was said that this seaweed was eaten by the green turtle.

Codium simulans Setch. & Gardn.
taca oomas 'triggerfish what-it-twined'
   "the cord that the triggerfish twined'
tacj oomas 'porpoise what-it-twined'
   "the cord that the porpoise twined"

The Seri described this seaweed as dark greenish, not a tall plant, with thick stems. They said it is common on the north and west sides of Tiburón Island and at El Desemboque during cool weather but not in the Infiernillo Channel.

This green alga commonly grows 10 to 20 cm in height. The branches are dark green, finger-like, cylindrical, less than 0.5 cm in diameter, and spongy (Figure 16.2).

Animal Food: Some people said that it was eaten by green turtles, particularly the ones found in places other than the Infiernillo Channel. However, others said it was not eaten by sea turtles.

Play: Children pretended that this fleshy seaweed was fish or turtle meat.

Colpomenia phaeodactyla Wynne & J. Norris
   [Phaeophyta; Scytosiphonaceae]
hast iti coteja 'rock on what-sways'
   "what sways on rock"

This plant is brown and has a cluster of hollow, finger-like, thin-walled elongated sacs (Figure 16.3). It is common in spring on rocks in the low intertidal zone and in beach drift.

Animal Food: Some people said the *moosni* (adult green turtle) fed on it, but that the *cooyam* (the young migratory phase of green turtle) did not.

Colpomenia tuberculata Saund.
hant iteja 'land its-bladder'
   "land's bladder"
xpeetc 'sea scrotum'

This unusual brown alga grows attached to rocks in the middle to lower intertidal zones and is common in beach drift in spring. It forms a hollow, irregular-globose structure with firm walls (Figure 16.3). There are several forms, ranging from globose to somewhat ovoid in shape.

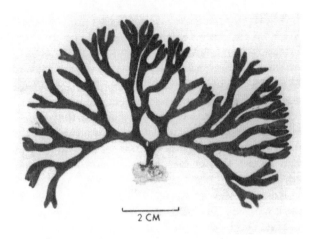

Figure 16.2. *Codium simulans*, a green alga. *AHH, 1975.*

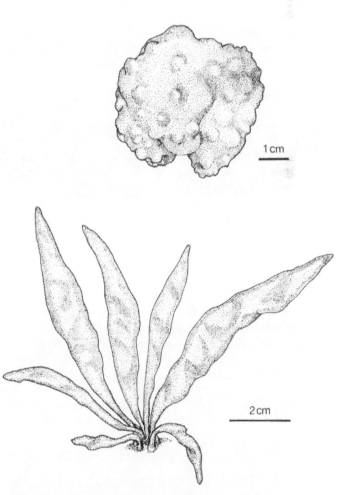

Figure 16.3. Two species of brown algae. Top: *Colpomenia tuberculata*; bottom: *C. phaeodactyla. SM.*

The Seri described it as a sort of bag of liquid, reaching the size of a baseball. It was said to occur primarily in the Infiernillo Channel, often in clusters in shell banks. They also said it grows in clusters alongside buried (dormant) green turtles and later washes ashore. When seen washed ashore at the end of March, it was said to be a sign that the *moosni hant coit* 'green-turtle land touch' (buried or dormant turtles) were leaving the sea floor and beginning to feed again. This seaweed was seen from December until June.

Animal Food: It was said that sea turtles did not eat it.

Play: Children played with it. They thought it was fun to drink the water inside the hollow seaweed, which they said tasted sweet. A piece of *Colpomenia* was sometimes used as a hat or cap on a seaweed doll (see *Sargassum herporhizum*).

Cryptonemia obovata J. Ag.
Halymenia coccinea (Harv.) Abbott
    [Rhodophyta; Cryptonemiaceae]
moosni ipnáil 'sea-turtle its-skirt'
    "sea turtle's skirt"

These attractive foliose (leaf-like) red algae (Figure 16.4) are almost impossible to distinguish without microscopic study. The blades are up to 25 cm in length. They grow on rocks in the subtidal zone and may be found in beach drift in spring. Rosa Flores said that there are different [kinds of] *moosni ipnáil*, and that they look similar but come in different colors. This *moosni ipnáil* was distinguished as being red; however, "red" does not seem to be part of the name.

Animal Food: It was said that the green turtle ate this plant.

Play: Little girls put the largest blades on their heads and wore them as scarves. Girls sometimes put a seaweed doll on a toy cradleboard made from a piece of sea turtle carapace and covered the doll with a piece of this seaweed as a blanket.

Cutleria hancockii Daws.   [Phaeophyta; Cutleriaceae]
hast iti coteja 'rock on what-sways'
    "what sways on rock"
xepe yazj 'sea its-membrane'

The fan-shaped blades of this brown alga superficially resemble those of *Padina* but differ in being

Figure 16.4. *Halymenia coccinea*, a red alga. The largest blade is 13 cm. long. *SM.*

very flaccid, slippery, and in having delicate, fringed margins. The blades are up to 10 cm wide. It is a cool-weather ephemeral.

The Seri said it grows at Tecomate and along most of the coast of Tiburón Island, except on the east shore, and that it also occurs in rocky areas north of El Desemboque. They pointed out that it grows on underwater rocks and sways with the movement of the water.

Animal Food: The Seri often observed green turtles with their tails pointing upward as they grazed on this seaweed. It was said that the *cooyam*, the young migratory phase of green turtle, did not eat it.

Dasya baillouviana var. *stanfordiana* (Farl.) J. Norris & Bucher   [Rhodophyta; Dasyaceae]
xpanáams caitic 'seaweed soft'
    "soft seaweed"

This delicate red alga has slender, feathery red branches. It occurs in the subtidal zone and less commonly in the low intertidal zone. In spring it is found in beach drift.

Dictyota flabellata (Coll.) Setch. & Gardn.
   [Phaeophyta; Dictyotaceae]
xepe zatx 'sea glochid'
ziix hant cpatj oohit 'thing land flat what-it-eats'
   "thing that the flounder eats"

The fronds are flat, brownish, irregular-dichotomously branched, and up to 15 cm tall. It grows on rocks from the middle to lower intertidal zones, and in spring occurs in beach drift.

Digenia simplex (Wulf.) C. Ag.   [Rhodophyta; Rhodomelaceae]
tacj oomas 'porpoise what-it-twined'
   "what the porpoise twined"
xepe oohit 'sea what-it-eats'
   "what the sea eats"
xpanáams coopol 'seaweed black'
   "black seaweed"

A common, coarse, and tufted red alga. In the northern Gulf of California it reaches 10 cm in height, but elsewhere in the world it is usually larger. This seaweed was not well known by the Seri. This seaweed, widely distributed in the tropical seas of the world, has been used as a vermifuge and food in the Orient since ancient times (Fenical 1983; Uphof 1968:179). However, we have no evidence of its use as a medicine or food by the Seri.

Enteromorpha acanthophora Setch. & Gardn.
   [Chlorophyta; Ulvaceae]
xpanáams coil 'seaweed green'
   "green seaweed"
xpanéezj 'sea membrane'

This bright green alga is tubular, hollow, branching from the base, and up to 20 cm tall. It is often very common on rocks in the intertidal zone during winter and spring.
Play: Girls played with dolls made from this seaweed.

Eucheuma uncinatum Setch. & Gardn.   [Rhodophyta; Solieriaceae]
taca-noosc 'triggerfish its-roughness'
   "triggerfish's papillae"
xepe oil caitic 'sea tuft soft'
   "soft sea fan"

This red alga has fronds which are elongated, up to 40 cm long, fleshy, branched, cylindrical, and covered with numerous papillae or spinose projections. This species is abundant during the cooler months. It grows on rocks from the low intertidal to shallow subtidal zones and occurs in beach drift.
Adornment: the thick, fleshy branches were sometimes cut into small pieces and strung as a necklace (Figure 16.5).
Animal Food: It was said that the green turtle ate it.
Play: While playing on the beach, children pretended that it was food. They broke the plant into small pieces and served it in seashell dishes, and playfully ate small pieces of it.

Galaxaura arborea Kjellm.   [Rhodophyta; Chaetangiaceae]
tacj oomas 'porpoise what-it-twined'
   "the cord that the porpoise twined"
ziix hant cpatj oohit 'thing land flat what-it-eats'
   "thing that the flounder eats"

This calcified red alga has flattened, pink-colored branches up to 15 cm long. It is common in beach

Figure 16.5. Seaweed necklace made by Raquel Hofer in 1975. The pendant is a piece of *Gigartina johnstonii* and the beads are *Eucheuma uncinatum*. The entire necklace is 60 cm in circumference, and the pendant is 9.5 cm long. *CMM.*

drift in spring. Both species of *Galaxaura* were apparently not well known by the Seri.

Galaxaura fastigiata Decaisne
hast iti coocp 'rock on what-grows'
   "what grows on rock"
nojóo ixpanáams 'spotted-sand-bass its-seaweed'
   "spotted sand bass's seaweed'
xepe oil 'sea tuft'
   "sea fan"
xpanáams ccapxl 'seaweed bitter'
   "bitter seaweed"

This calcified red alga has dichotomous, tubular, and purplish branches up to 10 cm in length. It occurs in beach drift in spring and grows on rocks in the low intertidal and subtidal zones. When bleached from the sun while drying in beach drift, it resembles the articulated coralline *Amphiroa van-bosseae*, and hence sometimes was called by the same name, *xpanáams ccapxl*. *Nojóo*, probably the spotted sand bass (*Paralabrax maculatofasciatus*), is a "small fish like *cabrillita*."

Gelidiopsis variabilis (Grev.) Schmitz   [Rhodophyta; Gracilariaceae]
tacj iha 'porpoise its-possessions'
   "porpoise's possessions"
xepe an impós 'sea in *Muhlenbergia*(or *Aristida-adscensionis*)'
   "small grass in the sea"

This red alga has dark (almost black) and wiry branches which are about 15 cm long and less than 1 mm in diameter.

Gigartina johnstonii Daws.
G. pectinata Daws.
   [Rhodophyta; Gigartinaceae]
tacj-anóosc 'porpoise its-roughness'
   "porpoise's papillae"

The name refers to the rough surface of the seaweed, resembling the texture and papillate nature of a tongue.

These red algae grow on rocks in the low intertidal zone and are common in beach drift in spring. The blades are reddish brown to dark greenish or reddish purple, up to 30 cm long, flat, and covered with numerous papillae, giving the appearance of a Turkish towel. The branches of *G. johnstonii* are up to 12 mm wide, and those of *G. pectinata* are about 6 mm wide.

Adornment: One necklace had a piece of *Gigartina* as a pendant (see Figure 16.5).

Animal Food: Green turtles were said to feed on it, particularly turtles from places other than the Infiernillo Channel.

Play: It was much used by children in play. They hung up pieces of the branches and pretended that the seaweed was fish or turtle meat. Girls used pieces of the plant as earrings or put it around their ears. They also used it as doll clothing.

Gracilaria textorii (Sur.) J. Ag. var. *textorii* [Rhodophyta; Gracilariaceae]
moosni yazj 'sea-turtle its-membrane'
   "sea turtle's membranes"

This red alga is common in spring and is generally found on rocks in the subtidal zone and growing on the carapaces of overwintering green turtles. The blades are membrane-like, up to 35 cm long, flat, branched in a sub-dichotomous pattern in one plane, and reddish in color.

Some people said this seaweed was a kind of *hast iti coteja* (*Cutleria*, etc.). It was said to grow up to about 30 cm tall on the carapaces of thin green turtles which were partially buried (dormant) during the winter in the Infiernillo Channel. It was stated that it did not grow on the carapaces of fat green turtles, even when they were buried. (It would seem that a turtle which had remained dormant long enough for seaweed to grow on it might indeed be thin.)

Gymnogongrus johnstonii (Setch. & Gardn.) Daws.   [Rhodophyta; Phyllophoraceae]
xepe an impós 'sea in *Muhlenbergia*(or *Aristida-adscensionis*)'
   "small grass in the sea"

The same name was applied to several other seaweeds. This red alga is especially abundant in spring. It occurs on rocks in the middle to lower intertidal zones. The plant is reddish brown, about 15 to 25 cm in size, densely branched, and with flattened branches each about 4 mm wide.

Animal Food: It was said that the green turtle ate it.

Halymenia—see *Cryptonemia*

Hypnea valentiae (Turn.) Mont.   [Rhodophyta; Hypneaceae]
xepe an impós 'sea in *Muhlenbergia*(or *Aristida-adscensionis*)'
  "small grass in the sea"
xepe an ihíms 'sea in its-fringe'
  "fringe of the sea"

This red alga was not well known by the Seri. It is brownish red, tangled but not densely branched, with branches up to 20 cm tall and 2 mm in diameter. It grows on rocks and can be epiphytic or entwined with other, larger seaweeds; sometimes it is entangled on gorgonians.

Laurencia johnstonii Setch. & Gardn.   [Rhodophyta; Rhodomelaceae]
xepe an ihíms 'sea in its-fringe'
  "fringe of the sea"
xepe an impós 'sea in *Muhlenbergia*(or *Aristida-adscensionis*)'
  "small grass in the sea"

This red alga reaches about 10 to 15 cm in height and is somewhat grass green in color (the green chlorophyll masks the red pigments). The slender branches (maximum diameter approximately 2 mm) are alternate and papillate and repeatedly branched from a main axis. It is common in spring on rocks in the intertidal zone and in the drift.

Lomentaria catenata Harv.   [Rhodophyta; Champiaceae]
xpanáams cquihöj 'seaweed red'
  "red seaweed"
xpanáams itojípz 'seaweed its-eyelashes'

This red alga occurs on rocks, mostly in the shallow subtidal zone. The branches are clumped, 8 to 14 cm tall, alternate to opposite, reddish, and terete.

Padina durvillaei Bory   [Phaeophyta; Dictyotaceae]
moosni ipnáil 'sea-turtle its-skirt'
  "sea turtle's skirt"
moosni yazj 'sea-turtle its-membrane'
  "sea turtle's mesenteries"

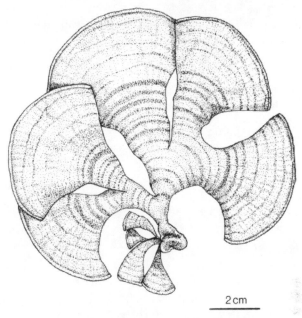

Figure 16.6. *Padina durvillaei*, a brown alga. *SM.*

"This is the common, coarse, fan-shaped brown alga along all the seaward reefs throughout the year (Dawson 1966:11)." Next to *Sargassum*, it is the most abundant seaweed in the Gulf of California. The blades are up to 25 cm wide, relatively thick and rigid, dark brownish, and with a heavy in-rolled margin (Figure 16.6). It was said to resemble the edge of the sea turtle's carapace, hence the name "sea turtle's skirt."

Plocamium cartilagineum (L.) Dixon   [Rhodophyta; Plocamiaceae]
xpanáams cquihöj 'seaweed red'
  "red seaweed"

This red alga grows up to 5 cm in height. It has a reddish, compressed axis. Branching is in a zig-zag manner and upper branchlets are pectinate.

Polysiphonia spp.   [Rhodophyta; Rhodomeliaceae]

These small red algae were known only as *xpanáams*, the general term for seaweeds. The branches are up to 5 cm long, flaccid, filamentous, and brownish red in color. These plants were often found on the carapaces of overwintering dormant green turtles. Two of these seaweeds, *P. paniculata* Montagne and *P. spherocarpa* var. *cheloniae* Hollenberg and J. Norris, are known only from the carapaces, heads, and flippers of overwintering green turtles taken in the vicinity of Tiburón Island.

Pterosiphonia dendroidea (Mont.) Falk.   [Rhodophyta; Rhodomelaceae]

This red alga has small, delicate, and reddish branches, 2 to 5 cm tall. It is usually found in the middle to low intertidal zone and in the drift. It was known only as *xpanáams.*

Rhodymenia divaricata Daws.   [Rhodophyta; Rhodymeniaceae]
moosni ipnáil 'sea-turtle its-skirt'
    "sea turtle's skirt"
ptcamn iha 'lobster its-possessions'
    "lobster's possessions"
xpanáams cheel 'seaweed red'
    "red seaweed"

This red alga is fleshy, dark red, firm, flat (in one plane), and dichotomously branched. The blades are 3 to 5 cm tall and only 2 to 3 mm wide, and do not have a stipe.

Rhodymenia hancockii Daws.
moosni iha 'sea-turtle its-possessions'
    "sea turtle's possessions"
xpanáams coozlil 'seaweed sticky'
    "sticky seaweed"
xpanáams cöquihméel 'seaweed purple'
xpanáams ool 'seaweed *Stenocereus-thurberi*'
    "organ pipe seaweed"

Although this red alga resembles the above species, its blades (up to 10 cm tall and 10 mm wide) have a distinct stripe.

Sargassum herporhizum Setch. & Gardn.
S. sinicola Setch. & Gardn.
    [Phaeophyta; Sargassaceae]
xpanáams 'sea in fringe'
    "fringe of the sea"
xpanáams isoj 'seaweed true(genuine)'
    "real seaweed"
sargaso, sargasso weed

Xpanáams and *xpanáams isoj* are synonyms. This seaweed was described as differing from other kinds of *xpanáams* (seaweeds) by being longer and taller, having thicker stems, wider leaves, and large "fruit balls." It was also described as being brownish, with brown "fruit" or "berries." It was said to be uncommon in the Infiernillo Channel, but common on the north and west coasts of Tiburón Island and north from there along the mainland coast. It was considered by some as "the dangerous seaweed" because people have been known to drown after being caught in it and also because the bodies of people who drowned have been cast ashore entangled in *Sargassum.*

Sargassum is the "ethno-type" or "model" of *xpanáams,* or seaweed. The term *xpanáams* is both a specific name for *Sargassum* and the general term for any seaweed.

The most abundant seaweed in the Gulf of California, *Sargassum* is most luxuriant and largest during spring, when the water is relatively cool. The entire plant is golden brown. It has long, cylindrical main axes and flat, leaf-like blades, with floating vesicles and globose fruiting bodies (Figure 16.7). The vesicles were called *tacj opjoj* 'porpoise its-blowing,' or "blown up (filled) by the porpoise."

S. herporhizum reaches 80 cm in length and *S. sinicola* is up to 1 m in length. A similar species, *S. johnstonii* Setch. and Gardn. has branches up to 80 cm in length and is undoubtedly included in the same ethnotaxon.

Animal Food: The triggerfish was said to feed extensively on this seaweed. Sea turtles, however, do not eat it.

Boats: Men used it to cushion their knees while paddling a balsa.

Play: Girls played with sargasso weed dolls (Figure 16.8). This custom was still common in the early 1980s.

Shelter: The plant was used as roofing for brush houses.

Sargassum sinicola Setch. & Gardn.
xpanáams caacöl 'seaweed large'
    "large seaweed"
xpanáams oaf 'seaweed waist-cord'

The two Seri names are apparently synonyms. This plant was said to be "the same as *xpanáams* [*Sargassum* spp., above], but with thicker stems, wider and longer leaves, and the plant is longer and taller." However, some people only recognized one kind of *xpanáams* (*Sargassum*).

Figure 16.7. Two species of sargasso weed. A) *Sargassum herporhizum*. B) *S. sinicola. AHH, 1975.*

Figure 16.8. Seaweed doll made of sargasso weed (*Sargassum* sp.) wrapped with strips of cloth. The cap is a piece of *Colpomenia*. This doll, 35 cm long, was made by Sara Villalobos in spring, 1974. *CMM.*

Spyridia filamentosa (Wulf.) Harv. [Rhodophyta; Ceraminaceae]
haxoj ano ihímz 'shoreline in its-fringes'
    "fringes of the shoreline"
xepe an ihíms 'sea in its-fringe'
    "fringe of the sea"
xepe an impós 'sea in *Muhlenbergia*(or *Aristida-adscensionis*)'
    "small grass in the sea"

This very common red alga is found throughout the year, but is most abundant from late spring to early fall. It grows up to 10 cm in height and has filamentous, red to pink branches, which turn pinkish white when cast ashore. It seems to be new to the Gulf of California since the 1960s (there are no records of it there prior to the 1970s).

The following seaweed names and information have not been keyed to identifiable specimens:

xpanáams hasít 'seaweed earring'
Animal Food: It was said that the green turtle ate it.
Play: Children played with this seaweed by hanging it over the ear as an earring.

xpanáams mojépe 'seaweed sahuaro'
"sahuaro seaweed"

xpanáams xaasj 'seaweed *Pachycereus*'
"*cardón* seaweed"

xpanéezj 'sea membrane'

This term was sometimes used in a distributive or general way for a number of seaweeds not otherwise distinguished by name. However, as with *xpanáams* (*Sargassum*), it also seems to be the specific name for a certain seaweed, but one which we have not been able to identify. However, it may be *Enteromorpha acanthophora*.

Xpanéezj occurs as floating sections of thin, leaf-like material, gelatin-like, and orangish in color. It dries in large sheets along the beach. It was said to be like *moosni ipnáil* (*Padina* or *Cryptonemia*, etc.) but thinner. It was known to occur primarily north of Tiburón Island and, to a much lesser extent, in the Infiernillo Channel.

Shelter: Dry *xpanéezj* was used as roofing for the brush house.

The Supernatural: A stone covered with *xpanéezj* was used to make the wind stop. It was collected at low tide, usually during a windy month, and a fire built on top of it to make the wind cease blowing.

Other Uses: When a person was on the run or fleeing from an enemy and had no other cover, dry *xpanéezj* might be gathered from the beach drift and used as a blanket.

xpanéezj cheel 'sea-membrane red'
"red sea membrane"

This seaweed was described as similar to *xpanéezj*, but reddish in color. It was said that sea turtles do not eat it, and that it occurs north of Tiburón Island. The same name was applied to *Rhodymenia divaricata*, which may be the same plant.

Non-vascular Land Plants

Fungi

Battarrea diguetii Pat. & Har.
Podaxis pistillaris (L.) Fr.
hant ootizx 'land what-it-peels-back'
"land's foreskin"

These species were not distinguished by the Seri. They knew of *hant ootizx* from the mainland and Tiburón Island.

These large, stalked puffballs, up to 30 cm in height, are common in the desert. They are not edible. *Battarrea* is worldwide, but most abundant in arid and semiarid regions. *Podaxis* (Figure 16.9), also worldwide, is generally confined to the warmer, more arid parts of the world.

Medicine: The dried, blackened mushroom was ground into powder and applied to cuts, sores, or burns. The efficacy might be as an antibiotic.

Ganoderma lucidum (W. Curt. ex. Fr.) Karst.
hehe iyas 'tree its-liver'
shelf fungus

The Seri knew of it growing on mesquite (*Prosopis*) and other trees. This is a wood-rotting fungus in the Polypore Family (Polyporaceae).

Tulostoma sp.
hacx cahóit 'somewhere what-causes-to-descend'
"what wastes it" or "what causes to be lost"

The name relates to the manner of spore release through the terminal pore, implying that the mushroom loses or "wastes" its dust (spores). It was distinguished from *hant ootizx* (*Battarrea* and *Podaxis*) by its small rounded cap with a terminal pore, as well as its overall smaller size. The Seri knew of it from the mainland and Tiburón Island. It is very common in the desert.

Medicine: The mushroom was crushed, and the dry, brownish powder (spores) spread on cuts, sores, or burns. It may have antibiotic properties.

Play: Children threw the mushrooms at each other, pretending that the one who was hit disappeared (wandered off) into the desert and got lost.

Algae, Lichens, and Mosses

Yamása is the general term for soil algae, lichens, and mosses. However, the term is used only with a modifying word describing the habitat, e.g., *hamt yamása* 'soil *yamása*,' *hast yamása* 'rock *yamása*,' and *iicj yamása* 'sand *yamása*.'

Cyanophyta, Blue-green algae
hamt ináil 'soil its-skin'
  "soil's skin"

These are the common bluish black algae which form a thin crust on the desert surface, especially on level coastal terrain (Figure 16.10). There are probably several genera, and they are probably not all blue-

greens. The Seri said that *hamt ináil* occurs where there is water, and is green when the soil is damp and black when dry. *Hamt ináil* and *hamt yamása* may not necessarily be different.

Rocella babingtonia Mont.
heecoj
orcilla

This rather large, gray, foliose lichen occurs as an epiphyte on desert wolfberry (*Lycium*) at Punta Sargento.

Medicine: Tea made from the plant was taken as a remedy for shortness of breath, and as a febrifuge. The plant was ground on a metate, squeezed through a cloth with a bit of water, and the resulting liquid placed on a burn or sore. The plant, ground with *casíime* (a kind of reddish clay) and water, was used as a febrifuge and to cure diarrhea. The ground plant, mixed in water, was used to bathe a child with a fever.

crustose and foliose lichens
hast yamása 'rock lichen'

Various lichens growing on rocks, including gray foliose and orange crustose species, were included in this ethnotaxon.

Medicine: Tea made from these lichens was taken as an emetic.

Figure 16.9. *Podaxis pistillaris*, a stalked puffball. *RSF, vicinity Campo Víboras, April 1983.*

Figure 16.10. Blue-green algae forming a thin crust on the soil at a shell midden in the vicinity of Punta Oona. *RSF, April 1973.*

Ferns and Fern Relatives

Notholaena standleyi Maxon
hehe quina 'plant with-hair'
  "hairy plant"
rock fern

A small desert fern, mostly among rocks on north-facing slopes, canyons, and higher elevations; mainland and Tiburón Island. The leaves remain tightly curled during rainless periods, giving the plant a dry, dead appearance. After a rain the fronds promptly unfold, presenting the appearance of a green, live fern. Even dry, dead leaves can unfold when dampened. The Seri reported that it occurs near the base of mountains.

Medicine: Tea made from the fronds was drunk by women to help bring on conception.

The Supernatural: Pieces of the fern were broken and stuffed into a small cloth bag and carried on the person to bring good luck in betting. One seeking good luck in gambling asked help of the plant. This bag was also sewn onto a cord and worn around the neck. It kept the wearer safe from a flood: "The flood water will not take you."

To bring rain one made a small cloth bag and filled it with *hehe quina* leaves. Dampening it with either fresh or sea water brought the rain.

Selaginella arizonica Maxon
hehe quina caacöl 'plant with-hair (*Notholaena*) large(plural)'
  "large rock fern"
spike moss

This moss-like plant forms dense mats on north- and east-facing slopes at higher elevations. We found it on Sierra Seri, and the Seri said it also occurs on Sierra Kunkaak on Tiburón Island. This plant was poorly known except by a few older people.

The plants usually appear dry, brown, and curled up. This aspect changes dramatically with rain, which transforms the ground into a green, mossy-looking carpet.

The Supernatural: Gathering the plant caused cloud formation.

17. Flowering Plants

Acanthaceae—Acanthus Family

Holographis virgata (Benth. & Hook.)
   T. F. Daniel subsp. *virgata*
   [= *Berginia v.* Benth. & Hook. var. *v.*]
hoinalca 'low hills'

A common small shrub, often forming low, dense mounds. It ranges through most of the Seri region and is often abundant on hills and mountains.

Justicia californica (Benth.) D. Gibs.
   [= *Beloperone c.* Benth.]
noj-oopis 'hummingbird what-it-sucks-out'
   "what hummingbirds suck out"
chuparosa

Noj is derived from *xeenoj*, a certain hummingbird, probably the black-chinned hummingbird. This common shrub occurs along dry washes and arroyos nearly throughout the Gulf Coast of Sonora, including Tiburón and San Esteban islands. The reddish orange flowers are pollinated by hummingbirds, as indicated by both the Spanish and Seri names. Various hummingbird-adapted flowers in Mexico are called *chuparosa*.

Ruellia californica (Rose) I.M. Johnst.
satóoml
rama parda

This small shrub has slender, brittle stems, and showy lavender flowers (Figure 17.1). It is common in the southern part of the Seri region and certain parts of Tiburón Island, especially on rocky bajadas and hills near the mountains.

R. peninsularis (Rose) I.M. Johnst. from the Baja California peninsula is similar and closely related to *R. californica*. Johnston (1924: 1171) states that "... *R. peninsularis* ... differs in having dull oily glandular-pubescent foliage and not glabrate foliage

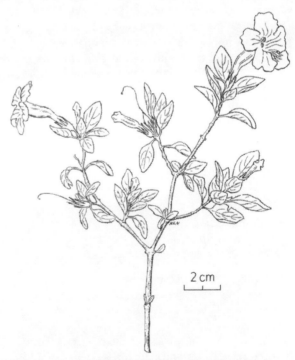

2 cm

Figure 17.1. *Rama parda* (*Ruellia californica*) from the mainland coast of Infiernillo Channel. *NLN, April 1974.*

which is glutinous and somewhat shiny." These characters might have an effect on the medicinal and smoking qualities of the plant.

Food: Nectar was sucked from the base of the corolla.

Medicine: Tea made from the leaves was taken for dizziness, and put on the face of one who was "tired out." The same liquid was used as shampoo to relieve a headache, and also as eyedrops. As another means to cure a headache, the leaves or leafy branches were put on moderately hot or warm coals; then the toasted leaves were put in an olla with warm water and the resulting tea was drunk. To counteract dizziness, fresh leaves were soaked in water, and the resulting yellowish water was used to wash the face. After the bark was removed, the root was boiled and the resulting tea was drunk to cure a cold or stuffy nose. Alfred F. Whiting (1957: #93) noted that it "is made into a poultice for headaches or the hair may be washed with it."

Smoking: Smoking the plant was said to "make one crazy" or cause hallucinations, although it was not as strong as *Datura*. It was smoked in a clay pipe. Some said that the leaves were smoked, others said the flowers were smoked. Rosa Flores broke the calyx and ovary off the flower and showed us that only the corolla was dried and smoked.

Agavaceae—Agave Family

Agave
maguey, century plant

Throughout the Seri region agaves are characteristically most numerous on upland rocky or hilly places, and some occur only at higher elevations. These populations tend to be densest on the less arid sites, such as north- and east-facing slopes, and absent from hot south- and west-facing slopes. Agaves played an important role in the indigenous cultures of Mexico and the southwestern United States (Castetter, Bell and Grove 1938; Felger and Moser 1970; Gentry 1942, 1972, 1978, 1982).

There are two subgenera: *Agave* (= *Euagave*) and *Littaea*. The species included in this region are listed below:

Agave
    A. *angustifolia*
    A. *cerulata*
    A. *colorata*
    A. *fortiflora*
    A. *subsimplex*
Littaea
    A. *chrysoglossa*
    A. *felgeri*
    A. *pelona*
    A. *schottii*

In the Gulf of California region subgenus *Agave* is characterized by a paniculate inflorescence (branched flowering stalk), flowers in clusters on lateral branches, and spiny leaf margins. *Littaea* in this region is distinguished by a racemose or spicate (unbranched) inflorescence and entire leaf margins (no marginal spines). Species in both groups bear a stout terminal leaf spine.

Certain species, particularly those in the subgenus *Agave*, provided a major food resource. Generally harvested in January and February at the end of the short winter, century plants provided an important sugar source. Eating this sweet vegetable was a joyful occasion. Youngsters especially awaited the event with eager anticipation.

Only plants showing a young emerging inflorescence were selected for harvesting. In the spring the large inflorescence develops rapidly and bears flowers in late spring. The Seri said that plants not showing signs of forming an inflorescence are bitter, whereas those that do are sweet. Obviously this high concentration of carbohydrate enables the plant to produce its relatively enormous inflorescence in such a short time. A plant that is going to flower produces progressively narrower and smaller leaves towards its center (Figure 17.2). The young emerging inflorescence, called *itöj*, usually less than 30 cm long, was sometimes cooked for food. The tall, mature inflorescence was called *icáp*. During years of severe drought we have observed substantially fewer plants producing inflorescences.

Methods of preparation are detailed for *A. subsimplex*. Agaves were still occasionally harvested in the early 1980s. The Seri reported that *A. schottii* is inedible. While the leaves of the other *Littaea* are bitter and not eaten, the hearts (meristematic tissue) of

Figure 17.2. The century plant (*Agave subsimplex*) on the left shows signs of initiating an inflorescence, as indicated by its narrowed and smaller central leaves. *RSF, 6 km northeast of El Desemboque, February 1969.*

the larger ones are edible. Both the leaf bases and hearts of plants of the subgenus *Agave* were eaten. Agaves must be thoroughly cooked in order to be edible, otherwise they are caustic and extremely dangerous to consume. Although agave flowers and nectar were consumed by other Indians (Bean and Saubel 1972:32–33; del Barco 1973:122; Gentry 1978:5), there is no indication that the Seri ate them. Agave hearts were roasted in a pit overnight or longer. Roasting time undoubtedly depended on plant size and quantity.

The Seri had a distinct name for each *Agave* species occurring in their region. Although various older people had not been in the Guaymas-Tastiota region for many years, if ever, they described plants which match species known to occur there. If the identity was certain, there was a one-to-one correspondence of Seri names to biological species. The Seri recognized the component species as a single group or folk genus. *Haamxö* (*A. subsimplex*), was sometimes used as a generic term for *Agave*. When asked—during a discussion of *A. subsimplex*—what other kinds of this plant were known to them, they listed the remaining species of agave occurring in the region. Most people also added *Hechtia* (Bromeliaceae) to this list.

The stars were created by *Hant Caai*, a principal Seri deity. The stars in the Milky Way are a kind of century plant which opens at night and gives off a light which is other than fire. It was said that a woman named Andrea learned about this in a dream.

Agave angustifolia Haw. var. *angustifolia*
[= *A. pacifica* Trel., *A. yaquiana* Trel.]
hamoc

The Seri distinguished this plant by its long, narrow leaves with marginal spines. They knew of it from a long north-south oriented mountain south of La Cienega, and at Cerros Los Mochos, north of El Desemboque.

Food: The heart was prepared in the same manner as for *A. subsimplex*.

Agave angustifolia Haw. var. *angustifolia*
coptoj

This century plant occurs north of Pozo Coyote, on the low hills several kilometers north of Los Mochos. The leaves are greenish, with small dark-colored marginal spines and a stout, dark terminal spine. It is stoloniferous. The plant resembles other populations of *A. angustifolia* from Sonora but with thicker, stouter, and greener leaves. It almost certainly is a northern population of the highly variable *A. angustifolia* as delimited by Gentry (1982: 559–563).

Food: The hearts and leaf bases were collected (Figure 17.3) and prepared in the same manner as for *A. subsimplex*.

Agave cerulata Trel. subsp. *dentiens* (Trel.) Gentry
heme
xica istj caitic 'things its-leaves soft'
   "soft-leaved thing"
maguey

The second, descriptive name given was a term apparently used by the San Esteban people.

This highly variable subspecies is endemic to San Esteban Island, and is the only agave occurring on that island. This medium-sized century plant is considerably larger and more massive than *A. subsimplex*, to which it is closely related. It forms extensive colonies on the coarse rocky slopes of San Esteban, and is one of the major landscape elements of the island (Figure 17.4).

The Seri correctly reported it does not occur on Tiburón Island. They also knew that some of the

plants are greenish, while others are distinctly gray in color (see Gentry 1982: 371).

Food: Unlike all other agaves in the Seri region, edible plants of this species could be found throughout the year. However, they are most flavorful during the latter part of January. The most savory ones, called *heme quims* "fringed *heme*," were described as being especially fibrous. A plant with a noticeably "oily" film on the leaves was said to be bitter and was not harvested.

To harvest the plant, the leaf tips were first cut off, so that one did not get stuck with the stout spines while working. Next, the developing inflorescence was cut out. Each leaf, starting with the inner ones, was then cut off near its base. The *hahéel* chisel–pry bar was used to sever the crown of the plant from the root.

It was cooked in much the same manner as for *A. subsimplex*. Seri terminology associated with the cooking of *A. subsimplex* was also applied to *A. cerulata*, as well as to other edible species. The hearts, cooked and mixed with sea turtle fat, were said to have a coconut-like flavor.

The San Esteban people dug huge pits "for cooking a hundred *heme* hearts at one time." Each agave heart in the pit carried the identifying mark of its owner. In 1983 there were still several large agave roasting pits on the island.

After the San Esteban people became extinct, other Seri went to the island from time to time to harvest *heme* (Quinn and Quinn 1956: 127 and 178). These harvest trips were made up until at least the 1940s. The first person to sight a returning boat would call out, "Here come the Mountain Travelers!" (San Esteban is mountainous), whereupon pandemonium would break loose. As their boat neared the beach, the men would stop paddling and begin throwing cooked *heme* hearts into the eager outstretched hands. Sometimes a spiny-tailed iguana called *heepni* (*Ctenosaura hemilopha*) was tossed ashore with the century plants, causing much screaming and laughter (although an iguana would be thrown as a joke, both iguanas and chuckwallas were brought from San Esteban as food).

Hair Care: Fibers from the leaves, tied in bunches, were used as hairbrushes.

Survival Water: An emergency source of potable liquid occasionally was obtained from this century plant. The margins and tips of each leaf were trimmed away, leaving a rectangular piece called *heme istj comtax* '*heme* its-leaf straight(plural).' These pieces were

Figure 17.3. Rosa Flores harvesting *Agave angustifolia*. RSF, vicinity Cerro Los Mochos, February 1968.

Figure 17.4. The San Esteban century plant (*Agave cerulata* subsp. *dentiens*). *RJH, San Esteban Island, 1980.*

roasted over a fire until the outsides were charred. The blackened exteriors were scraped away, and the remaining pieces cut up and placed in a sea turtle carapace. They were then pounded to extract a sweet liquid said to resemble pineapple juice. It was used only as a severe emergency measure.

Chico Romero and others told us about an incident concerning a small group of Seri who survived four or five days of summer heat by drinking agave juice prepared in the manner described. This event occurred at about the turn of the century, when Chico was a child. He told of five men and a boy who, arriving at San Esteban Island, left their wooden boat on the beach and went inland to harvest *heme*. Mexican fishermen happened along and set their boat adrift. Marooned on the island, they drank agave juice to stay alive while they constructed a raftlike boat or balsa from bundles of agave flower stalks. Fearing that the young boy, Juan Marcos (deceased ca. 1919), might not be able to survive much longer on the agave juice, the men sent him along with Manuel Encinas (deceased ca. 1931), who paddled the makeshift balsa to the opposite south shore of Tiburón Island for help. They landed at *Hant Copni* 'land carpenter-bee,' near *Cyajoj* (Arroyo Sauzal at the coast). Upon their arrival there, a boat was sent to rescue the stranded men.

Agaves were a common source of emergency liquid for the people living in the Central Desert of Baja California (Aschmann 1959:60).

Wine: Cooked leaves were cut into pieces and placed in a sea turtle carapace. They were then pounded with a rock to extract the juice. After standing for several days, the juice fermented. Warm water was added, and the mixture was ready for drinking. In 1922 Davis reported that the Seri went to San Esteban to gather *mescal* "to make liquor" to use at fiestas (Quinn and Quinn 1965:172).

Agave chrysoglossa I.M. Johnst.
hasot
   "narrow"
amole

The name *hasot* seems to be related to the word *coosot* 'narrow.' This century plant occurs in the Sierra Seri and Sierra Kunkaak. It is commonest at higher elevations. It is also on San Pedro Nolasco Island and the mountains north of Guaymas. The northern populations are probably a different subspecies from those in the Guaymas region. Plants in the northern populations have greener leaves, with prominent white margins and are profusely suckering

223

or stoloniferous. The leaves are about 30 to 40 cm long.

Food: The heart, or center, of the plant is edible although bitter. It was prepared by cutting off the leaves, including the leaf bases, and heating the heart thoroughly in an open fire. It was then pierced with a knife in different places and pit-baked in the same manner as for *Agave subsimplex*. Much of the bitter juice drained off during the cooking process. It was eaten by the people who lived in the interior of Tiburón Island.

It was said that it tasted bitter to a dull-minded or ignorant person, but that to an intelligent person it was sweet.

Agave colorata Gentry
A. fortiflora Gentry
haamxö caacöl 'Agave-subsimplex
   large(plural)'
   "large agave"
maguey

This century plant is distinguished by leaves 50 to 100 cm long and 10 to 12 cm wide. It occurs in the mountains from the vicinity of Guaymas northward to within 10 to 15 km of Tastiota.

Haamxö caacöl is the largest century plant known to the Seri. The Guaymas-Tastiota populations are *A. colorata* (Gentry 1982:431), and not *A. fortiflora*, as reported earlier (Felger and Moser 1976, Gentry 1972). Some Seri said *haamxö caacöl* also occurs in the mountains north of La Cienega, an area which corresponds with the known distribution for *A. fortiflora* (Gentry 1982:441). *A. fortiflora* superficially resembles *A. colorata* from the Guaymas region.

Food: The method of preparation was similar to that for *A. cerulata*. The large size of the plant must have made it an important food resource. However, it does not occur in as great a density as does *A. cerulata* on San Esteban Island.

Wine: The people of the Tastiota Region made wine from the cooked hearts. They were crushed with a grinding stone on a solid base, such as a depression in the ground lined with sea shells. Water was added and the mixture allowed to stand. The wine was periodically tasted until the sweetness was gone and it was slightly bitter. If not consumed it rapidly turned into vinegar.

In another method, water was not added after the cooked agave hearts were crushed. This wine remained sweet. It was probably comparable to the cactus fruit wine made without water and stronger than wine made with water (see discussion on wine under Columnar Cacti).

Agave felgeri Gentry

This small agave occurs in the southern part of the Guaymas-Tastiota Region, from San Pedro Bay south to Guaymas, and in the hills northwest of Hermosillo. It is superficially similar to *A. schottii* (Gentry 1982:109), and was probably used in a similar manner.

Agave pelona Gentry
inyéeno 'faceless'

This agave was said to occur north of Pozo Coyote and on Sierra Kunkaak. Some said it also occurs on Cerro Tepopa. It was described as having leaves shorter than those of *A. cerulata*, a very long pointed terminal spine and no marginal spines. However, information about this plant was not consistent. The very long terminal spine, entire leaf, and the locality "north of Pozo Coyote" indicates it is probably *A. pelona* (see Gentry 1982:169).

Food: The cooked heart and leaf bases were eaten.

Agave schottii Engelm.
icapánim 'what one washes hair with'
amolillo, amole

The Seri pointed out that this plant differs from *A. subsimplex* as follows: it lacks spines along the margin of the leaf; it has a long spine at the end of the leaf, narrow leaves like fingers, flower stalks about 2 m tall, small whitish flowers at the top of the flowering stalk similar in shape and size to those of ocotillo; and it is not edible.

It occurs northward and inland from El Desemboque and Puerto Libertad, where it is at its southern limit. The leaves are about 30 to 40 cm long and 1 cm in diameter. It superficially resembles *A. felgeri*.

This and related species have been widely used by Indian and Spanish-speaking peoples as a source of soap for washing clothes (Castetter, Bell, and Grove 1938:75). Rosa Flores said the plant was used by the Papago for soap.

Hair Care: The leaves were crushed in a container and water added. The resulting foamy mixture was considered excellent for washing hair. It was said to soften the hair, as well as make it grow long.

Weapons: The people of the Libertad Region, from near Puerto Lobos to El Desemboque, used the slender flowering stalk to make arrow shafts, since reedgrass (*Phragmites*) was scarce in their region.

Agave subsimplex Trel.
haamxö
maguey, century plant

This species is widespread and common along the coast, from the mountains near Puerto Libertad southward nearly to Kino Bay, and on the islands of Tiburón, Turners (Dátil), and Cholludo (a tiny island between Tiburón and Turners).

This is a relatively small century plant, and the leaves and spines are highly variable in size. The leaves are about 15 to 35 cm long and 3 to 6 cm wide. It is closely related to *A. cerulata* and *A. deserti* (Gentry 1982:405). Peak flowering is in late spring, usually May or early June (see *Ferocactus covillei*).

Plants of this species and two kinds of prickly-pear (*Opuntia phaeacantha* and *O. violacea*) were said to have been planted near the base of Punta Sargento many years ago by people of the Sargento Region. Isolated populations of these species occurring there in the 1980s may have been derived from these transplants. The Seri reported that the flowering stalks grow tallest at Cerro Tepopa.

Adornment: The flat, black seeds and flower buds were occasionally strung for necklaces (Figure 17.5).

Facepaint: A brown juice, which oozes from the base of the cooked heart, was used as facepaint by older people.

Food: Plants initiating an inflorescence were harvested in January and February (Figure 17.6). The plants were cut off at about ground level with the agave chisel—pry bar *hahéel*, made of catclaw (*Acacia greggii*), palo blanco (*Acacia willardiana*), palo verde (*Cercidium floridum* and *C. microphyllum*), *Colubrina*, or green ironwood (*Olneya*). The agave chisel—pry bar, hewn to form a cutting edge at one end, was pounded with a rock to sever the crown of the plant from the root. If the cutting edge became dull, it could be used to pry or otherwise uproot the plant.

The leaves were trimmed close with a knife, leaving a whitish core, or heart, with its broad white leaf stubs, or bases, firmly attached. This was the major edible portion of the plant. The green part of the leaf was discarded because it was bitter. A strip of leaf fibers was left attached to each heart. The strips from

two hearts were tied together to form a convenient handle for carrying them back to camp. One who harvests *haamxö* was called *coomxö* "he who goes for *haamxö*."

The method of cooking was similar for the various agaves throughout much of Mexico and the southwestern part of the United States (see Castetter, Bell, and Grove 1938; Gentry 1978). A deep pit was dug in the ground and a fire built in it. When it had burned down, the agave hearts were piled, top down, on the bed of coals. Flat rocks were placed over them and the rocks covered with earth to a depth of about 5 to 10 cm. Another fire was then built on top of that, the resulting bed of coals covered with more earth, and the hearts left to bake overnight.

On the following day the agave hearts or cores were removed from the pit. They were blackened from contact with the coals. Beneath the charred surface the pulp was brownish in color and rather firm in texture. These cooked agave hearts, called *haamxö ipxási* 'haamxö its-flesh,' were sweet and juicy.

Cooked agave was eaten in several ways. Flat cakes or patties, called *haamxö yapol* 'haamxö its-blackness,' were made from outer slices of the blackened heart. When dry, these cakes could be stored al-

Figure 17.5. Necklace of century plant (*Agave subsimplex*) seeds made by Andrea Romero in 1957. The entire necklace is 41 cm in circumference; the individual seeds are about 4 mm wide. *CMM*.

Figure 17.6. Ramona Casanova harvesting century plants (*Agave subsimplex*). A) Severing a plant with an *hahéel* chisel–pry bar. B) The severed plant. Note the pointed cutting edge of the chisel–pry bar. C) Trimming away the leaves. D) The cores with leaves trimmed. *RSF, 6 km northeast of El Desemboque, February 1969.*

most indefinitely. One simply dissolved the cakes in water and drank all but the pieces of charred pulp.

The inner portion of the agave heart was usually cut up and eaten with pieces of sea turtle fat (cooked or uncooked). Often it was dipped in turtle oil. Slices called *haamxö hacáscax* 'haamxö which-have-been-sliced' were sun dried. They could be kept almost indefinitely. It was said that agave "meat" made the

children fat and healthy and that it noticeably lightened the color of their skin.

Cakes made from the cooked leaf bases were called *hapátlc* 'what was pounded.' The leaf bases were pulled off the core (Figure 17.7). While still juicy, the leaf bases were pounded and shredded on a metate, and the resulting pulp was formed into cakes about 15 to 20 cm wide and several centimeters thick (Fig-

226

Figure 17.7. Ramona Casanova peeling away the leaf bases from the roasted heart of a century plant (*Agave subsimplex*). *RSF, El Desemboque, February 1969.*

Figure 17.8. Ramona Casanova shaping a cake with pieces of pounded agave (*Agave subsimplex*) leaf bases. These cakes were dried in the sun and stored for future use. *RSF, El Desemboque, February 1969.*

ure 17.8). When dried in the sun, they could be stored for a long time. These cakes were carried for food on long trips.

The cakes were prepared and eaten several ways. One way was to soften them in water. The cake released a flour, called *hapácösj* "what is shaken out," which was drunk with the water. A cake moistened with water was eaten by chewing and sucking out the sweet juice. The fibrous mass, still damp, was shaken over a basket to remove the remaining edible material which was often added to fish or sea turtle stew.

The cooked leaf bases were also torn into strips and chewed and sucked for their juice. The pieces of chewed leaves were opened up and spread out to dry. When dry they were shaken to remove the remaining finely powdered material, or flour. These dried strips of leaf base, as well as the flour shaken out of them, were also called *hapácösj*. The flour was eaten, and the fibers discarded.

Food Gathering: The inflorescence stalk, *icáp*, was sometimes used as a fruit-gathering pole for the large columnar cacti, *cardón* (*Pachycereus*) and sahuaro (*Carnegiea*).

Wine: It was apparently not used for wine making, at least not in the twentieth century.

Yucca arizonica McKelvy
hamat
dátil, Arizona yucca or banana yucca

This semi-arborescent yucca occurs in the inland desert about midway between Pitiquito and Puerto Libertad (Gentry 1972:166). The large flowers are borne on a branched flower stalk in April. The fruit, which develops in early summer, is fleshy and somewhat resembles a short, fat, green banana. When dry, the seeds rattle if shaken.

The older people stated that it was found "beyond Pozo Coyote." They said the fruit (on the flower stalk) looks like mullet strung on a stringer and that when dry it "sounds" or rattles. The young emerging flower stalk, or inflorescence, was called *iïöj*; after elongating it was called *icáp*; these terms were also used for agaves. The Seri did not consider *hamat* to be a kind of *haamxö*, or century plant.

Food: The ripe fruit was eaten fresh. The Seri did not know of any other use for this plant.

Aizoaceae—Carpetweed Family

Mollugo cerviana (L.) Ser.
hant iit 'land its-lice'
 "land's lice"
carpetweed

This diminutive summer ephemeral is widespread in the region. It is naturalized from the Old World.

Sesuvium verrucosum Raf.
spitj ctamcö 'Atriplex-barclayana males'
 "male coastal saltbush"
spitj caacöl 'Atriplex-barclayana large(plural)'
 "large coastal saltbush"
sea purslane

In this case "large" refers to many plants covering a large area of ground. The Seri distinguished *Sesuvium* by its narrow leaves and *Abronia maritima* (female *spitj*) by its rounder (broader) leaves. They have similar life forms and occupy similar niches, and both were considered to be "soft" plants.

This trailing perennial herb is glabrous, and has succulent leaves and stems (Figure 17.9). It forms low, spreading mats along the margins of mangroves and on upper beaches.

Food Preparation: A sea turtle carapace or basket was lined with *Sesuvium* or some other "soft" plant to provide a bed on which to place meat to keep it clean.

Trianthema portulacastrum L.
com-aacöl 'com-large(plural)'
verdolaga de cochi, horse purslane

This hot-weather ephemeral is common and widespread in the region.
Food: The seeds were ground and cooked as a gruel.

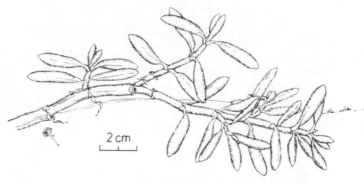

Figure 17.9. Sea purslane (*Sesuvium verrucosum*). FR.

Amaranthaceae—Amaranth Family

Amaranthus fimbriatus (Torr.) Benth.
ziim caitic 'ziim soft'
 "soft *ziim*"
bledo, quelite, fringed amaranth

The plant has relatively flexible herbage and thin, delicate leaves; the bracts are not prickly (Figure 17.10). These characters are indicated by its Seri name. This common hot-weather ephemeral occurs on the islands and throughout the mainland region. While it generally reaches maximum development along washes and arroyos, it may be seasonally common on open desert and rocky slopes. The seeds are dark brown or black, lens-shaped, and less than 1.0 mm in diameter.

There are three other kinds of *ziim: A. watsonii, Chenopodium* (Chenopodiaceae), and *Salsola* (Chenopodiaceae).

Food: The seeds were prepared in the same manner as for *A. watsonii.* Since *A. fimbriatus* is not as abundant as *A. watsonii*, we assume that it was of less importance. The seeds were stored in pottery ollas.

The leafy green shoots, when tender and young, were prepared as greens. The "leaves" (shoots or herbage) were cooked in water, and then the water squeezed out by hand, a handful at a time. These greens were sometimes cooked "in a bit of sea turtle oil" or mixed with honey.

Amaranthus watsonii Standl.
ziim quicös 'ziim prickly'
 "prickly *ziim*"
bledo, quelite, careless weed, pigweed, Watson amaranth

This hot-weather ephemeral commonly occurs in great abundance. In the relatively warm climate of

2 cm

the Gulf of California region, it often responds to both summer-fall and winter-spring rains. The plants are highly variable in size, depending on soil moisture and temperature.

The bracts are prickly and the stems stiff and coarse, as indicated by the Seri name. There are three kinds of *ziim* (see *A. fimbriatus*).

The fruit, like that of *A. fimbriatus*, is a circumscissle utricle, dry when mature, with a cap-like lid that falls away to free the single seed. The seeds are lens-shaped, dark brown or black, and about 1.0 mm in diameter.

Food: The seeds were an important food resource and were often stored in pottery ollas. It was said that the three kinds of *ziim* (excluding *Salsola*) produce many seeds, and "it is like eating the seeds of the columnar cacti." Apparently this species was seldom, if ever, prepared as greens.

The branches, held over a deer skin or cloth, were gently rolled in the hands, causing the seeds and flowers to fall. Winnowing separated the seeds from the dry flowers and other chaff. The seeds were toasted, ground into flour, and prepared as a gruel. The flour or gruel was often mixed with turtle oil. Children thought it was fun to stand near the seeds being toasted and anticipate their popping. *Ziim* seeds were said to have been a favorite food of the Giants.

Tidestromia lanuginosa (Nutt.) Standl.
halít an caascl 'head on what-causes-dandruff'
  "causes dandruff"
hierba ceniza

This summer-fall ephemeral is common through most of the Seri region. The plant has a silvery, speckled look, due to dense pubescence of branched hairs, and thus was said to cause dandruff.

Medicine: To relieve aching feet, the herbage was heated and placed under the feet. An infusion of the leaves was used for drawing out a thorn. The twigs were cooked and used as a shampoo to cure a headache.

Figure 17.10. Fringed amaranth (*Amaranthus fimbriatus*). This is a relatively small plant; individuals two to three times larger are common. Note the narrow leaves, dense axillary flower clusters, and fringed sepals of the pistillate flower. *FR.*

Apocynaceae—Dogbane Family

Vallesia glabra (Cav.) Link var. *glabra*
tonóopa, tinóopa
huevito

Medium to large shrub, small shiny green leaves, and translucent-white fruit about 1 cm long. Relatively copious fruit crops are produced up to several times per year. It is usually restricted to major arroyos and floodplains, and occurs on Tiburón Island and the mainland as far north as Arroyo San Ignacio.

Food: The fruit was eaten fresh.

Games: Sticks used to roll the hoops in the women's game of *hehe hahójoz cmaam* (see *Prosopis*) were sometimes made of *Vallesia* wood.

Medicine: The leaves were burned or toasted until blackened and then ground into a powder, which was rubbed on a part of the body inflicted with a rash or severe itching. It was said to relieve itching and associated pain, and was used for measles and other maladies.

Music: Violin tuning pegs were made from the wood.

Aristolochiaceae—Birthwort Family

Arisotolochia watsonii Woot. & Standl. [= *A. brevipes* Benth. var. *acuminata* S. Wats., *A. porphyrophylla* Pfeif.]
hatáast an ihíih 'tooth in where-it-is'
"what gets between the teeth"
hierba del indio, Indian root

The name refers to the plant's being placed on the tooth when used medicinally.

The Seri knew that this is an uncommon plant growing in arroyo beds on Tiburón Island and the mainland. It is a root perennial, with a carrot-shaped brown root. The shoot is facultatively drought deciduous.

Medicine: A decoction of the herbage cooked in water was held in the mouth to cure a toothache. The dry root was heated in a fire and placed over a cavity in a tooth ("tooth with a hole").

Asclepiadaceae—Milkweed Family

Asclepias albicans S. Wats.
white-stem milkweed
A. subulata Decne.
reed-stem milkweed
najcáazjc
mata candelilla, yamate

The Seri did not distinguish between these closely related species. *A. albicans* generally occurs on more arid sites than does *A. subulata*. *A. subulata* is the common reed-stem milkweed in the lowlands of the mainland and on Tiburón Island; it is absent from San Esteban Island. *A. albicans* is infrequently encountered, and is usually found on exposed mountain slopes; however, it is common on San Esteban Island. *A. albicans* is taller than *A. subulata* and there are significant floral differences. The stems of both species are slender, erect, and essentially leafless when mature (Figure 17.11).

Adornment: The stems, with the bark removed, were cut into pieces approximately 1 cm in length. These pieces were stained blue or red, left natural, or toasted black in a pan filled with sand, sometimes with animal fat added to aid in the toasting. The beads were then strung for necklaces, often with *Olivella* shells (Figure 17.12). A special necklace pattern made with alternating black or dark brown and natural colored pieces was called *hee yaháaho* "jackrabbit's path." The alternating dark- and light-colored beads were said to be reminiscent of the pattern of droppings left by a jackrabbit along its trail. Necklaces made of either this plant or *Baccharis salicifolia* are pictured by McGee (1898: 172, Figure 12; 173, Figure 13).

Hunting: The San Esteban Island people were said to have used white-stem milkweed in hunting *coof* (chuckwalla, *Sauromalus varius*). A hunter took one of the long slender branches to a rocky place where

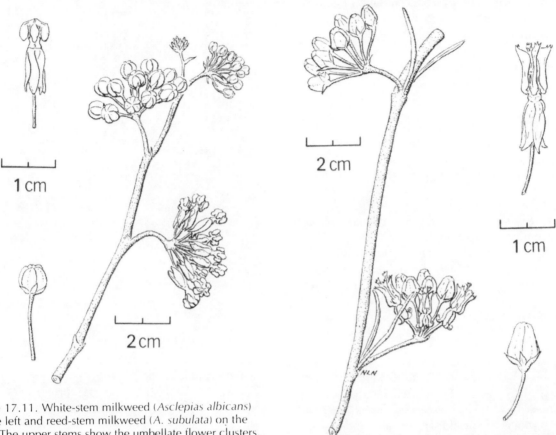

Figure 17.11. White-stem milkweed (*Asclepias albicans*) on the left and reed-stem milkweed (*A. subulata*) on the right. The upper stems show the umbellate flower clusters. Note the longer buds and elongated hood on the flower of *A. subulata*. *NLN.*

Figure 17.12. Necklace of *Olivella* shells, bleached white with the apical ends broken away, and reedstem milkweed (*Asclepias subulata*) beads toasted dark brown. Made in El Desemboque in 1957, it is 108 cm in circumference, and shows a pattern called "jackrabbit's road." *CMM.*

chuckwallas occurred, and whipped the branch through the air. The sound frightened the chuckwalla, which tried to hide in a different place. Hearing the large lizard move, the hunter dug it out. Piles of rocks were said to be evident where chuckwallas had been hunted on the island.

Medicine: Shampoo made by cooking the roots in water was used to cure a headache. The hair was then brushed to get rid of the headache.

To relieve a toothache, the root, together with the root of brittlebush (*Encelia*), and the entire plant of one of the small spurges (*Euphorbia polycarpa*/spp.)

were cooked in water, and the resulting liquid was held in the mouth. The liquid was also drunk as a remedy for heart pain.

Marsdenia edulis S. Wats.
xomée
talayote

Perennial vine, milky sap, often woody at the base, relatively large green leaves, and green ellipsoid fruit 8 to 12 cm long. Canyon and rocky slopes, generally in better-watered places on the mainland and Tiburón Island.

The Seri described it as follows: This plant has many green leaves which fill the tree upon which it climbs until the tree can't be seen. The pod-like fruit is longer than the fruit of *comot* (*Matelea cordifolia*) or *nas* (*Matelea pringlei*). It has more leaves than does *comot* and the fruit is thicker. The morphological comparisons with *Matalea* indicate a covert category in which generic relationships are recognized, although not named.

Food: The immature, tender fruit was eaten fresh. When older, it was cooked in ashes, peeled, and the inner white "skin" eaten. The seeds were not eaten.

Play: Years ago on Tiburón Island a person covered himself with dirt, except his arms and head. *Xomée* root was sprinkled over him, and this was said to attract the *naapxa* (a hawk), *col quiimet* (turkey vulture), and *hanaj* (raven). The person then grabbed the bird. Sara Villalobos told us, "Don't think that he would eat the bird! He released it and it flew away. This was just to play."

Poison: Pulverized bark from the root, mixed with drops of sap from *hierba de la flecha* (*Sapium*) and sprinkled on an enemy's food, was said to be fatal. This poison was used by the people of the Tastiota Region.

Matelea cordifolia (Gray) Woods.
comot

Perennial vine with ephemeral foliage; fruiting at various times of year but primarily during warmer months following rainy periods. Fruit elongate-ellipsoid, about 10 cm long (Figure 17.13). It is not common, and generally occurs along arroyos or washes.

The Seri distinguished this species from *nas* (*M. pringlei*) by the lack of spines on the fruit. Some people said it occurs on the "other" (east) side of the

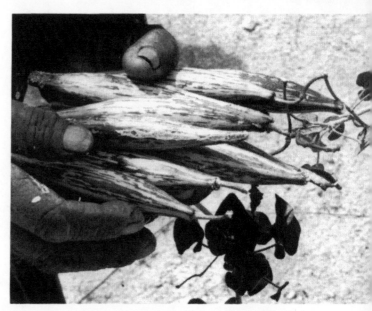

Figure 17.13. Fruit of *Matelea cordifolia*. RSF, Rancho Estrella, March 1983.

mountain called *Hast Eemla* in the Sierra Seri, and also knew of it by the road north of Campo Almond. They said it does not occur on Tiburón Island.

Food: The fruit was toasted in ashes. The entire fruit was eaten if young and tender; otherwise only the "skin" (ovary wall) was eaten.

Oral Tradition: An incident, humorous to the Seri, was told concerning this plant. Years ago a Mexican on horseback came upon a Seri and greeted him by saying, "*Como te va?* (How goes it with you?)." The Seri, not understanding Spanish, thought he heard reference to *comot*. Not wanting to be outdone, he answered by referring to the related plant *nas* (*M. pringlei*) and said, "*nas te va, ziix cooxo te va.*" With the words *ziix cooxo* 'everything,' the Seri was including all the plants with fruit similar to that of *comot*. This funny story was often retold and the greeting "*comot te va, nas te va*" was used in a joking manner.

Matelea pringlei (Gray) Woods.
nas
ziix is quicös 'thing its-fruit prickly' "prickly-fruited thing"

The Seri distinguished this plant from *comot* (*M. cordifolia*) by its shorter and thicker fruit. The fruit has short spine-like prickles, as indicated by the Seri name. It is an infrequently encountered vining or semi-vining perennial. It is generally found in better-watered or protected sites, such as arroyos and canyons, although it occasionally occurs in the open desert. Fruiting usually occurs in warm weather following sufficient rainfall.

Food: The fruit was cooked in ashes. The entire fruit was eaten if young and tender, otherwise only the "skin" (ovary wall) was eaten.

Play: When the people were relaxing in the desert, sitting around a fire, one diversion for the children was to make zigzag trails in the soil with their fingertips, and then fill the trails with "fuzz" (the silky appendages on the seeds) from the ripe fruit. One end was ignited with a burning ember and the fire burned through the zigzag trail.

Sarcostemma cynanchoides Decne. subsp.
  hartwegii (Vail) R. Holm.
hexe
huirote

This is one of the few common vines in the Seri region. It is often encountered sprawling over shrubs in floodplains, arroyos, and canyons. Flowers (Figure 17.14) may be produced at various times of the year.

Food: The flowers were picked and eaten fresh,

Figure 17.14. Flowers of *Sarcostemma cynanchoides*. RSF, Arroyo San Ignacio, March 1983.

often as a snack while walking through the desert. The flavor is faintly onion-like.

Medicine: To cure a severe headache, the head was washed in a decoction of the branches and leaves. Tea made by brewing stems and leaves was taken for the bite of a black widow spider. Eye drops were made by cooking the roots in water.

Bataceae—Saltwort Family

Batis maritima L.
xpaxóocsim, paxóocsim, xpacóocsim 'xpa-chew and spit out'
saltwort

Alfred F. Whiting (1957: #112) recorded the name as *paxho:ksim*. Succulent perennial, forming extensive colonies on tide flats and in mangrove esteros throughout the Gulf region (Figure 17.15).

Food: The roots were utilized for their sweetness. A large quantity of roots was crushed and cooked in water to make a beverage. The roots were also chewed and sucked for the juice. The root was peeled and used to sweeten coffee. Whiting (1957: #112) noted that the "roots are sweet and are sometimes chewed."

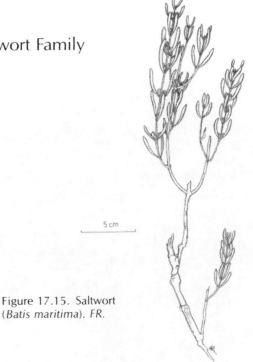

Figure 17.15. Saltwort (*Batis maritima*). FR.

Boraginaceae—Borage Family

Cordia parvifolia A. DC.
hehet ináil coopl 'plants their-skin black(plural)'
  "black-barked plants"
naz
vara prieta

Naz is probably the older name; the descriptive name refers to the dark-colored bark. Very common desert shrub, slender woody branches, relatively hard and flexible wood, and showy white flowers. Often associated with creosotebush (*Larrea*) on the mainland and Tiburón Island.

Food Gathering: *Icóopl* 'to string fruit,' the carrying stick for organ pipe (*Stenocereus thurberi*) fruit commonly was fashioned from a woody branch of this shrub. The cactus fruits were slipped onto the stick one by one like large beads. When carried in this manner they were called *hapápl* 'what are strung.' Men usually used carrying sticks for fruit, since they did not carry baskets or their modern counterparts (plastic buckets, etc.) when walking in the desert.

Toy: Boys made toy bows from the wood.

Cryptantha angustifolia (Torr.) Greene
C. maritima (Greene) Greene
Cryptantha spp.
hehe czatx 'plant stickery'
  "stickery plant"
hehe cotópl 'plant that-clings'
la peluda

Although not distinguished by name, the Seri pointed out that *C. maritima* was a larger plant than *C. angustifolia*. These winter-spring ephemerals are common and widespread in the region, occurring on the mainland, and Tiburón and San Esteban islands. The plants are covered with sharp, unpleasant spine-like hairs called *zatx* (Figure 17.16).

Heliotropium curassavicum L.
potács camoz 'future-*Allenrolfea* thinker'
  "what thinks it's an iodine bush"
hant otópl 'land what-it-sticks-to'
  "what the land sticks to"
hierba del torojo, alkali heliotrope

The name *potács camoz* signifies the plant "thinks" that perhaps it is *tacs* (*Allenrolfea*). Although these genera are quite different in gross morphology, both are perennial succulents and commonly occur on tidal flats and estero margins. Alkali heliotrope is also found on saline soils near water holes and as a farm weed (Figure 17.17).

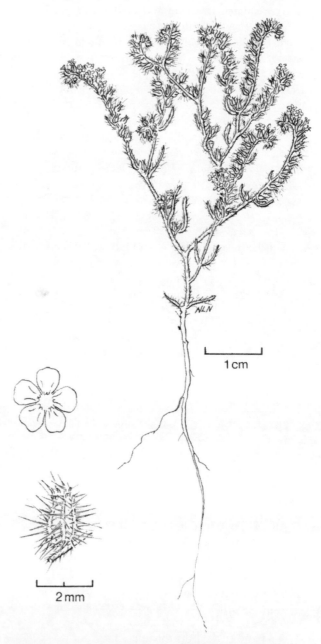

Figure 17.16. *Cryptantha angustifolia,* a common spring ephemeral. Note the glassy, spine-like hairs. *NLN, Infiernillo coast, April 1974.*

Tiquilia palmeri (Gray) A. Richardson
   [= *Coldenia p.* Gray]
hee xoját 'jackrabbit *Amoreuxia*'
   "jackrabbit *saiya*"

This plant was said to be the *xoját* (*Amoreuxia*) of the jackrabbit. It generally grows in sandy, gravelly soils and has long, thick, black roots.

Medicine: Tea made from the thickened portion of the root was taken as a remedy for stomachache or a cold.

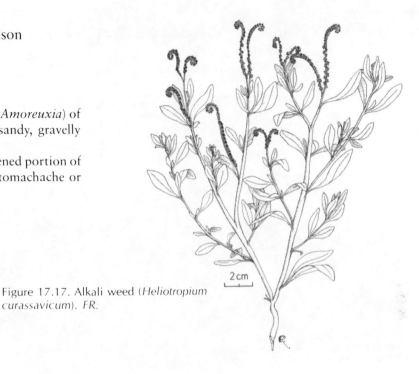

Figure 17.17. Alkali weed (*Heliotropium curassavicum*). FR.

Bromeliaceae—Pineapple Family

*Ananas comosus (L.) Merr.
haamxöii 'agave first'
   "trimmed agave" (see *Hechtia*)
piña, pineapple

Hechtia montana Brandegee
haamxöii 'agave first'
   "trimmed agave"

The name implies a century plant with the leaves already trimmed. It derives from the name for *Agave subsimplex* and the term for 'first' or 'already.' This plant was described as follows: Like a pineapple, with curled or twisted leaves and hooked spines on the edges (margin) of the leaves; found in the Tastiota region among the *haamxö* (*Agave*). The Tastiota region was often taken to include territory south of Tastiota, even as far as the vicinity of Guaymas. Also, in this case *haamxö* was used in a generic sense to include *Agave colorata*. This description was given by people living in El Desemboque who knew of *haamxöii* through oral history.

Hechtia occurs on the coast of Sonora from San Carlos Bay northward to San Pedro Bay. When we were collecting information about the various kinds of century plants (*Agave*), the Seri said that *haamxöii* was another of "that kind" of plant, indicating a covert or unnamed category for agave and agave-like plants.

Burseraceae—Frankincense Family

Bursera hindsiana (Benth.) Engler
xoop inl '*Bursera-microphylla* its-fingers'
   "elephant tree's fingers"
copalquín, torote prieto, red elephant tree

This large shrub or small tree has a thick woody trunk (Figure 17.18). It is readily distinguished from *B. microphylla* by its smooth reddish gray bark, larger simple or tri-foliate (occasionally 5-foliate) leaves (Figure 17.19) and somewhat harder wood; also, it is generally not as abundant. The wood is excellent for carving because it is relatively soft but firm, and does not split upon drying. It was preferred over other soft woods (such as *Bursera microphylla*)

Figure 17.18. Red elephant tree (*Bursera hindsiana*). *RSF, 8 km east of Estero Sargento, April 1983.*

because it was finer grained. It is found throughout most of the region including Tiburón and San Esteban islands.

Containers: Most adults owned a personal carrying box, called *hehe zamij an iqui ihácalca* 'wood palm in with where-one-has-belongings,' or "box to put belongings in" (see Palmae). These small portable boxes (Figure 17.20) were used for storing personal items, such as facepaint, buttons, fishing line, tobacco, etc. These were probably twentieth-century artifacts, acquired after the Seri became more sedentary. Originally these boxes were made from the wood of *Bursera hindsiana*, although later examples were often made from commercial wood. The ones made from red elephant tree wood were held together by pieces of creosotebush (*Larrea*) wood used as nails. These were pounded into holes bored with a piece of metal.

Personal storage boxes made from *Bursera* wood have not been used by the Seri for their own use since about the 1950s. Since about the 1960s these boxes have been made for sale to outsiders. Boxes made for sale were often decorated with blue and reddish Seri

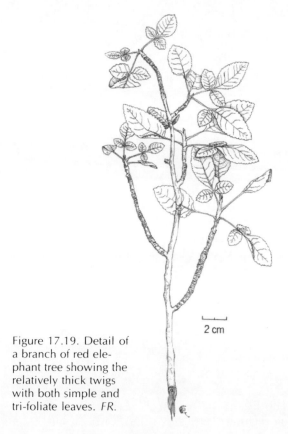

Figure 17.19. Detail of a branch of red elephant tree showing the relatively thick twigs with both simple and tri-foliate leaves. *FR.*

236

Figure 17.20. Red elephant tree (*Bursera hindsiana*) was used to make commercial carrying boxes. A) Jesús Morales bringing home a partially roughed out box. Note the large size of the wood. *MBM, El Desemboque, 1969.* B) The finished box, made by María Antonia Colosio. It has commercial nails and is 13.4 cm long and 9.7 cm wide. *CMM.*

designs, such as facepainting designs. When they were made for the Seri's own personal use, the boxes were not decorated or sometimes were painted a solid color.

Firemaking: A dry, dead branch attached to the living tree was sometimes called *xoop inl yapáain* 'xoop its-fingers what-caused-it-to-fall.' This special wood was a common material for firedrills.

Firewood: The dry wood was used for kindling and to keep a low fire going under a boiling pot.

Headpiece: The wood was used for the bird or knob affixed to the top of the *hehe hamásij* headdress (see *Jatropha cinerea*).

Medicine: Shavings from the wood were cooked in water with leafy branch tips of desert lavender (*Hyptis*), and the resulting tea was used to cure asthma or difficulty in breathing (*icayáax*).

Music: The box (body) of the one-string Seri violin was usually made of this wood. The wood was used also for the framework of a metal disk rattle called *ziix haquénla* 'thing sounded' (Figure 17.21) (Bowen and Moser 1970b: 187–188). It was played at fiestas, and to accompany lullabies, *icocóoxa*, which were sung only by grandparents and other relatives of their generation. As the instrument was shaken, the metal chips rattled, creating a sound much like that of a tambourine. It is similar to the *cenaso* instrument used by Yaqui and Mayo pascola dancers (see Densmore 1932: facing 27).

The Supernatural: According to Seri origin stories the first plant created was a *xoop inl* tree.

Fetishes were made from wood of this tree (Figures 17.22–17.24). The name for these fetishes is *icóocmolca* 'what (blame) is put on.' The name implies that it was the power of the fetish that caused good or evil to happen to a person; in other words, the shaman did not do the evil, the fetish did it. *Xoop inl* was used rather than *xoop* (*B. microphylla*) because the wood is finer grained.

These fetishes were carved only by a shaman and were rented or purchased from him as personal or household fetishes. Some were worn around the neck as amulets, while others hung from house rafters or

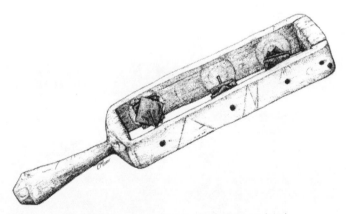

Figure 17.21. A disk rattle made of wood (*Bursera hindsiana*), nails, and flattened pieces of metal. This rattle, 29.8 cm long, was found on a dune in the vicinity of El Desemboque in 1958. *CMM.*

Figure 17.22. These fetishes, made of wood (*Bursera hindsiana*) painted with red and blue pigments, were made by Nacho Morales about 1960. The smiling (pointed) figure, 5.5 cm long, was found in the sand at El Desemboque in 1970 by Chapo Barnet and given to M. B. Moser. The discoid figure, 4.6 cm in diameter, was obtained from Nacho Morales at El Desemboque in 1969. *CMM.*

were used in vision quests and for shamanistic curing practices (Felger and Moser 1974b:418, Griffen 1959:51, Hardy 1829:294–295, Sheldon 1979:98, Xavier 1941).

There was a wide range of subjects, style, and personal meanings known only to the maker or owner. Some were for women, others for men, and some for both. Often abstract and of unique design, these small sculptures were a major aesthetic expression. Even a cursory examination of the few *icóocmolca* in museum collections and those still in use prior to about 1960 showed that wood sculpture is not a new artform among the Seri (see Chapter 14).

These fetishes undoubtedly represent a long tradition. During his visit to Tiburón Island, Hardy (1829:294–295) saw ". . . a wooden figure with a *carved hat*, and others of different shapes and sizes, as well also as leathern bags, the contents of which I was not permitted to explore."

Figure 17.23. These fetishes, made of red elephant tree wood, were collected in 1932 by Dane Coolidge and Mary Roberts Coolidge on Tiburón Island and "colored with native plants." A) This figure, 7.8 cm high (*ASM-5768x-1*), has black on the legs and hair, and blue on the upper torso, eyes, and notches on top of "hat." The rest of the face and the feet are natural wood. B) A small fetish, 5.8 cm in height (*ASM-5768x-2*), with natural wood and dark blue pigment. *CMM.*

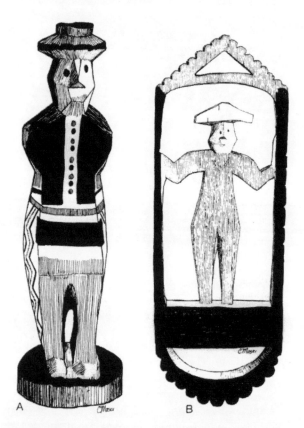

Figure 17.24. Fetishes made from red elephant tree wood. A) This figure, made by Nacho Morales in 1957, is 19.8 cm in height and decorated with red and blue pigments. B) A figure made by Jesús Morales in 1969. It has red and blue pigments and is 17.3 cm high. *CMM.*

Xavier's (1941:34–49) fine illustrations and notes to accompany her collection at the Arizona State Museum provide the best first-hand observations available. Our investigations essentially confirm her report. The following was excerpted from her unpublished manuscript (Xavier 1941:34–35). Annotations added by us are in brackets.

"Santos" of the Seri

These little painted and carved wooden figures the Seri call "santos" in Spanish, and in their own language "eh-eh-kok-*mus*-kaht" or "eh-eh kok-*meh*-keht" [*hehe icóocmolca*]. They are the most noticeable ceremonial or curative object around the Seri camps, for they may be seen worn around the neck, tied to the house doorways, and occasionally tied at regular intervals all around the inside structure of the house. Most Seri own one or more, and if they are not in evidence, they are carefully wrapped in cloth, and put away among their personal belongings in the wooden box in which so many keep their most personal things. At first they were reluctant to speak of them or show them, explaining that they were for curing, and offering no more information. Upon later visits, they became more communicative, and finally my daughter and I received several of them as presents, both from the medicine man and from friendly Seri. At the end several "santos" were offered for barter or sale. But others, some very interesting, their owners refused to part with, and they seemed to value them highly.

My first information about the "santos" came from old Santo Blanco, the medicine man, now dead. Most of the rest came from Jesús Ybarra, the best-known medicine man at present. There seems to be widespread feeling that the medicine man is the one who should be asked about "santos," for "he knows." Otherwise, "each person knows what his own santo is good for, but sometimes he does not wish to tell," as one woman put it. One person will not willingly attempt to explain another person's santo. The medicine men are the best source of information.

Santo Blanco offered some information when I found him carving and painting some of the little figurines, alone in his house one day. He said they were for curing or protective purposes, worn around the neck, or tied to the house as I had seen. When tied to the house it was to prevent sickness, evil or death from entering that house. Worn on the person, they were to cure a present illness. When a person was seriously sick, the figurines were taken to a cave where the spirits live. There are several caves, both on the mainland and the island of Tiburon which are used, and sometimes the medicine man sang to the santos there for a cure. The medicine man makes these santos for pay, and at times fasts and prays and sings for four or eight days before giving them to the patient. This makes them very strong for curing. He told me that anyone could make these santos, but by personal observation I have seen and known them to be made only by Santo Blanco or Jesús Ybarra, both medi-

cine men, and one of the Morales family, whom I suspect is learning to be a medicine man. It may be, however, that "anyone can make them." Certainly the ones made with fasting and prayer by the medicine man are considered more powerful, for a Seri upon offering one as a present or for sale will say, "The medicine man made this, fasting four days without food or water, it is very strong, muy fuerte." Some do not claim this extra power for their santos, however.

Jesús Ybarra, later, after Santo Blanco's death, made the same general statements, and offered some additional information concerning specific santos, and what they cured. He said also that certain types were for men, and others for women, and also mentioned santos which were "católicos" or catholic, and others which were "pura Seri." [To the Seri "católica" meant anything spiritual or powerful.] The birds, moon, sun, giant ray, snakes, and the representations of the doors of sacred caves were called "pure Seri." Where a cross appeared, or a man with a hat, he generally said "catolico." Yet there was no trace of catholicism in any of the explanations of the gods and their means of cure. The cross was generally explained as showing that the figurine was a "dios," a god, not just a "santo" or holy.

These figurines are of great variety, and it is evident even in this small collection and the short period of time I had for inquiry that much curing ritual, religious belief and symbol centers around them.

They seem to fall into several classifications. . . . A representation of the moon alone, generally the crescent moon, is associated only with the unmarried girls. No others wear or use them. But the rather frequently used santo in which a man is standing on the crescent moon is "a man's thing only," so I was told. Birds are usually explained as messengers, either to carry prayers, or to bring messages to the medicine man in a state of vision, to enable him to see "far off." . . . Each god has a special name, if the medicine man chooses to tell it. In a few cases he did. . . . There are one-legged gods and two-legged gods. The one-legged god has two legs at night and can step from heaven to earth. There is a two-faced god, who can see all ways, above the heavens and below the earth, frequently represented. The santos with the bottom shaped to a point or a narrow wedge, according to Jesús Ybarra, are thrust into the sand at night at the head of the sick person, and "the sickness will go." There are also santos which symbolize in various ways the doors of the sacred cave at Tiburon, where the spirits live, the "puertos" which are so important in the Seri mind, and so often represented in face-painting, santos, and drawings [the drawings obtained by Xavier], and which play a large part in their mythology. The santos seem to have a life of their own "at night,"—for example, one figurine representing the doors of the cave, with open slots, was said to "turn and turn" when all were asleep, the birds fly, and carry messages, the onelegged one grows two legs and moves about. . . . In the case of the figurine which turned at night, it was made clear that the figure itself did the turning. . . .

The figurines are always painted in the usual face-paint colors with much use of red, blue, and white, and

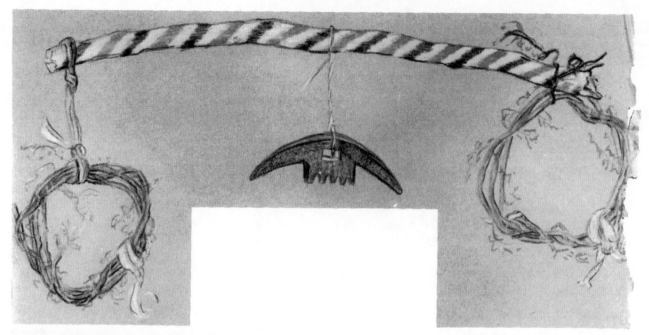

Figure 17.25. "Hidden cave" fetish collected by Gwyneth Harrington Xavier in July 1941 (*ASM-E966*). Her notes (Xavier 1941:49) indicate it was hung over a girl's head while she slept. "Then the moon, who is a señorita, will come in dreams to the señoritas, and speak with them." It was said also to be for curing. The two circles of elephant tree (*Bursera microphylla*) twigs, with "fragrant leaves" and tied with pieces of colored cloth, represent the full moon. The stick, 34.5 cm long, is ashy jatropha (*Jatropha cinerea*) painted with blue and reddish bands. The new moon fetish, carved from wood of *Bursera hindsiana*, is reddish on one side and blue on the other side. *From a color drawing by Xavier (1941:49); ASM.*

have highly formalized styles of decoration. The use of notches, alternately red and blue within the notch, is everpresent. Two sides of a santo are customarily painted in different and reversed colors—if one side is red with blue decoration, the other side is usually blue with red decoration. The notched edges are usually said to represent the edge of the earth, or rays of the sun and moon. Incised lines, painted within alternately red and blue, are frequent, and are known as "roads of the gods" or "caminos de dios." [The notched, scalloped edges were indeed representations of the edge of the earth, and the incised lines were considered moving roads or paths from heaven to earth.]

A special fetish called *zaaj hapesxö* 'cave hidden,' was hung at the entrance of a vision cave, and would probably also have been hung at other vision-seeking sites. It was made by a shaman, and included an *icóocmolca* as part of it. Xavier (1941:49) collected and illustrated a fetish design (for a young girl). A copy of

her illustration (Figure 17.25) was identified by several Seri as being that of a *zaaj hapesxö*. Her notes indicate that this one hung in a brush house; however, this observation does not preclude the possibility that it was also used in a vision cave. The Seri said they could not tell the meaning of this fetish because they did not make it; this confirmation supports Xavier's report that only the maker or owner of a fetish knew its full meaning.

Toys: The wood was used extensively for carving dolls (Figure 17.26) and toy boats. Boys often played with these toy boats in calm, shallow water in front of the camps or villages, or in the esteros. The hulls were carved from a single piece of *xoop inl*, and often skillfully outfitted with miniature sails, paddles, and sometimes even miniature wooden outboard motors. More often than not they were painted with the same striking combinations of bold colors as full size wooden boats.

Bursera laxiflora S. Wats. subsp. *laxiflora*
xoop caacöl '*Bursera-microphylla* large(plural)'
  "large elephant tree"
torote prieto

Large shrub or small tree, reddish brown bark, and fern-like, drought-deciduous foliage. Interior foothills and slopes, Tiburón Island, and on the mainland from the Sierra Seri southward and inland along major arroyos, such as at Rancho Estrella.

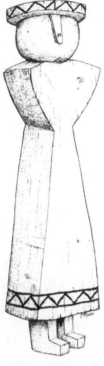

Figure 17.26. This girl's doll, obtained from Andrea Romero at El Desemboque in 1957, is 39 cm in height. It is made of wood (*Bursera hindsiana*) and decorated with blue pigment. *CMM*.

Medicine: Tea made by boiling strips or pieces of bark was taken as a remedy for a cold, sore throat, or cough. This remedy continued to be used in the early 1980s. Tea made from the dark portion of the bark was taken to calm the pain of a scorpion sting or black widow bite.

In June, 1951, while at Tecomate, on the north side of Tiburón Island, Alfred F. Whiting recorded the name as *hop ga'uk¹* (= *xoop caacöl*). His notes read:

> Brought in by a Seri woman, bark only. Comes from far into the interior of the mountains on the west side near the center of the island. Similar, as the name indicates, to *Bursera microphylla* (#38) but these leaves are said to be smaller, and grayer. Bark does not flake. Flowers similar, very small, yellow. Bark used as tea for dysentry, cold, and rarely as a substitute for coffee. The boiled liquid is red. Served with honey or sugar as a drink (Whiting 1957: #110).

Bursera microphylla Gray
xoop
torote, elephant tree

Both the Spanish and English common names subsume various pachycaulus desert and semi-desert trees and shrubs. However, the Seri name is specific for this species.

It is one of the most common and conspicuous desert shrubs or small trees in the region. The limbs and short trunk are fat and semi-succulent (Figure 17.27); the wood is pithy and very soft. The bark becomes papery in the late spring dry season. Foliage may appear at any time of year following even meagre rainfall (Figure 17.28). The sap contains turpines, and the leaves, when crushed, are highly aromatic. Flowering and fruiting likewise may occur at various times of the year, although flowering generally does not occur during the several colder months. The fruit is about 7 to 9 mm in diameter.

While walking in the desert, people occasionally chewed the hard fruit to quench thirst. It tastes like a bitter astringent and causes saliva to flow. The term for the gum was *xoop ooxö* 'elephant-tree excrement,' and for the dry wood *xoop cöcootij* "dry elephant tree."

Animal Food: The Seri said that old and thin male mule deer on Tiburón Island eat the bark of *xoop* and that these deer lack strength and fat. They said that the females and fat males do not eat the bark. Elephant trees with scarred trunks and limbs are common on the island, especially near permanent water holes (Figure 17.29).

Boats and Caulking: The ribs and vertical stem at the bow and at the stern of the first plank boats were carved from *xoop*. Juan Mata, maker of the first wooden boat around 1900, used this wood because it was very soft, and his homemade nails would not penetrate a hardwood, such as mesquite. These nails, fashioned from stolen fence wire, were about 4 cm long. *Bursera microphylla* was used rather than *B. hindsiana* because the trunk of the latter was crooked and not long enough.

Hocö ine "wood's mucus," the caulking compound or pitch used on these early boats, was sometimes made from elephant tree gum mixed with animal fat (see *Stenocereus thurberi*). It was also used to repair cracks and fill small holes in pottery vessels.

Facepaint: The blood-colored sap was squeezed from the inner "bark" or wood into a large clamshell (*xtiip*, *Laevicardium elatum*). A person's skin was said to become "white" with its continued use. Alfred Whiting (1957: #38) noted that "The bark is used in the preparation of face paint."

Firemaking: A dry stick or branch attached to the living tree was called *xoop yapáain* 'xoop what-caused-it-to-fall' (Figure 17.30). However, some said that the term *yapáain* should only be used with *hamísj* (*Jatropha cinerea*). *Xoop yapáain* was one of the most readily available materials for making the *caaa* firedrill. A crude but functional fireboard and drill tip could be carved from this wood in about five minutes

Figure 17.27. This medium-sized elephant tree (*Bursera microphylla*) was decorated with colored cloth during a leatherback fiesta at *Saps* in March 1981. The decoration was purely for fun. *RSF, December 1982.*

with an ordinary knife (see Figure 9.16). A firedrill of this material was usually made on the spur of the moment and was not one that a man kept in his quiver.

Firewood: The dry wood was used as kindling. It is very soft and burns quickly. The wood was often part of the fuel for cooking teddybear cholla stems (see *Opuntia bigelovii*).

Fishing: A belt made of the twigs was worn by a man spearing fish in waist-deep water to repel sharks.

Food Gathering: The dry wood was considered the best wood for smoking out bees.

Hair Care: The twigs were used with jojoba as a shampoo (see *Simmondsia*).

Headpiece: The leafy twigs were sometimes woven into a headband or wreath called *hehe yail it hapácatx*

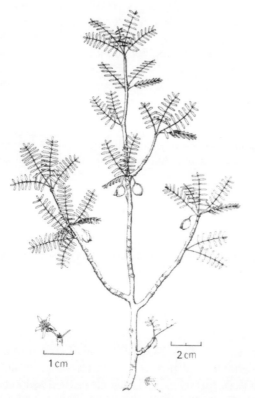

1 cm   2 cm

Figure 17.28. Details of a branch of elephant tree (*Bursera microphylla*) from the vicinity of Tastiota. *FR.*

Figure 17.29. During severe drought on Tiburón Island mule deer eat the bark of the elephant tree. The scars on the lower limbs of this tree were the results of extreme grazing pressure following the drought of summer 1982. *RSF, vicinity of Tinaja Anita, April 1983.*

'plant its-greenness on what-is-left-on.' It was worn in summertime by men and women for shade and to keep the hair in place (Bowen and Moser 1970a: 175).

Hunting: Temporary shelters or blinds for hunting mule deer were made from *xoop* branches because the highly aromatic foliage masked the odor of the hunter. These shelters were used in at least two hunting strategies. During summer, when deer came to eat organ pipe fruit, *xoop* shelters were built against these cacti (see *Stenocereus thurberi*).

The other method, called *hax cacóxaj* 'water what-is-with,' entailed hiding in blinds at a water hole on a moonlit night and waiting for the deer to come to drink. The blinds were usually placed on the leeward side of the water hole. If there were a number of hunters, they might build several blinds in a line at right angles to the line of approach of the deer to avoid shooting each other. It was said that four or five deer might be killed in a night and would provide enough meat for everyone.

Medicine: For a stingray wound the leaves were cooked in water with *Atriplex barclayana*, and the resulting liquid was used to bathe the wound. To cure a headache, the head was washed with a decoction of leaves from this species and *Stegnosperma*. A mother sometimes put the leaves on the stump of her baby's umbilical cord as it was drying.

To kill head lice, the crushed fruit was added to *Lippia* herbage which had been cooked in water, and the resulting mixture used as a shampoo.

The inner "bark" (reddish brown in color—probably the cortex) was mashed with water into a paste and applied to sores on a child's head. After the bark was mashed it often was squeezed into dark reddish brown balls called *xoop icotiixp* "squeezed *xoop*," which were stored for future medical use. The bark, boiled and taken as tea, was said to cure gonorrhea.

The sap was painted on a scratch or cut on the face to prevent the scarred area from becoming darker than the surrounding skin. The sap was also applied to head infections.

Painting: The red pulp (probably cortex) beneath the bark was kneaded with a bit of water until the mixture became foamy. It was then put into a clean cloth, the liquid squeezed into a container, and the container set aside until the liquid thickened. It was used to decorate pottery; after firing, the painted area was a dark color. It was also used to paint or decorate various other objects, particularly wooden ones, such as *icóocmolca* fetishes and violins; it produced a dark, reddish brown color (see *Bursera hindsiana*).

Figure 17.30. *Xoop yapáain*, a dead branch on a live elephant tree, was used for makeshift firedrills. *RSF, vicinity Hast Eemla, April 1983.*

The Supernatural: The *xoop* tree or shrub was considered to have a powerful spirit, and it featured prominently in religious practices. A handful of branches or a wand made from two or more twigs braided together and tied with a cloth or yarn (Figure 17.31) was used by a shaman in curing, and sometimes by those seeking power through visions (see *Vaseyanthus*). The bark was removed on the ones we have seen, and the twigs were dyed red or blue, or were left natural.

In one version of the vision quest, after three or four days of fasting in a vision circle, the supplicant reached out of the circle with a *xoop* wand. Power from *Icor* (a principal spirit) was picked up or transferred to the wand. By then touching the wand to himself, the supplicant received power from *Icor*. By this means one could receive his vision or enlightenment.

Griffen (1959:50) reported a version of the vision quest in which the aspirant built a hut or ramada with the framework of ocotillo (*Fouquieria splendens*) covered with *torote* and desert lavender (*Hyptis*). During the four-day fast the supplicant drank "only a minimum [amount of] water mixed with a little *torote verde*," which was probably leaves or leafy twigs of *xoop*. "Late in the afternoon of the fourth day he departs, taking with him some *torote verde*, and goes directly to the cave." In the cave he "lies down on the *torote verde* and waits for the *santos* to come from the ramada" (see *Bursera hindsiana*).

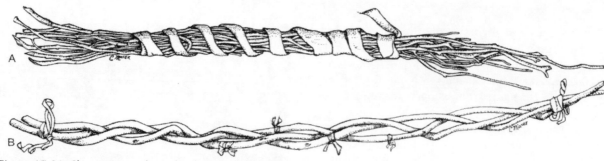

Figure 17.31. Shaman's wands made of twigs of elephant tree (*Bursera microphylla*) and cloth. A) This replica, made by María Antonia Colosio in 1964, is 57 cm long and decorated with pieces of colored cloth. B) A replica made by Jesús Morales in 1956. It is 58 cm long and decorated with orange, blue, and black cloth strips. *CMM.*

Figure 17.32. "Coyote's boat," fashioned from an elephant tree (*Bursera microphylla*) knothole, 11 cm long. Obtained at El Desemboque in 1956. *CMM.*

A shaman commonly held a *xoop* wand in his hands while practicing his curing powers. Griffen (1959:51) reported that during the curing procedure a shaman chewed *torote verde.* An intriguing, complex fetish collected by Gwyneth Harrington Xavier had two coils or circles of leafy *xoop* branches tied to it. These circles represented the full moon (see *Bursera hindsiana*; Xavier 1941:49).

During the four-day leatherback fiesta *xoop* branches were scattered over the giant turtle as offerings (see Figure 3.7; Smith 1974:141, 155).

Lowell's account of the legend of Lola Casanova includes a version of the Seri oral history surrounding Coyote Iguana (a Seri man also known as Jesús Avila). Part of this vision, given below, indicates the ritual use of *xoop.* The action in this story takes place shortly before 1850, and relates that some men went to *Hapis Ihoom* 'where tobacco occurs,' where a special *hapis*, or tobacco, grew (see *Nicotiana trigonophylla*).

When they didn't return soon the people back at camp were worried. A woman called Tola who had had visions of bows and arrows and fighting looked into the future by means of her power and sought to connect the spirits of the men still in camp with the spirits of the men who had made the trip. She walked before [back and forth in front of] the men there and sang as she walked. She put *xoop* [sacred Elephant Tree] branches in 8 different places and blew on the *hast icáh*, the shaman whistle, to call the spirits of the men who had left. If the men were still alive, there would appear some of the *hapis* plant, the green plants, near the branches of the *xoop* which she had laid out. Tola said that the plants [*hapis*] that her brother [who was on the excur-

sion] had cut would appear near the *xoop.* When she put out the *xoop* and sang, when she had sung 8 songs, then she said that if blood appeared on the *xoop* branches the men were dead. If the *hapis* appeared, they were alive. After she sang the 8 songs the people went to check the *xoop* branches. There were some *hapis* [green plants] in a piece of fresh horse hide (E. Moser in Lowell 1970:149).

One who dreamed a bad dream concerning another was supposed to reveal his dream to that person and then sweep the person with *xoop* twigs to prevent the dream from becoming reality. *Xoop* twigs were also used in the harpoon smoking ceremony (see *Atamisquea*). In 1921 Sheldon (1979:112) observed that, while hunting mule deer on Tiburón Island, one of the hunters threw *xoop* leaves in the fire and stated, "This is a Seri custom to bring good luck to the hunt."

Toys: The fruit was used for ammunition in a peashooter made of reedgrass (see *Phragmites*). A knothole-like growth on the tree resembled a tiny boat. It was called *oot icanóaa* "coyote's boat," and boys used them as toy boats (several measured 8 to 12 cm in length; Figure 17.32).

Weapons: Metal for making harpoon points was said to soften more readily (to facilitate shaping when pounded) if heated in a fire of dry *xoop* rather than ironwood (*Olneya*).

Buxaceae
(*see* Simmondsiaceae)

Cactaceae—Cactus Family

For the sake of convenience the columnar cacti are grouped as a unit. The remaining cacti are in alphabetical sequence following the discussion of the columnar cacti. The Seri have no terms for cactus or for larger groups of cacti, but rather refer to each species by name.

Columnar Cacti*

Columnar cacti (Tribe Cereeae) are dominant features of the desert landscape in the region. The six species in this region are:

Carnegiea gigantea, sahuaro
Lophocereus schottii, senita
Pachycereus pringlei, *cardón* or *sahueso*
Stenocereus alamosensis, sina
Stenocereus gummosus, *pitaya agria*
Stenocereus thurberi, *pitaya dulce*, organ pipe

Containers: The "boot" found in *cardón* and sahuaro is woody callus tissue lining a nest hole made in the stem by a woodpecker. After the plant dies and the fleshy parts of the stem decay, the boot and the woody ribs (vascular bundles) remain. The boot, *xcatnij*, was used to carry and store certain foods. Little girls used the boot for storing their dolls.

Fruit Harvest: The large-fruited species—*cardón*, organ pipe, *pitaya agria*, and sahuaro—provided major food resources for the Seri. McGee (1898:206) thought that cactus fruit was probably the most important Seri plant crop. *Senita* was infrequently harvested, primarily because of its small size, and *sina* fruit was occasionally eaten but not brought back to camp. Thus the following discussion concerns the four large-fruited species.

The fruit was a favorite food and in the 1980s still harvested each summer and fall. The fruit contains numerous small black seeds, *ics*, which are embedded in the sweet juicy pulp. The seeds are high in protein and in oil content. However, the seed coat apparently

must be broken or crushed before the contents can be digested. Whole seeds pass through human and animal digestive tracts undamaged. The fleshy pulp is high in sugar content. The pulp, along with the seeds, may be eaten fresh or variously prepared and stored.

Flowering generally occurs in spring and early summer. The fruit, usually produced in prodigious quantity, generally ripens at the height of the dry season, shortly before the onset of the brief summer monsoon. It is a time of very hot weather. *Pitaya agria* bears flowers and fruit later in summer and fall. The moon during which the columnar cactus fruit begins to ripen, called *imám imám iizax* "moon of the ripe fruit," corresponds closely to the month of June. *Imám imám* "ripe fruit" implies something cooked, and in this case refers to fruit cooked or ripened by the sun. In 1692 Padre Gilg gave *himamas* as the Seri term for "columnar cactus fruit."

As soon as the first buds were visible, a few were brought into camp. It was a welcome sign that fruit was soon to follow. However, mothers cautioned their first-born children never to touch any of the buds or their lives would be shortened.

The arrival of the first ripe fruit brought into camp was a joyful occasion. Bits of pulp from this first fruit collection were dabbed on the cheeks and the tip of the nose to bring good luck. Entire families often went into the desert to harvest the fruit, although women were the principal gatherers.

When a girl was nine or ten years old, she was ready to help gather fruit. She was told not to eat any of the first fruit she collected. If she did, she was said to be *inol quim* "one who swallows her arm." This phrase meant that she had "swallowed" her right arm and would therefore be lazy in gathering throughout her lifetime. For this reason her first fruit had to be given to her sister and her sister's husband; if she had no married sister, then it was for her family. The next fruit that she gathered had to be given to some elderly woman. Finally, she could eat fruit from her third gathering trip.

Fruit of sahuaro, *cardón*, organ pipe, and *pitaya agria* was harvested with three kinds of poles (Figure 17.33). The general term for the cactus fruit-gathering

*This section is revised from R. S. Felger and M. B. Moser, "Columnar cacti in Seri Indian culture," The Kiva (1974) 39 (3–4): 257–275. Used with permission.

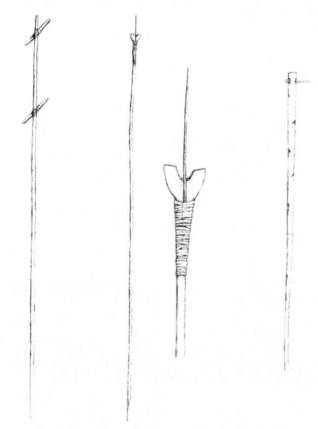

Figure 17.33. Cactus fruit-gathering poles, made in El Desemboque, c. 1956. *Left:* pole for *cardón* and sahuaro fruit is made from a sahuaro rib shaft with creosotebush cross-pieces. *Center:* pole for organ pipe fruit. The shaft is sahuaro rib, the wooden end piece is mesquite, the spike is a sharpened wire, and the bindings are strips of limberbush (*Jatropha cuneata*) stem. *Right:* pole for *pitaya agria* is made from a century plant (*Agave subsimplex*) flowering stalk with a creosotebush spike. *CMM.*

pole was *hehe imám ihapóc* 'plant its-fruit what-one-pries-off-with' or "stick to pry off fruit with." These poles were made from *xojíif*, the dry woody ribs from a dead sahuaro or *cardón*. Gathering poles were also fashioned from reedgrass (*Phragmites*), ocotillo (*Fouquieria*), or the flowering stalks of century plants (*Agave*). The specific construction and names for poles used to gather the fruits are detailed under each individual species.

Fruit gathering provided an opportunity for boys to display their skill at knocking off cactus fruit with improvised poles. Holding the pole with one hand, a boy tried to pry the fruit loose and catch it with his other hand. This was done only with sahuaro and *cardón* fruit, since the others were too spiny. If the

unharvested fruit had already burst open, the boy poked at it and tried to catch the falling juicy pulp in his mouth.

When larger quantities of fruit were gathered, probably primarily for wine making, the men sometimes helped carry the fruit back to camp. They carried it in large-mouthed pottery ollas or 20-liter (5-gallon) cans suspended from a pole over one shoulder. This method of transport, called *cooxop*, differed from the more common one that involved a carrying yoke.

When a man was walking through the desert alone and found ripe cactus fruit, he often cut a branch of a straight-stemmed bush, such as *Cordia*, and strung the fruit onto the stick like beads on a necklace for carrying back to camp. Or he might cut a stem of *Jatropha cuneata*, remove the bark, tear a long strand from the center portion of the flexible stem, and string the fruit on it. Fruit carried in this manner was called *hapápl* 'what is strung.'

A woman carried the fruit in a shallow basket on her head. If she had fruit of either organ pipe or *pitaya agria* from which the spines had not been removed, she was called *cahícös* 'who makes it spiny.' If she had removed the spines, or if she had peeled the fruit and had only the pulp in her basket (as was often the case with sahuaro and *cardón* fruit), then she and the load of fruit were called *queeolim*. Any juice that collected in her basket she drank. By the 1960s women carried the fruit in plastic or metal buckets.

When a woman was in the desert and found fruit but had no container in which to carry it, she made a pocket by tucking some of her skirt into the waistband in front and carried the fruit in it. She was then known as *isxamóni cooquim* 'outer-area-of-stomach who-puts.'

The fruit was commonly preserved by drying. The dry fruit was often stored in a bag made from a sea turtle stomach. Columnar cactus fruit that had dried on the plant was called *ixoj*, and was occasionally harvested.

Oral Tradition: The following narrative concerns the cactus boot, called *xcatnij*:

A mountain lion was looking for something to eat. He came to a cave in which some cottontails were hiding. He grabbed one, held it down with his paw, and said, "This cottontail's back is yellow." The cottontail said, "My back is yellow because I kill large male mountain lions and carry them on my back." Then he said to the other cottontails, "Bring out the lion's head over there.

I'll eat some of it, and then I'll go out and look for more lion tracks." From inside the cave the cottontails rolled out a cactus boot. When the lion heard the raspy sound of the dry rolling boot, he released the rabbit and fled. The cottontails ran the opposite way and into a thicket. Then the cottontail that had fooled the lion said to the others, "Let me get way in the back. It is because I fooled him that we got away" (adapted from E. Moser 1968:365–367).

Wine: Wine, called *imám hamáax* "fruit wine," was made from the fruit of *cardón*, organ pipe, *pitaya agria*, or sahuaro. *Pitaya agria* fruit produces the strongest wine. By the 1950s or earlier the Seri were no longer making cactus fruit wine.

The Seri stated that before they began fishing commercially (about 1925) the fruit of organ pipe and sahuaro was used almost exclusively for wine making. Every several days a new batch of wine was made. While the cactus fruit was ripe, family groups moved along the coast from camp to camp. It was a festive time. Mule deer, sea turtles, and mullet were plentiful during the summer months, and we were told "That's why we didn't have to eat the fruit [of organ pipe and sahuaro]."

Wine made with water was called *imám hamáax hax cöcaap* 'its-fruit wine water be-standing-with,' or "fruit wine with water." To make it, mashed fruit pulp was placed in a basket and water added. This mixture was poured into a pottery olla and placed in the sun. Fermentation occurred in several days. Further fermentation resulted in a sour wine. A scum which formed on the wine was removed and the wine poured into another olla. It was then ready for drinking. This kind of wine soured in a few hours, so it had to be consumed quickly. It was apparently rather low in alcoholic content, since the people said that one could consume a great deal of it.

Wine made without water was called *imám hamáax hax cöimáap* 'its-fruit wine water not-be-standing-with' or "fruit wine without water." Kneaded fruit pulp was placed in an olla and left for several days. It formed its own juice, which turned into a strong wine. This wine did not sour as did the kind made with water. It was said that one could consume only a moderate amount of it, presumably because of the higher alcoholic content. Once an olla had been used for wine making, it would not be used as a water container; it was stuck into bushes or otherwise put aside until needed again. A unique wine-making container was fashioned from a barrel cactus (see *Ferocactus wislizenii*).

Carnegiea gigantea (Engelm.) Brit. & Rose
   [= *Cereus g.* Engelm.]
mojépe
sahuaro

This giant cactus is common on the Sonoran mainland and on Tiburón Island; it does not occur on San Esteban Island. It generally reaches maximum density on sites inland from, but adjacent to, *cardón* (*Pachycereus pringlei*). It can readily be distinguished from *cardón* by its fewer branches which usually emerge more than one-third the distance above the ground. Flowering occurs in early summer and the fruit, *mojépe imám* "sahuaro's fruit," ripens shortly before the onset of the summer rains. The fruit is green and spineless on the outside; when fully ripe, it splits open to reveal succulent red pulp in which numerous shiny black seeds are embedded.

A smaller or stunted sahuaro, also with edible fruit, was called *mojépe zaac* "small sahuaro." Although the plant was shorter, it was said to always have larger fruit than the usual *mojépe*. A dry, woody rib of sahuaro or *cardón* was called *xojíif*.

Firemaking: The firedrill, *caaa*, was often made from a piece of dry sahuaro rib. The drill stick, *caaa ctam* 'firedrill male,' could be a single piece of wood (rather than spliced) because of the length of the raw material. A piece of sahuaro rib was often used to make the mainshaft or handle of a compound drill stick (see Chapter 9).

Food: The fruit-gathering pole used specifically for sahuaro and *cardón* (see Figure 17.33) was called *hacosáa*, which was also the name for the Big Dipper in the Ursa Major constellation (Kroeber 1931:12). The Big Dipper was said to resemble a sahuaro/*cardón* fruit-gathering pole. Most stars were formerly people, and certain stars near the *Hacosáa* constellation represent people holding the fruit-gathering pole. The Walapai and Havasupai of Arizona had similar concepts for the sahuaro fruit-gathering pole and Ursa Major (Kroeber 1931:41).

A typical *hacosáa* was about 5 m long. Two sticks of creosotebush (*Larrea*), each on the order of 10 to 15 cm long, were tied transversely to one end of the pole at acute angles. The top stick was called *ihactífa*. The bottom one, tied about 30 to 40 cm lower, was called *ihahíiquet* 'made a child,' i.e., child of the upper stick. With the cross pieces tied in such a manner, the fruit was dislodged by pushing up or pulling down the pole. The Papago used the same kind of pole to harvest sahuaro fruit (Castetter and Bell 1937:14).

The fruit of sahuaro, *pitaya dulce*, and *cardón* were all prepared similarly. Since *cardón* fruit was most used, fruit preparation is discussed under *Pachycereus*.

Games: Sahuaro fruit skins were used in the men's game called *cacómaloj* 'stick throwers.' The participants sat opposite each other, each with three pointed sticks and several fruit skins cut into narrow strips. The game consisted of taking turns throwing a stick at the opponent's fruit strips. If a player pierced one and returned it to his side without it dropping from the stick, he got to continue. The player who got all of his opponent's strips won that round. A previously determined number of rounds were played. The men played for large stakes: guns and knives. They became skilled at piercing very narrow strips called *hacóml* 'pierced things.'

Boys used a much wider strip called *haxz iipl* "dog's tongue." They played to win a bet of a certain number of cactus fruits which the losers had to go into the desert to gather. It was remembered that on one occasion the losers refused to pay their debt. The winners grabbed them, removed their shirts and tied the losers to trees, where they left them for a time in the summer heat. Then they released them and considered the debt paid.

Medicine: Rheumatism was treated by heating a slab of sahuaro stem (with the spines removed) in hot coals, wrapping it in a cloth and placing it on the aching part of the body.

Sealant: As reported by Bowen and Moser (1968), pottery vessels or ollas for making wine were sealed from seepage in the following manner: Leafy branches of creosotebush (*Larrea*) were packed tightly around the outside of the olla. Dry sahuaro flowers, called *mojépe ipolcam* "sahuaro's tails," were placed inside and ignited. Juices from the flowers and the creosotebush sealed the olla.

Shelter: The dry woody ribs sometimes were used for making wattle and daub house walls. Openings between the ribs were filled with mud.

The Supernatural: To stop rain, one built a fire against a sahuaro using desert saltbush (*Atriplex polycarpa*) for fuel. If the participant had enough faith the rain would stop.

The placenta of a newborn infant was buried at the base of a sahuaro or *cardón*. This insured a long healthy life for the baby (see *cardón*, *Pachycereus*).

The sahuaro was one of several common cacti which were once people (see *cardón*).

Tanning: Because of their high oil content the mashed seeds were used to soften deer skins.

Other Uses: Sahuaro ribs sometimes served as canes for the elderly or blind and as the steadying pole held by the foot drum dancer.

Lophocereus schottii (Engelm.) Brit. & Rose
    [= *Cereus s.* Engelm.]
hasahcápöj 'has-what is chewed'
    "chewed fruit"
hehe is quisil 'plant its-fruit little'
    "small-fruited plant"
sina, sinita, senita

Senita is widespread in the lowlands of Tiburón Island and the mainland in the Seri region (Figure 17.34). Flowering and fruiting occur in hot weather from late spring to fall, although peak flowering and fruiting tend to be in early summer. The Seri reported that *senita* in two areas—one near the center of Tiburón Island and the other on the mainland near Punta Sargento—flower and fruit twice a year (see *Stenocereus thurberi*). The spineless fruit is 2 to 3 cm in diameter and becomes red when ripe.

Food: The fruit was eaten fresh, although generally not harvested as a "crop." However, the people who lived in the interior of Tiburón Island ate much of the fruit. Since the fruit is mostly within easy reach, no pole was needed to collect it. Sometimes a fruit-bearing limb would be cut but not entirely severed, so that it swung down, allowing fruit in the center of the plant to be more accessible.

Games: Boys played a game called *hasahcápöj pte cjeaatim* 'senita together hitting' or "hitting each other with *senita*." They cut the spines off slabs of the cactus and then, choosing sides, threw them at one another.

Shelter: The wood was occasionally used in the walls of the wattle and daub house and as part of the brush house.

The Supernatural: *Senita* was one of the first plants formed (see *Allenrolfea*). The spirit of vegetation, called *Icor*, caused the *senita* to have a very powerful spirit. This spirit hovered over the cactus as a "hat" of vapor or fog and was called *xeele quionam* 'fog with-hat' or "hat made of fog."

Anyone might seek the aid of the spirit of the *senita* against an enemy. He could place a curse against

Figure 17.34. *Senita (Lophocereus schottii)* in the vicinity of Pozo Coyote. *HT, March 1969; ASM-21914.*

enemy's hair or a piece of cloth which had absorbed that person's sweat or saliva into the hole. The spirit of the cactus would cause the enemy to become weak and sickly and eventually die.

Should the supplicant suddenly become fearful of the spirit of the *senita* with which he was dealing, he must mash a small chunk of the plant and mix the juice with any one of the various native paints. This mixture was used to paint crosses on the face and thus avoid sickness that might otherwise affect him. One who painted his face for this reason was called *hehe ccactim* 'plant who-uses.' People who saw the crosses painted on his face might fear that he had been placing a curse on them.

A person who believed he had obtained good results from the spirit of the *senita* and continued to seek its help was called *hehe oziim* 'plant what-he-enjoys.' He might paint his face with the above-mentioned paint to bring good luck to himself and his family.

Good luck was solicited from the spirit of the *senita* by wedging clamshells, or sometimes twigs or other objects, into the stems of the cactus. This good luck cache was called *hasahcápöj heeyolca* 'senita what-are-given-to,' or "what are given to the *senita*" (see *Pachycereus*). It was believed that the *senita* was able to hear a conversation and people therefore often avoided it.

Toys: A section of the living stem was made into a toy. The fleshy part was removed from the woody central portion, leaving wheel-like discs at each end (Figure 17.35). The toy was pushed or rolled along with a forked stick. The same toy was also made from organ pipe stems (*Stenocereus thurberi*). The stem from which it was made could remain alive for many months—even a year or more.

someone whom he disliked, or seek help against coming danger—for example, against a curse which he believed someone had placed on him. However, one knew that he got involved with the *senita* at great personal risk. If too frequent use was made of the cactus in placing a curse on someone, that curse might backfire and affect him instead. The person was then said to be *hehe yaróocot* 'plant what-it-made crazy.' Also, if one began to fear the *senita* while seeking its power, he would become ill.

To place a curse, one cut a hole in a *senita* stem on the side facing the south wind and put a strand of the

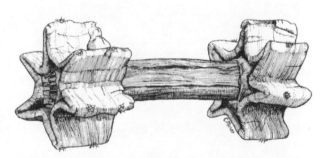

Figure 17.35. This wheel toy, 28 cm long, was made from a *senita (Lophocereus)* stem. It was found 3 km north of El Desemboque in April 1973. *CMM.*

Figure 17.36. *Sahuesos*, or *cardones* (*Pachycereus pringlei*). A) near Pozo Coyote. *ETN, March 1951.* B) On San Esteban Island. Note that the mainland plants tend to be larger and have more prominent trunks than those on San Esteban. *RSF, March 1965.*

Pachycereus pringlei (S. Wats.) Brit. & Rose [= *Cereus p.* S. Wats.]

xaasj
sahueso, cardón

Xaasj is also the name of a certain fish. *Cardón* is the largest cactus in the Sonoran Desert, and one of the most striking features of the landscape in the Gulf of California region (Figure 17.36). It is common throughout most of Baja California, the Sonoran coast from Guaymas northward to Puerto Lobos and on most of the islands in the Gulf, including Tiburón and San Esteban. The geographic distribution of *cardón* in Sonora closely approximates the original area of occupation of the various Seri groups (see Hastings, Turner, and Warren 1972:160).

The large white flowers usually appear in April and early May (Figure 17.37), and the fruit ripens in early summer. Fully ripe fruit often splits open to reveal the fleshy pulp and seeds.

Spine length (Figure 17.38) and color of the fruit pulp is variable, although an individual plant bears fruit of only one color and general spine length. Four types of fruit were classified according to color of the pulp:

1. red or red-purple, *cheel* 'red,' the most common color
2. white, *cöquimáxp* 'with whiteness'
3. light yellow-orange, *cöquimásol* 'with yellowness'
4. pinkish-white, *pti hicotaj ano cöcaaptim* 'together with(plural) in that-which-stands-with,' or "variable"

Three kinds of *cardón* were recognized and distinguished by the characteristics of their fruit:

1. *xcocni*, small fruit, without spines, rather bitter, edible but seldom eaten, always red
2. *xaasj quicös* 'cardón with-spines,' fruit very spiny, normal sized and good eating, all colors
3. *xaasj imícös* 'cardón without-spines,' fruit not spiny, normal sized and good eating, all colors

Toward the end of the "cool" season, probably in April, *cardón* buds sometimes drop off. The falling of the buds was said to be caused by *hant coopol* 'land black.' These are dark patches caused by winds, which come in over the land and cause the buds to fall. Among the columnar cacti this occurs only with *cardón*. It also happens to the fruit of the cliff fig (*Ficus petiolaris*). Not all of the buds and fruit fall, nor does

Figure 17.37. *Cardón* flowers in the vicinity of El Desemboque. *RSF, April 1983.*

it happen every year. The buds or fruit of any tree, including cactus, which fall prematurely were called *caactoj*.

A *cardón* with young fruit was called *cyaxa* 'with belly.' This is an old Seri word meaning "to be pregnant." The tiny, new fruit of such a plant was called *ihiyáxa* 'its offspring.'

Food: The fruit was gathered with the same kind of pole used for harvesting sahuaro fruit (Figure 17.39). Each *cardón* stem may bear numerous ripe fruits, so a woman usually placed her basket on the ground while collecting. She then sat in front of the pile of fruit, opened each one with a worn bone awl or knife, squeezed the pulp into the basket and threw

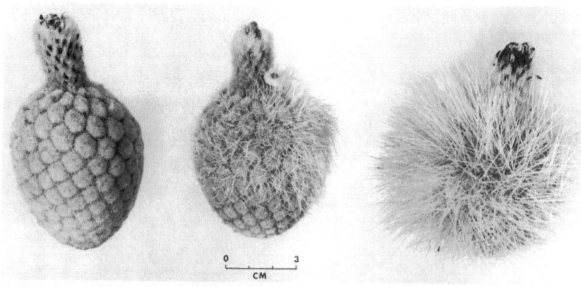

Figure 17.38. Variation of spine length on *cardón* fruit. Each fruit is from a different plant on the mainland side of the Infiernillo Channel. *RSF, July 1976.*

Figure 17.39. Coca Morales harvesting *cardón* fruit in the vicinity of El Desemboque. *MBM, July 1969.*

away the skin. Thus, she returned to camp with a basketful of pulp; sometimes women brought back the entire fruit. Certain individual plants consistently produced large fruit year after year, and were specifically sought at harvest time.

The fruit was eaten fresh or preserved. Pulp from ripe fruit was mixed with some pulp of unripe fruit. The mixture was then mashed and kneaded and the juice poured off into a pottery olla. The mashed fruit formed a sticky mixture that was patted into flat, round cakes and dried. When the Seri were camped by a *playa* (dry lake), the sticky mixture was poured directly onto the hardpan surface and left overnight to dry. It was then lifted or scraped up, broken into chunks, and stored in an olla. It would not spoil as long as it remained completely dry. This dry mixture of the green and ripe fruit was called *hanáij* 'what was grabbed.' To prepare *hanáij* for eating, water was added and the fruit was cooked and mashed. It was then called *haizj* 'what was mashed.'

Haizj cheel 'what-was-mashed red,' the fresh ripe fruit cooked and mashed, was made into fist-sized balls, which also could be dried and stored.

Haizj cooxp 'what-was-mashed white' was made from the seeds. The seeds were separated from the fleshy pulp by hitting the pulp onto the *icóasc* strainer (see Figure 15.18). The seeds fell through and the pulp remained in the strainer. The fresh seeds were cooked in water, mashed and salted. *Haizj cooxp* was eaten only when food was scarce, because it did not taste good.

Mothers mashed the seedless pulp, added salt, and fed it to their children. This seedless pulp was called *imám ináaj* 'its-fruit its-pulp' or "cactus fruit's flesh."

Toasted *cardón* seeds were ground on a metate and the resulting mash was called *haixa*. Since the seeds are oily, no additional fat or oil was needed to make a tasty dish, although it was often salted. Seeds remaining at the bottom of an olla during wine-making were saved to make *haixa*.

Xnois coinim 'eelgrass with-mix' was a mixture of the seeds of *cardón* and eelgrass (see *Zostera*). The toasted seeds were ground and water added to make a drink. The children were said to get fat on it.

According to oral tradition, a group of people known as *Xaasj It Coosot Quihízitam* 'cardón its-trunk slender who-inhabit,' or "the people of the slender-trunked *cardón* place," were great *cardón* seed eaters. It was said that they mourned when the fruit season ended. They were said to be the only Seri who ate *cardón* fruit before it was ripe and the only ones known to have extensively practiced recycling the seeds (second harvest).

Only the seeds of *cardón* were utilized for second harvest. After eating quantities of the fruit, the people defecated on flat rocky places, permitting the matter to dry in the hot summer sun. When it was completely dessicated, they returned, gleaned the seeds, and thoroughly cleaned and cooked them. The seeds were then prepared for food or stored in pottery vessels (see McGee 1898:209–213). Second harvest was also practiced by Indians in parts of Baja California (Baegert 1952:68; Clavigero 1937:94).

Hunting: During the intense heat of the summer a hunter often went from one *cardón* to another checking for mule deer lying in the shade. This strategy was most often employed on Tiburón Island. Of course, the hunter always hunted into the wind. Sometimes a specific *cardón* was identified with a particular hunter (Figure 17.40A). After a successful hunt, a man sometimes drove sticks into the stem of a *cardón* and hung meat on them in order to keep it away from animals (Figure 17.40B).

Medicine: Slabs of *cardón* with the spines removed were heated in ashes, then wrapped in cloth and placed on aching parts of the body.

Oral Tradition: "The Sea Horse's Entrance into the Ocean" tells that the sea horse and several common cacti were once people and relates how the *cardón* got its ribs. The free translation is given below.

Figure 17.40. This well-known *cardón*, about 3 km east of El Desemboque, was identified with José Torres. When it was hot and he was returning from hunting (around the 1940s), he often stopped at this cactus. However, others, such as Ramón Blanco and his son, also used this *cardón* because of its convenient location. A) Roberto Herrera at the base of the plant in 1983; although José Torres and Ramón Blanco were no longer living at that time, this *cardón* was still associated with them. B) José and other men hung deer meat from these sticks driven into the *cardón* stem as they were returning from a hunt. They were tired, and put the meat here so that people from El Desemboque would know where to come to get it. *RSF, April 1983.*

The sea horse, cholla (*Opuntia bigelovii*), *cardón*, and sahuaro used to be people and lived around here. One time those who are now the *cardón* and sahuaro were chasing the people who are now the sea horse and the cholla. They were chasing them across the desert and throwing stones and shooting arrows at them.

Going past the place called "Any Water (Tecomate, on Tiburón Island)," they came to the place now called "Where the Sea Horse Dodged." The sea horse jumped from side to side to avoid being hit. The rocks in that place look as though they were put there deliberately.

Then they came to the place called "Large Crucillo Bush Bay" (*Zizyphus*). The sea horse, arriving at the cliffs and about to go into the water, removed his sandals and tucked them into his waist band behind his back. He wrapped his braids around his head like anyone does when he is going to run fast. When he entered the water he turned into a sea horse, and now that is what he is called.

The cholla was full of holes from the stones and arrows that hit him, and one could see right through him. The *cardón* was covered with ridges made by the stones that the cholla and sea horse had thrown at him. All of these were ancestors and all were real people at one time. That is the end of it.

The Supernatural: On the occurrence of a miscarriage or the birth of a stillborn infant, the remains

Figure 17.41. An elevated burial in the limbs of a *cardón* in the vicinity of El Desemboque. The children are Cathy Moser and Mexican neighbors. *MBM, 1956.*

were wrapped in pieces of cloth, usually placed in a box, and then put on a platform of brush in the limbs of a *cardón* (Figure 17.41; Griffen 1959:28; M. Moser 1970b). If the body was buried in the ground, the mother would be burying all her future children with it. This elevated burial was called *ziix hacx cmiih cola hapáxquim* 'thing somewhere not-be high what-was-put,' or "the dead thing placed high." This type of burial has apparently not been practiced since about the 1960s.

The placenta of a newborn was buried at the base of a *cardón* or sahuaro. Five small plants of any species were buried with it. Ashes were put on top of the place of burial to keep coyotes from locating it. The cactus served to mark the spot. In later years one might visit the site of his placenta burial to put green branches of any plant on it for good luck. Most of the people did not know when they were born, but each knew the general area where his placenta was buried.

Luck could be sought from the spirit of the *cardón*. The supplicant marked four crosses on the bark and asked for the help desired. In addition, good luck was solicited by wedging seashells, usually clams (such as *Chione*), or sometimes twigs or other objects into the stem (Figure 17.42). This good luck cache was called *xaasj heeyolca* '*cardón* what-are-given to,' or "what are given to the *cardón*." This custom was practiced so that the spirit of the cactus would influence other people to give the supplicant material gifts. The clams usually remained in the cactus for a number of years, perhaps decades. Similar practices were followed with other columnar cacti—particularly *senita* (*Lophocereus*)—and also with ironwood (*Olneya*) and rock crevices.

To bring clouds or rain, holes were cut in the root and filled with water (Figure 17.43). This practice was performed when the supplicant wanted cloudy or rainy weather to alleviate the great heat during summer.

Each year brought a renewal of nature, and the first evidence of fruiting portended good fortune. For this reason, *cardón* buds were used as good luck charms.

Tanning: Because of their high oil content the mashed seeds were used to soften deer skins.

Tattoo: Juice of *cardón* fruit, mixed with charcoal, was used in tattooing (see section on tattooing in Chapter 12).

Other Uses: The dry ribs were used for house walls and as cactus fruit-gathering poles. The elderly and

Figure 17.42. Good luck caches in the vicinity of Estero Sargento. A) A *cardón* with numerous clams wedged into it. B) A clamshell (*Chione californica*) wedged into the trunk; note the scars where other shells have fallen out. *RSF, April 1983.*

Figure 17.43. Holes were cut in the root of a *cardón* and filled with water to bring clouds or rain to give relief from the summer heat. This one, 6 by 9 cm wide, was made by José Torres in the root of the *cardón* associated with him (see Figure 17.40). *RSF, March 1983.*

blind used the ribs for canes. A child often led a blind person around camp with a cactus rib, each holding one end of the cactus rib. The foot drum dancer steadied himself with a pole often made from a *cardón* rib.

Stenocereus alamosensis (Coult.) Gibs. & Horak
   [= *Rathbunia a.* (Coult.) Brit. & Rose; *Cereus a.* Coult.]
xasáacoj
sina

The boa constrictor, also called *xasáacoj*, occurs as far north and west as the vicinity of Hermosillo and San Carlos Bay; the Seri knew of it "near Tastiota." The term *xasáacoj* may be etymologically related to *xaasj caacoj* 'xaasj large' or "large *cardón*" (*Pachycereus*).

This plant somewhat resembles *pitaya agria*. *Sina* has small red, diurnal flowers, and more slender, snake-like stems than does the *pitaya agria*. *Sina* enters the periphery of the Seri region in the vicinity of San Carlos Bay, northwest of Guaymas, and at Siete

Cerros, between Hermosillo and Kino Bay. The Seri knew of its occurrence at Siete Cerros and near Hermosillo.

Food: The small fruit was sometimes eaten fresh, but it was considered bitter and consequently not harvested to be brought back to camp.

Stenocereus gummosus (Engelm.) Gibs. & Horak
   [= *Machaerocereus g.* (Engelm.) Brit. & Rose; *Cereus g.* Engelm.]
ool-axö 'Stenocereus-thurberi its-excrement'
   "organ pipe excrement"
ziix is ccapxl 'thing its-fruit sour'
   "sour-fruited thing"
pitaya agria

Ziix is ccapxl is the newer name for this cactus. When a young person had the same name as a plant or animal, or even a name which was very similar, that name became taboo or its use was said to be discontinued if the person died. *Oláöj*, the name for the juvenile *mero* (*Epinephelus itajara*) found in the esteros, was also the name for Ramón Montaño's son. He was given this name because he reminded people of the fish. The child died, and because both names (*ool-axö* and *oláöj*) sounded so similar, the name of the cactus was changed.

Pitaya agria is a large, sprawling cactus, the stems often arching and leaning, forming thickets of impenetrable spiny tangles commonly 2 to 3 m tall (Figure 17.44). It occurs throughout much of Baja California, on San Esteban and Tiburón islands, and along the Sonoran coast from the vicinity of El Desemboque southward nearly to Cerro Prieto near the north end of Kino Bay. It somewhat resembles *Stenocereus alamosensis*; however, the geographic ranges of the two species do not overlap and the flowers and fruit are very different.

The stems and fruit are covered with very sharp and brittle twisted spines (Figure 17.45). The large whitish flowers are nocturnal. The fruit is about the size of a small orange. When it is ripe the spines tend to fall away. Throughout most of its range the fruit is red when ripe, although in the vicinity of El Desemboque the Seri know of plants which bear yellowish fruit.

Flowering and fruiting generally begin later than for the other large-fruited species, and fruiting usually continues well into fall. However, flowering and

Figure 17.44. Pedro Comito next to a *pitaya agria* (*Stenocereus gummosus*) on the east side of Tiburón Island. *RSF, April 1983.*

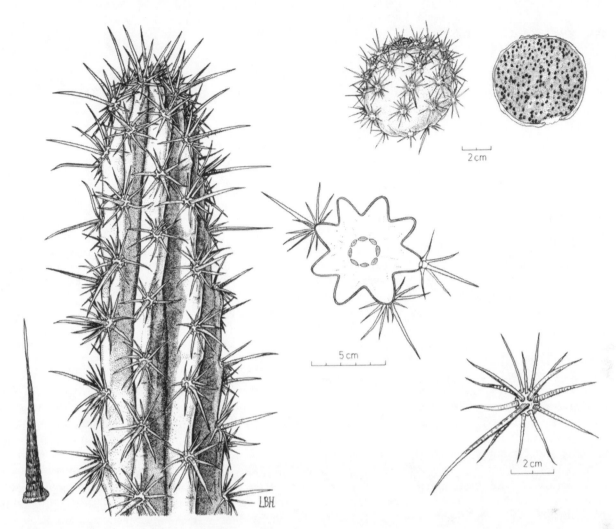

Figure 17.45. *Pitaya agria* (*Stenocereus gummosus*) from Tiburón Island. *LBH.*

fruiting may be reduced during years of low rainfall, and such reduction may yield scant harvests.

Caulking: *Pitaya agria* was occasionally used to make pitch (see *Stenocereus thurberi*).

Food: The fruit was highly esteemed for its sweet and tart flavor. It was eaten fresh or, if plentiful, often dried in the same manner as was organ pipe fruit. *Pitaya agria* fruit is very delicious, and it continued to be harvested in the 1980s.

The fruit-gathering pole for *pitaya agria* was called *hacozquíif*. At the end of the pole, which was about 2.5 m long, was inserted a pointed hardwood stick, such as creosotebush or *orégano* (*Lippia palmeri*) (see Figure 17.33). In modern times a stout nail served the same purpose, and the pole might be fashioned from a broom or mop handle. The gatherer pierced the fruit and used a twisting motion to pull it off the plant (Figure 17.46).

Figure 17.46. Rosa Flores harvesting *pitaya agria* fruit in the vicinity of El Desemboque. *RSF, October 1973.*

257

Stenocereus thurberi (Engelm.) Buxb.
  [= *Lemaireocereus t.* (Engelm.) Brit. & Rose;
  Cereus t. Engelm.]
ool
pitaya dulce, organ pipe

The term *ool* is also used as the name for the whale shark.

Organ pipe is common on Tiburón Island and the mainland from near sea level to peak elevation. It is rare on San Esteban Island. It has numerous, relatively slender branches arising from near the base of the plant (Figure 17.47). Maximum flowering and fruiting occur in early summer. Flowering and fruiting tend to occur over a longer span of time than for *cardón* and sahuaro. The fully ripe fruit generally splits open to reveal the fleshy red pulp. The spines tend to fall away from fully ripe fruit. According to the Seri, there is an area north of Pozo Coyote where certain of the organ pipes flower and fruit twice each year. Fruit from the second fruiting season—in fall—was called *queejam*. This term was applied to any out-of-season fruit.

Boats and Caulking: A black pitch or caulking compound called *hocö ine* "wood's mucus," was made with organ pipe and animal fat. *Pitaya agria* (*Stenocereus gummosus*) or elephant tree (*Bursera microphylla*) were sometimes substituted for organ pipe.

The dry cactus, apparently consisting mostly of stem cortex and some pith, was crushed, placed in a sea turtle carapace and pounded until almost like powder. It was then put into a pottery vessel made especially for this purpose (Bowen and Moser 1968: 121). Enough animal oil was added to thoroughly moisten the crushed pulp. The pot was placed on a slow-burning fire, usually of ironwood (*Olneya*), and the mixture cooked for several hours. It was stirred occasionally with a stick, and more oil added as needed. When the mixture attained the consistency of tar it was taken off the fire. The pitch was then ready to be spread along the seams of the boat with a stick flattened at one end.

The early twentieth-century plank boats were covered with this pitch (see Figure 10.8) but still leaked so badly that they needed constant bailing. Coolidge and Coolidge (1939:90) refer to a pitch made from the fat of birds and gum from a dead giant cactus. The *hocö ine* caulking was also used to fill cracks or small holes in pottery vessels.

In 1924 Edward H. Davis recorded the preparation of the caulking compound (Quinn and Quinn 1965:201):

> I witnessed the operation of making a certain kind of pitch or tar which the Seri use to make the seams in their canoes or boats tight. The women pound up the dried pulp next to the outside skin of a dead *pitahalla* cactus (organ cactus). This is very dry and brown in color and the pieces are pounded up to a dry powder with a wash or shore rock in a dried deer hide. This is placed in a *batello*, or flat basket, and manipulated so that the coarse pieces come to the top and are scraped off and thrown away. The fine powder is then put in a five gallon can and some porpoise or sea-lion oil or horse oil is poured on it and stirred into a thick gummy mass. When thoroughly mixed it is put over the fire, in an olla, to boil and continually stirred for an hour or two until it is the consistency and has the appearance of coal tar. While being worked it is kept heated and applied to seams inside and outside the boat. It hardens just like tar and answers the same purpose.

It is interesting to note Davis's report that the boat was painted with caulking inside as well as outside. His statement that porpoise oil was used is probably in error: they would have had no easy means to capture one.

Food: The organ pipe fruit-gathering pole was called *hactáapa* (see Figure 17.33). It was about 4 m long with a spatula-shaped piece of mesquite wood, *hacóipj* 'right angled,' fastened to the end. McGee (1898:230, Figure 36) pictures two of these, which he thought were awls. Extending beyond the *hacóipj* was a wire or sharp, pointed strip of metal to spear the fruit. This point, originally made of creosotebush, was called *hactáapa iti ihíip* '*hactáapa* on its-standing.'

The general term for organ pipe fruit was *'ool imám* "organ pipe's fruit." The term *cáahataj* was specific for nearly ripe fruit which had not yet split open. This fruit was picked for carrying on a trip, so that it would ripen in a day or so. Each stem bears at the most only a few ripe fruit at a time. As a woman walked around the cactus with a basket on her head she pierced one or two of them. Then, as she lowered the pole to retrieve the fruit, she ripped open the skins with the metal point and tossed the opened fruit up into her basket. When organ pipe fruit was scarce, the gatherer listened for the calls of the white-winged dove, for where the doves were, there would be fruit.

Fruit which had burst open but was not fully red, called *cmasij* 'burst open,' was considered the best tasting. The skin of organ pipe fruit is edible, which is not the case with other species of columnar cactus.

Figure 17.47. Edward H. Spicer in front of a large organ pipe (*Stenocereus thurberi*) in the desert east of El Desemboque. *ETN, March 1951.*

The seeds were not mashed or ground for food, presumably because they are too small.

The fruit can be preserved by drying. The woody ribs, *ool itajc* "organ pipe's bones," were crisscrossed on the ground for drying racks. The fruit was peeled and sliced in thin sections and placed on the organ pipe ribs to dry. It sometimes took several days for the fruit to dry, for humidity is often high at that time of year. During the drying process the slices were frequently turned. The dry fruit, *imám hapásj* 'its-fruit what-is-spread-out-to-dry,' could be stored well past the fruiting season. Water was added to the dried fruit and it was eaten either cooked or raw. Dry fruit of any cactus could be stored in a bag made from a sea turtle stomach.

Fruit nearest camp was harvested first. Prior to the 1920s, the collected fruit was generally not eaten because it was made into wine. Older children had to go far from camp to collect enough of it to eat. Little children seldom ate the fruit unless their mothers prepared some especially for them. In June, 1951, Alfred F. Whiting was at Tecomate and observed: "While I was on Tiburon the women of the family made an overnight expedition into the interior to bring back trays [baskets] full of these fruits (Whiting 1957: #67)."

The smell of organ pipe fruit, *ool imcáacas*, was pleasing to the Seri. The odor is usually not strong. However, organ pipe growing near Pozo Peña produced fruit with a strong, yet pleasant odor. When

259

Figure 17.48. Hunting torch made from dry ribs of organ pipe bound with strips of limberbush (*Jatropha cuneata*). This replica, made by Jesús Morales in 1965, is 74.5 cm long. *CMM.*

out in the desert, the people sometimes picked the flowers and ate the petals, but the flowers were not collected to bring back to camp.

In about the first decade of the twentieth century organ pipe fruit juice was used as emergency food for a newborn infant. María Luisa Chilión was among a small group of Seri fleeing from an approaching enemy attack when she gave birth to a son. Because of their desperate situation, she stayed behind and the others took the infant so that she might more readily recover and find her way to safety. The infant was fed on organ pipe fruit juice for about eight days, and survived in good health.

Food Preparation: Long pieces of the dry woody ribs of organ pipe were used as tongs to handle the stems of teddybear cholla (*Opuntia bigelovii*) during the process of cooking them (see Figure 6.11).

Games: The circle for the *camóiilcoj* game, played at a girl's puberty ceremony or leatherback fiesta, was marked with sets of five cross sections of the stems (see Figure 13.1).

Hunting: A hunter often sought mule deer tracks around an organ pipe during the fruit season. If there was evidence of deer in the area, he built small shelters of elephant tree (*Bursera microphylla*) branches, one against each of several organ pipes. To avoid alerting the deer with the scent of clothing, he hunted naked. In this manner, using the various shelters, he killed as many as three or four with bow and arrow at close range in a night (see *Bursera microphylla*).

Torches, called *heexoj*, were made of bundles of dry organ pipe ribs bound with strips of fresh *haat* stems (*Jatropha cuneata*; Figure 17.48). These were carried at night for hunting mullet, pelicans, and sea turtles (Quinn and Quinn 1965:198 and 213; Sheldon 1979:139).

A lantern was fashioned from a 20-liter (5-gallon) can with many holes punched into the sides with a knife. It was fueled by feeding organ pipe wood into the top of the can. The lantern was suspended from the prow of a wooden boat for hunting turtles at night during summertime.

Medicine: Fresh slabs of the cactus with the spines removed were heated in coals, wrapped in cloth, and placed on aching parts of the body.

Oral Tradition: There is a story concerning wine made from organ pipe fruit. *Moj coquépni* 'wine what-laps-up' is the name of a fish (*cabezón* or *pez baboso*) which has a red spot under the jaw. (*Moj* is derived from the word *hamáax* 'wine'). This tiny fish, found under rocks, was said to be a man drinking *ool* wine, lapping it up with his tongue. The Flood came upon him and he was changed into the fish with his chin still red. This fish is probably the male bay blenny, *Hypsoblennius gentilis* (see Thomson, Findley, and Kerstitch 1979:170–171).

Play and Toys: Small chunks of the fresh stem with the spines removed were thrown by boys in playful group fights. The stem was also fashioned into a boy's wheel toy (see Figure 13.6).

The Supernatural: To calm the wind, one went into the desert and cut "plenty of *ool*," removed the spines and skin, and cut it into eight cross-section slices. These were brought back to camp on a stick and put into a fire. The pieces were then thrown one by one into the sea as the supplicant called to the wind to stop.

Shelter: The dry ribs often were used for wattle and daub houses, and incorporated into the framework of the brush house (McGee 1971:354).

Signaling: When burned, the dry stems produced a black smoke which was used for smoke signals (Quinn and Quinn 1965:204; Sheldon 1979:107).

Other Uses: Dry organ pipe wood was used for smoking out bees. Pieces of the ribs were used to build up the sides of a basket in order to carry a larger load (see Figure 15.24). This arrangement was called *heet hapácax* 'sticks what-is-carried.'

Other Cacti

Echinocereus spp.
hant iipzx iteja caacöl 'land its-torn(place) its-bladder (*Mammillaria*) large(plural)'
  "large fishhook cactus" or "large bladder of the arroyo"

The several species of *Echinocereus* and *Mammillaria estebanensis* were not differentiated by name, although the Seri recognized them as different kinds of plants.

Four species of *Echinocereus* occur in the Seri region:

E. engelmannii (Engelm.) Rumpl. var. *nicholii*
L. Bens.
golden hedgehog cactus
This large hedgehog cactus is found in the Sierra Seri.

E. fendleri (Engelm.) Rumpl.
hedgehog cactus
Large clumps of this cactus are found on the bajada east of El Desemboque. It is seldom common in the Seri region.

E. grandis Brit. & Rose
giant rainbow cactus
This large rainbow cactus is endemic to San Esteban and San Lorenzo islands. Fresh specimens from San Esteban were immediately recognized as being from that island.
Fishing: The cactus was used from the shore in conjunction with the *hacaaizáa* 'hacáaiz true,' the single-pronged fish harpoon, to attract fish. Strips of "white meat" (cortex) from the plant were thrown into the sea. The fishermen then banged harpoons on the rocks and speared fish which came to feed.

E. pectinatus (Scheidw.) Engelm. var. *scopularum* (Brit. & Rose) L. Bens.
The Sonora subspecies of rainbow cactus occurs on rocky slopes in the Sierra Seri and southward to Guaymas, and on Tiburón Island.

Animal Food: The Seri reported that on the mainland these cacti were eaten by desert bighorn sheep and javelina.
Food: The fruits were eaten fresh, although there were seldom enough to suffice for more than snacks.

Ferocactus acanthodes (Lem.) Brit. & Rose
mojépe siml 'sahuaro barrel-cactus'
biznaga, barrel cactus

The Seri distinguished this barrel cactus from *siml* (*F. wislizenii*) by its taller stature, which makes it somewhat resemble a young sahuaro (Figure 17.49).
Mojépe siml occurs in sandy desert along the coast in the vicinity of El Desemboque and Kino Bay. It is generally taller, more slender, with longer and straighter spines than *F. wislizenii* (*siml*), and erect-

Figure 17.49. "Sahuaro barrel cactus," *Ferocactus acanthodes*. A) Roberto Herrera next to a large plant in the vicinity El Desemboque, April 1983. B) Detail of spine clusters, showing both stout and bristly spines. *RSF*.

growing rather than leaning southward. The spines, unlike those of any described variety of *F. acanthodes*, are dull colored.

In the Seri region smaller individual plants of *F. acanthodes* and *F. wislizenii* can be difficult to distinguish. Smaller plants of *F. acanthodes* were often identified as *siml* rather than *mojépe siml*. It seems that the name *mojépe siml* was applied to plants of *F. acanthodes* that were more than about 1 m tall. The flowers of the two species are also very similar.

Food: The seeds were prepared and eaten in the same manner as those of *F. wislizenii*. The flowers (Figure 17.50) were seldom eaten because they were

Figure 17.50. Flowers of *Ferocactus acanthodes* near Pozo Coyote. A) Collecting the flowers and buds. B) Close up of flowers. *RSF, April 1983.*

bitter. It was said that eating the pulp caused a headache and that the juice was not potable.

Ferocactus covillei Brit. & Rose
siml caacöl 'F.-wislizenii large(plural)'
    "large barrel cactus"
siml cöquicöt 'F.-wislizenii that kills'
    "killer barrel cactus"
caail iti siml 'dry-lake on F.-wislizenii'
    "dry lake barrel cactus"
siml yapxöt cheel 'F.-wislizenii its-flowering red'
    "red-flowered barrel cactus"
biznaga, barrel cactus

This stout, tough-spined barrel cactus (Figure 17.51) may become quite large. It occurs sporadically throughout most of the lowlands, including Tiburón Island. The spine clusters lack bristles and the flowers are usually red, occasionally yellow. The base of the emerging central spine is conspicuously thickened and deep rose-purple.

The Seri distinguished this barrel cactus by its high ridges or stem ribs, long thick spines, and reddish flowers. We were told that it starts budding one moon after the *haamxö* (*Agave subsimplex*) flowers, which is late spring.

Emergency Water: Emergency liquid or "water" was not extracted from this cactus because it was too strong and dangerous, causing upset stomach, diarrhea, aching muscles, and inability to walk. Sara Villalobos said any tall barrel cactus with red flowers was dangerous. Two cases of poisoning from it were remembered.

In the early twentieth century Guillermo Ortega's father was temporarily paralyzed from drinking juice of this cactus. "There was no part of him that was not paralyzed. Later he made a song concerning his paralysis. This (the cactus juice) did not kill him, for he later died at sea."

In the 1920s Carlos Ibarra was out in the desert near the estero at Kino Bay. There was no water. He drank barrel cactus (*caail iti siml*) juice, and ate meat with salt and *chiltepines*. Blood came from his nose, his eyes reddened and seemed as though they would burst, and his urine burned. He thought he would die. Some other Seri found him and brought him back to camp at Kino Bay. He subsequently recovered.

Facepaint: A deep rose facepaint was occasionally made from the newly emerging central spine. The spine was pulled out while immature and still flexible.

Figure 17.51. Coville barrel cactus (*Ferocactus covillei*) at Punta Santa Rosa. A) A medium-sized plant. B) The spine clusters have stout spines only. *RSF, April 1983.*

The base of the spine was macerated by chewing and the resulting juice dabbed on the cheeks.

Food: The buds and flowers were boiled and eaten. The fleshy fruit was sometimes eaten fresh. It has a tart, lemon-like flavor. The seeds were prepared in the same manner as those of *F. wislizenii.*

Medicine: A slice of the cactus with the spines removed was placed in coals, salted, and roasted until juice ran out of it. It was then wrapped in cloth and held on a sore place of the body for relief.

Ferocactus wislizenii (Engelm.) Brit. & Rose
siml
simláa 'true barrel cactus'
biznaga, barrel cactus

This species is widespread in the open desert, and on hills and slopes. It is very common on Tiburón Island but absent from San Esteban Island. The plant is stout, with both rigid and bristly radial spines (Figure 17.52). Larger plants commonly grow leaning to the south. The flowers are yellow to orange-red; the fruit is yellow.

Animal Food: The Seri said desert bighorn sheep and javelina eat this cactus, and that the bighorn tear it apart with their horns.

Containers: A unique container for liquids was made from this cactus. The plant was cut off at the base and the spines burned away with a fire made from brush. The top was cut off and the inside scraped out, leaving a shell at least 5 cm thick. A stick was put through the container below the rim and a loop of twine tied to the stick for carrying. In June, 1951, while camped on Tiburón Island, Alfred F. Whiting (1957:#67) noted that the spines were removed with a knife, the plant hollowed out from the base and hung in a carrying net, presumably from a carrying yoke. The barrel cactus container was especially used for carrying honey back to camp. After the honey was eaten the candied cactus was cut up and given to the children to eat. This container was also used to carry water (Whiting 1957:#67) and for wine-making.

Emergency Water: When fresh water was not available, liquid was extracted from this barrel cactus. Dry stems of limberbush (*Jatropha cuneata*) or other kindling were placed around the cactus and burned to remove the spines. It was then uprooted, or cut off just above the root, and chopped open. A thick clamshell, such as the Gulf cockle (*Trachycardium panamense*), or the lower jaw of a donkey was used to scrape out pieces of the juicy pulp (cortex and pith).

Another method, probably used in conjunction with the above method, was to poke the soft, fleshy pulp with sticks until it was somewhat mashed. Liquid was then squeezed by hand from the pieces of pulp into the resulting cavity in the cactus or into a 20-liter can or other suitable container. A fine, sandy-textured residue remained in the bottom of the can or container. The dried "meat" or pulp of the cactus with the liquid squeezed out was called *siml imínaj* "barrel cactus without pulp."

Alfred F. Whiting reported that the plant was

Figure 17.52. Common barrel cactus (*Ferocactus wislizenii*) on the east side of Tiburón Island. A) Pedro Comito next to a medium-sized plant. B) The clusters have both stout and bristle-like spines. *RSF, April 1983.*

". . . a common source of water. The adult cactus is pushed over, thorns burned off and the side opened with long rocks. The inside is pounded to a pulp and liquid collects in the cavity. It is palatable but tastes flat (Whiting 1957:#67)." A *siml* found leaning was said to contain more liquid than one growing erect. (Perhaps this was another means of distinguishing *F. wislizenii* from other barrel cacti.)

The juice caused some people to suffer pain in the "bones of their arms and legs." This pain occurred if one walked about a kilometer or more after drinking the juice. If one drank the liquid and remained still, the pain did not occur. Others were never affected by this pain. If one drank this cactus juice on an empty stomach (to the Seri this often meant not having eaten meat), he would usually get diarrhea, but not if he had recently eaten meat.

Use of barrel cactus juice as a substitute for fresh water was a common practice. It was especially important during the hot, dry months of late spring and early summer. Around 1928, about ten families were living at Las Víboras on the shore of the Infiernillo Channel. Fresh water was far away, and there were no roads. So instead of going for water, they used *siml* liquid. Each morning the men went out with cans to bring in scrapings from the inside of the cactus. Back at camp they squeezed the liquid from the scrapings and drank it. They used *siml* liquid for about a month, from March into April. Barrel cactus juice was occasionally used in making coffee.

In at least one instance *siml* liquid was used to bathe a newborn infant. The child was born about 1914 on the mainland side of the Infiernillo Channel. Because the nearest fresh water was in the Sierra Seri, the baby was bathed with barrel cactus liquid.

Facepaint: Yellow facepaint was prepared from the flowers. They were cooked until foamy, sugar or honey and white ashes were added, and the resulting mixture was used as facepaint. The flowers were also cooked with ground stem tips (herbage) of *Fagonia* spp. Barrel cactus petals were sometimes applied like facepaint.

Food: The flowers and buds were cooked in water, and sugar was added. Another method was to cook them in hot earth near a fire. The cooked flowers were a common food. (We found the taste to be similar to that of Brussels sprouts.) Children ate the fruit, although it is sour (like a lemon). The fresh pulp (cortex) was eaten as a survival food, or sometimes with honey. The seeds were collected from the fleshy fruit, ground and prepared as a gruel. Some of the seeds pop when heated. The hollowed out cactus was used as a honey container; after the honey was eaten, the cactus container was cut into long, thin slices like watermelon, and given to the children to eat.

The Supernatural: Many years ago the people of the Tastiota Region heard there would be a flood (a great, widespread disaster). They fled north, but the flood overtook them and they were turned into barrel cacti (*siml*).

Clouds come from all of the barrel cacti. The spirit power *Icor* causes them to form fog (*xeele*), which makes the clouds from which rain comes and gives life to all plants. "Fog and clouds have life; they are alive. You don't see clouds coming out of *siml* but there is a relationship." Once a *siml* sang a song. A shaman heard the cactus sing and learned songs from it. One of these songs is given in Chapter 13.

If one put *siml* juice on a *cardón* (*Pachycereus*), it was certain to bring rain. *Siml* was not considered a dangerous plant.

Wine: This cactus was used as a unique wine-making container. It was prepared in the same manner as the honey container described under Food. The hollowed-out cactus was filled with mashed fruit of columnar cacti, which was allowed to stand and ferment.

Mammillaria estebanensis Lindsay

hant iipzx iteja caacöl 'land its-torn(place) its-bladder (*Mammillaria*) large(plural)'
  "large fishhook cactus" or "large bladder of the arroyo"

This name was also applied to *Echinocereus*. This large fishhook cactus is endemic to San Esteban Island. A fresh specimen brought to El Desemboque was immediately recognized as being from that island (Figure 17.53).

Food: The orange-red fruit was eaten fresh as a snack.

Mammillaria microcarpa Engelm.

M. sheldonii (Brit. & Rose) Boed.
Mammillaria spp.
hant iipzx iteja 'land its-torn(place) its-bladder'
  "bladder of the arroyo"
fishhook cactus

Several small fishhook cacti described from western Sonora were not distinguished by the Seri. All bear small edible fruit about 1 to 2 cm long. Except for several distinctive species in the Guaymas region, most or all of the fishhook cacti in the mainland Seri region probably belong to one species complex.

Animal Food: The Seri reported that fishhook cacti were eaten by javelina.

Food: Children ate the fresh fruit.

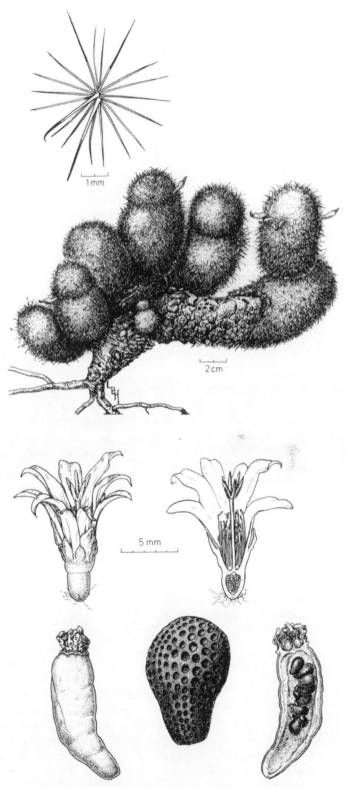

Figure 17.53. San Esteban fishhook cactus (*Mammillaria estebanensis*). *LBH.*

Medicine: As a remedy for earache, the spines were burned off, the plant cooked in water, peeled, and the "fiber" (vascular cylinder) crushed on a grinding stone. The resulting liquid was used as ear drops.

Neoevansia striata (Brandeg.) Sanch. Mejor.
[= *Cereus s.* Brandeg.; *C. diguetii* Web.; *Wilcoxia s.* (Brandeg.) Brit. & Rose; *W. diguetii* (Web.) Dig. & Gill.]
xtooxt

This slender-stemmed night-blooming cereus is fairly common and widespread in the region. Flowering is usually in late spring or early summer and the fruit matures later in the same summer and early fall. The fruit is red, succulent, 3 to 5 cm long, and produced sparingly along the pencil-thin stems. Each plant has a large cluster of potato-like roots.

Food: Some people ate the tuber-like root raw, but it was said that it may cause skin sores or rash. The fruit was eaten fresh, although it is not sweet like that of organ pipe (*Stenocereus thurberi*).

Medicine: The thickened root was prepared as a remedy for swellings. The raw tuber-like root was peeled and ground on a metate, salted, and heated with oil (such as olive oil) and placed on the swollen area. To harden a baby's fontanel, the tuber-like root was ground on a metate, salted, and placed on the baby's head.

Opuntia arbuscula Engelm.
siviri, pencil cholla

The Seri recognized two related kinds of cholla which seem to be *O. arbuscula*. It is a variable species and these may be modes or distinctive ecotypes, or populations. The glochid—the characteristic small spine of the genus *Opuntia*—was called *zatx*.

heem
heemáa 'true pencil cholla'

This plant was said to occur in the mountains and be similar to *heem icös cmasl* (see below) except that it does not have yellow spines. It was described as having few spines and "regular-sized" fruit, yellowish white when ripe. However, a few people applied the name *heem* to *O. versicolor* and the name *hepem ihéem* to *O. arbuscula*.

heem icös cmasl 'heem its-spines yellow-(plural)'
"yellow-spined *heem*"

The same name was applied to *O. cf. burrageana* from San Esteban Island. Yellow-spined *heem* was said to be more than a meter tall, and to have sour fruit with many large seeds. It was characterized by yellow spines, whitish-colored fruit, and large seeds; it had the most spines and the largest fruit of the three kinds of *heem* (includes *O. versicolor*).

Food: The fruits of the various *heem* were eaten fresh after the spines were removed, but were probably not a very important food resource.

Opuntia bigelovii Engelm.
coote, sea
cholla güera, teddybear cholla

This cholla is especially common on arid and exposed sites, often on hot south- and west-facing slopes (Figure 17.54). It occurs on the mainland and on Tiburón and San Esteban islands. It is not as common on San Esteban as *O. cf. burrageana*, the only other cholla on the island. The black gum is *coote ooxö* (see *O. fulgida*). The dry dead wood, or cholla "skeleton," was called *cotéexoj*.

Food: The younger stems ("joints") up to about one year's or season's growth were eaten. Dry brush, often from desert saltbush (*Atriplex polycarpa*) or *Colubrina*, was piled about 1 to 1.5 m high on top of a small shrub. On this were placed pieces of dry wood, usually from elephant tree (*Bursera microphylla*). By means of two pieces of the woody ribs of organ pipe (*Stenocereus thurberi*) used as tongs (see Figure 6.11), the young cholla stems were broken off and placed on top of the firewood (Figure 17.55). These tongs were often about 1 m long. The firewood was then ignited to burn off the spines and partially cook the stems. With the tongs the stems were lifted out of the embers and placed in a previously dug pit about 30 cm deep, then covered with earth and left for about one-half hour or so to continue cooking. The stems were removed from the pit, the soil cleaned off, and they were then ready to eat. The plant was an important food resource of the people living in the interior of Tiburón Island.

Medicine: The core of the root was boiled and the tea taken as a diuretic.

Oral Tradition: In "The Sea Horse's Entrance into

Figure 17.54. Teddybear cholla (*Opuntia bigelovii*). A) A dense stand of mature plants at Pozo Peña. B) Close up of plants near El Desemboque, showing new growth suitable for harvesting for food. *RSF, April 1983.*

Figure 17.55. Teddybear cholla stems on brush firewood prior to being cooked. *MBM, near El Desemboque, February 1964.*

the Ocean" this cholla (actually *cotéexoj*, the dry wood or "skeleton"), *cardón*, and sahuaro were once people. The holes in cholla wood came ". . . from the stones and arrows that hit him, and one could see right through him" (see *Pachycereus*).

The Supernatural: When a girl nearing puberty died, there was a four-day mourning period called *hant cacáxöla* 'place who-made-cry(plural).' Special singing and dancing took place, with crying, wailing, yelling, and laughing. The dead girl's personal property and food which would have been eaten at her puberty fiesta were supposed to be burned on a fire made only of *cotéexoj*, the dead wood of this cholla.

Opuntia cf. *burrageana* Brit. & Rose
heem icös cmasl '*Opuntia-arbuscula* its-spines yellow(plural)'
    "yellow-spined pencil cholla"

This is the common cholla on San Esteban Island. A fresh branch brought to El Desemboque was immediately recognized by Jesús Morales and others as being from San Esteban Island. Several Seri pointed out that it has longer and thicker stems than the *heem* (*O. versicolor*) from Tiburón Island.

Food: The fruit is edible, and was probably eaten by the Seri.

* *Opuntia ficus-indica* (L.) Mill.
heel cooxp 'prickly-pear white'
    "white prickly-pear"
heel cocsar yaa 'prickly-pear Mexican his-belonging'
    "Mexican's prickly-pear"
nopal, Indian-fig prickly-pear

This large, tree-like prickly-pear is often cultivated at ranches. It is widely grown in Mexico and elsewhere.

Opuntia fulgida Engelm.
cotéexet 'Opuntia bigelovii -exet'
cholla, jumping cholla

Coote seems to be the generic term for teddybear cholla (*O. bigelovii*) and jumping cholla. The concept of *cotéexet* includes both var. *mammillata* (Engelm.) Coult. and var. *fulgida*. Both varieties occur in the region; var. *fulgida* is so spiny that the stem is obscured, while var. *mammillata* has relatively fewer spines and the green stem is readily visible.

Cotéexet proliferates on the desert floor throughout most of the mainland region. It occurs on Tiburón Island, but it is not common throughout the island. Pendulous chains of fruit hang from the branches and may remain attached for several years. The fruit is fleshy, green even when ripe, and lacks spines but does have many glochids, called *zatx*. There is considerable variation in fruit size, which seems to be genetically controlled. Plants with extra-large fruit occur among both varieties in widely scattered places both inland and along the coast.

The fruit was called *tootöjc*. A gum which oozes from the stem and turns into black, fist-sized, dry nodules was called *coote ooxö* 'coote its-excrement.' It is an amorphous, tar-like substance. It is common and seems to result from an injury to the stem, such as that caused by certain large boring beetles. The powdered gum is very hygroscopic.

The fruit sometimes develops a hard dry "skin" said to be caused by heat waves. This hard-skinned fruit was called *hant ihajómjoj cmanim* "covered by the shimmering heat of the land."

Animal Food: The fruit was said to be eaten by desert bighorn sheep and javelina.

Food: The fruit was harvested throughout the year. It is tart but sweet, and continued to be harvested in the 1980s. It was a major food resource nearly throughout the region, and was highly esteemed by the people living in the interior of Tiburón Island. *Tootöjc* was selectively harvested from plants with extra-large fruit. The smaller, or ordinary-sized, fruit was not harvested because it does not have much "meat."

People removed the fruit by twisting it off the plant with their fingers, being careful to avoid the glochids, or by knocking it off the plant with a stick. To remove the glochids piles of the fruit were scrubbed, or swept, with branches of creosotebush (*Larrea*), although other desert shrubs, such as bur-sage (*Ambrosia dumosa*) and coast shrub (*Frankenia*), were also used. Sometimes the fruit was wiped with a wet cloth. The fruit was often rolled in the sand to remove most of the glochids, carried back to camp in buckets, and soaked in water to get rid of the rest of the glochids. The "skin" (thick ovary wall) was peeled with the fingernails or a knife, or sliced away with a sharp knife.

The fleshy green inner part, including the seeds, was eaten fresh—with salt or plain. It was also cooked in water and then mashed; honey was added for variety. The fruit was said to be very good roasted in hot coals for not more than about one-half hour. On rare occasions, when the people were in a hurry, they removed the glochids and cooked the entire fruit.

In April, 1983, we observed several women harvesting *tootöjc* fruit (Figure 17.56). They selected a site about 5 km east of Pozo Coyote because of the large fruit on many of the chollas. However, not all of the plants had large fruit. Each woman, after choosing a cholla with large fruit, walked around the cactus picking the better-looking and easily reached fruit. The women deftly twisted off the fruit with their fingers, avoiding the glochids, and usually tossed the fruit on the ground. Subsequently the fruit was picked up and put in a heap. They then broke off leafy creosotebush branches, put several branches together to make a small broom, and swatted and brushed the fruit on the sandy-gravelly ground to clean off the glochids. The outside of each fruit was trimmed away with a large, sharp, heavy-duty kitchen knife. Trimming was usually done while sitting on the ground. One woman used her sandal as a platform to clean and trim the fruit, and another stuck *tootöjc* fruit on a limb of the same cholla, impaling it on spines as she pared it down with her knife. The trimmed fruit was wrapped in a cloth to take home to El Desemboque. They frequently whetted their knives on any available rock. Some of the fruit was eaten right away. It was tart but sweet, with a very pleasant flavor.

While harvesting the *tootöjc* fruit, one of the women found some of the gum, *coote ooxö*. It was in a large, very spiny plant. She dislodged it with a stick and then used the stick to flip it out of the spines onto the ground. With her knife she cleaned away the spines and spider webs from the outside (Figure 17.57).

The gum was prepared in various ways. It was a much relished food. The softer, inside part of freshly gathered chunks was consumed raw or whole chunks were boiled and eaten. It was often toasted, ground, and mixed with water and drunk as a beverage. It was also eaten with honey or sugar. Sometimes it was

Figure 17.56. Lolita Blanco harvesting jumping cholla (*Opuntia fulgida*). A) Picking the fruit. B) Sweeping the fruit with creosotebush branches to remove the glochids. C) Trimming the fruit. D) Slicing off the outside of the fruit. E) The fruit ready to eat. *RSF, about 3 km east of Pozo Coyote, April 1983.*

Figure 17.57. Edible gum from jumping cholla, harvested about 3 km east of Pozo Coyote in April 1983. A) Aurora Comito cleaning the gum. *RSF.* B) Detail of the same piece of gum. *CMM.*

mixed with the sweet juice of cooked century plant (*Agave subsimplex*).

Medicine: The gum, ground and mixed with water, was taken as a remedy for diarrhea and also for shortness of breath. Children with persistent diarrhea were given the fleshy peel of the fruit cooked with a bit of the inside pulp and seeds, added for the tart flavor. The spines and "skin" were removed from the fresh stems, the fleshy green inside part boiled, and the resulting liquid was drunk as a remedy for heart pain or heart disease ("sickness"). For its use as a

Figure 17.58. Jumping cholla placed on a child's grave to discourage coyotes. *EHD, 1924; HF-23980.*

toothache remedy see *Anemopsis* and *Hyptis.* Inner bark from the root of this cactus and leaves of prickly poppy (*Argemone* spp.) were made into a tea taken to alleviate urinary problems, as a diuretic, and to "clear the urine."

The Supernatural: The gum was eaten by *Hant Hasóoma*, the principal spirit of the desert.

Other Uses: Pieces of the stem with the spines removed were thrown into a pool of muddy water. It was said that the water then became clear enough to drink. Cholla was placed in and on top of graves in order to discourage coyotes (Figure 17.58).

Opuntia fulgida Engelm. var. *fulgida*
sea icös cooxp 'Opuntia-bigelovii its-spines white'
    "white-spined teddybear cholla"

This appears to be a whitish-spined form of jumping cholla. The fruit was said to be smaller and with more spines than the common form(s) of jumping cholla.

Opuntia fulgida Engelm. cf. var. *fulgida*
sea cotópl 'Opuntia-bigelovii that-clings'
    "clinging teddybear cholla"

The characteristics given by the Seri indicate this is either an extra spiny form of var. *fulgida* or possibly a hybrid, such as one between *O. bigelovii* and *O. fulgida*.

This cholla was said to be widespread, have longer spines than those of *O. bigelovii*, and small fruit. However, some said it does not have fruit and is a low growth of *sea icös cooxp*, another kind of *O. fulgida*.

Food: The fruit was prepared in the same manner and given the same name, *tootöjc*, as the fruit of the common *O. fulgida*, although it was not eaten as frequently.

Opuntia leptocaulis DC.
iipxö
desert Christmas cactus

This slender-stemmed cholla is widespread and common and generally occurs on the desert floor rather than on rocky slopes. The fruit is reddish and about 1 to 2 cm long.

Food: The fruit was spread out on the ground, carefully brushed with branches of creosotebush (*Larrea*) or other soft brush to remove the spines. It was eaten fresh.

Opuntia marenae Parsons
O. reflexispina Wigg. & Roll.
xomcahóij

These are relatively rare plants. *O. marenae* occurs at widely scattered localities from Kino Bay northward to Caborca. *O. reflexispina* is known only from the vicinity of Tastiota. These closely related but distinct species are cholla-like plants with a cluster of tuber-like roots (Figure 17.59).

Medicine: The thick roots were cooked in ashes and eaten as a cure for diarrhea.

Opuntia phaeacantha Engelm. var. *discata* (Griff.) L. Bens.
[= *O. engelmannii* Salm-Dyck]
heel hayéen ipáii 'prickly-pear face what-it-is-done-with'
    "prickly-pear used for face painting"
nopal, Engelmann prickly-pear

Along the coast of the Seri region this prickly-pear is uncommon and highly localized, e.g., on Alcatraz Island and in sandy places near the base of Punta Sargento. It was one of several species planted at Punta Sargento by people of that region. The fruit was called *heel imám* 'prickly-pear its-fruit,' or "prickly-pear fruit."

Facepaint: The fruit was occasionally crushed and used as a reddish facepaint.

Food: The fresh fruit, with the spines removed, was occasionally eaten.

Opuntia versicolor Coult.
hepem ihéem 'white-tailed-deer its-pencil-cholla'
    "white-tailed deer's pencil cholla"
heem icös cmaxlilca 'heem its-spines stiffly-protruding(plural)'
    "stiff-spined pencil cholla"

This plant was said to be the *heem* (*O. arbuscula*) belonging to the white-tailed deer. The Seri told us it was common along the base of mountains. It was pointed out that *heem* is green, whereas this cholla has reddish stems.

It is generally absent or uncommon in the more arid and open desert situations, and becomes commoner in and around foothills and mountains. It flowers in late spring and bears fleshy green fruit which often remains attached to the plant the entire year. A few people identified *O. versicolor* as *heem*, and *O. arbuscula* as *hepem ihéem*.

Animal Food: Mule deer eat the fruit.

Food: When walking through the desert, the people occasionally picked the fruit. They scraped it on their long hair to remove the spines and ate the fruit fresh. The Seri were aware that the buds were eaten by the Papago (the Seri did not eat cholla buds). These buds were a major food among the Papago.

Opuntia violacea Engelm. var. *gosseliniana* (Web.) L. Bens.
heel
ziix istj captalca 'thing its-leaves wide'
    "wide-leaved thing"
saapom
duraznilla, purple prickly-pear

Fairly common on rocky slopes, particularly at higher elevations, from the Sierra Seri to the mountains north of Guaymas; occasionally found in lowland desert flats. The "pads" (stems) of this prickly-

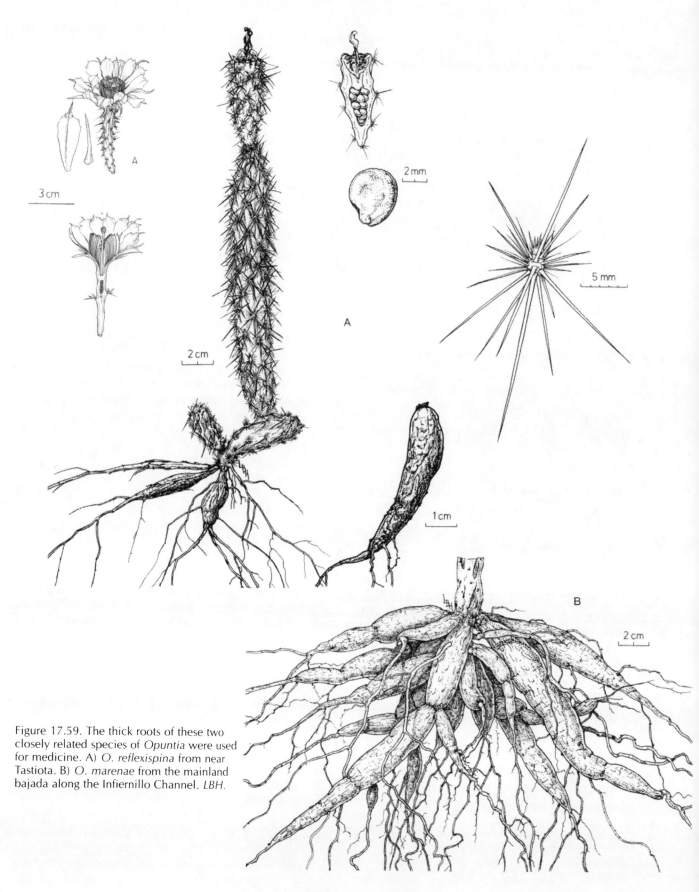

Figure 17.59. The thick roots of these two closely related species of *Opuntia* were used for medicine. A) *O. reflexispina* from near Tastiota. B) *O. marenae* from the mainland bajada along the Infiernillo Channel. *LBH*.

pear turn purplish during winter and early spring in response to cold or drought. It was one of several species planted near Punta Sargento by people of the Sargento Region.

Facepaint: A pink or pale red facepaint was occasionally made from the fruit. The spines were removed and the "skin" of the fruit rubbed in the palm of the hand to mash it. It was then applied directly to the face.

Food: The fresh fruit was occasionally eaten after the spines were removed. As with other *Opuntia* fruit, the spines were removed by rolling it in sandy or gravelly soil with a small branch. Some people said eating the fruit may result in a headache.

Campanulaceae—Bellflower Family

Nemocladus glanduliferus Jepson
xtamáaija oohit 'mud-turtle what-it-eats'
   "what mud turtles eat"
thread plant

The Sonoran mud turtle (*Kinosternon sonoriense*), inhabits fresh water, such as at the dam at Hermosillo and at Carrizal, north of Kino Bay.

This diminutive winter-spring ephemeral ranges through most of the region. It is often common in wet sand and mud along arroyo beds, as well as on the open desert floor.

Cannabaceae—Hemp Family

*Cannabis sativa L.
hapis coil 'what-is-smoked green'
   "green tobacco"
marijuana

Around the middle of the twentieth century a Seri man grew marijuana plants for his own use in the desert near El Desemboque. He carried water to his tiny, illicit garden. Running water had not yet been installed in the village, and with the exception of these few plants, the Seri were not at that time cultivating any other plants. A tube worm shell, *xtozaöj* (*Tripsycha tripsycha*), served as a pipe for smoking marijuana.

Capparaceae—Caper Family

Atamisquea emarginata Miers
cöset
palo zorrillo

A very common shrub through most of the Seri region, including Tiburón Island. Usually found along arroyos, bajada slopes, and bottom lands. The nearly evergreen foliage consists of small, linear, dark green, and leathery leaves. The wood is hard.

Firewood: The wood was used for fuel.

Oral Tradition: The dry wood was called *izáahoj quequéect* 'sunlight son-in-law' or "sunlight's son-in-law." This plant was once a boy who married the daughter of the sunlight. The plant was ashamed in the presence of its father-in-law and did not burn in the daylight, but rather gave off smoke.

The Supernatural: The wood had a powerful spirit and was used in several special ceremonies called *hamcáatxi* 'smoking.' One ceremony was performed to cure a fussy, crying baby. Any woman who had a gentle nature and was known to speak quietly and not be a gossip might have been asked to perform a special *hamcáatxi* ceremony for a baby which cried incessantly. M. Moser (1970a:208) described this ceremony as follows:

> During a ceremony which I observed, twigs of *xoop* (*Bursera microphylla*), *cöset* (*Atamisquea emarginata*), *seepol* (*Frankenia palmeri*) and the nest of a bird known as *ziic ano yait cöquiim* 'bird inside touch sleep-during,' i.e., bird that sleeps during the afternoon, were mixed together and burned in small fires built upon 4 piles of sand placed around the baby. As the smoke from each fire ascended and the curer sang appropriate songs, she called on the spirit of the bird of the burning nest and implored it to cure the baby of his crying. The spirit is said to come in the smoke and effect the healing. The fires were then extinguished. The unburned leaves and twigs were gathered together and given to another observer who was asked to take them to the baby's mother

who did not attend the ceremony. The piles of sand were smoothed out and a cross drawn on each spot. Picking the baby up, the curer drew a cross on the sand where he had lain and then returned the baby to his aunt who had brought him for the ceremony. She reminded the aunt to warn the mother not to throw the twigs away but to stick them somewhere high in her hut.

Another smoking ceremony was performed for the curing of a turtle hunter's harpoons. His female sponsor, *hamác*, made a fire from the wood of *palo zorrillo*, elephant tree (*Bursera microphylla*), and *Frankenia* or iodine bush (*Allenrolfea*). She added to the fire a bit of the outer surface of a sea turtle carapace and then, as he passed his harpoons and hands through the smoke, she called for his success in hunting.

Powdered *Atamisquea* wood was mixed with Seri Blue (see *Guaiacum*) and carried in a reedgrass (*Phragmites*) tube for good luck. It was used to make crosses on the face, signifying that the wearer was placing a curse or asking for luck. Others did not question the supplicant.

Cleome tenuis S. Wats.
cocóol

This name was also used for *Descurainia* and *Sisymbrium*. This non-seasonal ephemeral is widespread, although not abundant in the region.

Medicine: The seeds placed in water were given to a baby to drink as a remedy for stomachache.

Caryophyllaceae—Pink Family

Achyronychia cooperi Gray
hant yapxöt 'land its-flower'
   "land's flower"
tomítom hant cocpétij 'tomítom land circular-flat'
   "prostrate *tomítom*"
tomítom hant cocpétij caacöl 'tomítom land circular-flat large(plural)'
   "large prostrate *tomítom*"
frost mat, chaff-nail

The second name listed above implies a prostrate plant; this concept was usually applied to various herbaceous spurges. There are several variations of the term *tomítom* (see *Euphorbia polycarpa*). *Hant yapxöt* was also used for *Tillaea*.

It is a small, prostrate-growing, winter-spring ephemeral that is seasonally common on sandy soils through most of the region.

Celastraceae—Bittersweet Family

Maytenus phyllanthoides Benth.
   [= *Tricerma p.* (Benth.) Lundell]
cos
mangle dulce

A medium to large, multi-stem shrub with evergreen, succulent leaves (Figure 17.60). The small, fleshy red fruit ripens at the end of April. It occurs along the shores of the mainland and Tiburón Island. Impenetrable thickets of it are scattered along the shores of the Infiernillo Channel and at estero fringes. It is tolerant of seawater inundation and reaches maximum development on saline wet soils.

Firewood: *Cos* is one of the best firewoods found near the beach.

Food: The fruit was commonly removed from pack rat nests, where it was found in "slabs or hunks packed like raisins." It was also eaten fresh, but was considered to be sweet only when dark red and fully ripe.

Food Preparation: The leaves and soft, young, leafy branches were used to line a basket or turtle shell for holding meat. The wood was used to make a meat rack. A trunk or large branch about 2 m long with branches was brought into camp. The twigs or smaller upper branches were cut off, sticks laid across the branch forks, and on top of this were placed leafy twigs of "soft" plants which were not bitter. Meat was spread on it so the animals could not reach it.

Medicine: The bark was cooked in water and

Figure 17.60. *Mangle dulce* (*Maytenus phyllanthoides*) from the Infiernillo Coast. *FR.*

mashed, jojoba (*Simmondsia*) seeds added, and the liquid used as a gargle to cure a sore throat. Tea made from the bark of the root was taken to cure dysentery. This tea or one made from the leaves was used as a gargle for sore throat.

Shelter: Dry fallen leaves, gathered on the ground, were placed in thick layers on top of seaweed (*xpanáams*, Sargassum) as roofing for the brush house. This was said to keep out the rain. Because the branches are crooked, they were not used for construction.

Chenopodiaceae—Goosefoot Family

Allenrolfea occidentalis (S. Wats.) Kuntze
tacs
chamizo, iodine bush

This evergreen bush or small shrub has succulent stems with bead-like green or reddish green joints and an alternate branching pattern (Figure 17.61). It is abundant at the desert edge of marshy and saline wet places throughout the Gulf of California. The seeds are reddish brown, about 0.8 mm long, and are produced in great quantity in mid-winter. A container the size of an ordinary work basket can be filled in about ten minutes by shaking the seed-bearing branches into it.

Food: The seeds were toasted, ground, and cooked as gruel, or mixed with turtle oil. The seeds were said to pop when toasted.

Shelter: The branches were used to provide shade and roofing for the brush house.

The Supernatural: *Tacs* was considered to be one of the first plants formed. It was one of several plants used in the smoking ceremonies (see *Atamisquea*).

Tanning: Because it was a *caitic* 'soft' plant, it was used to cover the ground beneath a deer hide to keep it clean during the tanning process.

Figure 17.61. Iodine bush (*Allenrolfea occidentalis*) from Estero Sargento. Note the alternate, bead-like, and succulent stems. *CMM.*

Atriplex barclayana (Benth.) Dietr.
spitj
coastal saltbush

This semi-herbaceous, perennial saltbush is abundant along the shores and coastal regions of the Gulf of California and its various islands (Figure 17.62). It was considered to be a *caitic* 'soft' plant.

Medicine: The twigs and foliage were cooked in water with the leafy twigs of *Bursera microphylla*, and the resulting liquid used to bathe a stingray wound.

Shelter: The leafy branches were used in roofing of the brush house.

Tanning: The leafy stems were used to cover the ground beneath a deer skin being prepared for tanning.

Atriplex canescens (Pursh) Nutt.
A. linearis S. Wats.
hataj-isijc 'vulva immature'
    "immature vulva"
hataj-ixp 'vulva white'
    "white vulva"
chamizo, four-wing and narrow-leaf saltbush

These dense shrubs are common on coastal dunes and low-saline and alkaline soils. They have four-winged fruits. *A. canescens*, four-winged saltbush can be distinguished from *A. linearis* by its larger stature and larger fruits and leaves. The Seri did not distinguish between them, and also sometimes did not distinguish them from *A. polycarpa*.

Medicine: Tea made from the leaves was taken as an emetic.

Shelter: The leafy branches were used as roofing material for the brush house.

Atriplex polycarpa (Torr.) S. Wats.
hataj-ixp 'vulva white'
    "white vulva"
chamizo cenizo, desert saltbush

This common desert shrub extends over a wide range of arid habitats throughout the region, including Tiburón and San Esteban islands. It occurs on rocky slopes, desert flats, arroyos and dunes. The wood is fairly hard.

Firewood: It was used as kindling, for firing pottery, and for scorching the bark of *haat* (*Jatropha cuneata*) stems in preparation for basketmaking.

Hair Care and Washing: The leaves and twigs were mashed and added to water, and the solution used as shampoo or for washing clothes.

Shelter: The leafy branches were used in the roofing of the brush house.

Figure 17.62. Coastal saltbush (*Atriplex barclayana*). A) Plants on low beach dunes 2 km northwest of El Desemboque. *RSF, April 1983.* B) Detail of plants from Puerto Libertad: *left,* vegetative branch; *center,* branch from staminate plant; *right,* branch from pistillate plant. *NLN.*

The Supernatural: To stop rain, a fire of this wood was built against a sahuaro (*Carnegiea*). It was stated that one must have faith in the act if the rain was to stop.

Chenopodium murale L.
ziim xat 'ziim hail'
goosefoot

The name refers to the white, hail-like appearance of the seeds when they are popped. Hail rarely occurs in the Gulf of California. The Seri's report that the plant is abundant on Alcatraz Island confirms our observations. It is a winter-spring ephemeral (Figure 17.63) usually occurring in widely scattered but local pockets throughout the region, including Tiburón Island. It thrives as a weed around the camps and is abundant at El Desemboque. It is supposedly native

Figure 17.63. Goosefoot (*Chenopodium murale*) from Alcatraz Island, spring 1970. *FR.*

to the Old World and now widely naturalized in the New World.

Food: This important seed crop was harvested in late spring. The seeds were toasted, ground and made into gruel. The seeds also were stored in pottery vessels. While several other kinds of seeds may pop somewhat when heated (e.g., *cardón*), this is the only one said to open like popcorn.

Salicornia bigelovii Torr.
[= *S. europea* L. of western American authors]
xnaa caaa 'south-wind what-calls'
"what calls for the south wind"
pickleweed

This is the only annual *Salicornia* in the region. It is common in esteros and salt marshes in the northern Gulf of California. The stems are green and succulent, with an opposite branching pattern.

The Seri said that from a distance the newly sprouting plants look like young, green wheat, that it grows in the esteros, and may be found with mangroves, usually between the mangroves and the beach.

Food Preparation: It was sometimes used to line a basket or sea turtle shell to keep meat clean.

*Salsola kali L.
ziim caacöl 'ziim large'
    "large *ziim*"
chamizo volador, Russian-thistle, tumbleweed

This common agricultural weed has invaded disturbed sites along the Infiernillo coast. Other kinds of *ziim* are *Amaranthus* and *Chenopodium*.

Suaeda esteroa Ferren & Whitmore
sipöj yanéaax 'cardinal what-it-washes-its-hands-with'
    "what the cardinal washes its hands with"

The cardinal (bird) was once a person. After harpooning a sea turtle his hands were covered with blood, so he wiped them on this plant.

The plants are about 50 cm tall. The succulent leaves are awl-shaped and flattened on top (Figure 17.64*A*). The foliage is often reddish. It is common in mangrove esteros.

Shelter: The leafy branches were used for roofing for the brush house.

Suaeda moquinii (Torr.) Greene
    [= *S. torreyana* Wats.]
hataj-ipol 'vulva black'
    "black vulva"
sosa, sea blite

A large, dense hemispherically-shaped perennial shrub with succulent leaves and thin brittle stems (Figure 17.64B). There is only one species of large, shrubby *Suaeda* in the region. It is common along beaches, saline and alkaline flats, and lower bajada slopes near the shore. It occurs on the mainland and Tiburón and San Esteban islands.

Dye: A black dye, called *haat an ihahóopol* 'Jatropha-cuneata in to-blacken,' was occasionally made with sea blite. The stems and leaves were pounded on a grinding stone, mixed with water, placed in a pottery olla or pan, and cooked slowly for about 24 hours. Then a few pieces of pomegranate peels, if available, were added, along with several pods of *Pithecellobium confine* and a handful of desert lavender (*Hyptis*) leaves. After the mixture boiled for only a few minutes, basketry splints (see Chapter 15) previously dyed reddish brown with *cósahui* (*Krameria grayi*) were coiled in the solution. In a short time the splints were dyed black. Black dye was also made by boiling sea blite roots with those of *Jatropha cuneata* or with mesquite (*Prosopis*) bark.

Medicine: Tea made from the roots was taken to relieve a cold.

Shelter: The leafy branches were used as roofing for the brush house.

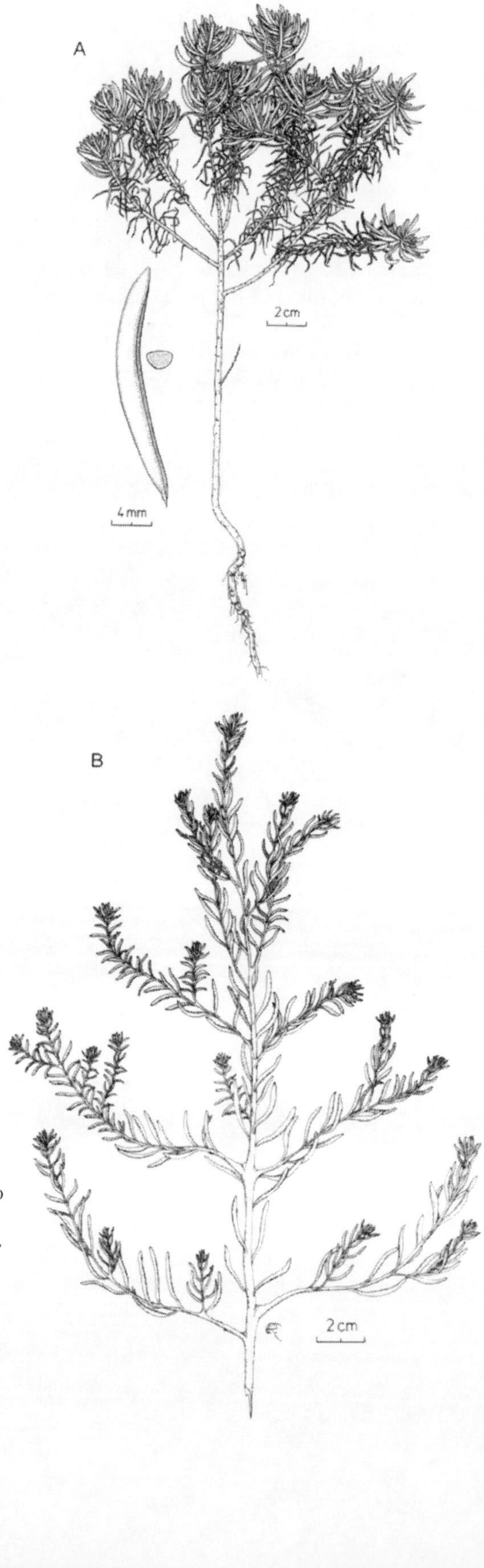

Figure 17.64. The two species of sea blite from the Seri region. A) A mature plant of *Suaeda esteroa*, from Estero Sargento. A single leaf and cross-section are shown on left. *LBH*. B) A vegetative branch of the shrubby species, *Suaeda moquinii*, from Kino Bay. *FR*.

Cochlospermaceae—Cochlospermum Family

Amoreuxia palmatifida DC.
xoját
saiya

This root perennial (Figure 17.65) responds to summer rains and is dormant during the rest of the year. The tuber-like, succulent roots are commonly 20 to 30 cm long. The flowers are bright orange-yellow, ca 5 cm across; the capsule is ovoid, greenish, 3 to 4 cm long, and fleshy until the seeds ripen, at which time it dries and splits. The seeds are kidney-shaped, black, and about 4 to 5 mm long. It is found on rocky slopes on mountains and foothills from near El Desemboque to Guaymas, and on Sierra Kunkaak. Often locally abundant. A noteworthy population occurs with wild teparies on steep, rocky, north-facing slopes near the tepary gathering camp on Tiburón Island (see *Phaseolus acutifolius*).

Adornment: The seeds were strung for necklaces (Figure 17.66; also see McGee 1898:172)

Food: The root was an important food resource and was especially significant to the people living in the interior of Tiburón Island. It was dug with the *hapoj* digging stick. The root was eaten raw, boiled with meat, or toasted lightly in the fire. The root was savored with honey, or mule deer or turtle fat. The uncooked root was said to taste like *jícama* (a common root vegetable of Mexico). Bark from the root was cooked like beans, with mule deer or green turtle fat. The buds, flowers, and tender young fruit also were eaten fresh or cooked with mule deer fat and bones.

In times of hunger, such as when walking in the desert, a mother sometimes gave slices of the dried strung root to her child to eat (Figure 17.67).

Song: This song was said to be sung by the *xoját* about itself. The yellow flowers are *xoját* flowers, and the blue ones are *mahyan* flowers.

> *hehe yapxöt coil ano hamíticol*
> *hehe yapxöt coil ano hamíticol*
> *ano hamíticol,*
> *yapxöt coox caail iti yomásol.*
> *hai he tap xoonoj.*

> We are among the blue flowers
> We are among the blue flowers
> We are among them,
> The dry lake is all yellow with flowers.
> The wind touches us and roars.

Figure 17.65. *Saiya* (*Amoreuxia palmatifida*) from northern Sonora. A) A newly emerging stem with the first flower of the season. The flower is about 5 cm across. B) The roots and base of the same plant. *RSF, July 1983.*

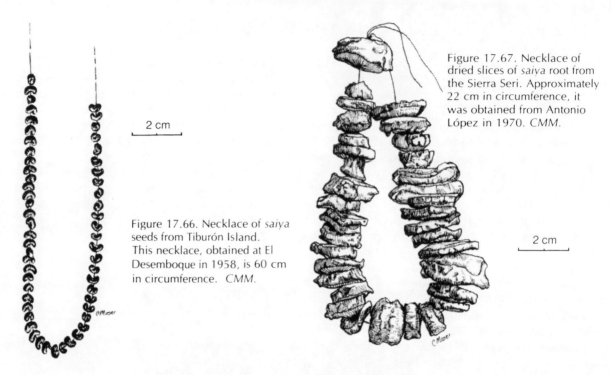

2 cm

Figure 17.66. Necklace of *saiya* seeds from Tiburón Island. This necklace, obtained at El Desemboque in 1958, is 60 cm in circumference. *CMM.*

Figure 17.67. Necklace of dried slices of *saiya* root from the Sierra Seri. Approximately 22 cm in circumference, it was obtained from Antonio López in 1970. *CMM.*

2 cm

Combretaceae—Combretum Family

Laguncularia racemosa (L.) Gaertn. f.
pnaacoj hacáaiz 'mangrove harpoon'
mangle, white mangrove

Pnaacoj is the generic term for mangrove. This is one of the three mangroves in the Gulf of California (see *Avicennia* and *Rhizophora*). It is a large evergreen shrub or small tree with relatively hard wood, and shiny green leaves (Figures 17.68, 17.69). It occurs in esteros, bays, and inlets, from Punta Perla on Tiburón Island and Estero Sargento southward into the tropics.

Boats: White mangrove wood was often used to make the handle of the single-blade paddle for sailboats. Although it was not very strong, sometimes the wood was also used for a sailboat mast.

Firewood: It was considered inferior as firewood, but was used if nothing else was available, as often was the case along the shore.

Hunting: Harpoon shafts were fashioned from straight branches. Young men decorated them in the following manner. Designs were cut into the bark, and bark not within the outline of the design was excised. The shaft was then scorched, and the remaining bark removed, producing a natural, light-colored ("white") design. Each man created his own design.

Shelter: The branches with their dense foliage were extensively used for roofing on the brush houses and ramadas. The wood was used for house posts, beams, and roofing of the wattle and daub houses and the brush houses.

Compositae (Asteraceae)—Composite or Daisy Family

Ambrosia ambrosioides (Cav.) Payne
    [= *Franseria a.* Cav.]
tincl
chicura, canyon ragweed

Common bushy perennial with slender stems and elongated, triangular, rough-surfaced leaves. It usually occurs in canyons and arroyos, and ranges throughout the region, including Tiburón Island.

Medicine: Tea made from the roots was given to a woman near parturition.

Pigment: The pounded root bark was sometimes

Figure 17.68. White mangrove (*Laguncularia racemosa*) at
Estero Sargento. A) An old, isolated tree, April 1974.
B) The aerial roots, April 1983. *RSF.*

281

Figure 17.69. A vegetative branch of white mangrove (*Laguncularia racemosa*) from the Estero at Campo Víboras, January 1972. *LBH.*

substituted for bur-sage (*Ambrosia dumosa*/spp.) in the manufacture of Seri Blue (see *Guaiacum*). It gave the pigment a distinct aroma.

Smoking: The leaves were dried and smoked.

Ambrosia cf. *confertifolia* DC.
paxáaza
ragweed

This perennial herb is a characteristic weed on farmlands, roadsides, and other disturbed sites. We found it at Rancho Miramar near El Desemboque.

Ambrosia deltoidea (Gray) Payne
A. divaricata (Brdge.) Payne
A. dumosa (Gray) Payne
A. magdalenae (Brdge.) Payne
    [= *Franseria* spp.]
xcoctz 'old'
an icoquétl
cmajíic ihásaquim 'women what-one-brushes-hair-with'
    "what women brush their hair with"
chamizo, bur-sage, burro-weed

The Seri did not distinguish among these small, bushy plants. They are common desert shrubs with grayish foliage and small bur-like fruits. *A. deltoidea* and *A. dumosa* are especially common and widespread in the northern Gulf of California region, including Tiburón Island. These species do not occur on San Esteban Island.

Pigment: The pulverized root bark was used in the manufacture of Seri Blue (see *Guaiacum*).

Ambrosia ilicifolia (Gray) Payne
    [= *Franseria i.* Gray]
tincl
xcoctz 'old'
hollyleaf bur-sage

The same names were used for *A. ambrosioides* and *A. deltoidea*/spp. Although recognized as a distinct plant, *A. ilicifolia* was not given a distinct name. Low, spreading bush with stiff, holly-like, scabrous leaves. In the Seri region it is abundant on San Esteban Island and also occurs on the northwest coast of Tiburón Island.

Baccharis salicifolia (R.& P.) Pers.
    [= *B. glutinosa* Pers.]
caaöj
batamote, seep willow

Wispy, leafy shrub with long, straight, slender stems. The young herbage is resinous-glutinous. Localized populations occur in widely scattered riparian situations on the mainland and *Xapij* (Sauzal) on Tiburón Island.

Adornment: After the bark was removed, the stem was cut into pieces approximately 1 cm long and the beads strung as necklaces. Sometimes the beads were toasted to color them black (see *Asclepias subulata*) or dyed reddish orange.

A special necklace worn by men, called *hant xec* 'land *xec*,' consisting of three or four strands of these beads, was worn across the front of a blouse or shirt called *coton*. The necklace was pinned or tied to both sides of the *coton* with colored cloth strips. The *coton* was worn by both men and women in the nineteenth and twentieth centuries. It had a round neck, sometimes a small opening in front, and extended to within about 10 cm of the natural waistline (see Figure 1.5).

Medicine: Tea made from the leaves was taken to help one lose weight, and as a contraceptive. This same tea also was used to stop loss of blood after childbirth. The leaves, heated over coals, were placed

282

on the head as a remedy for headache or on a sore area of the body.

Baccharis sarothroides Gray
casol caacöl ‘*casol* large(plural)’
   “large *casol*”
escoba amarga, romerillo, desert broom

The term *casol* subsumes a number of other species of composites with aromatic, glutinous, or resinous herbage, e.g., *Haplopappus, Hymenoclea*, and *Pectis*, and also hop bush (*Dodonaea*).

Tall bushy shrub with green stems and twigs, and reduced leaves. Highly localized populations occur here and there at the margins of the dry lake beds south of El Desemboque and elsewhere, and along arroyos and floodplains of major streamways.
Medicine: Tea made by cooking the twigs was taken as a remedy for a cold. The same tea was rubbed on sore muscles for relief.

Baileya multiradiata Torr.
caháahazxot ‘what causes sneezing’
desert-marigold

This is an onomatopoetic name, and sounds like a sneeze. The name is also applied to *Dyssodia*.
A spring wildflower; common from the vicinity of Puerto Libertad northward, and infrequently found southward.

Bebbia juncea (Benth.) Greene var. *juncea*
sapátx
sweet-bush

A scabrous desert bush, common and widespread on rocky slopes, canyons and arroyos, and across the desert floor. The Seri liked the fragrance of the flowers.
Shelter: The branches were used for roofing of the brush house. It was one of the plants used to make the semi-circular roofless windbreak (see *Encelia*).
Water Transport: Full cans or containers of water were covered with branches of this plant to prevent spilling. It also gave the water a pleasant aroma.

Brickellia coulteri Gray var. *coulteri*
comíma

This small bush has leaves somewhat similar to those of *Eupatorium sagittatum*, also called *comíma*.

Brickellia is common along washes and arroyos in the mountains and inland, such as at Rancho Estrella.
Adornment: The fresh leafy stems were put in little cloth bags on necklaces because of their pleasant aroma. For this reason the plant was also rubbed on the clothing and put in one’s shirt or blouse pocket.

Chaenactis carphoclinia Gray
xtamóosn-oohit ‘desert-tortoise what-it-eats’
   “what desert tortoises eat”
cacátajc ‘what causes vomiting’

Delicate winter-spring ephemeral with whitish flowers; common in the northern part of the Seri region. The same names were also applied to *Fagonia* and *Pectis*.

Dyssodia concinna (Gray) Robinson
caháahazxot ‘what causes sneezing’
fetid marigold

The same name was also given to *Baileya*. Some Seri did not recognize the plant by name. This small winter-spring ephemeral is often seasonally common in the northern part of the region.

Encelia farinosa Torr.
cotx ‘acrid smell’
hierba del vaso, incienso, brittlebush

The name is derived from *ccotxta* ‘acrid smell.’ However, *cotx* refers only to brittlebush.
Perennial bush with whitish to grayish green foliage; becoming leafless during drought. Bright yellow, daisy-like flower heads on long stalks are produced at various times of the year, generally in spring and again with the summer rains. It is one of the most common and conspicuous wildflowers in the Sonoran Desert and seasonally turns the landscape yellow.
Two varieties occur in the region: var. *farinosa* with all yellow flower heads and var. *phenicodonta* (Blake) I.M. Johnst. with a brownish disk and yellow rays. The latter is more common.
Adhesives, Sealants, and Gums: The *cotx icsípx* ‘brittlebush its-resin’ from this plant is not common, and exudes near the “roots” (apparently meaning it is produced on the stem near the ground). It is dry and hard, often yellowish or dark in color. This resin was heated and used as a glue. It was sometimes used for

hafting, such as for the point and shaft of a sea turtle harpoon.

The Seri distinguished another dried sap or resin called *cotx itájc* 'cotx its-saliva.' It is somewhat gummy and was collected from the upper stems. It was used for sealing pottery vessels and as violin rosin. "A glob of resin . . . is often attached to the reverse side of the pegbox. The bow string is rubbed on this substance after every few songs (Bowen and Moser 1970b:180)."

Medicine: A green twig with the bark removed was heated in ashes. It was then placed in the mouth, and one would bite into it to harden a loose tooth. The root was used with *Asclepias* and *Euphorbia* for toothache and heart pain (see *Asclepias*). The resin, *csipx*, was ground and sprinkled on sores.

Shelter: The semi-circular windbreak, *hacáiin hant hapáhtolca* 'windbreak down what-is-put(plural)' was often constructed with upturned brittlebush.

Eupatorium sagittatum Gray
comíma

A weak-stemmed bush, generally in low and seasonally wet places in mesquite thickets. The same name was given to *Brickellia*, which it somewhat resembles in foliage. The two species often occur together. Concerning this plant it was said, "It just smells good."

Franseria, see Ambrosia

Haplopappus sonoriensis (Gray) Blake
casol cacat 'casol bitter'
  "bitter *casol*"
casol ziix ic cöihíipe 'casol thing with make-better'
  "medicinal *casol*"

The same names were also applied to *Baccharis sarothroides* and *Hymenoclea salsola*. It is a small shrub, generally in flat places near the coast.

Hofmeisteria fasciculata (Benth.) Walp.
taca imas 'triggerfish its-body-hair'
  "triggerfish's body hair"

This small, globose, succulent perennial is common on sea cliffs and steep rocky canyon walls along the coast. In the Guaymas region it is replaced by *H. crassifolia* S. Wats.

Hofmeisteria laphamioides Rose
hamt inóosj 'soil its-claw'
  "soil's claw"
hast yapxöt 'rock its-flowering'
  "rock's flower"

This small shrub is common nearly throughout the region, where it grows on steep rock slopes and cliffs. Seri names for the plant were not used consistently nor was it always recognized by name.

Hymenoclea monogyra Torr. & Gray
casol cacat 'casol bitter'
  "bitter *casol*"
casol coozlil 'casol sticky'
  "sticky *casol*"
casol ziix ic cöihíipe 'casol thing with make-better'
  "medicinal *casol*"
jécota, burro brush

The same names were also variously applied to *Baccharis sarothroides*, *Haplopappus sonorensis*, and *Hymenoclea salsola*. The name *casol cacat* became taboo because it was the favorite medicine of a woman who died. However, with the passing of time the original name came back into use.

It is a common, wispy shrub 2 to 3 m tall found along arroyos and floodplains of major drainage-ways, and as a weed on farmland. The herbage is resinous-viscid. It has many upright stems which are seldom more than 5 cm in thickness; the wood is rather brittle and moderately soft.

Firewood: It was occasionally used for firing pottery.

Medicine: The twigs (or herbage) were mixed with *Koeberlinia* twigs and boiled, and the tea was taken to cure skin rash. Tea made from the stems (or herbage) was used to relieve pain in the lungs and trachea and to reduce swellings. A remedy for rheumatism was made from this plant and *hierba de manso* (see *Anemopsis*).

Shelter: When camped at Pozo Coyote, people used the branches for the roofing and framework of brush houses.

Hymenoclea salsola Torr. & Gray
casol itac coosotoj 'casol its-bone thin(plural)'
  "thin-stemmed *casol*"

casol coozlil 'casol sticky'
  "sticky *casol*"
jécota, burro brush

Casol coozlil refers to the resinous-glutinous twigs which stick together when crushed. This name was sometimes also applied to *H. salsola*; also see *Baccharis sarothroides*.

Machaeranthera parviflora Gray
zaah coocta 'sun watcher'
hehe cacátajc 'plant that-causes-vomiting'
hee imcát 'jackrabbit what-it-doesn't-bite-off'
  "what the jackrabbit doesn't bite off"

The name *hee imcát* was also used for the rock daisy (*Perityle emoryi*). Various wildflowers were called "sun watcher" because the leaves and flowers turn to follow the path of the sun through the day. *Hehe cacátajc* was said to be poisonous.

Ephemeral or annual; fairly common through much of the lowland desert region. The flower heads are blue and yellow.

Palafoxia arida Turner & Morris var. *arida*
  [= *P. linearis* (Cav.) Lag., in part]
moosni iha 'sea turtle its-possessions'
  "sea turtle's possessions"
moosn-oohit 'sea-turtle what-it-eats'
  "what the sea turtles eat"
Spanish needles

This common ephemeral or annual occurs nearly throughout the region, and is especially common along the coast.

Medicine: The plant was ground and used to kill insect larvae in a dog's sores.

Pectis palmeri S. Wats.
casol heecto caacöl 'casol small large(plural)'
  "large small *casol*"

This species resembles *P. papposa* but differs in that the entire plant is larger, including the flower head. It occurs in the southern part of the Seri region and on the southern end of Tiburón Island.

Pectis papposa Harv. & Gray var. *papposa*

casol heecto 'casol small(plural)'
  "small *casol*"
casol ihasíi tiipe 'casol its-odor is-good'
  "fragrant *casol*"
cacátajc 'what causes vomiting'

This is one of the most abundant summer-fall wildflowers in the northern Gulf of California region. It has bright yellow flower heads and pungently aromatic foliage. Its flowering and relative abundance are reliable indicators of hot weather and soil moisture.

Pitaya agria (*Stenocereus gummosus*) was said to bear fruit when *casol heecto* was in flower. Thus, when fall rains and warm weather extend into early December, the *Pectis* is in flower, and fruit may still be found on the *pitaya agria*.

Perityle emoryi Torr.
hee imcát 'jackrabbit what-it-doesn't bite-off'
  "what the jackrabbit doesn't bite off"
hehe cotópl 'plant that clings'
rock daisy

It was said that jackrabbits know the plant has tiny spines and tastes terrible and, therefore, do not eat it (Figure 17.70). This is one of the most abundant winter-spring ephemerals in northwestern Sonora. The flower heads have yellow centers and white rays, although plants with all-yellow flower heads are sometimes also present.

Perityle leptoglossa Harv. & Gray subsp.
  leptoglossa
hee imcát caacöl 'jackrabbit what-it-doesn't
  bite-off' (*Perityle emoryi*) large(plural)'
  "large rock daisy"

This small perennial usually grows in rock crevices in the mountains on the mainland and on Sierra Kunkaak on Tiburón Island. It somewhat resembles *P. emoryi* but has larger and all-yellow flower heads and stouter, slightly woody stems.

Porophyllum gracile Benth.
xtisil
hierba del venado

This delicate perennial bush has pungently aromatic herbage. It is widespread on arid slopes and the

Figure 17.70. Rock daisy (*Perityle emoryi*) from near El Desemboque, spring 1970. *FR.*

desert floor on the mainland and Tiburón Island.

Medicine: Tea made from the stems was taken as a remedy for colds and also during a difficult delivery. When the mother was given this tea the baby was said to dislike the plant's "bad odor" and was born quickly. Tea made from the roots was used to cure toothache and diarrhea. One woman said, "The thing it is not good for does not exist."

Senecio douglasii DC. var. *douglasii*
tee caacöl 'Eriogonum-inflatum large(plural)'
    "large desert trumpet"
hast ipénim ctam 'rock what-it-is-splattered-with male'

Hast ipénim is the name for *Camissonia californica*, *Eschscholzia*, and *Kallstroemia*. Many Seri did not know a name for this plant, and the names given were not used consistently. This winter-spring ephemeral has bright yellow flower heads. It ranges only into the northern part of the Seri region, where it is seldom common.

Medicine: Tea made by boiling the roots was used as a remedy for a cold.

Stephanomeria pauciflora (Torr.) A. Nels.
hehe imixáa 'plant rootless'
    "rootless plant"
posapátx camoz 'future-*Bebbia* thinker'
    "what thinks it's a sweet-bush"
wire lettuce

The plant seems to have little or no root and can easily be knocked over, thus the name. The second name indicates the plant "thinks" that perhaps it is *Bebbia juncea*.

Bushy perennial, often forming a hemispherical mound on beach dunes.

Shelter: The plant was used as shading material on a roof frame of the brush houses.

Trixis californica Kell. var. *californica*
cocazn-ootizx 'rattlesnake what-it-peels-back'
    "rattlesnake's foreskin"

A very common small shrub with brittle stems, green foliage, and attractive yellow flower heads. It is widespread in the region, including Tiburón Island.

Medicine: Tea made from the roots was taken by a woman before delivery to hasten the birth.

Smoking: The leaves were smoked like tobacco.

Verbesina palmeri S. Wats. subsp. *oligocephala* (I.M. Johnst.) Felger & Lowe
cocazn-ootizx caacöl 'rattlesnake what-it-peels-back (*Trixis*) large(plural)'
    "large *Trixis*"

This small shrub grows on Sierra Kunkaak on Tiburón Island and in the mountains on the opposite mainland.

Viguiera deltoidea Gray
hehe imoz coopol 'plant its-heart black'
    "black-hearted plant"

The name derives from the heartwood, which is said to be blackish. It is a common desert bush or small shrub with brittle stems, triangular leaves, and daisy-like yellow flower heads. It is widespread in the northern Gulf of California.

Medicine: The roots were mashed and cooked in water, and the tea taken as a contraceptive.

Xanthium strumarium L.
cözazni caacöl 'tangled (= *Cenchrus*)
   large(plural)'
      "large sandbur"
cadillo, cocklebur

The name refers to the bur-like fruit, which gets tangled in clothing and hair. It is not native, and occurs as a weed mostly on abandoned farmland. We found it at Rancho Miramar near El Desemboque. It is an annual.
   Medicine: Tea made from the bur was taken to relieve kidney pain.

Zinnia acerosa (DC.) Gray
   [= *Z. pumila* Gray]

saapom ipémt 'Opuntia-violacea what-it-is-rubbed-with'
      "what purple prickly-pear is rubbed with"
cmajíic ihásaquim 'women brush-hair-with'
      "what women brush their hair with"
mojépe ihásaquim cmaam 'sahuaro what-one-brushes-hair-with female'
      "female sahuaro hairbrush"
desert zinnia

This dwarf shrub has showy white ray petals. It is infrequent in the northern part of the Seri region.
   Food Preparation: It was said to often grow near *Opuntia violacea* and was used to clean the spines and glochids from the fruit of this prickly-pear.
   Medicine: The Seri reported that the Papago used this plant to make medicine to cure diarrhea.

Convolvulaceae—Morning Glory Family

Cressa truxillensis H.B.K.
ziix atc casa insíi 'thing testicle putrid not-smell'
alkali weed

This small, wiry perennial is silvery-pubescent. It is a halophyte and often found at the edge of mangroves on Tiburón Island and the mainland. The Seri considered it to have a very objectionable odor and a "very ugly name." The name refers to a camp and a territory (*ihízitim*) on the southeastern shore of Tiburón Island named for a man who long ago suffered from a diseased testicle.

Cuscuta leptantha Engelm.
C. corymbosa Ruiz & Pav. var. *grandiflora* Engelm.

hamt itóozj 'soil its-intestines'
      "soil's intestines"
dodder

These very similar-appearing species were not distinguished by the Seri. *C. leptantha*—with orange, thread-like stems—is a common, widespread parasite on *Amaranthus watsonii* and *Euphorbia polycarpa*. *C. corymbosa*, a larger dodder, is parasitic on shrubs such as *Colubrina*, *Lippia*, and *Mimosa*. It occurs near the northwest base of Sierra Kunkaak.

Ipomoea sp.
hatáaij 'what is spun (like a top)'
hehe quiijam 'plant that-curls-around it'

   Summer ephemeral; delicate vine. The flowers were said to be like those of *Ruellia californica*.

Crassulaceae—Stonecrop Family

Dudleya arizonica Rose
hast yapxöt 'rock its-flowering'
      "rock's flower"

The name is descriptive of a plant growing among rocks. In Sonora this thick-leaved succulent occurs in coastal mountains as far south as Cerro Tepopa, but it is not common in the region. It usually grows wedged in crevices of cliffs or canyon walls.

Tillaea erecta Hook. & Arn.
hant yapxöt 'land its-flowering'
      "land's flower"
xat 'hail'

   Hant yapxöt was also applied to *Achyronychia*. During favorable years this tiny winter ephemeral is common in the northern part of the Seri region.

Cruciferae (Brassicaceae)—Mustard Family

Descurainia pinnata (Walt.) Britt.
cocóol
tansy mustard

A winter-spring ephemeral, sometimes very common. It occurs throughout the Seri region including Tiburón Island. The same name is applied to *Cleome* and *Sisymbrium.*

Gambling: Plants of this species and *Lotus* spp. were rubbed on the person to bring good luck in certain gambling games.

Medicine: An infusion of the seeds was used as eyedrops for sore eyes.

Dithyrea californica Harv.
hehe imixáa 'plant rootless'
   "rootless plant"
moosni iha 'sea-turtle its-possessions'
   "sea turtle's possessions"
hant istj 'land its-leaf'
   "land's leaf"
spectacle pod

The names given for this plant varied, apparently reflecting its non-utilitarian and relatively inconspicuous nature. The first two names were also used for *Stephanomeria* and *Palafoxia.*

It is a spring ephemeral with fragrant white flowers found on sandy soils and dunes in the northern part of the Seri region.

Draba cuneifolia Torr. & Gray var. *sonorae* (Greene) Parish
cocóol cmaam 'cocóol (*Descurainia*) female'
   "female tansy mustard"

It is interesting to note that its relationship with other mustards was recognized. A small winter-spring ephemeral with a basal rosette of broad, spreading leaves. It is fairly common in the northern part of the Seri region and on Tiburón and San Esteban islands after periods of sufficient rain.

Lepidium lasiocarpum Nutt.
coquée
isnáap ic is 'its-side with its-fruit'
   "whose fruit is on one side"

queeto oohit 'Aldebaran what-it-eats'
   "what Aldebaran eats"
pepper-grass

Coquée has become the generic term for chile peppers and black pepper. The second name was usually applied to *Bouteloua.* The last name given implies that this plant was the food of the star *queeto,* or Aldebaran, described as a red star visible in October and said to cause plants to flower and fruit out of season (see *Stenocereus thurberi*). However, to many Seri this plant was not well known and not recognized by name. It is a winter-spring ephemeral, generally most common along major drainageways, such as Arroyo San Ignacio and in the mesquite bosques between Kino Bay and Tastiota.

Food: The plant has a spicy, chile-like flavor, particularly the inflorescence, both in flower and seed. The whole plant or just the flowering or fruiting tops were used fresh and dried. It was said to have been eaten like chile, or *chiltepín,* to flavor meat, such as green turtle, leatherback turtle, and horse meat. It was said that the ancestors, particularly the people from the Tastiota region, made extensive use of *coquée* and ate meat flavored with *coquée* and salt.

Lyrocarpa coulteri Harv. var. *coulteri*
ponás camoz 'future-*Matelea-pringlei* thinker'
   "what thinks it's a *Matelea pringlei*"

As indicated by the name, this plant thinks that perhaps it is a *nas* (*Matelea pringlei*), which it somewhat resembles in life-form, size, and habitat. It is widespread on the mainland and Tiburón Island. The only comment concerning this plant was, "It just grows."

*Sisymbrium irio L.
cocóol
London rocket

The same name was also applied to *Descurainia. Sisymbrium* occurs along major arroyos, such as Arroyo San Ignacio near El Desemboque. It is a naturalized Old World species.

Medicine: An infusion of the seeds was used as eye drops.

Unidentified Cruciferae
hasoj an hehe 'river area plant'

The plant was generally not recognized. Two people said it was a plant which grows near rivers (washes), and their name for it was merely descriptive. It is a winter-spring ephemeral found in the northern part of the Seri region.

Cucurbitaceae—Gourd Family

Brandegea bigelovii (S. Wats.) Cogn.
hehe iti scahjíit 'plant on let's-fall'
  "let's fall on the plant"
hant caitoj 'land creeper'

The first name implies it grows on another plant, and describes a vine. There is no other Seri concept or word for vine, and in a rather vague manner this name might be considered a life-form term. Vines are relatively few or even absent in deserts throughout the world (Raunkiaer 1934). When called *hant caitoj* it was associated with the vision quest.

This rank-growing vine is generally annual or ephemeral, with a fleshy, carrot-shaped white root, seasonally luxuriant foliage, and small white flowers. It usually occurs in larger arroyos and washes, such as Arroyo San Ignacio, where it often covers shrubs and small trees like a green curtain. It is common around the head of the Gulf of California and ranges into the northern part of the mainland Seri region, but does not occur on the islands. On the islands and in the Guaymas area it is replaced by *Vaseyanthus*.

Hair Care: The roots were pounded on a metate, and the fibrous pulp used as a shampoo.

The Supernatural: Some people said the root was used for the vision quest (in the same manner described for *Vaseyanthus*).

*Citrullus lanatus (Thunb.) Mats. & Nak.
ziix is coil 'thing its-fruit green'
  "green-fruited thing"
coxi ixám 'dead-one his-squash'
  "dead man's squash"
sandía, watermelon

*Cucumis melo L.
meróon 'melon'
iicj ano meróon 'sand from-in melon'
  "sand melon"
melón, cantaloupe

Cucurbita digitata Gray
ziix is cmasol 'thing its-fruit yellow'
  "yellow-fruited thing"
chichi coyote, coyote gourd

This sprawling perennial vine is common in sandy places, abandoned farmland, and disturbed sites. The round gourds, about 8 cm in diameter, are yellowish when ripe and persist long after the vine withers.

Music: The gourd rattle, *hexez*, described by Kroeber (1931:14) was made either from this gourd or *xjii* (*Lagenaria*). The gourd was dried, a hole punched in it, and the dry insides scraped loose with a stick and shaken out. Pebbles were dropped in, and a stick for a handle was inserted into the hole. The gourd rattle was replaced by the can rattle (see Figure 13.10) before the mid–twentieth century (Bowen and Moser 1970b:187).

Play: Children used the yellow gourds as toys.

*Cucurbita mixta Pang. and/or *C. moschata* (Duch.) Duch. ex Poir.
xam
squash

Xam is the generic term for cultivated squash and gourds.

*Cucurbita moschata (Duch.) Duch. ex Poir.
tziino ixám 'Chinaman his-squash'
  "Chinaman's squash"
sehualca, butternut squash

This is one of the common squashes cultivated in Sonora. The word *tziino* is derived from the Spanish *chino*.

*Cucurbita pepo L.
xam cozalc 'squash ribbed'
  "ribbed squash"
calabaza, pumpkin

Cucurbita pepo L. var. *ovifera* (L.) Alefeld
caay ixám 'horse its-gourd'
   "horse's gourd"
calabacilla, ornamental gourd

Caay is derived from the Spanish word for horse, *caballo*. This is the hard, non-edible, decorative gourd.

Lagenaria siceraria (Mol.) Standl.
xjii
guaje, *bule*, calabash, bottle gourd

The bottle gourd has been cultivated by neighbors of the Seri since ancient times. The Seri obtained it from nearby ranches and towns. Some people said that *xjii* came from Caborca. However, it was generally not well known by the modern Seri.

Music: The gourd rattle, *hexez*, was made from *xjii* or *Cucurbita digitata*. When used as a gourd rattle it was called *xjii heetni* 'xjii that-which-is-tapped.'

Water: A photograph taken by Sheldon in 1921 shows a dipper made from a bottle gourd (see Figure 6.1). Several Seri identified the dipper in this photograph as *xjii*, and said that it would have been obtained from a ranch.

Maximowiczia sonorae S. Wats. var. *sonorae*
   [= *Ibervillea s.* Greene]
hant yax 'land its-belly'
   "land's belly"
cow-pie plant

This cucurbit has a large, above-ground, swollen caudex from which emerge the tendrilled vining stems. The shoots appear with summer rains and the fruit ripens in late summer and fall.

The Seri described it as follows: It has a bulging stem, or trunk, like a ball on the ground which is about 1/3 m in height. The stems are about 1.5 m in height, slender, vining, and with leaves. The fruit is elongated, red, and soft. The fruit is so pretty that people want to eat it, and some do. But it is so bitter it is considered inedible. The pulp of the trunk is also bitter. There is much of it far north of Pozo Coyote by the big ranches.

Vaseyanthus insularis (Wats.) Rose
hant caitoj 'land creeper'

Sprawling vine with ephemeral growth and a fleshy tap root, probably annual (Figure 17.71). It is common near the sea in the southern part of the Seri region and on Tiburón and San Esteban islands.

Hair Care: Liquid shampoo, prepared by cooking the green leaves in water, was said to cause abundant hair growth.

The Supernatural: The roots were broken up in water and allowed to stand for several days. The bitter drink was used in the vision quest, during which four sips were drunk many times over three or four days of fasting. It was said to make one "like a drunk person." This liquid, called *hant caitoj iix* 'land creeper liquid,' was kept in a pottery vessel called *hehe an ihamáax* 'plant area its-wine,' or "wine for the uninhabited desert" (see *Brandegea*).

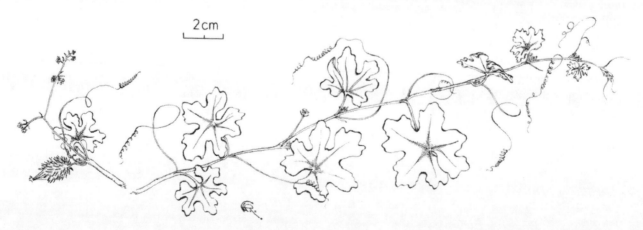

Figure 17.71. *Vaseyanthus insularis* from near Guaymas. FR.

Euphorbiaceae—Spurge Family

Acalypha californica Benth.
queejam iti hacníix 'bearer-out-of-season on what-is-dumped'
"what out-of-season fruit is dumped on"
hierba del cáncer

Queejam indicates a plant which bears flowers or fruit out of season. After collecting the very last columnar cactus fruit of the year, the people of the Libertad Region were said to have spread branches of this shrub on the ground and dumped the fruit out of their collecting baskets onto the branches.

This small, drought-deciduous shrub ranges through most of the region, including Tiburón and San Esteban islands.

Shelter: The branches were used as roofing for the brush house.

Andrachne ciliato-glandulosa (Millsp.) Croiz.
hasoj an hehe 'river area plant'

An inconspicuous winter-spring ephemeral; widespread on the mainland and islands.

Medicine: To cure a headache, the plant was put in water and the head washed with the infusion.

Argythamnia lanceolata (Benth.) Muell. Arg.
A. neomexicana (Torr.) Muell. Arg.
   [= *Ditaxis l.* (Benth.) Pax & Hoffm., *D. n.* (Muell. Arg.) Hell.]
hehe czatx caacöl 'plant stickery large(plural)'
   "large stickery plant"

The Seri did not distinguish these two species by name. The small, spine-like hairs, called *zatx*, are probably the reason for the name (see *Cryptantha* and *Opuntia fulgida*).

A. lanceolata is a weak-stemmed perennial with silvery pubescent foliage; *A. neomexicana* is herbaceous and facultatively ephemeral to perennial (Figure 17.72). Both are common over a wide spectrum of desertscrub through the Gulf region, including Tiburón and San Esteban islands.

Cnidoscolus palmeri (S. Wats.) Rose
coap
mala mujer, ortiguilla

This small shrub grows in rock crevices on steep

2 cm

Figure 17.72. *Argythamnia neomexicana.* Note hispid pubescence, or small, spine-like hairs, called *zatx. FR.*

granitic slopes and cliffs, usually at higher elevations (Figure 17.73). It occurs on hills and mountains from near El Desemboque to the Sierra Seri, Sierra Kunkaak on Tiburón Island, Turners Island, near Guaymas, and in Baja California.

It has thick, crinkled leaves with milky veins, and stout, stinging hairs which inflict a painful but short-lasting rash. This rash is known as *oozxlim.* The sap is milky and contains rubber. Each shrub produces several dozen fleshy and swollen tuber-like roots 5 to

20 cm or more in length. For the sake of convenience, these tuber-like roots are hereafter referred to as tubers. The tubers are potato-like, but often compressed as they grow wedged between rocks. Several or more tubers develop at irregular intervals along each slender lateral root which may extend 1 to 2 m from the center of the plant. This slender connecting root was called *inz* 'spinal cord.' Several tubers may coalesce to form an elongated, thickened root 0.5 to 1 m in length. Such a root was called *iscám* 'his balsa' because it was long and slender like a balsa (see *Phragmites*). The terms *inz* and *iscám* were not applied to any other plant.

Food: *Coap* was a particularly important food resource because it could be harvested at any time of year. The collector had to remove the surrounding rocks to expose the roots and dig the tubers. They were dug and pried out from rocks with the *hapoj* digging stick.

In December, 1981, Rosa Flores located and harvested *coap* at about 150 m elevation on a granitic mountain about 20 km south of El Desemboque. Known as Cerro Dos Amigos (= Cerro Dos Hermanos), or *Iisc* 'its-grayness,' it is several kilometers inland (east) from the coastal fishing camp called *Saps*.

She chose a large shrub, about 1.5 m tall, on a steep slope. She began by removing rocks more than one meter below the crown of the plant. Surface rocks up to about a meter in length were pulled away by hand and tumbled downhill. When finished, she had removed rocks and dug tubers from an area extending out two meters below the shrub. She worked for 50 minutes, obtaining 14 tubers (Figure 17.74) weighing 2 kg (fresh weight). These represented only a fraction of the tubers, since she excavated only a single, trench-like area; large boulders and solid rock precluded excavating a larger area. She used a small shovel instead of a digging stick, which she convincingly pointed out would have been a better tool for the job. During the digging process the leaves, with their stinging hairs, were treated with considerable caution and carefully avoided. Towards the end of the excavation the plant toppled over because of rock removal and she pulled it out and threw it downhill. While digging she paused several times to peel and eat a few of the smaller tubers: they were crisp, succulent, white like a raw potato, and bland tasting.

Figure 17.73. Rosa Flores and José Astorga beside a shrub of *coap* (*Cnidoscolus palmeri*) on a rocky slope east of Estero Sargento. *RSF, March 1981.*

Figure 17.74. Rosa Flores with part of the tuber-like roots dug from a single *coap* shrub on Cerro Dos Hermanos, about 5 km east of Estero Sargento. *RSF, December 1981.*

She peeled off the thin but tough, brown bark (*coap ináil* 'coap its-skin') with her fingernails.

In March, 1983, at the same location, Rosa Flores and her husband José Astorga harvested 5.5 kg (fresh weight) of tubers from a single shrub. They used a steel crowbar to excavate the tubers in 1 hour and 35 minutes (Figure 17.75).

When harvesting *coap*, the majority of the tubers were taken back to camp whole. They could easily be kept for a number of days or even weeks. The tubers were roasted whole in hot ashes as well as on top of

Figure 17.75. Rosa Flores and José Astorga harvesting the roots of *coap* on Cerro Dos Hermanos. A) Digging for the roots of this plant is very difficult and requires moving large rocks from the surface. B) One of the larger tuber-like roots. It is flattened because it grew wedged between rocks. *RSF, March 1983.*

the coals or on a grill. The blackened bark or skin was removed, and the inside eaten, often with honey. The young tubers were said to be best for eating. Honey was commonly obtained in the same area as *coap*, so there was ample opportunity to eat *coap* with honey.

Coap was one of the food resources resorted to when it was too windy for the men to go to sea for turtles. At such times the men sought terrestrial game, particularly mule deer, and the women went to the mountains to harvest *coap* and whatever other food might be available.

Croton californicus Muell. Arg.
hacáiin cooscl 'what-is-gathered-for-windbreak drab'
    "drab windbreak"
moosni iti hatépx 'sea-turtle on what-is-rested-on'
    "what sea turtle meat rests on"
sand croton

While on Tiburón Island in 1951, Alfred F. Whiting (1957:#40) obtained the same names for this species which he recorded as *agai'inko:'sk*[l] and *mosnitiatep*[x].

Perennial bush, silvery-gray herbage and slender stems (Figure 17.76). Common along the shores of the northern Gulf of California including Tiburón Island; particularly abundant on coastal dunes.

Dye: Natural basketry splints were dyed yellow-orange by placing them in boiling sea water with sand croton, or the basket was rubbed with the leafy flowering stems. The color was not long lasting but was said to be pretty until it faded.

Food Preparation: Since the plant has no spines or unpleasant odor, it was spread on the ground or used to line a basket on which sea turtle (Whiting 1957:#40) or other meat was to be placed. This kept the meat clean while it was being cut up and until it was cooked or consumed.

Freshly gathered shellfish—such as oysters, clams, horse mussel, sea pen, and others—were spread out on a flat surface of sand or dirt. They were covered with green sand croton brush, which was then burned in order to steam the shellfish.

Poison: The green plant was crushed and thrown into estero water to poison mullet.

Shelter: The leafy branches were used to form a brush windbreak and for roofing of brush houses.

Figure 17.76. Sand croton (*Croton californicus*) from Kino Bay. *FR.*

Croton magdalenae Millsp.
hehe pnaacoj 'plant mangrove'
    "mangrove plant"

This is the name for *Sideroxylon leucophyllum* on San Esteban Island. The term *hehe pnaacoj* seems to have been applied to *Croton magdalenae* because of resemblance of its foliage to that of *Sideroxylon*.

C. *magdalenae* is a shrub about 1.5 m tall which has large, greenish white, fuzzy leaves. It occurs in canyons on the southwest side of Sierra Kunkaak on Tiburón Island.

Croton sonorae Torr.
hoinalca 'low hills'

The same name was also applied to *Holographis*. This small shrub is usually found in the hills and mountains.

Euphorbia eriantha Benth.
pteept, taapt, teept
desert poinsettia

A common and widespread ephemeral, occurring throughout the region.

Euphorbia misera Benth.
hamácj 'fires'

Small- to medium-sized shrub with knotty stems (Figure 17.77), often on north-facing slopes. Superficially resembling *Jatropha cuneata* but readily distinguished by its copious milky sap. Coolidge and Coolidge (1939:200) reported that "*ah-mahk*" was used in the making of blue clay, or Seri Blue, but this practice was not verified by the Seri.

Medicine: Tea made from the roots was used as a remedy for stomachache, dysentery, and venereal diseases.

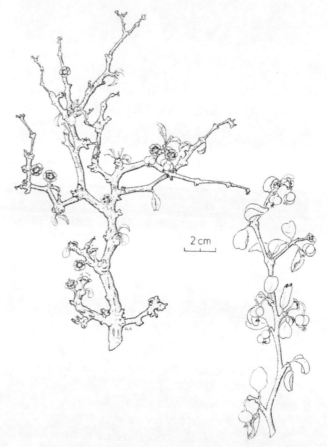

Figure 17.77. *Euphorbia misera* from Cerro Tepopa, April 1974. *NLN.*

Euphorbia polycarpa Benth.
E. petrina S. Wats.
E. setiloba Engelm.
tomítom hant cocpétij 'tomítom land circular-flat'
  "prostrate *tomítom*"
also *hamítom hant cocpétij*
  xomítom hant cocpétij
golondrina, spurge

As the name indicates, these are prostrate, mat-like herbs. The Seri did not distinguish among the several species of small *Euphorbia* in the subgenus *Chamaesyce* occurring in their region. They are abundant over a wide range of desert throughout the Gulf of California region. *E. petrina* and *E. setiloba* are ephemeral, while *E. polycarpa* is facultatively ephemeral or perennial (Figure 17.78).

Medicine: The green leaves (herbage) were mashed, salt and oil added, and the mixture applied as a poultice to a swollen area. The plant was also used in a remedy for toothache or heart pain (see *Asclepias*).

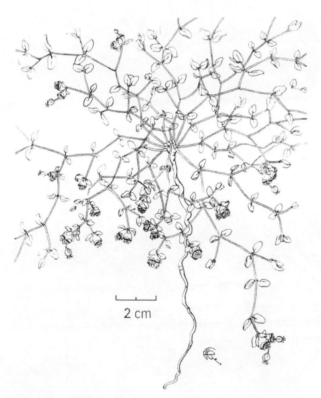

Figure 17.78. *Golondrina* (*Euphorbia polycarpa*) flowering in its first season. *FR.*

Poison: The plant, cooked with the liver of a certain puffer fish (*tzih*), cinnamon and salt, was used on one occasion to poison a man who had killed two others.

Euphorbia tomentulosa S. Wats.
tomítom hant cocpétij caacöl 'tomítom land circular-flat large(plural)'
xomítom hant cocpétij caacöl
  "large prostrate *tomítom*"

The name signifies a large kind of *E. polycarpa*/spp. This small shrub is common on hillslopes and bajadas through most of the Seri region.

Euphorbia xanti Boiss.
hehe iix cooxp 'plant its-liquid white'
  "white-sapped plant"

This sparsely branched shrub has tall green stems and short-lived, drought-deciduous foliage. It is common along the broad bajada bordering the mainland side of the Infiernillo Channel. As the name indicates, it has milky sap, a generic character of *Euphorbia*.

Animal Food: It was reported that mule deer browsed the plant.

Jatropha cardiophylla (Torr.) Muell. Arg.
heecl
sangrengrado, limberbush

Trunkless shrub, about 1 m tall, flexible stems, reddish bark, and shiny, heart-shaped leaves present only during summer rainy season. Common north and east of the Seri region.

The Seri described this plant as resembling *hamísj* (*J. cinerea*) but differing in having straighter and more flexible stems, longer internodes, reddish rather than gray bark, and in usually being leafless. It was also pointed out that the stem tips of *J. cardiophylla* are straighter than those of *J. cinerea*, that when the former does bear leaves the bark is even more red, and that the fruits each contain two seeds.

Elvira Valenzuela and several other women showed us a single, nearly mature *heecl* shrub in Arroyo San Ignacio at Rancho Miramar, about 2 km north of El Desemboque. We were told it is common on the large ranches north and east of El Desemboque, such as at Rancho Estrella. These women believed that this single, extralimital shrub resulted from a seed carried down the riverbed with floodwaters, i.e., *hax-exl* 'water-what-it-took,' or "what the water took."

Basketry: Elvira Valenzuela said that when she was a girl her family had travelled in the northern part of the Seri territory, and that they had seen Papagos using stems of this shrub for making baskets.

Dye: The bark from the root, crushed on a metate and sprinkled with water, was squeezed onto the inner side of a deer skin in order to stain it red.

Jatropha cinerea (Ort.) Muell. Arg.
hamísj
sangrengrado, ashy limberbush

Multi-stem shrub (Figure 17.79), with thick roots, smooth gray bark, flexible branches, and very soft wood. The relatively large, rounded leaves appear after sufficient rainfall and quickly fall with drought. This is one of the more conspicuous shrubs in the open desert throughout the coast of Sonora and on Tiburón Island. When cut, the roots and lower stems exude copious blood-like sap, hence the Spanish name, which is a corruption of *sangre de drago* 'blood of dragon.'

The dry wood was called *oot iquéöjc* "coyote's firewood," because it is nearly worthless as a firewood. A dead, dry branch remaining attached to the living shrub was called *hamísj yapáain* 'hamísj what-caused-it-to-fall.' Some said that the term *yapáain* should only be used with *hamísj* (see *Bursera hindsiana* and *B. microphylla*).

Cradleboard: The curved frame was sometimes made from the straight stems of this shrub (see Figure 10.16). Since the 1950s the wood has been used for the framework of cradleboard dolls made for sale (Figure 17.80).

Firemaking: The fireboard and drill tip of the *caaa* firedrill were often made from *hamísj yapáain*. It generally was used for making a firedrill on the spur of the moment, and the men usually did not keep an *hamísj* firedrill in their arrow quivers.

Games: The stems were used to make decorated stick counters, called *hapéxz* (Figure 17.81). These were used in the *camóiilcoj* circle game played at a

Figure 17.79. Ashy jatropha (*Jatropha cinerea*) near El Desemboque, April 1983. *RSF.*

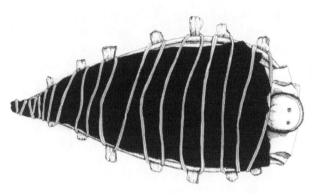

Figure 17.81. Stick counters, *hapéxz*, from a *camóiilcoj* circle game played at a girl's puberty fiesta in El Desemboque, 1963. The sticks are made of *Jatropha cinerea* with blue (commercial bluing) and red bands and spots, and multi-colored cloth strips. The longest one is 61 cm. *CMM.*

Figure 17.80. Cradleboard doll made by Sara Villalobos for Cathy Moser in 1973 at El Desemboque. The doll is 46 cm long and made of wood (*Jatropha cinerea*), cloth, and yarn. *CMM.*

girl's puberty ceremony or the leatherback turtle fiesta (see *Phragmites*).

Headpieces: The wood of this shrub was used to make two kinds of crowns described in detail by Bowen and Moser (1970a:169–171) and in Chapter 12. The *hehe hamásij* 'wood that-is-opened-up' was a type of crown with a framework consisting of slats and hoops of ashy jatropha wood fastened with sinew (later often replaced by string or nails). The crown terminated in a peak to which was fastened a tuft of feathers, often the white breast feathers of a gull. Later the feather tuft was commonly supplanted by a knob or bird carved from red elephant tree (*Bursera hindsiana*; Figure 17.82).

The *hehe hamásij* crown derived supernatural significance through its association with coyote men. During the nineteenth century many men wore these crowns, although the association with supernatural properties eventually diminished. During the 1920s and 1930s they were still worn, but their popularity was declining. By the late 1940s the crowns had all but disappeared from use (Hayden 1942:24); after that time they were made for sale to outsiders. Some specimens from the 1940s and 1950s are artistically intriguing (see Figure 12.3) compared with later versions. One fine example in the collections of the Amerind Foundation (AF-2132) is decorated with strings of small olivella shells alternating with *Viscainoa* seeds. In the early 1980s men performing the foot-drum dance and other pascola-like dances at fiestas usually wore one of these crowns (see Figure 13.9).

The other headdress, called an antenna headpiece by Bowen and Moser, was known as *ano cojozim*

Figure 17.82. Crown, *hehe hamásij*, 40 cm in height. The bird is carved from red elephant tree (*Bursera hindsiana*) and the rest of the crown is made from slats of ashy jatropha (*Jatropha cinerea*). This replica, made by Nacho Morales in 1961, is decorated with red (hatching) and blue (solid) pigments and white gull feathers. *CMM.*

quih itáamalca 'inside who-flees the his-horns' (see Figure 12.4). It usually consisted of a cloth headband into which were stuck two knobbed sticks carved from ashy jatropha wood, with each stick about 30 cm tall. It was used in the dance called *ziix coox cö-*

coila 'thing all who-dance-about,' or "they who dance about everything," the men's circle dance performed at girls' puberty ceremonies. The Seri said Juan Mata initiated the use of this headpiece (it would have been around the turn of the century). In a dream or vision he was instructed by *cama*, the manta ray, to introduce it into the *ziix coox cöcoila* dance.

The antenna headpiece was worn by each of two men who were the leaders of this dance. They stood opposite each other in a circle of men who sang and played split reed instruments, *xapij hamánoj* (see *Phragmites*), as they danced counter-clockwise. The pace was fast and the dancers jumped and stamped their feet. As the two leaders danced with their antennae-like headpieces bobbing, they watched each other closely. When they signaled—by dipping their headpieces—they crossed the circle and exchanged places. The antenna headpiece was said to hold no special significance other than to identify the dance leaders.

Hunting: During fawn season the leaf was used to call in mule deer. The leaf was held to the mouth and air sucked through it to produce a shrill sound imitating the cry of a fawn. Adult deer would come to investigate, and the hunter could get a close shot at them.

Medicine: The roots of small plants with the bark removed were mashed and made into tea used to cure dysentery. "When a baby's mouth becomes sore from nursing, it may be treated with rattlesnake oil or with sap from *hamísj* . . . (M. Moser 1970a : 207)."

Shelter: Brush houses were apparently occasionally built with *hamísj* branches. Rosa Avila, mother of Ramona Casanova, was called *hamísj quih ano cama* "she who lived in an *hamísj* hut."

Weapons: The sap was used as a poison for arrow points.

Other uses: Stuffed animal heads were used in certain dances and as decoys in hunting. In the taxidermy of a mule deer or jackrabbit's head (see *Larrea*), strips of *hamísj* or mesquite root (*Prosopis*) were sewn along the margins of the ears until they dried and stiffened. The strips were later removed (Bowen and Moser 1970a : 173).

Jatropha cuneata Wigg. & Roll.
haat
matacora, torote, sangrengrado, limberbush

This shrub is abundant and widespread in the Gulf of California region, including most of the major is-

lands. It ranges from the Guaymas area northward into southwestern Arizona, and extends through most of the peninsula of Baja California. It is characteristic of arid places, and is often abundant on hot, dry, south- or west-facing slopes. It is common throughout all of the territory of the various Seri groups.

It is a semi-succulent and multiple-stemmed shrub, commonly reaching 1 to 2 m in height (Figure 17.83). Each shrub has a number of large, radiating, and carrot-shaped roots, up to about 10 cm in diameter. The roots and lower stems exude copious blood-like sap when cut. The stems are flexible, the wood very soft, and the foliage drought-deciduous (Figure 17.84). The sensitivity of this species to freezing weather limits its northern range. Relatively rapid stem growth may occur with the summer-fall rains. After being cut, such as for basketmaking, the stems seem to regenerate without any significant damage to the plant.

Basketmaking: All Seri baskets have been made of splints prepared from *haat* stems (Figure 17.85); basketry is discussed in detail in Chapter 15.

Dye and Pigment: Reddish brown paint or dye was obtained from the roots. See *Suaeda moquinii* for basketry dye. The sap stains clothing and was said to be indelible.

Firewood: The dry, brown tips made the best kindling.

Headring: The headring, *hatxíin*, was almost always made with *Jatropha* stems, although mesquite (*Prosopis*) was sometimes substituted. In 1692 Gilg illustrated a woman carrying a basket on a headring (see Figure 1.4). All of the twentieth-century headrings we have seen were wrapped or bound with variously colored fabric such as cloth, scraps of blankets or clothing, canvas, or yarn. There was considerable variation in quality of construction and aesthetic appearance; some were bound with two or three selected colors of cloth or yarn incorporated into bold patterns. A piece of cloth deftly looped into a ring on the head served as a makeshift headring.

The *hatxíin* was in daily use by women to balance and cushion a wide array of objects carried on the head, such as loaded baskets or their replacements (cans and buckets), firewood, and water vessels—earthen or, later, metal ones. Although it was indispensable in earlier times, the headring declined in use during the middle of the twentieth century.

In 1968, María Antonia Colosio demonstrated the making of a headring. María chose *Jatropha* stems

Figure 17.84. Details of limberbush (*Jatropha cuneata*) from Santa Rosa peninsula, October 1965. A) Leafless dry season aspect. B) Short shoots with foliage following a rainy period. C) Rapid, primary growth following summer-fall rains. Note the larger, lobed leaves and relatively long internodes. *LBH*.

Figure 17.83. Limberbush (*Jatropha cuneata*). A) A shrub at the base of *Hast Eemla* in the Sierra Seri. *RSF, April 1983*. B) Jesús Morales preparing the thick roots to make dye for a tanned hide. *MBM, El Desemboque, 1969*.

which were not long and straight enough to be useful for basket making. She scorched the stems in a fire and then removed the outer and inner bark. The remaining portion of stem was flattened by chewing until it was about 1 cm or more in width. Each chewed stem was further flattened by gently pounding it with a mano on a metate. The flattened stems were soaked in water for an hour or so. Next, María

split each strip into two layers. She wound the smaller of these strips around her hand to form a ring (ca 9 cm in diameter) and bound them with a few more slender strips. The outer edge of this coil was then built up with thicker and wider strips. Around this ring she wound long strips of green inner bark which had earlier been peeled off the stem from beneath the papery, black (scorched) outer bark. The headring was then allowed to dry. To finish the ring she bound it with yarn (Figure 17.86).

Tattoo: A strip of the green stem was pulled back and forth over a bleeding, freshly tattooed area of skin to stop the bleeding and help it heal.

Weapons: The sap was used on arrow points for

Figure 17.85. Limberbush (*Jatropha cuneata*) stems collected by Rosa Flores at Rancho Estrella. These stems were cut for use in basketmaking. *RSF, March 1983.*

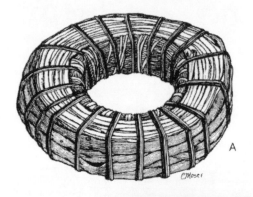

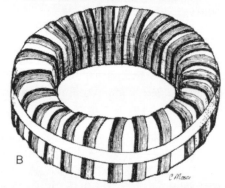

Figure 17.86. Headrings made by María Antonia Colosio. The foundation materials are basketry splints prepared from stems of *Jatropha cuneata*. A) An unfinished headring, 15.5 cm in diameter, made in 1964. B) A finished headring, 14.5 cm in diameter, bound with red and blue cloth and yarn, made in 1962. *CMM.*

hunting and warfare. However, some said it was not poisonous.

Other Uses: Flexible strips from living stems were used to bind or tie a wide variety of objects. Women used *haat* strips to tie the framework of the brush house. Men used them to bind a torch (see *Stenocereus thurberi*) and tie a desert tortoise to a pole to carry it back to camp (Felger, Moser, and Moser 1983). Men sometimes carried columnar cactus fruit into camp strung on a loop of *haat* wood. A stem was cut, the bark removed, and a long strand torn from the center portion of the flexible stem in order to string the fruit. Fruit carried in this manner was called *hapápl* 'what was strung.'

* Ricinus communis L.
hehe caacoj 'plant large'
  "large plant"
higuerilla, castor bean

A common weed around farmland; found at Rancho Miramar north of El Desemboque. The Seri knew the seeds are poisonous.

Medicine: Mashed seeds were used on sores on the head.

Tanning: In dressing a deer skin, the seeds were mashed, mixed with salt, and rubbed into the hide. The salted castor bean mixture was left on the stretched skin for about two days.

Sapium biloculare (S. Wats.) Pax
hehe coanj 'plant poison'
  "poisonous plant"
hierba de la flecha, Mexican jumping bean

This medium to large shrub has shiny green or greenish red foliage and milky sap. It occurs on the mainland and Tiburón Island in arroyos, on rocky slopes and upper bajadas. It is generally not common in the open desert.

Cradleboard: The wood was used for the curved frame of the cradleboard and was said to give an infant a long life.

Hunting and Warfare: The sap was put on arrow tips used in warfare and hunting. When a deer was shot with a poisoned arrow, the meat surrounding the arrow wound was discarded. McGee (1898:256) likewise reported that the portion around the arrow wound was thrown away. The dreaded arrow poison of the Seri undoubtedly was made from this plant, although the potency of the poison may have been exaggerated.

Poison: The plant was feared and avoided because of its poisonous qualities (see *Dodonaea*). People of the Tastiota Region used it to poison their enemies. The sap, mixed with pulverized root bark of *Marsdenia* was sprinkled on the victim's food.

Tragia amblyodonta (Muell. Arg.) Pax & Hoff.
hehe cocóozxlim 'plant that-causes-rash'

An inconspicuous vining or semi-vining perennial, found in better-watered places along arroyos. The plant was avoided because stinging hairs on the foliage cause a skin rash.

Fouquieriaceae—Ocotillo Family

Fouquieria columnaris (Kell.) Curran
    [= *Idria c.* Kell.]
cototaj
cirio, boojum tree

The boojum is one of the most unusual plants in the world, looking like an upturned carrot 5 to 10 m tall (in Sonora), with scraggly little secondary branches (Figure 17.87). It occurs in Baja California and there is an isolated population on Angel de la Guarda Island; in Sonora it is found in the granitic coastal mountains between El Desemboque and Puerto Libertad. In Sonora the boojum tree "forests" are mostly on north-facing slopes. The Seri did not know that boojum trees occur in Baja California or on Angel de la Guarda Island.

Oral Tradition: In order to escape a great flood, a group of Giants from the south fled northward to the mountains between El Desemboque and Puerto Libertad. There the flood overtook them, and they were changed into boojum trees. The tall ones were men and the short fat ones were pregnant women. In another version, those who were turned into boojum trees were people from the Tastiota Region, and there was no mention of whether or not they were Giants. Some say the flood overtook them just as they passed

Figure 17.87. Boojum trees (*Fouquieria columnaris*) at Punta Cirio, about 10 km south of Puerto Libertad. *RMT, January 1975.*

the crests of the mountains and were descending, and that is why the boojum trees are on the north sides of the mountains.

The Supernatural: These plants have much power from the spirit *Icor* and, if touched or harmed, will cause strong winds and rains. The branches were thrown into the sea to attract fish, but there was fear that this could result in bringing strong wind.

Fouquieria diguetii (Van Tieghem) I.M. Johnst.
xomxéziz caacöl 'Fouquieria-splendens large-
    (plural)'
    "large ocotillo"
palo adán, Baja California tree ocotillo

This plant resembles ocotillo (*F. splendens*) but is more tree-like and the flowers are darker red. It occurs from near Tastiota to the Guaymas region.

Fouquieria splendens Engelm. subsp. *splendens*
xomxéziz
ocotillo

This unique desert plant is one of the most common and conspicuous elements of the open desert landscape in the northern Gulf of California region, including Tiburón Island (Figure 17.88). Ocotillo does not occur on San Esteban Island. The long, wand-like, spiny branches arise from the base of the plant. The wood is soft and burns rapidly. Red-orange flowers are produced from the branch tips in spring, usually in March and April.

Adornment: The buds, flowers, and dry capsules were strung for necklaces (see Figure 12.5*c*).

Firewood: Although a poor firewood, ocotillo was a common fuel because it was often the only one available. It was useful for feeding into a fire to keep it burning.

Food: When out in the desert, the people would sometimes suck the nectar from the flowers.

Food Preparation: Stout ocotillo poles were used to thresh eelgrass (*Zostera*). A horizontally placed pole—such as an ocotillo branch or rope or line—used to dry meat or fish was called *hant iti icotíin* 'place on to-dry-meat-on' (see Figure 6.8).

Games: Bats for softball were occasionally made of ocotillo wood.

Hunting: Clubs made of ocotillo wood were used for killing pelicans, which were hunted by torchlight.

Music: The violin bow was occasionally made from the wood. "If the resin [from brittlebush, *En-*

Figure 17.88. Ocotillo (*Fouquieria splendens*) near Pozo Coyote. *RSF, April 1983.*

celia farinosa] is unavailable, the friction of the bow string can be increased by rubbing it across the bark of a green ocotillo (Bowen and Moser 1970b: 180)."

Shelter: The framework for the brush house was more often than not made from the flexible branches of ocotillo (Figure 17.89). Ocotillo frame shelters continued to be made in the 1980s when temporary camps were set up away from the villages.

The soft wood soon disintegrates, and evidence of the brush house may disappear within several years. However, sometimes branches stuck in the sand form roots and grow into a new plant, and years later the site of a former house is outlined by a row of ocotillo shrubs. One such colony, on a high dune at the north end of Tiburón Island, resulted from brush houses said to have been abandoned about 1900 (Figure

Figure 17.89. Ocotillo poles form the framework of this abandoned brush house at Campo Almond. *TGB, 1967; ASM-13524.*

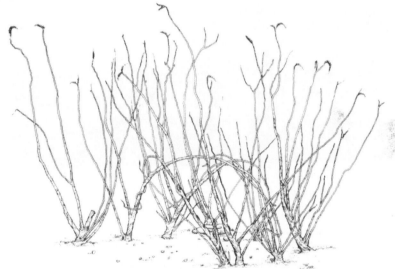

Figure 17.90. These plants grew from ocotillo poles set in the ground for a brush house on the crest of a high dune on the north side of Tiburón Island. The poles subsequently took root and formed new plants, outlining the original brush house. This camp was probably abandoned when balsa boats were phased out at the beginning of the twentieth century. *CMM, from a photo by MBM in 1977.*

17.90). When the people began using wooden boats, the camp was moved from the high dune to lower terrain next to the shore because, unlike balsas, wooden boats were too heavy to carry above the beach.

Ocotillo poles were used in the walls and framework of wattle and daub houses and the framework of tar paper shacks.

To provide shade in the desert, the branches were bent over towards the center of the plant, and then covered with brush. One would sit on the shady side.

The Supernatural: One who was seeking spirits or a vision sometimes built an *hasóoma* (ramada) with an ocotillo framework and covered it with branches of such plants as desert lavender (*Hyptis*) or elephant tree (*Bursera microphylla*; Griffen 1959:50). Although the elephant tree and desert lavender had religious significance, the ocotillo was probably merely of practical consideration.

Other Uses: A stout ocotillo pole was often used as a makeshift carrying pole. For example, a large fish or sea turtle was suspended by a rope from an ocotillo pole carried across the shoulders of two men. Ocotillo was a common, functional, but poor quality substitute for a number of less common stronger woods.

Frankeniaceae—Frankenia Family

Frankenia palmeri S. Wats.
seepol
saladito

Dwarf woody shrub with gray foliage of tiny evergreen, beadlike leaves (Figure 17.91). It is abundant in coastal situations over a wide variety of soils, from the head of the Gulf south to Punta Baja and on Tiburón Island.

Adhesives: A kind of resin or gum is occasionally found at the base of the plant. It was heated and melted and used as a glue for hafting harpoon and arrow points and knife blades, and for sealing pottery vessels. It was called *seepol icsípx* 'Frankenia its-resin.'

Firewood and Food Preparation: The dry wood was usually the first choice as fuel for cooking breast meat of a sea turtle (see Figure 3.11). After the turtle was killed, it was turned on its back, cut open at the base of the neck and the intestines and stomach removed. Then a fire of *Frankenia* brush was built on its plastron. After the fire burned down the ashes were cleaned off, and the plastron removed with a thin layer of cooked meat clinging to it. The plastron meat was generally eaten immediately. The remaining meat—after the plastron had been removed—was divided among the recipients, who cóoked it at their own hearths.

Medicine: Tea made from the root was taken as a remedy for a cold.

Shelter: *Frankenia* brush was used as roofing.

The Supernatural: To bring the south wind, *xnai*, four cross-shaped canals (each on the order of 10 to 20 cm in size) were dug below the high-tide line and filled with *Frankenia* branches. The supplicant dipped more branches into the sea and, as he stood on one of the cross canals, he waved the branches in the direction of the south wind to cause it to blow. The south wind was often desired because it is a warm wind.

In another version, *seepol* was used to bring wind for sailing. A cross was marked in the sand at the edge of the sea and filled with pieces of the plant. The supplicant stood on the cross and turned in the direction of the wind he wanted to call. He took more *seepol* in his right hand, dipped it in the sea, waved it in the desired direction, and yelled for the wind to come.

According to a common version of the Seri origin myth, *Frankenia* was one of the first plants to be created. It was used in smoking ceremonies (see *Atamisquea*).

Figure 17.91. A branch of *Frankenia palmeri* from Kino Bay. *FR.*

Gramineae (Poaceae)—Grass Family

Aristida adscensionis L.
impós
six-weeks threeawn

Impós was also the name for *Muhlenbergia microsperma*. Common and widespread ephemeral, occurring at any time of year following adequate rains.

Aristida californica S. Wats.
conée csai 'grass hairbrush'
Mohave threeawn

Perennial grass, forming small, dense clumps in sandy soils, generally near the shore. Relatively thick, tough, almost wiry roots. It is common from Kino

Bay northward and on the east and north coasts of Tiburón Island.

Boats: Men used pads of this grass to cushion their knees when paddling a balsa (see *Phragmites*).

Hair Care: The hairbrush, *csai*, was usually made from roots of this grass. Several plants were pulled out and severed close to the rootcrown. The roots were tied into a bundle—with yarn, strips of colored cloth, or sinew (Quinn and Quinn 1965 : 160)—and evened off by burning (Figure 17.92; also see McGee 1898 : 226; and Sheldon 1979 : 141, Figure 6-6).

Firemaking: Pieces of this grass mixed with rabbit dung were used as tinder.

The Supernatural: If one used the *csai* brush at night his spirit would wander off toward *coaxyat*, the place of the dead. If one dared to use it a second time at night, his hand would swell. Such a person was then known as *csai himo cöhaalajc* "the brush user

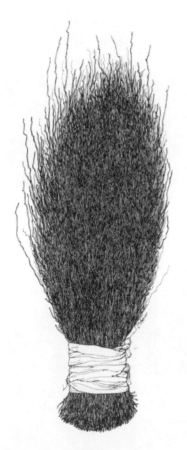

Figure 17.92. A hairbrush made by Aurora Colosio in 1962. The brush, made of the roots of *Aristida californica*, is bound with twisted red, blue, and brown cloth strips and is 31.5 cm long. *CMM.*

who finishes last." A discarded hairbrush or a brush used by one who had died was considered dangerous.

Tanning: The grass was mixed with animal brains and made into patties which were dried for storage. These were moistened and rubbed into a deer hide to cure it.

* *Arundo donax* L.
xapij
xapij-aacöl '*Phragmites* large(plural)'
  "large reedgrass"
carrizo, giant reed

Arundo donax is native to the Mediterranean region, but is now widely cultivated and naturalized in the New World. It closely resembles *Phragmites*, or reedgrass, from which it is distinguished by characters of the floret and the swollen or enlarged base of the leaf blade; it also tends to be larger and more robust (Figure 17.93).

Xapij-aacöl was found south of La Ciénega. It was brought to the shore in bundles suspended from carrying yokes, and used for making balsas (see *Phragmites*). *Xapij-aacöl* may be *Arundo*.

The *Pazj Hax* water hole on Tiburón Island and perhaps several others contain *Arundo* (Figure 17.94) and not *Phragmites*. *Arundo* may have replaced *Phragmites* from bird-introduced seeds. There is no evidence that it was introduced by the Seri. Artifacts made from *Arundo* are indistinguishable from those made from *Phragmites*, and are described under *Phragmites*. At least some of the model balsas made by Jesús Morales in the 1960s were fashioned from *Arundo* collected by him at *Pazj Hax*. This *Arundo* was identified by the Seri as *xapij*.

Bouteloua aristidoides (H.B.K.) Griseb.
aceitilla, six-weeks needle grama

B. barbata Lag.
navajita, six-weeks grama

B. repens (H.B.K.) Scribn. & Merr.
navajita, slender grama

isnáap ic is 'its-side with its-fruit'
  "whose fruit is on one side"

The Seri did not distinguish between these three species by name. They regarded *isnáap ic is* to be a kind of *conée* (the generic name for grass) with "seeds" on only one side of the stalk. One-sided

Figure 17.93. *Carrizo*, or giant reed (*Arundo donax*), from Tinaja Anita on Tiburón Island, November 1969. Note the swollen base of the leaf blade. *FR.*

Figure 17.94. A colony of giant reed (*Arundo donax*) at the Tinaja Anita water hole on Tiburón Island. Jesús Morales has two bundles of *carrizo* suspended from the ends of his carrying yoke. *RSF, November 1969.*

(secund) spikes are a key character in the genus *Bouteloua*. Pepper-grass (*Lepidium*) was also sometimes called *isnáap ic is*.

Bouteloua aristidoides and *B. barbata* are common, widespread summer ephemerals. *B. repens*, a perennial, occurs at scattered localities in the mountains.

Food: During favorable years *B. barbata* may form extensive stands on the desert floor, such as in the Agua Dulce Valley on Tiburón Island. The people collected "lots of it, like *xnois*" (eelgrass, *Zostera*). The grain was threshed, winnowed, and ground into flour.

Brachiaria arizonica (Scribn. & Merr.) S. T. Blake
  [= *Panicum a.* Scribn. & Merr.]
conée ccapxl 'grass sour'
  "sour grass"

Annual grass commonly appearing with summer rains. The same name was used for *Eragrostis* and *Erioneuron*.

Cenchrus palmeri Vasey
cözazni 'tangled'
huizapol, sandbur

The name refers to the spiny "fruit," which gets tangled in hair or clothing. The fruit is an obnoxious bur with sharp, stiff spines (Figure 17.95). It is a common, non-seasonal ephemeral, usually in sandy soils near the shore throughout the region, including Tiburón Island.

Medicine: The bur (fruit) was crushed and cooked in water, and the tea used as a diuretic and for kidney pain. The bur was also crushed in water with a fireworm (*Eurythöe complanata*). The resulting mixture was then squeezed through a cloth, and the liquid taken to cause cessation of the menstrual flow.

Digitaria californica (Benth.) Henrard
  [= *Trichachne c.* (Benth.) Chase]
cpooj
punta blanca, California cottontop

Cpooj was also used for *Sporobolus cryptandrus*. This perennial grass is common in rocky places on the islands and mainland.

Eragrostis sp.
conée ccapxl 'grass sour'
  "sour grass"
love grass

This name was also applied to *Brachiaria* and *Erioneuron*. The Seri knew of this annual grass near ranches and dry lake beds.

Figure 17.95. Sandbur (*Cenchrus palmeri*). *FR.*

Erioneuron pulchellum (H.B.K.) Tateoka
  [= *Tridens p.* (H.B.K.) Hitchc.]
conée cosyat 'grass spines'
  "spiny grass"
conée ccapxl 'grass sour'
  "sour grass"
fluff grass

Low, tufted perennial grass, often common on dry rocky slopes and elsewhere in the desert. The plant is stiff, bristly, and almost spinescent.

Jouvea pilosa (Presl.) Scribn.
cocásjc, xojásjc

This coarse, spinescent saltgrass grows in beach sand near the high tide line at several sites along the mainland side of the Infiernillo. It is perennial and forms dense clumps; the roots are coarse and wiry. The fishermen know this grass because they see it when they beach their boats. They know of it at only three sites: Punta Chueca, Campo Oona, and Dos Amigos near Estero Sargento. We saw it also on a sandy point 10 km north of Punta Chueca.

Hair Care: The roots were used to make the *csai* hairbrush in the same manner as for *Aristida californica.*

Monanthochloe littoralis Engelm.
cötep

Wiry-stemmed perennial saltgrass, with short, stiff, sharp-pointed leaves (Figure 17.96). Generally forming dense colonies at esteros and tidal flats along the mainland and on Tiburón Island.

Figure 17.96. *Monanthochloe littoralis. FR.*

Beliefs: One who felt pain when walking barefoot on this grass was said to be stingy.

Shelter: Used as roofing for the brush house, it provided protection against light rain.

Signaling: It was considered one of the best plants for making smoke signals.

Muhlenbergia microsperma (DC.) Kunth
ziizil
impós

Delicate, non-seasonal ephemeral, widespread and generally commonest in better-watered niches, such as arroyo margins. The light-colored grain is 1.5 to 2.0 mm long. Dark-colored, cleistogamously developed grain commonly occurs near the base of the plant.

This grass, often not given a specific name, was generally called "a kind of *conée*," the generic name for various grasses. The Seri described it as having tiny leaves and no thorns, and indicated a height of about 30 cm.

Food: The grain, described as being yellowish, was toasted, ground, and cooked as a gruel.

*Oryza sativa L.
xica coosotoj 'things thin(plural)'
   "thin things"
xica potáat cmis 'things maggots what-resemble'
   "what resemble maggots"
arroz, rice

The second name is the older of the two names for rice.

Panicum arizonicum—see *Brachiaria*

Phragmites australis (Cav.) Steud.
   [= *P. communis* Trin.]
xapij
carrizo, reedgrass, common reed or cane

The bamboo-like leafy culms, or stems, are 2 to 5 m tall. New shoots emerge from the thick, perennial rootstock in early spring and rapidly attain full height. Reedgrass formed dense thickets at fresh water seeps and water holes in widely scattered and highly localized sites on the mainland and Tiburón Island. *Phragmites australis* is worldwide in distribution.

Arundo donax occurs at some of the water holes,
even on Tiburón Island, where it has replaced the native *Phragmites*. The Seri do not distinguish between them, and the following discussion of *xapij* covers both species as they occur native and naturalized in the region. *Phragmites* closely resembles *Arundo* but the stems of the former are generally smaller. Places named below are shown on Map 6.1.

Xapij does not occur near the coast north of the Kino Bay region, and the people of the northern mainland region had to go far inland to obtain it. They went to *Cpaaija* 'that which falls by itself(plural)' (Rancho Félix Gómez) and Pozo San Ignacio, which is on a hill south of Rancho La Ciénega. It was a journey of several days to go inland, cut the cane, and carry it to the coast.

Xapij was found at the three major permanent water places near Kino Bay: *Hax Cáail* 'water wide' (Pozo Carrizo), *Haspót Hax* 'new-mesquite water,' and *Xapij An Hax* 'reedgrass inside water.' *Xapij* occurs at four water holes on Tiburón Island: *Xapij*, *Sopc Hax*, Xacácj, and *Pazj Hax*. The most extensive *xapij* stand in the entire Seri region is at the camp and water hole called *Xapij* (Arroyo Sauzal) at the south end of the island, where great thickets of it extend nearly 1 km along shallow water in canyon bedrock. It was the source of *xapij* for the San Esteban people. In the late 1970s *carrizo* was planted at Tecomate at the north end of Tiburón Island by personnel of the Mexican federal wildlife department.

The people of the Tastiota region came north to obtain *carrizo*. They got it at a water hole called *Haxásol* 'water -asol.' *Carrizo* grows at the edge of a small playa (dry "lake") at San Pedro Bay between San Carlos Bay and Tastiota. However, the stand is relatively small and probably would not yield enough to have warranted harvesting. Neither the Seri nor we know of any other locality for *xapij* in the Tastiota–San Carlos Bay region.

At small, bedrock, spring-fed sites—such as *Pazj Hax* on Tiburón Island—the *carrizo* tends to fill the entire water hole with an impenetrable thicket. In such places the men cut a channel through the *xapij* to let the water flow more freely and to give them access to the water. A photograph taken on the McGee expedition in December, 1895, shows a pool of water more than several meters across just below the area of *xapij* (a copy of the photo is at the University of Arizona Library, Special Collections). The pool no longer existed in the early 1960s.

The tender, new shoot-buds of reedgrass (*Phragmites*) are edible, and Indians of the Colorado River region and elsewhere harvested a sweet exudate from the leaves called honeydew (del Barco 1973:136, Heizer 1945, Jones 1945, Medsger 1939:226–227, and Tanaka 1976:547). However, the Seri had no knowledge of reedgrass being used as food. Perhaps it was too valuable to have been used for food.

Boats: The Seri balsa, or reed boat, *hascám*, was made from *xapij* stems (see Figures 10.6, 10.7). The men cut the cane and carried it to the shore. Each man carried two large bundles suspended from the ends of a carrying yoke. The bundles were tied with mesquite root cord. The leaves were then removed from the long, bamboo-like stems (or the leaves may have been stripped off before being transported to the shore). The ends of the stems were intertwined to form long bundles which were lashed together with mesquite (*Prosopis*) root cordage. Three large bundles were finally bound together to form the balsa. Griffen (1959:10) says that a balsa "could be made within two or three days, possibly less." When the owner of a balsa died, it was burned by his burial sponsor, *hamác* (Griffen 1959:28).

The balsa was used for transport to the islands and coastal camps, and for hunting and fishing. It was a double-ended craft with graceful lines, varying in length from about 3 to 10 m. The smaller ones carried one man and were probably used primarily for hunting sea turtles and fishing. The people of San Esteban Island had two-man balsas which could carry as many as six or seven large turtles, whereas the usual one could carry about four. The larger balsas could carry three families (e.g., six adults and nine or ten children) or six to seven large turtles.

McGee (1898:216–219) described and illustrated a balsa his party "collected" at a Seri camp after the inhabitants fled (see Figure 10.6). He reported it as "measuring barely 4 feet abeam, 1½ feet in depth, and some 30 feet in length over all (p. 216)" and estimated it to weigh 113 kg dry, and about 126 kg when wet, "so that it could easily be picked up by three or four, or even two, strong men and carried ashore (p. 217–218)." He pointed out it was an efficient and safe craft even under severe conditions, capable of carrying twice its weight. Since the tapered ends of the craft rose out of the water when it was heavily loaded, water drag was substantially reduced.

Cargo was secured with spikes, called *icócaöj*, driven into the side of the balsa. These spikes were made of hardwood, such as catclaw (*Acacia greggii*) or creosotebush (*Larrea*). McGee (1898:217) showed two *icócaöj*, which he called marlin spikes. When a

turtle was harpooned and brought alongside the balsa, it was killed, because a live turtle would have been difficult to transport. Slits were cut through the front flippers in order to secure it to the balsa with the *icócaöj* spikes.

Neither people nor cargo on a balsa remained dry (Hardy 1829:292). The craft was paddled by men in a kneeling position, with the knees cushioned on pads of grass (*Aristida californica*) or *xpanáams* (*Sargassum* spp.). When the balsa was loaded, the knees were in the water. Commander Dewey wrote, "They kneel in these canoes when paddling, the water being at the same level in the canoe as outside it (U.S. Hydrographic Office 1880:145)." The paddle generally was made of ironwood (*Olneya*), because it would easily enter the water due to its weight.

Early descriptions of balsas, beginning with the sixteenth century, are similar to those of McGee and others more than three and a half centuries later. In a vague reference probably concerning the Seri in December, 1535, Alvar Núñez Cabeza de Vaca mentioned "fish that they catch in the sea from balsas (Núñez 1555:45)." In March 1539 Fray Marcos de Niza saw Indians who reached their island home by means of balsas; however, these were probably Yaqui or Mayo rather than Seri (Bandelier 1890:119).

In 1539 Francisco de Ulloa saw the balsa in use by the Cochimí along the Gulf Coast of Baja California:

It was made of canes tied in three bundles, each part separately, and then all tied together, the middle section being larger than the lateral. They rowed it with a slender oar, little more than half a fathom long, and two small badly made paddles, one at each end (Wagner 1929:22).

In 1692 Padre Adamo Gilg described the Seri balsa as follows:

My Seris do not build their boats of planks, but of three bundles of reed bound together, which are joined in a narrow point behind and before, and are widely separated from each other in the middle, forming a hollow bilge. In place of an anchor, they throw a good sized stone into the depths when they want to stop (Di Peso and Matson 1965:50–51).

During a campaign against the Seri in late March, 1700, the soldier Juan Bautista de Escalante described balsas used by the Seri to cross the Infiernillo Channel:

The Seri cross in balsas composed of many slender reeds, disposed in three bundles, thick in the middle and narrow at the ends, tied together, up to five or six varas in length. These balsas sustain the weight of four or five persons, and cut the water easily. Their oars are two varas in length with a paddle [blade] at each end (Alegre 1960:167; one vara = 0.83 m, or 33 in.).

Later reports further document the manner of construction and use of the balsa (e.g., Hardy 1829: 291–292).

The Seri frequently crossed from Tiburón Island to San Esteban by balsa, and on occasion they also travelled to San Lorenzo and Angel de la Guarda islands and even to the opposite shore of Baja California. In 1824 Fray Francisco Troncoso wrote that the Seri crossed the Gulf in balsas to rob and pillage the mission at Loreto (McGee 1898:82). During the late nineteenth century small parties of Seri men journeyed from the Tiburón Island region to Guaymas on balsas or on foot to barter pelican skin robes for liquor, cloth, guns, and other goods (see Figure 1.5).

There was probably at least one balsa for each family living along the coast: the Seri said each man owned a balsa. Hardy (1829:291) saw 15 or 20 balsas in the Infiernillo Channel. During a military campaign against the Seri in 1844, Tomás Espense circumnavigated Tiburón Island and captured 384 men, women, and children, and burned 64 brush houses and 97 balsas (Velasco 1850:169).

After the Seri began making wooden boats around the turn of the century (see *Stenocereus thurberi*), they started phasing out balsas. In the 1930s Alfredo Topete (personal communication) saw an abandoned balsa rotting in the Santa Rosa estero south of Punta Chueca. This was probably the last balsa used by the Seri. In the 1940s and 1950s William Neil Smith obtained several replica balsas which are in museum collections in Arizona and California, and in the 1960s Jesús Morales and others made a number of model balsas for sale (see *Arundo*).

In 1922 Edward H. Davis obtained one of the last functioning balsas. Davis stated that he "made a trade with Blanco for his balsa, promising him sufficient lumber, nails and tools, to make a boat large enough to carry five men (Quinn and Quinn 1965:171)." The Seri told us that Ramón Blanco was still using his balsa when Davis bought it, and that "it is at the University in New York." (It is in the Museum of the American Indian.) Davis's description of a one-man balsa used for turtle hunting was probably based on this same balsa:

> When a Seri goes in his native *balsa*, which only holds one man, he straddles the middle part on his knees, resting them on seaweed. The craft is light and is a good sea craft, but it is low in the water and the man is wet all the time. He uses a double-bladed paddle, and his harpoon rests against pegs set in the *carrizos*. A rock which serves to quiet the turtle, two pointed stakes and a heavy fish line complete the equipment. He has to paddle across to the other side [of the Infiernillo Channel] in that cramped position, say eight or ten miles, very early in the morning, then watch his chance to harpoon. If successful he brings up the turtle, knocks him on the head with the rock, pulls him on the end of the craft by skillful manipulation, cut[s] slots in the front flippers, put[s] stakes through and into the *carrizo*, pinning him firmly in place, then paddle[s] back (Quinn and Quinn: 166).

Containers: A container made from a piece of the hollow stem was used for carrying and preserving powdered or granular materials (Figure 17.97), such as cattail (*Typha*) pollen for facepaint, powdered ochre (*xpaahöj*) for paint, and creosotebush (*Larrea*) lac (*csipx*). One of these containers has been illustrated by McGee (1898: facing 261).

Fishing: The entire culm, or stem, and the root, called *xapij icáapöquij* 'reedgrass caller,' was used to "call" fish. The stalk was held by the top, and the root end plunged up and down on the water to attract fish to the surface, where they were then harpooned. For example, fishermen living on Tiburón Island hunted leopard grouper, *Mycteroperca rosacea* (*tatcö*), with this method. They went to rocks or tide pools where groupers occurred, taking along a bundle of reedgrass to stir up the water. When a grouper

Figure 17.97. This *carrizo (Arundo or Phragmites)* container with a wooden stopper is 16 cm long and decorated with blue pigment. It was obtained in 1961 at El Desemboque. *CMM.*

stuck its head out to see what all the commotion was about, the fisherman harpooned it. It was said that the larger fish thought that the movement was another, smaller fish. This technique was used in conjunction with the *hacaaizáa*, the single-prong fish harpoon.

Food Gathering: Columnar cactus fruit-gathering poles were occasionally made from reedgrass culms.

Games: The scoring for the *camóiilcoj* circle game, played by women and children at puberty and leatherback turtle fiestas, was done with game sticks, called *hemot*, which were used as dice (Figure 17.98). The *hemot* were made from reedgrass stems. The game was played with a set of three game sticks, each of which was decorated or banded on only one side.

Different combinations obtained from throwing the *hemot* dice determined how far a player moved the *hapéxz* stick counter forward or backwards around the *hamóiij* circle. The first player to move her counter a complete round plus four spaces won the game. Griffen (1959:14) observed bets consisting "of pins, thread, and the like."

Xapij caanlam 'reedgrass enclosed(plural),' was a men's gambling game played with four reedgrass tubes and 100 counting sticks (Figure 17.99; Griffen 1959:14–15, Kroeber 1931:15). It was played with two teams of four men on a side. The reedgrass tubes were open at one end. A twig or sliver—often reedgrass—or a coral bean seed (see *Erythrina*) was put in one tube, and all four were filled with sand. The filled tubes were passed to the opposing side. A member of this team emptied them one by one and tried to guess the one with the twig on the third attempt.

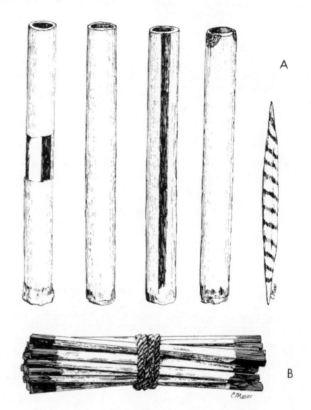

Figure 17.99. *Carrizo* tubes and counting sticks for the *xapij caanlam* game. A) These tubes and wood sliver are decorated with blue pigment (shown as shaded areas). They were made by Nacho Morales about 1962. *From the left: imac coopol* 'its-middle black'; *comcáii* 'old woman'; *cöhatáxca* 'striped'; *ilit coopol* 'its-head black.' The longest tube is 14 cm and the wood sliver is 10 cm long. B) Counting sticks made by Lupe Comito in 1958. The longest is 12 cm; the ends are dyed blue and the bundle is tied with commercial string. *CMM*.

Figure 17.98. These *hemot* game sticks for the *camóiilcoj* circle game are made of *carrizo*. Made by Sara Villalobos in 1962, they are each 14 cm long. One side has red and blue pigments; the other side is plain. *CMM*.

Griffen (1959:15) reported that score was kept as follows: to find the twig on the first try, lose ten points; second, lose six; third, lose turn; and fourth, lose four points. The game was played for a score of 100 points, counted out with 100 markers of shells or sticks placed between the two teams. The counting sticks were often made from pieces of reedgrass. The ends were commonly dyed reddish, from *Kramaria grayi*, or blue, from Seri Blue (see *Guaiacum*). On each throw the number of points was subtracted from the counters in the center pile, or, when these were exhausted, from those in front of the opposing team. The counters were placed in front of the group losing the points, unless the twig appeared on the third attempt, in which case only the turn was lost. Play was continued until all counters were on one side, and the corresponding team had then lost the game. The turns for emptying and filling the reeds were rotated within each team, and each person had to place a bet in order to play.

Xapij caanlam continued to be a popular game at fiestas in the 1980s (see Figure 13.2). The men sat on the sand at the beach or on a dune. On one occasion, in 1983, they sat on benches facing a wheelbarrow filled with sand. A blanket was spread over the sand in front of them. A scorekeeper was at each end of the bench with a bundle of counting sticks. The men on opposite sides took turns hiding a matchstick in one tube while keeping the tubes hidden under the blanket. The tubes were filled with sand underneath the blanket to disguise the one which had the matchstick. Men on the other side attempted to guess by elimination which tube contained the stick, which had to be the third tube upended in order to win. In 1983 they played for 1,000-peso stakes. This game was widespread in southwestern North America (e.g., Kroeber 1931:41; Russell 1908:176–177).

A game called *xtapácaj caahit* 'xtapácaj he-who-caused-to-eat,' or "he who feeds the tower shell" was played with a section of reedgrass and a tower shell (e.g., *Turritella gonostoma*). The tower shell, called *xtapácaj*, was flipped with the fingers into the open end of the reedgrass tube, called *xapij ccaiit* 'reedgrass he-who-shoots.' In order to win, one had to flip the shell into the piece of cane ten consecutive times. Two people at a time played, and the winner challenged the next contestant. It was played by men and women, at any time, but probably it was primarily a men's gambling game played at fiestas. Some men were experts at this game.

Griffen (1959:15) reported boat races with balsas and wooden boats. There was usually only one man to a balsa, or occasionally two. Each wooden boat had a team of six or seven men, each one paddling.

Hunting: It was discovered that the split reed musical instrument (see Music, below) *xapij hamánoj* 'reedgrass rolled-between-hands' could be used for a non-musical purpose. Juana Necia, a woman spoken of by many of the adults, devised a novel method for hunting jackrabbits. On moonlit nights during jackrabbit mating season she went out into the desert alone. Sitting down, she covered herself with vines and rolled the split reed instrument between her hands. The sound was said to be similar to that of male jackrabbits fighting. Juana also made lip trills and smacked her lips against her hand to help lure a rabbit in close. When a curious jackrabbit hopped up close enough she would club it. Juana killed several jackrabbits on each outing, and for a time other Seri could not figure out how she hunted them (Bowen and Moser 1970b).

Music: The following discussion of reedgrass instruments is revised from a previous report.[*] The Seri mouth bow, called *xapij haahi* 'reedgrass what-is-caused-to-sound,' was constructed from a piece of cane up to about 1.7 m long and 2.5 cm thick (Figure 17.100). The twisted sinew or mesquite root twine which served as the string was tied to one end of the cane. At the other end, the string was wound around the tip of a hardwood tuning peg inserted in a hole in the cane.

This instrument, remembered only by the older Seri, was played while the musician was lying on his back (see Figure 13.14). A right-handed musician held one end of the cane between his teeth, with the peg end extending out to his left and the string facing away from him. The instrument was strummed with the right hand and fretted with the fingers of the left. The mouth bow was an instrument to be played at home. Probably this was the "one-string viol" mentioned by the Coolidges (1939:209).

The split reed instrument, *xapij hamánoj* 'reedgrass rolled-between-hands,' (Figure 17.101) was rolled between the palms to produce a rattling sound (Griffen 1959:17). A short section of the cane was split lengthwise several or more times, leaving all the slats fastened at the base. The men used it in two dances. The *ziix coox cöcoila* "dances concerning all things" was a men's circle dance performed during a girl's puberty ceremony, and included the use of the antenna headpiece (see *Jatropha cinerea*). The *oot cöicóit* "dance concerning coyote" was performed by a man, usually when he was inebriated. It included songs about the coyote and fox, and was done purely for fun to amuse the children. The Yaqui have used a similar instrument for their coyote dance, held before dawn on the Day of the Dead and on December 12 at Potam, Sonora (Fernando Murillo Rendón, personal communication, 1982).

The shaman's flute, *xapij an icóos* 'reedgrass in what-one-sings-with,' was made from a completely hollowed out section of reedgrass about 30 cm long (see Figure 13.15). A slit was cut at the middle of the

[*] Revised from T. Bowen and E. Moser, "Material and functional aspects of Seri instrumental music," *The Kiva* (1970) 35(4):178–200. Used with permission.

Figure 17.100. Mouth bow made from a *carrizo* stem and mesquite root twine; the tuning peg is hardwood (*Larrea*). This replica, made by Jesús Morales in 1962, is 1.5 m long. *CMM*.

tube. This instrument was used only by a shaman who blew through one end to summon the spirits.

Xapij an iquípl 'reedgrass in what-one-licks-with' was a toy flute used by young boys. It was made from a section of hollowed cane about 30 cm long. Finger holes spaced at equal intervals were sometimes cut into the tube. Instruments with finger holes were capable of producing different pitches but were not used for playing tunes.

A whistle, *xapij iti hacázx* 'reedgrass on what-was-slit,' was manufactured from a section of cane usually somewhat less than 30 cm long. It was hollowed except for the distal end, which was left plugged. The open mouth end was whittled so that it terminated in two points on opposite sides of the tube. The tube was then split down either side almost to the base. This whistle was a toy used by boys to signal to their companions. The sound could be heard for a considerable distance.

Oral Tradition: In 1960 two Seri men said that about ten years earlier they saw coyote people on a dry lake bed near Kino Bay. They did not go too close to them. They had thin, hairy, and yellow faces, and wrinkled foreheads. There were men, women, and children. One old man had a bow. He shot at a tree trunk, and the arrow went all the way through, skidded on the ground, stopped, and formed a *xapij* plant with leaves on it.

The Supernatural: A shaman made special powder which he put in small tubes of reedgrass called *xapij hehe ano yaii* "reedgrass to put plant powder in" (Fig-

Figure 17.101. This split reed rattle, or musical instrument, is made of polished *carrizo*, 14 cm long, with bands of Seri Blue near the base and at the tip. This replica was made by Jesús Morales in 1962. *CMM*.

ure 17.102). The powder was obtained with the aid of the spirit power *Icor*. The tubes were rented to people to bring good luck, and one could use them to cure himself of illness.

Shelter: The stems were often incorporated into the framework of the brush house. In April, 1922, Edward H. Davis saw brush houses on the east shore of Tiburón Island with ". . . the frames of carrizo reed and ocotillo" (Quinn and Quinn 1965:151). At 5 to 10 km to the southeast of *Pazj Hax* (= Tinaja Anita) McGee (1894:127) found a ". . . fairly recent ranchería of half [a] dozen houses of [the] usual type except that cane is used largely in construction."

Smoking: A segment of the stem, called *xapij an icóopis* 'reedgrass in what-one-sucks-with' served as a cigarette holder and as a kind of pipe. It was heated in a fire and filled with tobacco (*Nicotiana trigonophylla*). While it was being smoked, saliva was put on the cane to prevent it from burning.

Toys: A reedgrass gun, known as *xapij iti caacni* 'reedgrass on bowed,' shot slivers of reedgrass a distance of several meters (Figure 17.103).

A peashooter, *xapij an icóotp* 'reedgrass in what-one-spits-with,' used elephant tree (*Bursera microphylla*) fruit as ammunition. A squirt gun, *xapij iqui icajöámz* 'reedgrass with what-one-squirts-with', was made with a small section of reedgrass fitting into a larger one, loaded with fresh or salt water. Boys made toy harpoons from reedgrass for spearing crabs (*Callinectes*) and small stingrays along the shore.

Weapons: *Hasménelca*, the mainshaft of the Seri arrow, was made from reedgrass because it made the arrow light.

Other Uses: Small knives were fashioned from the stem. McGee (1898:190) reported that women removed pelican skins by making an incision "either with the edge of a shell-cup or with a sharp sliver of cane-stalk taken from an injured arrow or a broken balsa-cane."

Sections of the stem were used as drinking straws called *xapij hax ipási* 'reedgrass water what-it-is-drunk-with.' When the men returned to camp with ollas filled with water, the children ran with their "straws," jammed them into the ollas, and drank their fill. Sometimes these straws also were used to drink cactus wine.

In 1692 Gilg stated that the Seri wove ". . . mats of palm and reed" (Di Peso and Matson 1965:55). *Carrizo* (*Arundo* and/or *Phragmites*) would have been readily available at the Seri mission at Pópulo and elsewhere along the Río San Miguel. Gilg illustrated a woman carrying a woven mat (see Figure 1.4) which could have been made from *carrizo* or palm leaves.

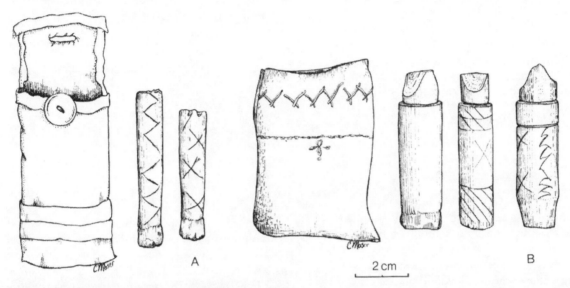

Figure 17.102. Nacho Morales's shaman paraphernalia. A) *Carrizo* tubes (wooden stoppers missing) and a blue cloth bag with a red button and red and white stitching; obtained in 1960. B) *Carrizo* tubes with wooden stoppers and a blue cloth bag with decorative stitching of red thread; given to M. B. Moser by Lupe Comito in 1968. *CMM.*

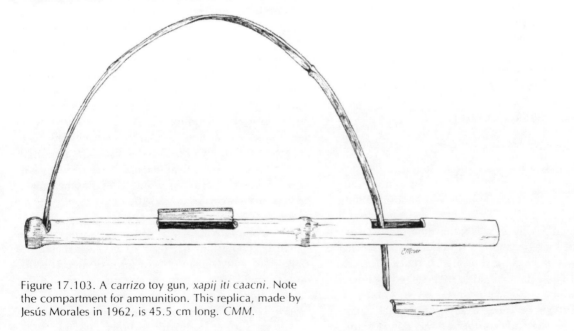

Figure 17.103. A *carrizo* toy gun, *xapij iti caacni*. Note the compartment for ammunition. This replica, made by Jesús Morales in 1962, is 45.5 cm long. *CMM.*

Saccharum officinarum L.
xapij coatöj 'reedgrass sweet'
  "sweet reedgrass"
caña azúcar, sugar cane

Setaria liebmannii Fourn.
ziizil

Summer-fall ephemeral grass, usually found beneath large legume trees; mainland and Tiburón Island.

Food: The small seeds were parched, ground, and cooked in water and eaten as a gruel.

Setaria macrostachya H.B.K.
hasac
xica quiix 'things globular(plural)'
  "globular things"
plains bristlegrass

Perennial with stems 1.0 to 1.5 m tall, and dense, spike-like panicles. The grain is about 2.5 m in diameter and globose. Usually found at higher elevations on north-facing slopes among rocks, and infrequently in widely scattered lowland places, generally in better-watered situations. This is the only perennial *Setaria* in the Seri region.

The Seri said it was abundant in the area of Rancho Miramar near El Desemboque, at La Ciénega, and in the mountains north of Hermosillo. They said it was once common along the Arroyo San Ignacio, but was no longer found there (meaning that it is not common there). The term *xica quiix* seems to be applied to larger plants of this species.

Food: The grain was toasted, ground into flour, cooked with or without green turtle oil, and eaten as a gruel. It was apparently a fairly important food.

Sporobolus cryptandrus (Torr.) Gray
cpooj
sand dropseed

This name was also used for *Digitaria californica*. This perennial clumping grass is seldom common and usually occurs on sandy soil, such as dunes and middens near the sea.

The Seri said this grass is taller than *conée caacöl* (unidentified grass), and has roots which are thicker and not as straight as the roots of *Aristida californica*.

Animal Food: It was said to be browsed by horses and cattle.

Hair Care: The roots were used to make hairbrushes when *Aristida californica* was not available.

Sporobolus virginicus (L.) Kunth
xojásjc

A perennial saltgrass, abundant along beaches and the inland margins of mangroves on the mainland and on Tiburón Island. It has stiff leaves and forms numerous shoots from long underground rhizomes.

Food: Children enjoyed eating the new, emerging shoots.

Trichachne californica—see *Digitaria*

Tridens pulchellus—see *Erioneuron*

Triticum aestivum L.
caztaz 'Castilian *Zostera*'
  "Castilian eelgrass"
trigo, wheat

Just as many people have named wheat after their most important grain or grain-like crop, the Seri seem to have named wheat "Castilian eelgrass." The term *caztaz* was probably derived from *cazt* 'castellano' (Castilian) and *eaz*, the term for the detached eelgrass plant as it is found floating or washed ashore in spring with seeds (see *Zostera*). It is common among Indian languages in Mexico to use such phrases as "Castilian tortilla" for "bread" or "Castilian *guajolote* (= turkey)" for "chicken" (Lyle Campbell, personal communication, 1983).

Food: Wheat, introduced from the Old World, has been cultivated by neighbors of the Seri since early Spanish colonial times. It was one of the important crops grown by missionized Seri at Hermosillo and Pópulo during the seventeenth and eighteenth centuries. During the nineteenth century some Seri camped near ranches and often worked for daily rations of food, including wheat, maize (corn), cheese, and milk. Wheat flour was called *hapáha quis* 'what-is-ground-up raw.' *Pinole*, the flour used to make *atole* (gruel), was called *hapáha*. *Hai cöquisj* 'wind spray,' was the term for winnowing a grain, such as wheat, by throwing it into the air with a threshing fork.

After winnowing it was called *caztaz ano hahóóca* "wheat in what-has-been-scraped.' It was then moistened and rough ground on a flat metate. The flour was cooked in water to which animal fat (e.g., sea turtle, horse, or beef) was added, then salted and eaten. During the twentieth century wheat flour was made into atole, fry bread, and tortillas.

Zea mays L.
hapxöl
xica caacöl 'things large'
maíz, maize or corn

Hapxöl is a Seri word. *Xica caacöl* is a term from the Tiburón Region people, and has been in use in modern times, although used less commonly than the term *hapxöl*. Field corn is often called *hapxöláa* 'true corn.'

Corn has been grown by neighbors of the Seri since ancient times. In 1645 Perez de Ribas (1944, II:148) reported that "They sustain themselves by hunting, although during the time of maize harvest, they go with deer hides and salt which they gather from the sea to trade with other nations." During Spanish colonial times they further became familiar with it through the missions. In the nineteenth and early twentieth centuries some Seri worked on ranches, harvesting corn, beans, and other crops (Quinn and Quinn 1965:191).

Food: Popcorn, *hapxöl cmaptx* 'corn burst,' was made from common field corn. The kernels were put in a large basket, and embers put on the corn. A woman shook the basket continuously in an up-and-down motion until the corn was popped. The hot, popped corn and embers were poured into another basket or pottery vessel, the embers removed, and the corn eaten. It was also popped in a pottery vessel or olla with hot sand.

In 1921 Sheldon (1979:137) observed that the Seri had metal kettles, and that when they could get corn they parched it in these kettles by shaking it with hot sand. The dry, hard kernels were ground on a metate. This coarse-ground corn was cooked in water, and animal fat (e.g., sea turtle, beef, or mule deer) was added. When it thickened it was removed from the fire. The gruel was drunk from a large clamshell (*Laevicardium elatum*) or other suitable container. Children used a smaller clam or mussel shell. An ear of green corn (*elote*) was called *hapxöl coil* 'corn green.'

Unidentified grasses

conée caacöl 'grass large(plural)'
   "large grass"

The Seri said this grass, found near Siete Cerros, was similar to *zai*, but it did not grow as tall, and it had crooked rather than straight roots.

xapij-aas 'reedgrass (*Phragmites*)-aas'
bamboo?

The source of this plant was not known, since it was found only as driftwood. We have not seen it.

Cradleboard: Crossboards for a cradleboard sometimes were made of this plant.

Other Use: It was said that on rare occasions it was used as a mast for the Seri sail boat; such use indicates it is stronger and more rigid than *Arundo* or *Phragmites*.

zai

This was said to be a tall grass with straight roots. It grows near the Agua Zarca ranch near Kino Bay.

Hydrophyllaceae—Water-leaf Family

Nama hispidum Gray
hohr-oohit 'donkey what-it-eats'
   "what donkeys eat"

The name *hohra* 'donkey' is derived from the Spanish word *burro*. It is a low-growing, winter-spring ephemeral with lavender corollas. During favorable years it is abundant.

Animal Food: The name indicates a plant eaten by donkeys.

Phacelia ambigua Jones
najmís

caháahazxot ctam 'what-causes-sneezing (*Baileya*) male'
   "male desert marigold"
heliotrope phacelia

Caháahazxot is based on the onomatopoetic word for sneeze. Contact with the plant often causes an allergic reaction. This winter-spring ephemeral has lavender petals and coiled (scorpoid) flower stalks. It is widespread and in some years very common in the northern part of the Seri region, including Tiburón Island. Some Seri did not recognize the plant by name.

Juncaceae—Rush Family

Juncus acutus L. var. *sphaerocarpus Engelm.*
caail oocmoj 'dry-lake what-it-wears-on-waist'
  "dry lake's waist cord"
junco

This giant rush grows along margins of dry lakes, "like the dry lake's waist cord."

This unusual plant forms a hemispherical, pin-cushion-like, perennial clump of spine-tipped shoots about 1.0 to 1.5 m. tall. It grows in low wet places and thrives in highly alkaline conditions. In the Gulf region it occurs in Baja California and in the vicinity of Kino Bay.

The Seri said it occurred north of Kino Bay, near Pozo Carrizo, growing with *xapij* (see *Phragmites*), had spines and green stems but no leaves, and that the stems were thin, like wires, with many stems sticking up about 1 m.

Games: Men made gambling chips from the stems (culms). Women used the dry, grain-like fruit for their gambling.

Koeberliniaceae—Crucifixion-thorn Family

Koeberlinia spinosa Zucc.
xooml
crucifixion thorn

Shrub or small tree (Figure 17.104), mainland coast south nearly to Tastiota and on Tiburón Island. The wood is exceedingly hard and produces copious oily black smoke when burned. The plant was said to be poisonous or toxic and to cause the flesh to swell when one was pricked by the thorny branches.

Medicine: Tea made from the flowers was taken as a remedy for dizziness and a certain intestinal disorder, or diarrhea (Whiting 1957: #61), known as *xica cmasol* "yellow stuff." The twigs were combined with *hierba de manso* as a remedy for rheumatism (see *Anemopsis*).

The wood was burned in a hut to drive away disease during epidemics, probably measles. During the nineteenth century, after excursions to Hermosillo to sell and trade, the Seri came home and burned *xooml* wood to disinfect their houses against diseases they had encountered in the city.

Figure 17.104. Crucifixion thorn tree (*Koeberlinia spinosa*) near Kino Bay. *RSF.*

Krameriaceae—Ratany Family

Krameria grayi Rose & Painter
heepol
cósahui, white ratany

Low-spreading shrub, spine-like branch tips, and small drought-deciduous leaves (Figure 17.105). The deep magenta flowers, ca 1 cm in diameter, are crowded in small clusters and may appear at various times of the year. The bur-like fruit, ca 1 cm in diam-eter, resembles a miniature sea urchin. The thick roots radiate from a point about 15 cm below the soil surface.

It is a common desert plant, found on the mainland and Tiburón Island. South of Kino Bay it is replaced by a similar, although distinct, species—*K. sonorae* Britt., which develops into a shrub 1.5 to 2 m in height.

Dye: The most common basket dye color and, ac-

Figure 17.105. A shrub of *cósahui* (*Krameria grayi*) surrounding a young *cardón* on the coastal bajada near the north end of the Infiernillo Channel. *RSF, March 1983.*

cording to Seri oral history, the only color traditionally used was a reddish brown dye. It was generally made from *cósahui* root. Johnson (1959 : 12) described it as a "rich burnt sienna color." With considerable effort the lateral roots (not the main root) were dug up with a stick, seashell, knife, or metal spoon. One or two kilograms of roots might be gathered at one time and carried back to the village or camp.

The greater bulk of the root consists of thick water-storing bark from which the dye was prepared (Figure 17.106*A*). The dirt was scraped off the root with a blunt instrument, such as a spoon. The bark was then removed (Figure 17.106*B*) by pounding the root on a metate with a mano (grinding stone), and the mashed bark was placed in boiling water. Rolls of basketry splints (from *Jatropha cuneata*) were then placed in the dye, and allowed to steep in the boiling brew until they reached the desired color saturation

Figure 17.106. Bark from the roots of *cósahui* (*Krameria grayi*) is used for basketry dye. A) Rosa Flores with the thickened roots obtained from digging up part of one shrub along the coastal bajada near the north end of the Infiernillo Channel. *RSF, March 1983.* B) Removing the bark by pounding the roots on a metate. Note the light-colored roots with the bark removed, just in front of the metate. *MBM, El Desemboque, 1968.* C) The freshly mashed bark is spread out on a plastic bag to dry. The pan at center right holds the *Krameria* dye in which basketry splints are steeped. *RSF, near El Desemboque, March 1983.*

319

(Figure 17.106 C). Prolonged simmering was essential so that the red-brown hue would not later fade (Johnson 1959). Unbleached muslin was sometimes dyed in this manner.

The statement that reddish brown was the only color traditionally used for basket designs was borne out by examination of Seri baskets in numerous museums and private collections. *Krameria* dye continued to be the predominant color in most baskets in the 1980s. Two other plants, *Hoffmanseggia* and *Melochia*, were occasionally prepared in a similar manner to produce red-brown dye. In rare instances *Krameria* roots were mixed with those of *Hoffmanseggia*. In addition, various objects, such as game sticks and beads, were dyed with *Krameria* dye.

Medicine: Tea made from the flowers was taken to cure an upset stomach and diarrhea. Tea made from the stems with the bark removed was said to make the blood very red. The stems, dried and ground into powder, were sprinkled on a skin sore that was slow in healing. It was said to help the healing and prevent infection.

Krameria parvifolia Benth.
haxz iztim 'dog its-hipbone'
"dog's hipbone"

The same name was applied to *Hoffmanseggia* and *Calliandra*. This low, stiff-branched shrub is not common in the Seri region. It occurs on the mainland and Tiburón Island. Most of the Seri were unsure of the name of this plant.

Labiatae (Lamiaceae)—Mint Family

Hyptis emoryi Torr.
xeescl
desert lavender

This common desert shrub occurs throughout the Seri region, including Tiburón and San Esteban islands. It grows on rocky slopes, as well as across the desert floor and along dry washes and arroyos. The multiple, slender stems reach 2 to 4 m in height. The small, simple leaves are covered with grayish white hairs and the tiny, lavender-blue flowers are fragrant. The wood is somewhat flexible and moderately hard. In spring, when the desert lavender was in flower, women and girls picked handfuls of *xeescl* to take home because of the pleasant fragrance.

Dye: For use as a basket dye, see *Suaeda moquinii*.

Fishing: The shaft for *hacaaizáa*, the single-pronged fish harpoon, was often made from this shrub.

Hunting: Desert tortoises (*Gopherus agassizi*) were usually hunted by women. During the cooler months of the year, the tortoises were known to crawl into small caves or burrows extending beneath large rocks. The usual method of hunting was to search in canyons for likely caves or burrows. The women then fashioned a long pole made from a stout branch of desert lavender. *Hyptis* was used because it was frequently common where tortoises were found and it has tall stems. With a wire hook fastened to one end of the pole, they pushed it into the cave to feel for tortoises. When they felt one, they poked at it until it was hooked in the posterior part of the carapace and then pulled the animal out (Felger, Moser, and Moser, 1983).

Medicine: Tea brewed from the leaves, with cinnamon added, was taken to cure a cold. As a remedy for asthma and rapid breathing (*icayáx*) a tea was made with the shavings of *Bursera hindsiana* and *xeescl iyat* '*Hyptis* stem-tips.' To relieve a toothache, *Hyptis* leaves, a piece of the pith from the stem of *cotéexet* (*Opuntia fulgida*), and a piece of *comáanal* root (*Anemopsis*) were brewed together in water and the liquid held in the mouth over an aching tooth. A few leaves were placed in the ear to help one hear better.

The Supernatural: Some considered *xeescl* to be one of the first plants formed. Griffen (1959:50) indicated that temporary huts for a vision quest were sometimes covered with elephant tree (*Bursera microphylla*) and desert lavender branches. While curing, a shaman often held a bundle of the leafy stems and blew through them (see Figure 7.6), chanting, "*zop, zop, zop. . . .*"

Salvia columbariae Benth.
hehe yapxöt imóxi 'plant its-flowering what-doesn't-die'
"plant whose flower doesn't die"
chia

This small mint is widespread in arid regions in the southwestern part of the United States, and in Sonora it extends south into the northern part of the Seri region as a winter-spring ephemeral. The Seri name is based on the dry flower stalks, with their old clusters of dry calices, which remain long after the plant has withered. Chia seed was an important food resource for Indians and early settlers in parts of the southwestern United States. Apparently, it was too scarce in the Seri region to be significant, and was not used by them.

Teucrium cubense Jacq.
hehe itac coozalc 'plant its-bone ribbed(plural)'
　　"ribbed-stem plant"

The Seri name, also applied to *T. glandulosum*, refers to the sharp-angled, or square, stems characteristic of the mint family. It is a delicate annual or ephemeral, usually found in low, wet places and poorly drained soils.

Teucrium glandulosum Kell.
hehe itac coozalc 'plant its-bone ribbed(plural)'
　　"ribbed-stem plant"
hasoj an hehe 'river area plant'

The first name was also used for *T. cubense*, and the second was a descriptive name applied to various mustards. Some people did not recognize or have a name for the plant. It is a large mint found in highly localized populations at the edges of dry lakes and other seasonally wet places.

Leguminosae (Fabaceae)—Legume or Bean Family

Acacia constricta Benth.
oeno-raama
vinorama, white-thorn

The Seri name seems to be derived from the Spanish name for the plant, or perhaps from *buena rama*. A shrub with straight, white spines of variable size, it closely resembles *Acacia farnesiana*, from which it is distinguished by its pods, which are smaller, more slender, more numerous, and have constrictions between each seed. The Seri likewise distinguished it from *A. farnesiana* by its smaller pods. It generally occurs along arroyos and washes, and is widely distributed in the Seri region.

Medicine: The seeds and leaves, mashed and cooked as a tea, were used as a remedy for upset stomach or diarrhea.

Acacia farnesiana (L.) Willd.
poháas camoz 'future-mesquite thinker'
　　"what thinks it's a mesquite"
huizache, sweet acacia

The name implies this shrub "thinks" that perhaps it is a mesquite or *haas* (*Prosopis*). This name was also used for *Desmanthus covillei*. The Seri pointed out that it differed from *oeno-raama* (*A. constricta*) by its larger fruit, and that the fruit was similar to the mesquite pod.

Sweet acacia is a large shrub with straight white spines. The pod is thick, like that of the mesquite, although shorter.

It occurs near Rancho Costa Rica and at other ranches east of the coastal mountains. At Pozo Coyote it grows along the floodplain but is rare, and was said to be a "plant taken by the water," implying that the seed was brought by floodwater from upstream.

Acacia greggii Gray
tis
uña de gato, catclaw

Catclaw is a shrub or small tree with irregular branches and trunk, and small, sharp, recurved spines (Figure 17.107). The wood is strong and hard, and the heartwood reddish. It ranges nearly throughout the Seri region, including Tiburón Island, and is common along arroyos and other drainageways. It does not occur on San Esteban Island.

Facepaint: See *Cercidium microphyllum*.

Food Gathering: The wood was used as the *hahéel* chisel—pry bar for the various century plants (*Agave*) and for the *hapoj* digging stick for *saiya* (*Amoreuxia*), etc.

Hunting and Fishing: Before metal was available the preferred material for fish and turtle harpoon points was catclaw wood. The generic term for an entire harpoon or just the harpoon point, wooden or metal, was *tis*.

Music: *Tis* was one of several kinds of wood used

Figure 17.107. Catclaw (*Acacia greggii*) along the trail to Tinaja Anita on Tiburón Island. *RSF, April 1983.*

to make the violin bow. The musical rasp, *hexöp*, was occasionally made from catclaw instead of ironwood (*Olneya*).

Weapons: The wood was used to make the bow, *haacni* (Whiting 1957: #19).

Other Uses: The wood was used for the *peen* carrying yoke and for *icócaöj* spikes to hold cargo on the balsa (see *Phragmites*).

Acacia willardiana Rose
cap
palo blanco

This thin, wispy tree is spineless and has paperlike, peeling, white bark (Figure 17.108). The trunk and limbs are usually straight, slender, and flexible, and the wood is hard. It is very common on rocky slopes, cliffs, bajadas, and arroyos near mountains. *Palo blanco* occurs in coastal mountains from the Sierra Seri to Guaymas and on Tiburón Island. It is endemic to western Sonora.

Food Gathering: The wood was used to make the *hahéel* chisel−pry bar and the *hapoj* digging stick.

Hunting: The detachable foreshaft of a sea turtle harpoon was often made from *palo blanco* wood.

Shelter: The trunks commonly were used in the framework of brush houses and as posts in wattle and daub houses.

Other uses: Uses of the wood included: the *heectim* club for killing fish and sea turtles; the *iqui iquéexoz* 'with what-one-scrapes with,' a pole on which a deer skin was scraped; and the *peen* carrying yoke (Sheldon 1979: 101). This was often made on the spot after killing a deer, in order to transport the meat back to camp (Sheldon 1979: 114, 115, 118, Figure 5−18).

Astragalus magdalenae Greene
iix casa insíi 'his-water putrid who-doesn't smell'
"who doesn't smell his putrid water"
magdalena loco-weed

Annual or perennial with silvery foliage, lavender petals, and characteristic hollow loco-weed pods. Found on beach dunes on the mainland and Tiburón Island.

Figure 17.108. Pedro Comito next to *palo blanco* (*Acacia willardiana*) trees along the trail to Tinaja Anita on Tiburón Island. *RSF, April 1983.*

The name refers to an incident which occurred long ago. There was a man who hoarded his drinking water and did not share it with others. His water lasted a long time, but it became stale and putrid, although he didn't mind the smell. In fact, he said it did not smell at all. Subsequently, the children liked to ask an unsuspecting person to smell this plant. If one said that the plant had no odor (which it doesn't) great hilarity overcame the children, who said that he must be stingy, like the man who hoarded his water.

Caesalpinia palmeri S. Wats.
hap oácajam 'mule-deer what-it-hits'
   "what mule deer flay antlers on"
cmapöjquij 'what bursts open'
piojito

The first name signifies that mule deer use the branches to flay or scrape velvet off new antlers. The second name refers to the manner of dehiscence: the dry, mature pods burst open explosively and fling the seeds away from the parent plant.

This large, spineless shrub has drought-deciduous foliage and small, bright yellow, cassia-like flowers. The trunk and branches are slender and the wood is hard, but flexible and resilient. It is found in the foothills and mountains from the Sierra Seri southward and on Tiburón Island.

Fishing: The wood was often used to make the shaft of *hacaaizáa* the single-pronged fish harpoon.

Pigments: The red sap was used for painting arrows, and for decorating various other objects. To obtain the sap, the shrub was not cut or bruised: "One must just look for places from which it is oozing."

Weapons: It was often the preferred wood for the bow, *haacni.*

Other uses: It was one of the preferred woods for making the carrying yoke and the outer, curved frame of the cradleboard.

Calliandra eriophylla Benth.
haxz iztim 'dog its-hipbone'
   "dog's hipbone"
huajillo, fairy duster

The same name was also used for *Hoffmanseggia* and *Krameria parvifolia*. It is relatively rare in the Seri region and generally occurs at higher elevations.

Cassia covesii Gray
hehe quiinla 'plant that-rings'

The name refers to the dry pods with rattling seeds. Perennial herb about 0.5 m tall with pale yellow flowers; widely distributed on the mainland and Tiburón Island

Medicine: Tea made from the roots would give one an appetite and also "clean out the stomach." This tea was also used in the treatment of measles, as kidney medicine, and to help bring about conception. Tea made from the leaves and stems was said to be good for the liver and was given to one afflicted with chicken pox.

Cercidium floridum Gray subsp. *floridum*
ziij, iiz
blue palo verde

Iiz is also the name for the gafftopsail pampano, or *pez pampano* (*Trachinotus rhodopus*).

This is one of the few common desert trees in the region which provides ample shade (Figure 17.109). Blazing yellow masses of flowers appear in spring and a prodigious crop of pods usually ripens in May and June. It is characteristic of major arroyos and drainageways on the mainland and Tiburón Island.

Ziij was often considered to be a different tree from *iiz*. *Ziij* was distinguished as spineless and with edible flowers, while *iiz* flowers were said to be bitter and not easily gathered because of its spiny branches. The *ziij* may be of hybrid origin, because some hybrids are characterized by the lack of spines (see Jones 1978).

The Seri distinguished this species from foothill palo verde (*C. microphyllum*) as follows: blue palo verde (*ziij* and *iiz*) is found in arroyos, while foothill palo verde occurs in the mountains and hills. Blue palo verde is a taller tree. Otherwise they are the same and are similarly used.

Adornment: The seeds were strung as necklaces (Figure 17.110).

Facepaint: See *C. microphyllum.*

Firewood: It was used as fuel in the same manner as was *C. microphyllum.*

Food: The pods usually were gathered from pack rat nests (see *C. microphyllum*). The seeds were sun dried, toasted, and finely ground, and the flour was mixed with water. The mixture was then cooked and eaten with honey or sea turtle oil. It was a fairly important food and the flour was stored in pottery ves-

Figure 17.109. Blue palo verde (*Cercidium floridum*) in an arroyo on the southeast side of Tiburón Island. *RSF, April 1983.*

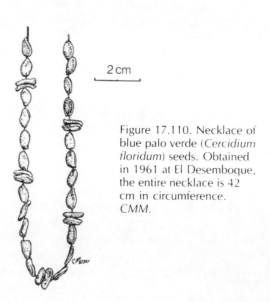

2 cm

Figure 17.110. Necklace of blue palo verde (*Cercidium floridum*) seeds. Obtained in 1961 at El Desemboque, the entire necklace is 42 cm in circumference. *CMM.*

sels. The green pods were cooked with meat. *Ziij* flowers were stripped off the thornless twigs with the fingers. The flowers were prepared in the same manner as for *C. microphyllum*, but were not used as extensively.

Food Gathering: The wood was used to make the *hahéel* agave chisel–pry bar.

Cercidium microphyllum (Torr.) Rose & Johnst.
ziipxöl
foothill palo verde

This large shrub or small desert tree occurs on hills and mountains, bajada slopes, and minor streamways across the desert floor (Figure 17.111). It ranges throughout the mainland region and on Tiburón Island. It is one of the more common and characteristic species of the Sonoran Desert. Masses of pale yellow flowers are produced in spring and the pods ripen in late May and June (Figure 17.112).

Adornment: The seeds were strung into necklaces (Figure 17.113).

Animal Food: Jackrabbits fatten on the fallen flowers.

Facepaint: Reddish facepaint was made with powdered ochre, *xpaahöj*, mixed in the palm of the hand with honeycomb containing flowers of palo verde (*C. floridum* or *C. microphyllum*) or catclaw (*Acacia greggii*).

Firewood: Although very soft, the wood sometimes was used as fuel because it was readily available.

Food: The pods generally were gathered from pack rat nests. They were spread out in the sun to dry, then rubbed together in a cloth to free the seeds. After winnowing, the seeds were toasted and ground, and the flour was prepared as a gruel called *haaztoj*. Honey or mesquite flour, *haas copxöt* 'mesquite loose,' sometimes was added to it. It was an important food and the flour was often stored. The seeds were an especially important staple for the desert Seri who did not fish or have ready access to the sea, such as the people of the Interior Tiburón Region. The green pods were cooked with meat. The sweet, green seeds were eaten fresh. The flowers were eaten fresh or cooked in water, and the juice squeezed from them. This liquid was mixed with sea turtle oil or honey, added to the crushed flowers, and eaten.

Food Gathering: The wood was used for the *hahéel* agave chisel–pry bar.

Pottery Sealant: To seal a pottery olla for transporting water, the seeds were ground and the flour

Figure 17.111. Aurora Colosio and Lolita Blanco collecting desert lavender (*Hyptis emoryi*) and other plants near a foothill palo verde (*Cercidium microphyllum*) in April 1983, 3 km east of Pozo Coyote. *RSF*.

2 cm

Figure 17.112. Flowers and pods of foothill palo verde (*Cercidium microphyllum*). Note the constrictions between the seeds. *NEW*.

Figure 17.113. Necklace of foothill palo verde (*Cercidium microphyllum*) seeds. The entire necklace is 45 cm in circumference. *CMM*.

boiled. A scum that formed was skimmed off and rubbed into the exterior of the olla to waterproof it.

Cercidium praecox (Ruiz & Pav.) Harms
maas
palo brea

This low, spreading tree has nearly horizontal branches, pale pea-green bark, and bright yellow flowers in spring. It is wide ranging in Latin America, and extends northward along the coast to Tastiota and Siete Cerros.

Medicine: The bark, ground with sea turtle oil and salt, was used as a lotion for bruises and sprains. Tea made from the bark was taken to aid in the expulsion of a torn placenta.

*Cicer arietinum L.
paar icomítin 'padre his-ironwood-seed'
   "padre's ironwood seed"
garbanzo, chick pea

The name indicates that the padres, or Catholic priests (probably Jesuits), introduced garbanzos to the Seri. Ironwood (*Olneya*) and garbanzo seeds are somewhat similar in appearance, especially when cooked.

Coursetia glandulosa Gray
hehe ctoozi 'plant resilient'
   "resilient plant"
cmapöjquij 'what bursts open'
   "bursts open"
sámota

Hehe ctoozi is the usual name for this shrub. On Tiburón Island *Lysiloma divaricata* was called *hehe ctoozi*, at least on occasion, and *Coursetia* was called *cmapöjquij*, the name usually applied to *Caesalpinia palmeri*. The term *cmapöjquij* refers to the explosively dehissive nature of the pods, a common feature of this *Caesalpinia* and *Coursetia*.

This multiple-stem shrub is common on north-facing rocky slopes and along arroyos and canyons (Figure 17.114). *Hehe ctoozi* refers to the resilient branches, which are tough and hard to break. The stems are sometimes encrusted with orange lac from a scale insect, *Tachardiella*. The lac, called *csipx*, was an important and highly prized material.

Adhesive: The lac from the stem was used for haft-ing harpoon heads and other objects. It has the same properties and uses as the lac from creosotebush (*Larrea*).

Fishing: The shaft for *hacaaizáa*, the single-pronged fish harpoon, was often made from *sámota* wood.

Other uses: The wood was also used for:

haacni, bow
hapoj, digging stick for *saiya* (*Amoreuxia*) roots, etc.
hasítj, cradleboard (outer frame)
hehe iti icáipj 'plant on what-one-places-crosswise-with,' a makeshift gun rest for aiming
icazóoj, walking stick
iquéete, ramrod for guns

Dalea emoryi—see *Psorothamnus e.*

Dalea mollis Benth.
hanaj itámt 'raven its-sandals'
   "raven's sandals"
silky dalea

This low-growing, winter-spring ephemeral is widespread in the region, including Tiburón Island.

Medicine: The plant was heated in coals and used as a poultice on swellings. The poultice was held in place with a rag or cloth band.

Dalea parryi—see *Marina p.*

Desmanthus covillei (Brit. & Rose) Wigg. ex Turn.
poháas camoz 'future-mesquite thinker'
   "what thinks it's a mesquite"

This name was also applied to *Acacia farnesiana*. Slender shrub about 1.5 m tall, with delicate stems and drought-deciduous, filmy foliage. Found in mountains and foothills, often in canyons in Sierra Kunkaak and Sierra Seri.

Desmanthus fruticosus Rose
hehe casa 'plant putrid'
   "putrid plant"

Very slender shrub about 2 to 3 m tall with bipinnate leaves. Found on upper bajadas, foothills, and mountains on the mainland and on Tiburón and San Esteban islands.

Adornment: We were told that the ancient Seri

Figure 17.114. *Sámota* (*Coursetia glandulosa*). A) Angelita Torres collecting stems of the shrub. B) Angelita Torres demonstrating the meaning of the Seri name for *Coursetia*—"resilient plant." C) Lac insects encrusting a stem of *sámota*. RSF, Rancho Estrella, April 1983.

strung the flowers and leaflets for necklaces.

Medicines: Tea made from the roots was held in the mouth to cure mouth sores.

The Supernatural: Crosses made from the twigs were hung on a cord as a necklace to give the wearer protection from sickness. This plant was brought by The Flood (see Chapter 7).

Errazurizia megacarpa (S. Wats.) I.M. Johnst.
coxi ihéet 'dead his-gambling-sticks'
   "dead man's gambling sticks"

This small, mound-shaped shrub has whitish foliage and small flowers with yellow petals. It is a characteristic species of the Central Gulf Coast of Sonora vegetation region.

Adornment: The seeds were strung for necklaces.

Erythrina flabelliformis Kearney
xloolcö
chilicote, coral bean

Shrub or small tree, often with a very thick trunk; trunkless at higher elevations where exposed to freezing temperatures, and leafless except during summer rainy season. The bright red seeds are 12 to 18 mm long. It occurs in the eastern margin of the Seri region and in the mountains near Guaymas. The Seri knew of this plant from near Rancho "Aldipo" (= El Dipo) and Rancho Félix Gómez towards Hermosillo.

Games: It was believed that a player would have better luck in playing the four-tube hiding game, *xapij caanlam*, if he used a seed of this plant instead of a sliver of reedgrass (see *Phragmites*).

Medicine: Tea made by cooking the seeds was used to cure diarrhea.

Hoffmanseggia intricata Brandegee
haxz iztim 'dog its-hipbone'
   "dog's hipbone"

The name was also applied to *Calliandra* and *Krameria parvifolia*.

This woody perennial, less than 0.5 m tall, has attractive yellow flowers and reddish buds. It is common on the mainland and on Tiburón and San Esteban islands.

Dye: The roots were mashed and boiled to yield a reddish brown dye. It was used as an alternative source of reddish dye when *cósahui* (*Krameria grayi*) was not available, and was prepared in the same manner. Reddish brown dye was also made by combining roots of *cósahui* and *Hoffmanseggia*. This mixture was used if the *cósahui* dye was not strong enough. The resulting dye was said to look like "the red dye you buy from a store."

Lotus salsuginosus Greene
L. tomentellus Greene
hee inóosj 'jackrabbit its-claw'
   "jackrabbit's claw"
hant ipépj 'land *ipépj*'

Winter-spring ephemerals with tiny, pea-like, yellow flowers, and prostrate or spreading stems. Common during favorable years on the mainland and Tiburón Island. These superficially similar species are not distinguished by the Seri.

Games: One of these plants, together with tansy mustard (*Descurainia*), would bring good luck when rubbed on a person who planned to play certain gambling games.

Lupinus arizonicus S. Wats. subsp. *lagunensis* (M.E. Jones) Christ. & Dunn
zaah coocta 'sun watcher'
Arizona lupine

Various wildflowers were called sun watcher because the leaves turn through the day towards the sun, or "watch the sun." During favorable years this lupine is a common winter-spring ephemeral from El Desemboque northward.

Lysiloma divaricata (Jacq.) J.F. McBride
hehe ctoozi 'plant resilient'
   "resilient plant"
mauto

The Seri name refers to the long, slender, flexible branches. This was the name usually given to *Coursetia* on the mainland. When *hehe ctoozi* was used to identify *Lysiloma* on Tiburón Island, then *Coursetia* was called *cmapöjquij*.

This small tree grows on Sierra Kunkaak on Tiburón Island. It has many long, straight, and slender branches arising from the base. The wood is very hard and the stems relatively flexible.

Fishing: The long slender poles were often used to make the shaft of the *hacaaizáa* single-pronged fish harpoon.

Weapons: It was a preferred wood for making the bow, *haacni*. For this purpose it was considered better than *Coursetia* because the wood is harder while still being flexible.

Other uses: It was one of the preferred woods for making the carrying yoke (*peen*) and the outer, curved frame of the cradleboard.

Marina parryi (Torr. & Gray) Barneby
   [= *Dalea p.* (Gray) Torr. & Gray]
hanaj iit ixac 'raven its-lice their-eggs'
   "nits of the raven's lice"

The name derives from the tiny leaflets that are glandular-punctate (having dot-like glands).

This common winter-spring ephemeral sometimes persists as a short-lived perennial. It has small flowers with dark blue petals. It is widespread through the region, including Tiburón and San Esteban Islands.

Medicine: The plant was cooked in water and the tea given to the youngest child of a pregnant woman. Since the child, called *caal* "companion-child (of the expected baby)," received relatively little nourishment from the mother, it often suffered from malnutrition. This medicine, although said to be very bitter, was supposed to give the *caal* strength (see *Triteleiopsis*).

A child who reached walking age, but showed no ability to do so, was bathed in water with this plant. This would help the child to walk.

Melilotus indica (L.) All.
haacoz
sour-clover

Annual, alfalfa-like plant with tiny, pea-like, yellow flowers. Common farm weed through much of North America; naturalized from Europe. The Seri knew of it only from the vicinity of Puerto Libertad, and we have not seen it elsewhere in the Seri region except as a farm weed east of Kino Bay.

Adornment: Because of its fragrance it was put into small cloth bags which were sewn on a cord and worn as a necklace.

Medicine: Tea made from the plant was taken to cure a stomachache.

Mimosa laxiflora Benth.
hehe cotázita 'plant that-pinches'
garabatillo

The Seri name, indicating a plant that grabs hold or pinches, refers to the sharp, recurved spines. This shrub grows along canyons, rocky slopes and adjacent bajadas.

Olneya tesota Gray
comítin
palo fierro, ironwood

Ironwood is one of the most conspicuous landscape features of the Sonoran Desert. It characteristically develops into a small tree with several massive trunks, shredding gray bark, dense gray-green foliage, and short, paired, sharp spines (Figure 17.115). The tree is long lived and slow growing. Masses of pale, violet-pink flowers are usually produced in May and develop into short, thick pods in June and July. Each pod bears one, two or occasionally three or four seeds. The wood is extremely hard, does not float, and burns with a hot flame. Record and Hess (1943:300) describe the wood as follows:

Heartwood rich dark brown, somewhat variegated; sharply demarcated from the rather thick, yellowish white sapwood. Fairly lustrous. Without distinctive odor or taste when dry. Very hard, heavy, compact, and strong, but brittle; sp. gr. (air dry) about 1.15; weight 72 lbs. per cu. ft.; texture medium coarse; grain irregular; not easy to work, but finishing very smoothly with a high natural polish; is very durable.

Ironwood trees are characteristic elements of the low semi-riparian gallery "forests" along arroyos, floodplains, and drainageways of even minor size. Smaller individuals also occur on rocky slopes. It is common on the mainland and Tiburón and San Esteban islands. In recent years ironwood stands on the mainland have been over-harvested for Seri wood sculptures.

The following terminology was associated with ironwood:

comítin, the tree, seed, and lighter-colored wood —including sapwood
hesen, dead wood "from the old trees," the dark heartwood
comítin coxi 'ironwood dead,' a young tree no longer living
comítin an ihís 'ironwood inside its-fruit,' the pod
comítin cöcootij, 'ironwood dry,' the drier green wood

Uses for the green wood, comítin, included:

hapát, meat rack
peen, carrying yoke
tis it ihíip, harpoon point

Uses of the dark heartwood, hesen, included:

hahéel, chisel—pry bar for century plants (*Agave*)
hascám iquéelx, balsa paddle
haspéen, a large pestle
heectim, club to kill fish and sea turtles

Animal Food: Mule deer were said to feed on the green, leafy twigs.

Boat Paddle: The dark heartwood, hesen, was preferred for making the blades of the double-bladed balsa paddle because the wood entered the water easily due to its own weight.

Firewood: Ironwood is one of the finest firewoods. The preferred fuel was hesen, the dead wood, although comítin cöcootij, the drier green wood was also used. Hesen gives such a hot fire it was thought that continued use would injure the lungs. Waste

Figure 17.115. Ironwood trees (*Olneya tesota*). A) Rosa Flores next to a large ironwood in a riparian habitat at Rancho Estrella. Note the rough bark. *RSF, April 1983.* B) Ramona Casanova and Roberto Herrera next to a medium-sized ironwood in the open desert along the former trail between Pozo Peña and the coast. This tree was a much used and well known resting place along the trail. *MBM, 1972.*

chips from sculpture making were used for cooking fires.

Fishing and Hunting: A mash made from the seeds was used to still or calm wind-roughened water surfaces in an estero (see Oral Tradition, below). The fisherman carried seeds for this purpose, chewed them and spat the mash into the water. The roughened water surface was said to still quickly, so that one could see through the water to spear crabs or fish in the shallow water. The *heectim* club for killing fish and sea turtles was made from *hesen* (Quinn and Quinn 1965:166).

In 1980 an ironwood harpoon shaft was brought up in a turtle net. The Seri believed it to be at least one-half century old. It was about 110 cm long, 3.5 cm in diameter and flattened at one end. The Seri pointed out that it was the distal end of the shaft for a double-pronged fish harpoon. It was found at the *Sacpátix* turtle hole or "home" (*iime*) near Punta Chueca. This kind of harpoon had not been used for many years.

The detachable foreshaft of a sea turtle harpoon was often made from ironwood. Harpoon points were sometimes made from *comítin*, and projectile points were occasionally made from *hesen*. In summer hunters expected to find mule deer in the shade of larger ironwood trees.

Since a sea turtle *iime* 'home' was a good place to hunt turtles, the men of the Tiburón Region made some artificial turtle "homes," probably during the early twentieth century. It was said they made at least several of these artificial "homes" by dropping dead ironwood logs into the Infiernillo shallows in likely places. The turtles utilized these "homes" and the men hunted them there in the summer, although only during high tide. In time the ironwood disintegrated and the practice was not repeated.

Food: The ripe pods were gathered in early summer. The pods, cached by pack rats, were also gathered during subsequent months from the nests.

The dry, mature seeds were cooked in water, drained, and cooked for a longer time in a second water until soft. The seeds were cooked in a second water to get rid of the unpleasant smell (*ccon* 'stink'). The seed coats floated to the surface and were discarded. The cooked seeds were eaten whole or ground and salted. The Seri said the ground mixture was oily and tasted like peanut butter. The seeds were also boiled with meat. This was the only food which the Seri cooked in a second water. The seeds contain canavalin (Rosenthal 1977), a mild toxin also known to occur in jack beans (*Canavalia ensiformis*).

Food Gathering and Preparation: The *hahéel* chisel−pry bar for century plants (*Agave*) was made from *hesen*, as was the large pestle *haspéen*, for pounding mesquite pods in bedrock mortars (see *Prosopis*, Food). Branched poles of *comítin* were often used to make a meat rack, *hapát*.

Games: A solid ironwood ball, about the size of an ordinary softball, was used in the men's football race called *hehe hahójoz* 'wood what-is-made-to-flee.' The ball was kicked and the race run over a considerable distance, probably on the order of 4 to 10 km. It was run along the beach where cactus and brush would not interfere. There were four sides, each with a runner kicking an ironwood ball. The men on a side relieved each other in informal relays, and usually only one or two finished (Griffen 1959:15; Kroeber 1931:16). The ball was also made of mesquite (*Prosopis*).

Kroeber (1931:16) stated that, "the Seri once went north to play the Papago this game, but lost." Football relay races with a wooden ball were widespread in the region, the most famous being those of the Tarahumara (Pennington 1963:167−173).

Medicine: Sapwood was placed overnight in water and about one-half liter of the liquid drunk as an emetic. The same liquid was also taken to prevent one from "breathing hard" when running.

Music: The following information is revised from a previous treatise on Seri musical instruments.* The violin bow, *henj ica ihoom* 'violin with what-lies-with,' was sometimes made from the dark heartwood (*hesen*).

The musical rasp, *hexöp*, was normally made of ironwood (occasionally from catclaw), sawed and filed to shape. Size was quite variable, from approximately 0.5 to 1 m long. The unnotched portion of

the handle was sometimes decorated with a carved design. This instrument was a popular item of purchase by tourists. One end of the musical rasp was placed on an inverted basket which acted as a resonator. Probably because of the scarcity of baskets for home use since about the 1960s an inverted 20-liter lard can often sufficed. A rasping sound was produced by sliding a stick rapidly back and forth across the notches with a rhythmic flick of the wrist (see Figure 13.13).

The musical rasp was widely distributed among the neighbors of the Seri, and the Seri believed that the idea came from the Yaqui. The Seri musical rasp was nearly identical in form to those characteristic of the Yaqui (see Densmore 1932: plate 28). The musical rasp was used by the Seri for the deer dance, also said to have been learned from the Yaqui.

Oral Tradition: A Giant chewed ironwood seeds into a mash and blew it out of his mouth while he was on his balsa on a windy day. The water calmed and he was able to see through it to harpoon fish.

The Supernatural: A drink used in the vision quest for power was made from a piece of green ironwood when *Vaseyanthus* was not available. The wood was cut from the region of junction of the green branch and the main trunk or old wood. After the wood soaked in water for several days the liquid was ready for drinking.

Bullroarers, *hacáaij* 'what is twirled,' were carved from *hesen* (Figure 17.116). Each side was usually painted with red and blue zigzag lines, although some were undecorated. Two of them were tied to a cord, one at each end, although only one at a time was used. It was swung in a wide circle around the head so that sound was generated four times. Bullroarers were used by shamans and vision seekers to summon the spirits (Griffen 1959:16, 50).

Sea shells and other objects were stuck into ironwood bark for the same reasons as for *cardón* and other cacti (see *Pachycereus*). This practice was called *hesen heeyolca* 'ironwood what-are-given-to.'

Sculpture: Since the 1960s the Seri have become famous for their ironwood sculpture (see Chapter 14).

Shelter: Ironwood poles were occasionally used for houseposts.

Tanning: Seeds collected from ironwood trees or pack rat nests were used to tan deer hides. The seeds were removed from the pods by soaking them in water. The seeds were mashed and rubbed into the skin, which was then allowed to dry for two to three days. Next, the skin was soaked in salt water for a

*Revised from T. Bowen and E. Moser, "Material and functional aspects of Seri instrumental music." The Kiva (1970) 35(4):178−200. Used with permission.

Figure 17.116. Bullroarers, 18.2 and 17.8 cm long, made of ironwood, buckskin, and commercial thread. These replicas were made by Pedro Comito in 1969. *CMM.*

Figure 17.117. This replica of an ironwood knife for fighting, made by Jesús Morales in 1960, is 54 cm long. *CMM.*

day and then kneaded. Mashed green twigs, *hap ináil ihacáitic* 'mule-deer its-skin what-it-is-made-soft-with,' were also used for softening deer skins.

Tattoo: The ashes were used in tattooing practices.

Weapons: Long knives made of *hesen* were said to have been used for fighting. The replica shown in Figure 17.117 is clearly adapted from a machete or European-style knife. McGee (1898:260) mentions a machete made of *palo blanco* wood. However, the Seri said *palo blanco* (*Acacia willardiana*) was not so used, and McGee probably confused it with iron-

wood. Projectile points occasionally were made from *hesen.*

Parkinsonia aculeata L.
snapxöl
bagota, Mexican palo verde

This large, green-barked tree is commonly cultivated or semi-naturalized on farms and ranches.

Medicine: Tea made from the twigs was taken to relieve stomach pain.

Phaseolus acutifolius Gray var. *latifolius* Freem.
haap
tépari del monte, wild tepary

This native, wild tepary grows with late summer and early fall rains and is found near the center of Tiburón Island. It is an ephemeral vine climbing through desert shrubs and small trees. The pods are commonly 5 to 5.5 cm long, and each contains several seeds (Figure 17.118). The ones we found have dark, nearly black seeds which are relatively large for a wild tepary. *Haap* seeds range from 46.5 to 89.0 mg in weight, with an average weight of 73.7 mg (n = 35, SE = 9.1). The pods of wild teparies, unlike those of domesticated ones, burst open explosively when dry and ripe, and the seeds are smaller (Nabhan and Felger 1978).

Long before we saw this plant it was described to us as follows: It grows with the summer rains and occurs in only one area near Sierra Kunkaak on Tiburón Island. It is a vine with a single stem, and may grow on *Colubrina,* ironwood (*Olneya*), desert wolfberry (*Lycium andersonii*), or *Sapium.* Each plant may have about 250 pods, and each pod contains six or seven seeds. The pod is called *an ihís* 'inside place-of-fruit.' The seeds are gray, yellow or brownish, or black. They are like a small pinto bean or cultivated tepary.

The site where it is known is called *haap caaizi quih yaii* 'tepary who-do(plural) the where-they-are' or "tepary users' place." We finally saw *haap* plants in 1976, when we were guided there by Rosa Flores, who had not visited the area for 34 years. The site consists of expansive, mostly north-facing slopes of blackish volcanic scree. The plants grow up through several layers of largely barren loose rocks.

Food: The pods, which ripen at the end of the short summer monsoon, were gathered in the early morning while it was still relatively damp and not

Figure 17.118. Pods and seeds of the wild tepary (*Phaseolus acutifolius*) from Tiburón Island. *RSF.*

too hot. If harvested at midday the pods would dry quickly and burst open, scattering the seeds. Rosa Flores related that women picked mature pods and gathered them into their skirts to carry back to camp. When the pods were dry the women shelled them by rolling them in their hands over blankets or deer skins.

The seeds were boiled alone or with meat, if it was available—especially deer meat or bones. They were said to taste like the Mexican pinto bean. Concerning *haap*, Sara Villalobos said:

> They are boiled just like regular beans [pinto beans]. You do not grind them at all. They cook right away—regular beans take a long time to cook because their skin is thick. You don't use a lot of water [to cook them]. You don't salt them because the water is sweet.

Haap seeds were an important food staple of the people who lived in the interior of Tiburón Island. The renegade Yaqui, who lived with the Seri for a short time on Tiburón Island prior to 1904 (García y Alva 1905) made extensive use of this wild tepary.

* *Phaseolus acutifolius* Gray
teepar 'tepary'
teepar cooxp "white tepary"
teepar cmasol "yellow tepary"
teepar coopol "black tepary"
teepar coospoj "spotted tepary"

These are the domesticated, cultivated teparies. The Seri name is derived from *tépari*—the Sonoran Spanish term for this bean. Since ancient times teparies were cultivated in areas near the Seri region (Nabhan and Felger 1978). Many color variations, or cultivars, were known (Freeman 1918). The names *teepar cooxp* and *teepar cmasol* were elicited when we showed white and brown seeds respectively to several elderly people. We were then told that there were two other kinds, the black (*teepar coopol*) and spotted (*teepar coospoj*).

Phaseolus filiformis Benth.
haamoja iháap 'pronghorn its-tepary'
  "pronghorn's tepary"

This delicate ephemeral bean can appear at various times of the year following sufficient rainfall. It has twining stems, pink, pea-like flowers, and small pods. The Seri said it is the *haap*, or wild tepary (*P. acutifolius*), belonging to the pronghorn. The term *haamoja* is an old word from the dialect of the people from the Libertad Region. The modern name for pronghorn is *ziix itx cooxp* 'thing its-rump white.'

Animal Food: It was eaten by quail.

* *Phaseolus lunatus* L.
hapats imóon 'Apache his-beans'
  "Apache's beans"
Lima bean

* *Phaseolus vulgaris* L.
ziix is cquihöj 'thing its-seeds red'
  "red-seeded thing"
xica is cheel 'things its-seeds red'
  "red-seeded thing"
frijol, pinto bean, common bean

Although the common bean has been widely cultivated in Mexico and southwestern United States since pre-Columbian times, the Seri were apparently not familiar with it until Spanish-speaking farmers settled in Sonora.

Food: In the nineteenth and early twentieth centuries the Seri prepared the beans (seeds) by toasting them in pottery vessels containing hot coals. The toasted beans were ground on a metate and then winnowed to remove the hard seed coat and debris. The bean flour was mixed with water and eaten as gruel. It was said to taste like packaged "*frijol al minuto*" (instant beans).

Pisum sativum L.
paar icomíhlc 'padre his-what-are-smooth'
  "padre's mesquite seed"
chícharo, garden pea or green pea

Comíhlc 'what-are-smooth' is the fresh green seed
of the mesquite (*Prosopis*). The Seri name apparently
refers to the introduction of this European food by
the early Catholic priests.

Pithecellobium confine Standley
heejac

Large shrub with stiff, zigzag twigs, sharp spines,
and very hard wood. The woody pods are persistent
and become black with age. Found in Sonora along
the Infiernillo coast, on Tiburón Island, and in Baja
California.

Dye: The dry pods were used in making a black
basketry dye (see *Suaeda moquinii*).

Firewood: It was a valued firewood but not extensively used because it is generally not common in the
vicinity of the camps.

Medicine: Tea made from the whole, dry pod was
taken to cure a cold, cough, or sore throat.

**Pithecellobium dulce* (Roxb.) Benth.
camótzila
guamúchil

Camótzila is derived from the Spanish term *guamúchil*. This large tree has sweet, edible pods often
sold in market places. It is now widely planted and
semi-naturalized in the warm arid parts of the world.
It is cultivated in Caborca, Hermosillo, and elsewhere
in the lowlands in Sonora. It was described as looking
like mesquite but with darker leaves.

Oral History: Manuel Molino planted *guamúchil*
and dates (*Phoenix*) at one or more camps along the
coast south of Cerro San Nicolas. (This would have
been in the mid–nineteenth century.) According to
some versions he planted them at camps known as
Xat Icahéme 'hail deserted-camp' and *Hax Imáma*
'water it-doesn't-flow.' Other versions state that it
was at a camp known as *Hacac Ihíin* "close to
Hacac." Some people said that he planted dates and
caramochi, which seems to be another name for
guamúchil. *Caramochi* was described as a tall tree
with black seeds and pod-like fruit.

Prosopis glandulosa Torr. var. *torreyana* (L.
  Bens.) M.C. Johnst.
haas
mezquite, western honey mesquite

There was extensive terminology associated with
haas. Both the tree and the pod were called *haas*; the
term for the root was *caaz*. Mesquite root driftwood,
satómatox was said to be carried downstream by
floodwaters and deposited on the coast in beach
drift.

Mesquite was the single most important resource
to the diverse peoples of the arid and semi-arid regions of southwestern North America. It served as a
primary resource for food, fuel, shelter, weapons,
tools, fiber, medicine, and many other purposes for
hunters and gatherers, as well as agriculturalists (Bell
and Castetter 1937; Felger 1977). The trunks are
usually crooked or irregular and the branches strong
but somewhat brittle, while the roots are pliable
(Hardy 1829:296–297). The root is softwood, while
the wood of the trunks and branches is hard, heavy,
strong, and very resistant to decay (Record and Hess
1943:318). The following discussion is revised from
our earlier report on Seri use of mesquite.*

Three species of mesquite occur in western Sonora:
P. articulata S. Wats., *P. velutina* Woot., and *P. glandulosa* var. *torreyana*. In western Sonora these mesquites are generally segregated by geography and
habitat. All three occur in the Guaymas region. *P. articulata*, rare elsewhere in Sonora, is distinguished by
a relatively thin endocarp and very little or no mesocarp in the pod. *P. velutina*, the velvet mesquite, is
the common mesquite of the desert and semi-desert
regions inland and north of Guaymas extending into
Arizona and elsewhere.

The western honey mesquite (*P. glandulosa* var.
torreyana; Figure 17.119) is the only mesquite we
have found in the coastal areas from Tastiota northward to Puerto Lobos and on Tiburón and San Esteban islands. It is the common mesquite tree of the
subtropical scrub south and southeast of the Sonoran
Desert in Sonora. In the desert it ranges northward
from Guaymas along the coast (including the Gulf islands) to the Colorado River and into the lowland
deserts of southeastern California, western Arizona,

---

*Revised from R. S. Felger and M. B. Moser, "Seri use of mesquite
(*Prosopis glandulosa* var. *torreyana*)," The Kiva (1971), 37:53–
60. Used with permission.

Figure 17.119. Western honey mesquite (*Prosopis glandulosa* var. *torreyana*). A) A medium-sized tree at Pozo Coyote. B) An older tree with rough bark. The fallen dead branches were much used for firewood. *RSF, April 1983.*

and southern Nevada. It is common along the drainageways on Tiburón Island, but on San Esteban Island it is too scarce to have been of significant value.

Honey mesquite is readily distinguished from other mesquites in the region by its jugate leaves, with numerous long, narrow, individual leaflets that are commonly more than five times longer than wide, glabrous, and widely separated from each other (Figure 17.120). The twigs commonly bear stout, straight, stipular spines which are highly variable in size even on the same plant; spines are occasionally absent. The Seri pointed out that the larger trees or shrubs generally have small spines. The mesquite spine was called *xpeemoja*.

In the better-watered areas, such as floodplains and major arroyos, honey mesquite develops into a small tree; with age it forms a massive trunk. In drier locations, such as coastal dunes, it forms a broad, spreading shrub. Great stands of mesquite, called *mezquital*, once occurred in aggregated masses along the coastal plains between Kino and Tastiota Bays. Land clearing, wood cutting, cattle grazing, and general environmental degradation subsequently greatly reduced the extent of the *mezquital*. In the early 1980s impressive mesquite forests still lined the channels and floodplains of the Arroyo San Ignacio and the Bacoache.

Since its deep roots can tap underground waters, mesquite does not depend solely upon rainfall for its moisture supply. New foliage appears in spring when the weather turns warm. Major flowering follows the appearance of new foliage, and the fruit develops in early summer. A second, lesser crop of flowers and fruit may occur in late summer or fall. Foliage is gradually shed through late fall and winter. However, during exceptionally dry years mesquite may locally fail to produce pods. Some Seri claimed that the major mesquite groves produced much greater quantities in earlier times.

Boat Building: Ribs of modern Seri wooden boats were usually made from mesquite.

Cordage: Extensive use was made of strong cordage fashioned from mesquite roots (Figure 17.121; also see McGee 1898:228–229). Cordage was made from the roots of young trees with roots about the thickness of a finger or slightly larger. The roots, with the bark removed, were chewed to break down the fi-

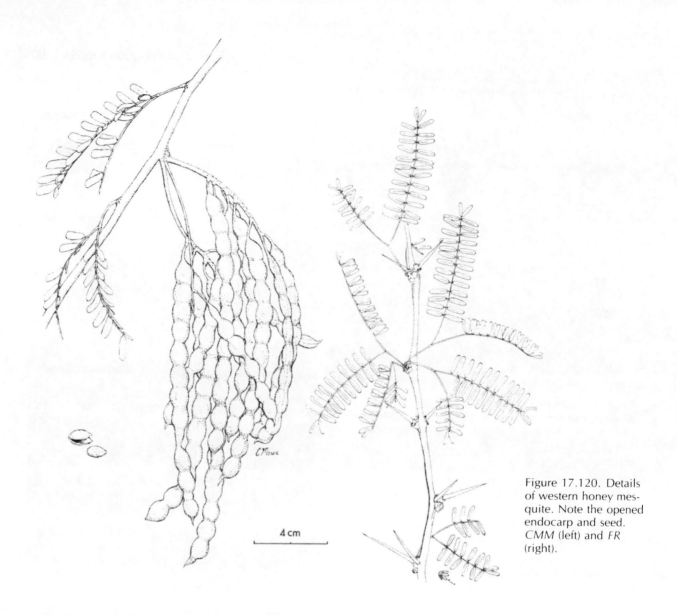

Figure 17.120. Details of western honey mesquite. Note the opened endocarp and seed. *CMM* (left) and *FR* (right).

4 cm

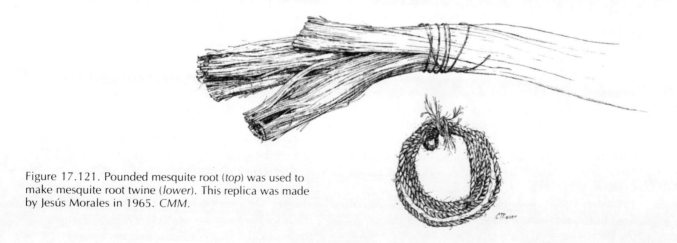

Figure 17.121. Pounded mesquite root (*top*) was used to make mesquite root twine (*lower*). This replica was made by Jesús Morales in 1965. *CMM*.

bers, which were then twisted or spun into twine. The twine was spun by rolling the fibers on the thigh while one was sitting. Terminology for the different kinds of cordage includes the following:

hapámas 'what is spun,' twine

caaz poosj 'mesquite-root twine,' double-strand twine or rope

hapámlc 'what is twisted,' double-strand rope

xepe poosj 'sea twine,' double-strand rope for use at sea

ano hapáxquim 'inside what-is-put,' three-strand rope

icáazlca 'what one strings up with,' cordage used in connection with hunting or fishing

Rope was made from two or three twisted strands of twine that were plied together. The term for three-strand rope refers to the fact that the third strand was plied, or twisted into, or "put inside" the other two. *Icáazlca* was mesquite cordage which the men kept specifically for use during hunting and fishing to tie up meat to be carried back to camp.

Mesquite cordage was used for harpoon lines for hunting sea turtles; it was also used to bind together the pieces of the harpoon pole and to bind harpoon points to the shaft. A certain fish line, *haxáapa*, was used for stringing fish as they were harpooned. The fisherman fastened one end of the line around his waist. Wading through the water, he speared fish and strung them on the line with a creosotebush point. The fish trailed several meters behind him in the water, so that if sharks were attracted, the man might not be attacked.

Mesquite cordage was used to lash together bundles of *carrizo* for the reed boat or balsa (see *Phragmites*). A story was told concerning a man seeking vengeance. He cut almost through the mesquite bindings of his enemy's balsa. The result was that the balsa broke apart at sea and the enemy drowned. Temporary rafts were made from driftwood logs tied together with mesquite cordage. After using the raft, the hunter dismantled it in order to save the cord, which was regarded as a valuable possession.

The carrying net, *cool*, made of mesquite twine, played a prominent and important role in Seri culture. Water was brought to camp by the men. A man usually carried two large water-filled ollas, each suspended in a carrying net (see Figure 6.1) from the end of the carrying yoke (*peen*). In the twentieth century the ollas were replaced by 20-liter cans. The carrying net was also used for transporting a wide range of other objects. On occasion, a small child placed in a shallow tray basket was carried across the desert in one of these nets swinging from a carrying yoke, with the other end balanced with a water-filled vessel or cargo.

The waist cord, *hapáaf*, was sometimes made of mesquite twine.

Cradleboard: Slender mesquite stems or poles were said to be superior material for the curved frame of the *hasítj* cradleboard.

Dye, Facepaint, and Paint: Black paint was made from mesquite bark. Pieces of the bark with pitch (the partially hardened, oozing black sap) on them were slowly cooked in water. Sugar was added to help darken the mixture. When sufficiently thick, the black mass was allowed to dry for one or two days. It was then formed into cakes or patties called *heep*, which were stored until needed. When *heep* was rubbed into a bit of water on a stone, a paste resulted which was used as a black paint or pigment. It was used as a facepaint and basket dye and to decorate fetishes and other objects.

Dry *heep* cakes were placed in boiling water to make a black basket dye. Strips of natural basketry splints (see Chapter 15) cooked in this dye were stained a light gray. However, black was preferred to gray. To obtain black, basketry splints which had been dyed reddish (from the root of *cósahui, Krameria grayi*) were dyed a second time with *heep*. Some basketmakers did not bother to make the *heep* patties but simply boiled pieces of mesquite bark covered with the characteristic oozing black sap (pitch). The bundles of basketry splints were then placed in the resulting dye. Black basketry dye was also made by boiling *haas ihasnáilc* 'mesquite outer-bark' with the roots of sea blite (*Suaeda moquinii*).

Firemaking: Mesquite root driftwood, called *satómatox*, was a prime material for making the firedrill, *caaa* (both the fireboard and the drill tip). It was easier to start a fire with this wood than most other firedrill woods. The firedrill that a man kept in his arrow quiver was usually made from *satómatox* or cliff fig (*Ficus petiolaris*). The term *satómatox* is derived from *coomatox* 'he who makes fire by friction.' *Satómatox* was usually found as driftwood on the coast near the mouths of major arroyos or rivers, such as the Río San Ignacio several kilometers north of El Desemboque. The Seri told us that mesquite was torn out by flood waters and carried downstream to the coast; this process cleaned off the bark.

Firewood: Mesquite was the most important

cooking fuel, and continued to be the preferred fire-wood for cooking in the 1980s.

Food: Bell and Castetter (1937:21) were of the opinion that "mesquite was one of the most important wild plant staples of the Southwest. . . ." McGee (1898:206–207) surmised that it was probably second only to the giant cacti as the most important plant food of the Seri, although he incorrectly regarded all plant foods as being of relatively minor significance. We concur that mesquite was one of the most important Seri food staples. However, it varied in significance among the different geographic groups. For instance, the people living on San Esteban Island and the southern and interior portion of Tiburón Island had relatively little mesquite available to them in their environment. While the uses of mesquite described here were little practiced by the mid–twentieth century, it was still occasionally harvested.

There were eight named stages of growth of the pod:

> *isman* youngest stage, less than ca 2.5 cm long
> *ihazíinz* 'its fringes,' young pod, longer than *isman*
> *cocpémelx* 'twisted,' firm but unripe pod
> *cocpátyax* 'striped,' the pod, almost ripe, striped with color

haas coil 'mesquite-pod green,' the pod picked green from the tree
cpasi 'what is wrinkled,' the ripe pod
haas cöcootij 'mesquite-pod dry,' the dry pod on the tree
cahtóopj 'what falls,' the fallen pod, edible until it gets wet and spoils or molds.

The young pods, *ihazíinz*, were tied into small bundles and cooked with meat. The green pods, *haas coil*, were mashed in a mortar formed in a bedrock (Figure 17.122) or hard earth. These pounding holes were called *hacala*. The pounding club or pestle, *haspéen*, was made of either mesquite or ironwood and was nearly 1 m long and 7 to 10 cm in diameter at the base (Figure 17.123). After the green pods were mashed, they were cooked in a pot. Mesquite pods prepared in this manner were called *haas coil hapámöjc* 'mesquite-pods green what-is-mashed.' The ripe pods, *cpasi*, were prepared in the same manner as the green pods.

The dry pods were gathered in early summer, usually from mid-June to late July. Whole, dry pods were also gathered in considerable quantities later in the year from pack rat nests.

Although men sometimes helped with the mesquite harvest, the dry pods were usually gathered off the ground beneath mesquite trees by women. They transported the pods in baskets carried on their heads. A large load of pods could be accommodated by building up successive layers with the ends of the outside pods intertwined to bind them together. This oversize load, called *hasái*, could also be held together by tying it with strips made from long mesquite twigs. The capacity could also be increased by extending the sides of the load with vertically placed sticks held in position by the pressure of the load.

Figure 17.122. Elvira Valenzuela (*right*) and her daughter Angelita Torres demonstrating the use of ironwood pestles for pounding mesquite pods in bedrock mortars at Pozo Coyote. *RSF, April 1983.*

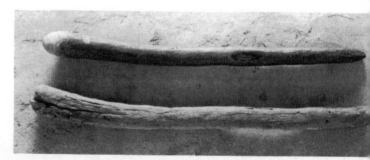

Figure 17.123. Ironwood pestles for pounding mesquite pods. Found near El Desemboque in 1968, the pestles are 87 and 85 cm long. *RSF, March 1983.*

The dry fallen pods, *cahtóopj*, were the most commonly used form of the fruit. In one manner of preparation they were pounded in a mortar and chewed; the juice was then swallowed and the pulp discarded. Mesquite pods prepared in this manner were called *quis hapámöjc* 'what-is-raw mashed.'

The most common way to prepare the dry pods was to toast them before pounding. The ground was cleared, a fire burned on the cleared area, and the coals subsequently removed. The dry pods were then piled on the hot earth. At the same time, four piles of sand were placed around the area, and fires set on each pile to heat the sand. The pods were then sprinkled with the hot sand, and the heat from the fires plus the hot sand toasted them. The moon, or month of the year, known as *icóozlajc iizax* 'to-sprinkle moon' derived its name from the sprinkling of hot sand on the mesquite pods. This moon approximates the month of July and for the Seri was the beginning of the new year.

After the pods were toasted, they were carried to a bedrock mortar to be pounded. A large pile of pods was poured in the mortar, and more placed on the ground surrounding the hole. Several women might be pounding at the same time, each at a mortar. As the pods were crushed, more were added from those piled around the hole. One man said that the rhythmic sound of the pestles falling upon the mesquite pods was a sound he liked to recall. After the pods were mashed they were placed between deer skins to prevent spoiling in the hot July wind. The pounding continued until all the pods were mashed.

The women then placed the pestle across the mortar hole. Mashed pods or pulp were put in a basket and gently winnowed by tapping the basket against the pestle. Flour from the mesocarp, or pulp, of the pod fell into the mortar hole; the "seeds" (seeds and endocarp) and pieces of fiber and shell or pod (exocarp) remained in the basket and were set aside on a skin. The flour, *haas copxöt* 'mesquite loose,' was winnowed again until pure. It was then placed in a pottery vessel to keep it dry and could be stored for a "long time" (probably weeks or months), retaining its smell and taste. One man estimated that two women, working with a man who kept them supplied with mesquite pods, were able to prepare about 40 kg of mesquite flour in a day.

This flour was commonly boiled in water to make *atole*, a gruel or thin porridge. Sometimes flour from palo verde seeds was added to it (see *Cercidium microphyllum*). The flour was also put into a large

basket, mixed with water, and kneaded into a dough called *haas copxöt hax cöhapáca* 'mesquite loose water what-is-mixed-with.' The dough was shaped into *hahéelcoj* 'capsule-shaped things' which were rolls about 20 cm long and 5 cm thick, or into round cakes called *haas hatócnalca* 'mesquite what-are-made-round.' The rolls and cakes were dried immediately so that they would not spoil. When dry they could be stored in pottery vessels for a long time. Seri families often had two or more large vessels filled with mesquite rolls hidden in caves for times of need.

The mesquite seed is enclosed in a tough, leathery pit or endocarp, which resembles a hard outer seed coat. The specific term for the mesquite seed and its endocarp was *ihij*. The actual seed, called *comíhlc* 'what-are-smooth," or "smooth things," is inside the endocarp. When fresh and green it has a thin, shiny, smooth seed coat enclosing an embryo with large, soft, dark green cotyledons. In the first pounding the *ihij* were separated from the rest of the pod. The remaining pod fibers were called *ipóosj* 'its-twine.' After separation, a second pounding—this time of the *ihij*—broke open the hard endocarp to free the seed. Winnowing separated the seed from the endocarp. Following this separation, the seeds were ground on a grinding stone. The resulting flour was mixed with water, some dry flour (*haas copxöt*) from the pod was added, and the mixture drunk. Apparently the mesquite seed was infrequently used elsewhere in southwestern North America, although the pod (mesocarp) was widely used (Bell and Castetter 1937; Felger 1977).

Ipóosj, the fibers of the pods resulting from the first pounding, were chewed for the sweet flavor and then discarded. Sometimes the fibers were pounded a second time, water added, the fibers sucked, and the pulp discarded. This chewed pulp was sometimes saved, mixed with sugar, toasted and then added to water to make a drink called *haas icóocsim* 'mesquite chewed-pulp.' Hardy wrote of being near the mouth of the Colorado River in 1826 and seeing

an old woman . . . engaged in chewing a kind of salad herb, which she took from an *earthen* bowl, and put into her mouth with her shrivelled hand; and after chewing it sufficiently, she spat it out again. . . . I afterwards examined the mixture, and found it to be the pod of the Mesquite-tree steeped in water. It is naturally very saccharine, and from its sour *smell*, must have undergone fermentation. The inhabitants of the island of Tiburow [sic] have the same dirty practice (Hardy 1829:334, 337).

Sometimes, after the first pounding, a quantity of *ihij* (endocarp with the enclosed seed) was put into a pottery vessel with water, weighted down with a stone and left until the water became sweet. This juice was a special treat to the children. Sometimes it was prepared by men, who let it ferment for several days before drinking it.

The dry pods were also placed in a vessel of water, weighted down, and cooked. The pods were then chewed (the inedible portions were spat out), and the broth drunk. Mesquite pods prepared in this way were called *haas hapáznij* 'mesquite that-is-cooked.' When children ate these pods they often did not spit out the *ihij* ("seeds"), but swallowed them. The "seeds" caused a specific kind of indigestion, and one suffering from it was called *quifj.* The Seri said that food from the mesquite made the children fat and their skin lighter in color.

Mesquite gum, *heep cooxp* 'sap white,' is bitter but children liked to chew it, spit it out, and then drink water. They said that the water then tasted sweet.

The difficulty of separating the true seed from the encasing inedible hard pit, or endocarp, apparently limited the use of the seed in southwestern North America and elsewhere. The prehistoric inhabitants of the Pinacate region in northwestern Sonora devised stone gyratory crushers, which were essentially metates with a hole through the center, although they were worked on both sides of the hole (Figure 17.124; Hayden 1969). Similar stone implements have been found elsewhere in Sonora, as well as from ancient times in the Old World (Hole, Flannery, and Neeley 1969). Hayden (1969) suggested that the stone gyratory crusher was developed to grind mesquite pods, and later investigation has shown that it is ideally designed not only to grind large quantities of mesocarp, but also to break open the endocarp to free the seeds with relative ease (J. Hayden, personal communication; Felger 1977).

Jim Hills (1973:103) found a gyratory crusher on Tiburón Island, Cathy Marlett found one in a sand dune near El Desemboque, and pieces of several others have been found in the region. The Seri did not recall having used this device. However, the Giants were said to have strung large perforated stones together and slung one set over each shoulder bandolier-fashion for use as a chest armor. The name for a string of such stones was *xapcoj iya* 'metates his'; *xapcoj* is an archaic Seri word. A Seri man shown a photograph of the Tiburón Island gyratory

crusher said he had never before seen one, but he immediately identified it as a *xapcoj iya* stone (Bowen 1976:79, 105).

Reasons for the disuse of gyratory crushers remain unknown, although Hayden (1969) hypothesized the disappearance of mesquite forests in northwestern Sonora. Perhaps, in the millennia following the extinction of the late Pleistocene megafauna grazers, there was a change in composition of the mesquite pod. It can be presumed that mesquite pods were well adapted to extinct Pleistocene grazing animals (Janzen and Martin 1982; Cornejo et al. 1982).

Food Gathering: The wood was used to make the *hacóipj,* a carved, spatula-shaped piece which was part of the gathering pole for organ pipe fruit (see *Stenocereus thurberi*).

Food Preparation: Both kinds of skewers for barbecuing meat were often made from mesquite: *hehe icaméquet* 'wood heat-with,' a single-pointed stick, and *hamquéecol quih coocj ano coofija* 'cooking-fork the two in what-passes-through,' a two-pronged skewer for holding a large piece of meat.

Games: Women had a race in which they rolled hoops made of mesquite root strips (Figure 17.125). Each player had a stick made of *Melochia* or *Vallesia* used to roll a hoop. This game, similar to the football

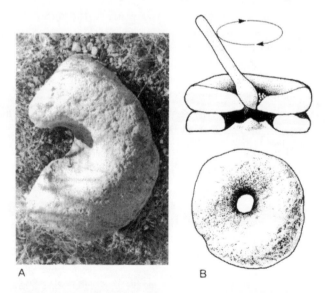

A   B

Figure 17.124. Gyratory crushers, occasionally found in the Seri region, were probably used in ancient times to process mesquite pods and separate the seeds from the leathery endocarp ("pit"). A) A broken gyratory crusher found on Tiburón Island. *RJH, 1973.* B) Diagram of a gyratory crusher from the Pinacate Region. *Drawing based on specimen 57 cm in width and Hayden 1972. KM.*

Figure 17.126. Mesquite club for killing fish and sea turtles. Found on the beach near El Desemboque by M. B. Moser in 1958, it is 51 cm long (*ASM-E8645*). The "x" was the owner's identification mark. This kind of club was probably adapted for use with wooden boats and commercial fishing. *CMM.*

Figure 17.125. A hoop for the women's hoop rolling race, made of mesquite wood bound with pelican sinew and decorated with bands of blue pigment. This replica, made by María Antonia Colosio in 1957, is 14 cm across. *CMM.*

race of the men (see *Olneya*), was known as *hehe hahójoz cmaam* 'wood what-is-made-to-flee female.' The men's football race was played with a solid ball of green mesquite wood or ironwood (*Olneya*).

Hunting: The mainshaft of the turtle harpoon was often made of wood from mesquite roots. Hardy (1829:296–297) described the use of a turtle harpoon "about 12 to 14 feet in length, the wood of which is the root of a thorn called mesquite." The detachable foreshaft of the turtle harpoon was commonly made from the hardwood of a trunk or branch.

The *heectim* club for killing fish and sea turtles was usually made from mesquite (Figure 17.126). The local Spanish name for this club is *macana*.

Medicine: Tea made from mesquite leaves, called *haas istj haizj* 'mesquite leaves made-pulp,' was drunk as an emetic. Bark of green or young branches was cut into long strips, tied into rolls, and soaked in water; this liquid was taken as a laxative. *Haas ooxö* 'mesquite excrement,' a milky sap hardened like resin, was dissolved in water and used as eye drops. The use of mesquite sap for eye drops was widespread in North America (Felger 1977:166).

Music: The violin bow was sometimes made of mesquite root wood. The one-string violin was occasionally made from mesquite root driftwood, called *satómatox*. It gave the best tone of the several woods used in violin making (Bowen and Moser 1970b; see *Bursera hindsiana* and *Ficus petiolaris*).

Oral Tradition: A story relates that coyote was hurriedly making mesquite rope (*hapámlc*). A person who was present said he was going somewhere, but then vacillated and decided not to go. Because he delayed, the coyote was able to hang him with the mesquite rope. When someone changed his mind and decided not to go somewhere, he was said to be hung by the coyote: *oot cola ocái* 'coyote high what-it-hung' or "what the coyote hung high."

Shelter: The corner posts and roof beams of ramadas and wattle and daub houses were often made of mesquite, and it was sometimes incorporated into brush houses. In June 1951, at Tecomate on the north shore of Tiburón Island, Alfred Whiting (1957:#22) noted that mesquite "trees are trimmed and pulled into weird shapes to form part of a house or shade. This is often used for it is one of the few readily available leafy trees in the area at this season." Whiting observed brush houses and ramadas.

Songs: The mesquite tree was said to sing this song:

yapxöt tazíim
caail iti miticol
caail imac cöhayoom
yapxöt he mooxp

Pretty flowers
They are on the dry lake
I lie in the middle of the dry lake
The flowers are white on me.

Other Uses: Mesquite wood was commonly used for making a carrying yoke, *peen* (Figure 17.127; also see McGee 1892:184, Figure 16) and a pestle for pounding mesquite pods, which was called *haspéen*.

The headring was occasionally made from strips of mesquite wood, probably the roots, if *haat* stems were not readily available (see *Jatropha cuneata*). Strips of the root were used in the taxidermy of a mule deer or jackrabbit head (see *Larrea*). To support the ears, strips of mesquite root or *hamísj* (*Jatropha*

Figure 17.127. This carrying yoke was found near El Desemboque in 1958. Made of mesquite, it is 107 cm long. It has reddish bands of paint, which was probably made from elephant tree (*Bursera microphylla*). CMM.

cinerea) were sewn along the margins; the strips were removed after the ears had dried and stiffened. Rings of mesquite root were sewn onto the base of the head (Bowen and Moser 1970a : 173).

McGee (1898 : 198) claimed that mesquite gum was used for hafting. He undoubtedly confused it with lac from creosotebush (*Larrea*).

Psorothamnus emoryi (Gray) Rydb. var. *emoryi* [= *Dalea e.* Gray]
xométe
Emory indigo bush

This low, spreading shrub has grayish foliage which imparts a yellow stain when rubbed. It is common along coastal dunes and sandy places on the mainland from Tastiota northward and on Tiburón Island.

Dye: The flowering stems and foliage were boiled in salt water and the resulting liquid used to dye basketry splints a pale yellowish color (see Chapter 15). However, basket splints stained prior to weaving soon became soft and broke. Therefore the leafy twigs and flowers were rubbed directly on a plain basket to stain it yellow. The color easily rubbed off, and this method of staining was seldom employed.

Facepaint: The twigs were used as brushes for applying facepaint.

Shelter: It was incorporated into roofing of brush houses because it is spineless.

Tephrosia palmeri S. Wats.
cozi hax ihapóin 'Condalia water what-it-is-closed-with'

Perennial or annual with silvery-green foliage. Common on Tiburón Island and the mainland.

Water Transport: As indicated by the name, the leafy stems were put in the tops of ollas or water cans being carried to camp to help prevent spilling and to impart a pleasant smell to the water.

* *Vigna unguiculata* (L.) Walp. subsp. *unguiculata*
yori imóon 'Mexican his-beans'
"Mexican's beans"
xica ihíijim coopl 'things its-sight black(plural)'
"black-eyed things"
black-eyed pea

The Seri have two names for black-eyed peas; the second one given above is the older name. *Ihíijim* is a form of an archaic term for 'see.' *Yori* is used to indicate any non-Seri, and *moon* is a Yaqui and Piman word for bean.

Liliaceae—Lily Family

* *Allium cepa* L.
hehe ccon 'plant that-reeks'
cebolla, onion

Allium haematochiton S. Wats.
mojet oohit 'desert-bighorn-sheep what-it-eats'
"what desert bighorn eat"

This dwarf onion occurs in the coastal mountains between El Desemboque and Puerto Libertad. The slender bulb is covered with onion-like maroon scales. Foliage and flowers appear in spring.

It was described by the Seri as having large red bulbs (as compared with *Triteleiopsis*) and flowers similar to the cultivated onion.

Food: The bulb was used to flavor meat.

Allium sativum L.

haaxo

ajo, garlic

The Seri name is derived from the Spanish.

Brodiaea pulchella (Salisb.) Greene
[= *Dichelostemma p.* Salisb., *B. capitata* Benth.]
caal oohit caacöl 'companion-child what-he-eats (*Triteleiopsis*) large'
"large blue sand lily"
wild hyacinth

This species may be distinguished by its slender flower stalk bearing several blue flowers, and small single bulb. It flowers in spring. We found it at higher elevations in the mountains southeast of El Desemboque. The plant was not well known by the Seri.

Triteleiopsis palmeri (S. Wats.) Hoover
caal oohit 'companion-child what-he-eats'
"what the companion child eats"
blue sand lily

The small blue flowers are borne in an umbellate arrangement on a slender stalk arising from a cluster of tiny underground bulbs (Figure 17.128). The foliage, consisting of several leaves, and flowers appear in spring. In the early 1980s some of the women and girls at El Desemboque dug up *caal oohit* plants and transplanted them to their gardens. They grew them in cans, buckets, and in the ground.

Food: *Caal* "companion child" was the name given to a young child whose mother was pregnant. Since the mother was often unable to provide sufficient milk, the child was likely to be undernourished. The tiny bulbs were said to be survival food for the companion child. The bulbs were cooked in water until they became soft mush. This mush was one of the soft foods fed to young children when they started eating solids.

In addition, children enjoyed looking for the plants on dunes in springtime and eating the bulbs fresh. For this reason it was often difficult to find mature plants near El Desemboque. The children often brought handfuls of them back to the village.

Figure 17.128. Blue sand lily (*Triteleiopsis palmeri*) from the coastal dune on the north side of Cerro Tepopa. A) The flowers are born in umbrella-like clusters. B) The plant proliferates by small bulbs—seeds are unknown. *RSF, April 1983.*

Loasaceae—Loasa Family

Eucnide rupestris (Baill.) Thomps. & Ernst
  [= *Sympetaleia r.* (Baill.) Gray]
zaaj iti cocái 'cliff on what-hangs'
  "cliff hanger"

This ephemeral has bright, shiny green leaves covered with large, sticky hairs. It grows on cliffs and rock canyon walls.

Mentzelia adhaerens Benth.
hehe cotópl 'plant that-clings'
hehe czatx 'plant stickery'
  "stickery plant"

Names used with this plant were also applied to *Mentzelia involucrata* and *Cryptantha*. These are all small herbs with "clinging leaves" due to barbed and spinescent pubescence.

This ephemeral may appear after rains at any time of the year. It is seasonally abundant on the mainland and on Tiburón and San Esteban islands. The petals are orange.

The plant is said to have a bitter taste. It is so abundant in one area of Tiburón Island that women's skirts got covered with the leaves.

Mentzelia involucrata S. Wats.
hehe cotópl 'plant that-clings'
zaaj iti cocái cooxp 'cliff on what-hangs white'
  "white *Eucnide*"

Spring ephemeral, pale silvery-yellow petals, and "clinging" foliage; mainland and Tiburón Island.

Loranthaceae—Showy Mistletoe Family

Phoradendron—see Viscaceae

Struthanthus haenkeanus (Presl.) Standley
xcocoj

The name derives from *quiix cocoj* 'globular rounded.' *Xcocoj* is also the name for the common housefly.

The plant was described by the Seri as a parasite growing on mesquite, with leaves longer than those of *Phoradendron diguetianum* and a larger fruit. The Seri knew of it in mesquite trees in the vicinity of Kino Bay.

We brought some of the plant to El Desemboque. Rosa Flores said, "It came from this side of Tastiota, didn't it? There is a lot of it around Kino. It sits up in mesquite and has red fruit. The fruit is not eaten."

Medicine: Tea made from the twigs was taken to cure dysentery. The fruit was broken and the juice painted on the nipple to make one fat.

Malpighiaceae—Malpighia Family

Echinopterys eglandulosa (Juss.) Small
hap oacajam 'mule-deer what-it-hits'
  "what mule deer flay antlers on"

This plant was poorly known among the Seri. The name was usually applied to *Caesalpinia palmeri*, but was also used for *Thryallis*. Small shrub with showy yellow flowers; mostly at higher elevations and not common.

Janusia californica Benth.
J. gracilis Gray
hehe quiijam 'plant that-curls-around-it'

Small desert vines bearing delicate yellow flowers, commonly growing on shrubs and small trees. Both species are widespread and common in the region. These closely related and similar-appearing species were not differentiated by the Seri.

Mascagnia macroptera (Ses. & Moc.) Nieden.
haxz oocmoj 'dog what-it-wears-on-waist'
  "dog's waist-cord"
gallinita

A bushy perennial, becoming a vine in better-watered areas with denser vegetation. Found in arroyos and mountains on the mainland from Arroyo San Ignacio southward and on Tiburón Island. It has yellow petals and unusual fruit with large papery wings.

Headpiece: A crown or wreath woven from the leafy stems, sometimes with the flowers included, was worn by men and women during summertime for shade and to keep the hair in place. It was called *hehe yail it hapácatx* 'plant its-greenness on what-is-left-on' (see Figures 1.5, 12.2; Bowen and Moser 1970a: 175–176). Women and girls continued to use it in the 1980s.

Medicine: Tea made from the roots was taken as a remedy for a cold and for diarrhea. It was also said to help a woman gain strength after parturition.

Thryallis angustifolia (Benth.) Kuntze
hap oacajam 'mule-deer what-it-hits'
    "what mule deer flay antlers on"

This name was usually applied to *Caesalpinia palmeri*, and sometimes to *Echinopterys*, which superficially resembles *Thryallis*. This small, perennial subshrub is generally restricted to higher elevations. It was poorly known among the Seri.

Malvaceae—Mallow Family

Abutilon californicum Benth.
hant ipásaquim 'land what-it-is-swept-with'
    "broom"

Sparsely branched shrub with thin stems and drought-deciduous foliage. It is common in arroyos, canyons and other brushy places on the mainland and Tiburón Island. According to the Seri, there is much of it along the mountains south of Tepopa and on Tiburón Island.

The stems were tied together and used as a broom.

Abutilon incanum (Link) Sweet
hasla an ihoom 'outer-ear in where-it-lies'
    "ear is its place"
caatc ipápl 'grasshopper what-it-is-strung-with-(plural)'
    "what grasshoppers are strung with"

Small shrub with slender stems and drought-deciduous foliage; widespread and common on the mainland and Tiburón Island.

Adornment: As indicated by the first name given, small pieces of the stem were put through pierced ear holes to keep them from closing.

Food Preparation: Several mallows apparently were once used to string grasshoppers (see *Horsfordia*).

Abutilon palmeri Gray
caatc ipápl 'grasshopper what-it-is-strung-with-(plural)'
    "what grasshoppers are strung with"

Perennial bush with relatively large, velvety leaves and orange petals; found in arroyos on the mainland and the islands.

Food Preparation: The name indicates that the Seri used to capture and string grasshoppers for food (see *Horsfordia*).

Medicine: The bark was removed from the root, and the root then cut into pieces and put in cold water; the liquid was drunk to alleviate a sore throat.

* *Gossypium* spp.
mooj
algodón, cotton

Cotton is an important agricultural crop in Sonora. The cotton boll is *mooj is* "cotton's fruit."

Hibiscus denudatus Benth.
hepem ijcóa 'white-tailed-deer its-*Sphaeralcea*'
    "white-tailed deer's globe mallow"
rock hibiscus

The name indicates relationship with globe mallow (*Sphaeralcea*). This small desert perennial is widespread and common in western Sonora.

Horsfordia alata (S. Wats.) Gray
hehe coozlil 'plant sticky'
    "sticky plant"
caatc ipápl 'grasshopper what-it-is-strung-with-(plural)'
    "what grasshoppers are strung with"

The first name derives from the stems being sticky, gummy, or slimy beneath the bark. The second name was also used for *Abutilon incanum* and *A. palmeri*. It is a spindly shrub 2 to 3 m tall with sparse foliage. Mainland, and Tiburón and San Esteban islands. Identification of this plant was sometimes confused among the Seri.

Food Preparation: The second name indicates it was used to string grasshoppers, presumably to carry them to camp (see *Abutilon palmeri*). In the seventeenth century Adamo Gilg reported that the Seri ate grasshoppers (Di Peso and Matson 1965:55). However, the modern Seri claimed no knowledge of grasshoppers being used for food.

Medicine: A thick liquid made by soaking the outer bark of the root and the lower part of the stem was used as a medicine for sores in the mouth or eyes.

* Malva parviflora L.
malva, cheeseweed

Common weed, native of Eurasia; it is found on farmland and is apparently a relatively recent invader in El Desemboque and Punta Chueca. The Seri said that it is a Mexican plant (implying one not native to their region), and they used the Mexican common name for it.

Adornment: The green capsules were occasionally strung as necklaces.

Sphaeralcea ambigua Gray var. *ambigua*
S. coulteri (S. Wats.) Gray
jcoa
mal de ojo, globe mallow

Orange petals and scurfy foliage: *S. ambigua* is a small perennial bush; *S. coulteri* a winter-spring ephemeral.

Jcoa was a generic name for *Sphaeralcea*. Some people distinguished *jcoa ctam* "male *jcoa*" from *jcoa cmaam* "female *jcoa*." Generally the taller plants were considered male and the stouter or shorter ones female.

Animal Food: Dogs were said to sometimes eat globe mallow when out in the desert hunting lizards and rabbits.

Medicine: The bark was removed from the roots by pounding them on a metate. The resulting somewhat frothy and slimy pulp was made into a tea taken as a remedy for diarrhea and sore throat, and was also used as eye drops.

Martyniaceae—Unicorn-plant Family

Proboscidea altheifolia (Benth.) Decne.
xonj
cuernitos, devil's claw

The fruit was called *xonj itáast*, '*xonj* its-tooth.' The slender underground stem arising from the thick root was called *ipaöjam inol* 'its-tail its-finger.' This term was used only with this species. In contrast with *P. parviflora*, this species was said to be common on sandy places near the sea.

The dry devil's claw capsule can be found at any time of the year, usually in small sandy arroyos and floodplains. Fruit of this species is much smaller than that of *P. parviflora*. When the summer rains begin, leafy green shoots rapidly develop from a large tuber-like root and bear snapdragon-shaped yellow flowers. The green fruit develops quickly and splits apart to form the unique devil's claw capsule as it dries. Common on the mainland and Tiburón Island.

Adornment: The seeds were occasionally strung for necklaces.

Animal Food: The root was said to be eaten by javelina and desert bighorn sheep.

Food: The relatively large, fleshy root was peeled and the outer portion (cortex) beneath the bark eaten fresh. The inner part (pith) was said to be bitter. It was a relatively important staple for the people who lived in the interior of Tiburón Island.

Proboscidea parviflora (Woot.) Woot. & Standl.
xonj caacöl 'P.-altheifolia large(plural)'
    "large devil's claw"
cuernitos, devil's claw

Summer ephemeral, commonly in sandy soils on the mainland. The fruit is considerably larger than that of *P. altheifolia*. The Seri recognized it from dry fruit found in winter. They described it as a plant with "hardly any root" (an ephemeral) and pink flowers, whereas *xonj* (*P. altheifolia*) has a large root and yellow flowers.

The Seri apparently have no knowledge of devil's claw fibers used in basketry designs by other people in southwestern North America. When shown long-clawed fruit of the domesticated form cultivated by the Papago, they merely identified it as *xonj caacöl* and offered no additional comments.

Menispermaceae—Moonseed Family

Cocculus diversifolius DC.
com-ixaz 'com-rattle'
snail seed

There is no apparent meaning for *com*, but *ixaz* is based on *quixaz* 'what rattles.'

In the Seri region we know of this slender, perennial vine only from Arroyo San Ignacio (several kilometers east of Pozo Coyote), and the Seri confirmed our observations.

Moraceae—Mulberry Family

Ficus padifolia H.B.K.
nacapule

This tree occurs in the mountains, generally near water, between San Carlos Bay and San Pedro Bay (= Ensenada Grande) in the Guaymas region. It produces substantial quantities of small edible figs, and the people of the Guaymas-Tastiota Region probably harvested the fruit. *Nacapule* is the local Spanish name for it.

Ficus petiolaris H.B.K. subsp. *palmeri* (S. Wats.)
    Felger & Lowe
    [= *F. palmeri* S. Wats.]
xpaasni
tescalama, cliff fig

This is the only fig in the desert north of the Guaymas–San Carlos Bay region and is the largest-leaved tree in the region (Figure 17.129). It grows on sea cliffs, sheer canyon walls, and mountain rock, generally in sheltered locations (Figure 17.130). The roots grasp the rock and cascade down over the surface, and may reach riparian soils, such as a canyon floor. The bark is whitish and the leaves relatively large and heart shaped. It is generally a large shrub, although where the roots obtain sufficient moisture it develops into a tree. In the interior of Tiburón Island there are *Ficus* trees over 10 m in height, and a few are about 20 m tall. These are the largest trees of the northern Gulf of California. The fruit is a miniature fig about 1.5 cm in diameter. The trees often produce crops in early spring and late fall. The sap is milky and the wood relatively soft and light colored. It occurs on Tiburón and San Esteban islands, although it is relatively rare on San Esteban, and from the Cerros Anacoretas and Sierra Seri southward.

There is a Seri saying that if one sees a flower of

Figure 17.129. A branch of cliff fig (*Ficus petiolaris* subsp. *palmeri*) from San Pedro Martir Island in the Gulf of California, January 1963. *FR.*

Figure 17.130. Antonio López beneath a cliff fig about 5 km southeast of El Desemboque. *RSF, April 1983.*

xpaasni he will become rich. As in all figs, the minute flowers are enclosed within the fig (receptacle), and would not be recognized as flowers by a non-botanist.

During cool weather, probably in April, the fruit sometimes falls off. This was said to be caused by *hant coopol* 'land black,' described as dark patches caused by winds which come in over the land and cause buds (and small fruit) to fall (see *Pachycereus*).

Adornment: The fresh bark, which was considered to have a very pleasant odor, was cut into small pieces and tied on a cord in tiny bundles as a necklace (see Figure 12.5B). It was said that it gave the wearer a long life. Girls especially wore these necklaces because of the pleasant aroma, which was said to last one or two years. However, the bark can cause skin irritation on some people.

Firemaking: Cliff fig was one of the preferred woods for the fireboard and drill stick or drill tip of the *caaa* firedrill (see Figure 9.14). A branch of green wood was cut, carved while still fresh, and then allowed to dry. The firedrill that a man kept in his arrow quiver was usually made from cliff fig wood or *satómatox* (a special kind of mesquite driftwood; see *Prosopis*). Cliff fig was a preferred wood because the fire caught right away when a little bit of jackrabbit dung was used as punk.

Food: In favorable years the fruit was collected in late spring and fall. It was cooked in water with sea turtle oil until soft. It was also eaten fresh, but if one ate too much it was said to cause a headache.

Music: The one-string Seri violin was sometimes made from the green wood (see *Bursera hindsiana*).

Concerning a Seri harp, Bowen and Moser (1970b:184) stated:

> The Coolidges mention . . . chordophones (1939:209–210) including an 11 string Seri harp. This may well have been the single harp-like instrument seen by Chico Romero many years ago and believed to have been made by Luis Torres. Constructed of wood from the wild fig tree, the frame was pointed at one end and strung with the sinew-like fibers stripped from sea turtle intestines. The instrument was called *heenj hatítöj* 'what-is-played what-is-pinched' (the present name for guitar). It was very possibly an adaptation of the Yaqui harp. Probably very few were ever constructed.

Water Transport: While camped at Tecomate, on the north shore of Tiburón Island, in June 1951 Alfred Whiting (1957:#8) collected a specimen of "*pa'sni*" which consisted of "leaves from another section of the island which were put into a can of drinking water to keep the liquid from splashing during a boat trip."

Ficus radulina S. Wats.

This species occurs with *F. padifolia* in the mountains north of San Carlos Bay, but it is locally more restricted in distribution. In this region it occurs only near water and develops into a relatively large tree. It produces edible figs, and it is likely that the people of the Guaymas-Tastiota Region harvested the fruit.

Musaceae—Banana Family

* *Musa sapientum* Kuntze
xonj itáast cmis 'Proboscidea its-tooth what-resembles'
　"like devil's claw's tooth"
plátano, banana

The banana was thought to resemble the unripe green fruit of the devil's claw before it dries and splits open to form the claws.

Nyctaginaceae—Four O'clock Family

Abronia maritima S. Watts
spitj cmajíic 'Atriplex-barclayana females'
"female coastal saltbush"
coastal sand verbena

This prostrate-growing perennial has thick succulent leaves and bright magenta flowers (Figure 17.131). It forms dense mats on upper beaches and beach dunes throughout the region, including Tiburón Island (see *Sesuvium*).

Medicine: Tea made by cooking the outer portion ("bark") of the root was taken to aid in the expulsion of a torn placenta.

Abronia villosa S. Wats.
hant-oosinaj 'land-*oosinaj*'
sand verbena

During years with favorable rains this winter-spring ephemeral is common from the vicinity of Cerro Tepopa northward. It characteristically occurs on sandy soils, often on coastal dunes.

Girls enjoyed bringing the flowers home because they were so pretty and fragrant; they placed the bouquets in cracks or holes in the wall inside their houses (see *Oenothera*).

Allionia incarnata L.
hamíp cmaam 'Boerhaavia-coulteri female'
"female spiderling"
trailing four o'clock

The plant spreads over the ground. For this reason it is considered a female plant.

It is a perennial herb with trailing stems dying back to the roots in severe drought. It has lavender flowers several times larger than those of *Boerhaavia coulteri*. It is common over a wide range of habitats on the mainland and Tiburón and San Esteban islands.

Medicine: The plant was cooked in water and the tea drunk to cure diarrhea.

Boerhaavia coulteri (Hook.f.) S. Wats.
hamíp
spiderling

A common summer-fall desert ephemeral; mainland and Tiburón and San Esteban islands. The small, pinkish flowers are on slender racemose inflorescences. The Seri said it is like tall hay, but with small leaves.

Food: The green herbage was cooked, mashed and eaten. This was one of the few greens used by the Seri (see *Amaranthus fimbriatus*).

Water Transport: It was one of several kinds of plants used to cover an olla or 20-liter can to prevent water from splashing or spilling.

Boerhaavia erecta L.
hamíp caacöl 'B.-coulteri large(plural)'
"large spiderling"

Summer-fall ephemeral, resembling *B. coulteri* but having flowers in tiny umbellate clusters. Some Seri

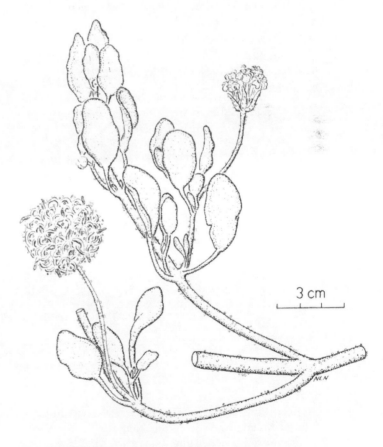

3 cm

Figure 17.131. Coastal sand verbena (*Abronia maritima*) from Kino Bay, April 1974. Note the sand grains adhering to the stem. *NLN*.

said it was the same as *B. coulteri* but that it had grown taller.

Animal Food: The edible caterpillar of the sphinx moth (*Hyles lineata*) was said to be found more often on *hamíp caacöl* than on other kinds of plants.

Medicine: The stems were cooked in water, and the resulting tea taken to combat measles.

Mirabilis bigelovii Gray
hepem isla 'white-tailed-deer its ear'
    "white-tailed deer's ear"

Herbaceous perennial; generally in mountains among rocks. This species was not widely recognized.

Onagraceae—Evening-primrose Family

Camissonia californica (T. & G.) Raven
hast ipénim 'rock what-it-is-splattered-with'
    "splattered against rock"

Winter-spring ephemeral, often very common across the desert floor on the mainland and Tiburón Island.

Medicine: The flowers and leaves were cooked in water and used to wash the head as a remedy for falling hair and hair with split ends.

Camissonia chamaenerioides (Gray) Raven

No Seri name was given for this plant, but we were told, "It looks like *tee* (*Eriogonum inflatum*) but isn't." It is a winter-spring ephemeral.

Camissonia claviformis (Torr. & Frem.) Raven
hant-oosinaj ctam 'land-*oosinaj* (*Abronia villosa*) male'
    "male sand verbena"

Common winter-spring ephemeral in the northern part of the region. Many Seri did not have a name for this plant.

Oenothera californica S. Wats. subsp. *arizonica* (Munz) Klein
hant-oosinaj cooxp 'land-*oosinaj* (*Abronia villosa*) white'
    "white sand verbena"
evening primrose

This winter-spring ephemeral has large white flowers which open at dusk and remain open through the morning on cool days. Generally found on sandy soils, it is common on dunes on the mainland and Tiburón Island.

To the Seri it is the "same" as sand verbena (*Abronia villosa*) except for the difference in flower color. Both species occur together on beach dunes and sandy soils.

Girls brought home bouquets of the flowers which were tied together and hung from the walls of their huts.

Orobanchaceae—Broom-rape Family

Orobanche cooperi (Gray) Heller
matar
broom-rape

According to Seri oral history, a Papago Indian named Matar lived for a short time with a northern group of Seri, hence the name. The derivation of Matar seems to be from *maw'o tatk* 'head root,' the Pima-Papago term for broom-rape (Amadeo Rea,

personal communication, 1983; also see Curtin 1949:49). *Orobanche* is generally more common to the north, in Papago territory.

The thick, succulent stems appear in spring, and are parasitic on various species of bur-sage (*Ambrosia* spp.). It occurs near El Desemboque and northward.

Food: The people of the Libertad Region ate the "roots" (thickened base of the stem and inflorescence) after they were thoroughly cooked in ashes.

Palmae (Arecaceae)—Palm Family

Three species of fan palms are native to the coastal mountains between San Pedro Bay and San Carlos Bay. Seri reference to "wild dates" growing "south of Tastiota" pertains to these palms. *Zamij* is the modern generic term for any palm tree.

The people of the Tastiota-Guaymas region probably made extensive use of these palms. Some hint of earlier use of palms is found in the terms for a wooden box, *hehe zamij* 'wood palm,' and the personal carrying box, *hehe zamij an iqui ihácalca* 'wood palm in with where-one-has-belongings' (see *Bursera hindsiana*). Perhaps these containers were once made from woven palm leaves.

The seeds, fruit, and terminal leaf buds were eaten by other people in southwestern North America (e.g., Bean and Saubel 1972:146; Castetter and Bell 1951:206; Gentry 1942:68; Pennington 1980:234; Standley 1920:72). Since all of these palms produce substantial quantities of fruit, it may be presumed that they were important food resources. The terminal bud, or tender young leaves were eaten fresh or roasted in hot coals or pit-baked. However, the Seri had no information on use of the leaf bud for food. Like *Phragmites*, these palms were probably too valuable and relatively limited in numbers to have been used in this manner, since the young palm is killed when the terminal leaf bud is harvested.

Writing from Pópulo in 1692, Adamo Gilg mentioned that the Seri wove ". . . mats of palm and reed" (Di Peso and Matson 1965:55; see Figure 1.4). *Sabal uresana* is common in the region.

Brahea aculeata (Brdg.) H.E. Moore
[= *Erythea a.* Brdg.]
zamij cmaam 'palm female'
  "female palm"
hesper palm

This palm is common in the coastal mountains from the vicinity of San Carlos Bay to San Pedro Bay. It is closely related to *B. armata*, from which it is distinguished by its somewhat smaller fruit and more slender trunk.

The Seri told us that the fruit was not edible. We attribute *zamij cmaam* to *Brahea* because this palm is usually not as tall as *Sabal* or *Washingtonia*, and the seeds are not edible unless prepared. (The Cocopa

pit-roasted the seeds of *Brahea armata* to destroy the bitter flavor [Felger and Rea, unpublished notes]).

Headpiece: The people of the Tastiota Region made a kind of hat woven from the tender inner leaves to provide protection from the sun. It was called *zamij yeen haonam* 'palm its-face hat.' "Its face" refers to the inner leaves.

Brahea armata S. Wats.
[= *Erythea a.* S. Wats.]
Blue hesper palm

This palm occurs in central and northern Baja California and on Angel de la Guarda Island. An immature leaf—presumably of this species—was brought to us by a Seri man, who claimed he obtained it from a single plant on the west side of Tiburón Island. Other Seri also told us about this palm, but we have not been able to find it.

*Phoenix dactylifera L.
dátil, date palm

Date palms, brought to Sonora by the Jesuits during Spanish colonial times, have long been known to the Seri.

Oral Tradition: Manuel Molino planted dates and *guamúchil* at one or more camps south of Cerro San Nicolás (see *Pithecellobium dulce*). This would have been in the mid-nineteenth century.

Sabal uresana Trel., and/or Washingtonia robusta Wendl.
zamij ctam 'palm male'
  "male palm"

Sabal uresana is a large palmetto with huge, dull green, tough leaves. The petiole is spineless. It occurs in arroyo and canyon bottoms, generally with standing water, and along the beach north of San Carlos Bay. It is very common between Ures and Rayón in the lower Río San Miguel Valley near Pópulo, the Jesuit mission for the Seri during the seventeenth and eighteenth centuries.

Washingtonia robusta, or desert fan palm, has shiny green leaves with prominent spines along the petiole and generally develops a tall, slender trunk. In Sonora it is restricted to canyon streamways with permanent water near San Carlos Bay.

Food: The Seri said *zamij ctam* grows wild "south of Tastiota" and that the fruit was eaten.

The seeds and the thin, fleshy, sweet pericarp of *Washingtonia* were eaten by various people in southwestern North America (Bean and Saubel 1972:146; Castetter and Bell 1951:206; Standley 1920:72). The fruit of *Sabal uresana* was eaten fresh by people in the Río San Miguel valley and elsewhere in Sonora (Felger, unpublished notes).

Papaveraceae—Poppy Family

Argemone pleiacantha Green
[= *A. platyceras* Link & Otto, in part]
*A. mexicana L.
xazácöz
cardo, prickly poppy

The Seri did not distinguish between these two species. *A. pleiacantha* is common along arroyos and floodplains and as a weed around farms and ranches. It is a perennial herb, bearing large flowers with white petals and numerous large, yellow stamens. *A. mexicana* is a smaller plant with yellowish petals. We found it on abandoned farmland northeast of El Desemboque. Prickly poppy is rare in the undisturbed natural desert in the Seri region.

Medicine: To relieve kidney pain a tea was made from the leaves wrapped in cloth and steeped in water. This tea was also used to make a woman "lose bad blood" after parturition. Tea made from prickly poppy was taken to cause expulsion of the remaining portion of a torn placenta.

A medicinal tea made from prickly poppy leaves cooked in water with the inner bark from the root of *cotéexet* (*Opuntia fulgida*) was taken for urinary problems, as a diuretic, and to "clear the urine." It was said to "rest" the kidneys and relieve the pain.

Argemone subintegrifolia Ownby
xazácöz caacöl 'Argemone large(plural)'
  "large prickly poppy"

This unusual spiny-leaved shrub is known from Baja California, Angel de la Guarda Island, and San Esteban Island. A fresh specimen collected on San Esteban Island and brought to El Desemboque was quickly recognized as being from that island.

Eschscholzia parishii Greene
hast ipénim 'rock what-it-is-splattered-with'
  "splattered against rock"
desert poppy

In years with favorable winter-spring rainfall, this spring wildflower is found on the mainland in the northern part of the Seri region. The flowers are golden yellow. It is replaced by *E. minutiflora* S. Wats. in the Sierra Seri and on Tiburón Island.

Passifloraceae—Passion Flower Family

Passiflora arida (Mast. & Rose) Killip var. *arida*
oot ijöéne 'coyote its-*Passiflora-palmeri*'
  "coyote's passion vine"

It was said to be one of coyote's plants because of the inferior quality of the fruit as compared with the other passion vine in the region.

This small passion vine occurs in scattered localities across the desert on the mainland and Tiburón Island. It flowers and fruits at various times of the year following periods of sufficient soil moisture. The leaves are whitish lanate (wooly).

Food: The fruit was eaten fresh. Since about the middle of the twentieth century it was usually eaten only by children. It was of minor dietary importance. Some said it had a disagreeable taste, while others claimed it was good eating. The fruit was said to stink and to smell like guava (*guayaba*).

Medicine: The roots were cooked in water and the tea drunk for intestinal disorders.

Passiflora palmeri Rose
jöéne

Small perennial with semi-vining stems, similar in

habit to *P. arida*, but with brownish or yellowish brown viscous hairs.

The Seri described it as follows: It is a vine which covers the ground like an apron, with just a thin, narrow stem. The fruit of this plant is about the size of that of *hant yax* (*Maximowiczia*). The fruit is yellowish, smells like cantaloupe, and has stripes when it is ripe. It has a pleasing odor and is edible.

Food: The fruit was eaten fresh or sometimes cooked with meat. It was said to be sweet.

Phytolaccaceae—Pokeweed Family

Phaulothamnus spinescens Gray
an icös 'inside its-thorns'
    "thorny inside"

The name refers to the fact that the inside, or interior part, of the shrub is thorny. The Seri described this plant as a thorny shrub with tiny, white fruit that is found on Tiburón Island and the mainland. It is common and widespread, resembling *Lycium* in habit.

Medicine: The outer bark was removed and the "inner bark" burned. The blackened area was then scraped, water added to the scrapings, and the resulting liquid used as eye drops to improve eyesight.

Stegnosperma halimifolium Benth.
xneejam-siictoj 'xneejam red'

This plant is not "claimed" or "owned" because the seeds are hard and cannot be used in necklaces, as are those of *Viscainoa*, called "*xneejam* seeds claimed." *Siictoj* is an archaic word for red.

It is a common shrub along the shores of the Gulf including the major islands, and often ranges inland through arroyos and washes. The nearly evergreen foliage consists of simple, semi-succulent leaves (Figure 17.132). At various times of the year it bears fragrant, star-shaped flowers and juicy red fruit.

Medicine: To relieve a headache, the leaves were cooked with leaves of *Bursera microphylla* and the mixture used as shampoo.

It was said that women going for firewood should carry leaves of this plant in their clothing or tied in their head scarves. If bitten by a rattlesnake, they were to chew the leaves to make a paste, then rub it on the bite. It was said that one would not die if treated in this manner.

Oral Tradition: This plant features in the story of two brothers who became angry with their people and went to Baja California (E. Moser 1963b). The brothers were found by two Giant women who turned them into Giants by rubbing their feet, hands, and heads with whale brains. After a long time one brother said, "Brother, the end of the year has come. The *xneejam-siictoj* is bearing fruit [the fruit is ripe]. Let's return home." This story was well known among the Seri. The end of the year would have been June, and their home was Tiburón Island.

Pigment: The bark of the root was sometimes substituted for bur-sage (*Ambrosia dumosa*/spp.) in the making of Seri Blue (see *Guaiacum*).

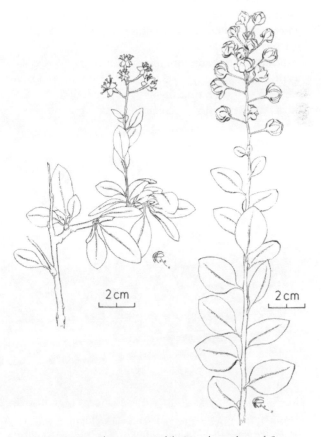

Figure 17.132. Flowering and fruiting branches of *Stegnosperma halimifolium*. FR.

Piperaceae—Pepper Family

Piper nigrum L.
coquée coopol 'chile black'
  "black chile"
pimienta, black pepper

Plantaginaceae—Plantain Family

Plantago insularis Eastw. var. *fastigiata* (Morris) Jeps.
hataj-en 'vulva inside'
wooly plantain, Indian wheat

This winter-spring ephemeral often carpets the desert following times of sufficient rainfall. It generally occurs on level terrain of the desert floor on the mainland and Tiburón Island. The size of the plant and number of flowers and seeds varies with the amount of soil moisture. Under favorable conditions each plant develops a half-dozen or more slender spikes 10 to 15 (max. 30) cm tall, each of which bears numerous yellowish to reddish brown seeds 2 to 2.5 mm long.

Food: The seeds, often stored in pottery vessels, were an important food. In the 1980s a common method of harvesting involved the following procedure. Whole plants with mature seeds were pulled up and gathered in a pile (Figure 17.133). While sitting on the ground, the women rolled the fruiting spikes in their hands to free the seeds and let the seeds and chaff fall onto their outspread skirts. They then picked up handfuls of the seed and chaff and wind-winnowed it (see Figure 6.12).

The whole seeds were mixed with water and allowed to sit for about one-half hour, and then eaten; they were not cooked. The soaked seeds became an edible, gelatin-like mass. Sugar was commonly added, and with water added the mixture was often consumed as a refreshing drink. It was a much relished food. In 1978 Sara Villalobos, after eating a dish of it, said, "This is the best food of all."

Medicine: Prepared in the above-mentioned manner, it was a much-used remedy for stomachache. It was considered excellent for children with stomach ailments.

Figure 17.133. A handful of freshly picked Indian wheat (*Plantago insularis*), 3 km east of Pozo Coyote. *RSF, April 1983.*

Polygonaceae—Buckwheat Family

Chorizanthe brevicornu Torr.
xtamóosn-oohit 'desert-tortoise what-it-eats'
  "what desert tortoises eat"
brittle spine-flower

Several other wildflowers were given the same name, e.g., *Chaenactis* and *Fagonia*. *C. brevicornu* is a spring ephemeral.

Chorizanthe corrugata (Torr.) Torr. & Gray
tee

The Seri name may be derived from the Spanish word *té* (tea).

A winter-spring ephemeral found in the mountains between El Desemboque and Puerto Libertad, and northward. It was said to be like *tee* (*Eriogonum inflatum*) but small.

Chorizanthe rigida (Torr.) Torr. & Gray
xomxéziz caacöl 'Fouquieria-splendens
  large(plural)'
  "large ocotillo"

This little winter-spring ephemeral dries into a persistent skeleton with stout spines, looking somewhat like a miniature ocotillo. It is common across the desert in the northern part of the region.

Eriogonum inflatum Torr. & Frem. var. *inflatum*
tee
desert trumpet

The Seri name may be derived from the Spanish word *té* (tea).

Perennial herb with several or many slender, erect-growing, branched flower stalks. The main axis is sometimes hollow and inflated. Commonly found among rocks on the mainland and on Tiburón and San Esteban islands.
Medicine: Tea made from the plant was taken as a remedy for a cold.

Eriogonum trichopes Torr.
tee cmaam 'E.-inflatum female'
  "female desert trumpet"

Female here refers to the relative size and shape of the plant as compared with desert trumpet (*E. inflatum*), a taller and more slender plant. *E. trichopes* was not always recognized; for example, one person said, "I don't know its name, but I would call it *tee*." Winter-spring ephemeral, common on open, sandy desert.

Punicaceae—Pomegranate Family

* *Punica granatum* L.
hehe is quiixlc 'plant its-fruit globular(plural)'
  "round-fruited plant"
granada, pomegranate

Pomegranates have been cultivated in Sonora since Spanish colonial times.
Dye: The peels were used occasionally in the preparation of a black basketry dye (see *Suaeda moquinii*).

Rafflesiaceae—Rafflesia Family

Pilostyles thurberi Gray

This tiny plant is parasitic on *Psorothamnus emoryi*. Purplish brown buds, flowers, and fruits, each 2 to 4 mm in diameter, dot the stems of the host plant. In the Seri region it occurs only on the high beach dunes encroaching the north side of Cerro Tepopa. The Seri had no specific name for it, and considered it to merely be a kind of animal life, *ziix ccam* 'thing with-life.'

Resedaceae—Mignonette Family

Oligomeris linifolia (Vahl.) Macbr.
tomáasa, xamáasa, xomáasa

Ephemeral, often abundant and in dense stands in late spring (Figure 17.134). It occurs on the mainland and Tiburón Island. These small, slender plants have tiny capsules containing numerous shiny black seeds about 0.4 mm in diameter.

Food: The seeds, worked out of the dry plant onto deer hides, were toasted, ground, and mixed with water to make a gruel. Substantial quantities of the seed could be obtained during favorable years.

Medicine: Tea made from the roots was used as a remedy for measles.

Oral Tradition: In Seri mythology, Giant girls gathered the plant for the seeds. One of them, who was in love, strung the little seeds on a necklace and gave it to her lover (in Seri mythology the Giants often used tiny objects).

Figure 17.134. *Oligomeris linifolia.* A) Mature, seed bearing plants near El Desemboque. B) The chaff and tiny, black seeds. *RSF, April 1983.*

Rhamnaceae—Buckthorn Family

Colubrina viridis (Jones) M.C. Johnst.
  [= *C. glabra* S. Wats.]
ptaact
granadita, palo colorado

One of the most common shrubs in the Seri region, it occurs on the mainland and on Tiburón and San Esteban islands. The bright green foliage of thin, rounded leaves falls quickly with drought and reappears promptly with each rainy period. The branches are stiff and the wood is very hard (Figure 17.135).

Caulking: The leaves, mashed in the hand with a bit of water, were used to plug small holes in pottery vessels. It was said to be as good as *hocö ine,* "wood's mucous," the pitch made from organ pipe (*Stenocereus thurberi*).

Firewood: An excellent fuel, it was much used.

Food Gathering: This was one of several kinds of wood used to make the *hahéel* chisel–pry bar for century plants (*Agave*).

Hunting: The detachable foreshaft of the turtle harpoon often was made from *ptaact* wood.

Figure 17.135. *Palo colorado (Colubrina viridis)* near Pozo Peña. A) Pedro Comito next to a mature shrub. B) The dry dead wood, much used for fuel. *RSF, April 1983.*

Condalia globosa I.M. Johnst.
cozi
crucerilla, bitter condalia

Shrub with very stiff, woody branches and twigs ending in sharp thorns (Figure 17.136). The wood is very hard. The shrub usually does not exceed 1.5 to 2 m in height, although some in the interior of Tiburón Island exceed 5 m in height. It has small spatulate leaves, and small brownish or black fruit bitter to the taste. It generally occurs in desert washes and upper bajada slopes and is widely scattered, although seldom common, throughout the region.

Firewood: It is an excellent firewood, said to be as good as ironwood.

Tattoo: The sharp, thorn-tipped twigs and mashed leaves were used in tattooing practices (see section on tattooing in Chapter 12).

Figure 17.136. *Condalia globosa* at Rancho Estrella. A) A mature shrub about 3 m tall. B) The thorn-tipped twigs. *RSF, March 1983.*

Zizyphus obtusifolia (T. & G.) Gray var. *canescens* (Gray) M.C. Johnst.
[= *Condaliopsis lycioides* (Gray) Sues. var.
c. (Gray) Sues.; *Condalia l.* Gray var. *c.*
Gray]
haaca
xica imám coopol 'things its-fruit black'
"black-fruited things"
bachata, white crucillo, lotebush

The name *haaca* became taboo because a child
called *Haaca Ixái* "*haaca* roots" died. However, it
was not an absolute taboo, since we found both plant
names in use.

This thorny, scraggly shrub often forms dense
thickets on the mainland and Tiburón and San Esteban islands. Fruiting occurs at various seasons, although generally during warmer months following
rainy periods. The ovoid fruit, about 7 to 9 mm in diameter, has a thin, fleshy pericarp.

Hair Care: The roots, pounded and with water
added, were used for shampoo.

Food: The fruits were collected from the shrub
and eaten fresh. They were also commonly collected
by robbing pack rat nests. Whiting (1957: #54) also
noted that the fruit was eaten.

Medicine: The dry roots were ground into a powder which was put on skin or scalp sores.

Rhizophoraceae—Red Mangrove Family

Rhizophora mangle L.
pnaacoj-xnazolcam 'mangrove crisscrossed'
"crisscrossed mangrove"
pnaazolcam (contraction of above)
mangle, red mangrove

Pnaacoj is the generic term for mangrove. "Crisscrossed" refers to the pattern of the stilt roots (see
Figure 2.6). Red mangrove is distinguished from the

other Gulf of California mangroves (*Avicennia* and
Laguncularia) by its prominent stilt roots, shiny
green, semi-succulent leaves, and unique fruit (Figure
17.137). The fruit, which germinates while still attached to the parent plant, forms an embryo with a
thickened green "root" about 20 to 30 cm long before falling. Fruit is produced throughout the warmer
months of the year.

There are impenetrable thickets of red mangrove

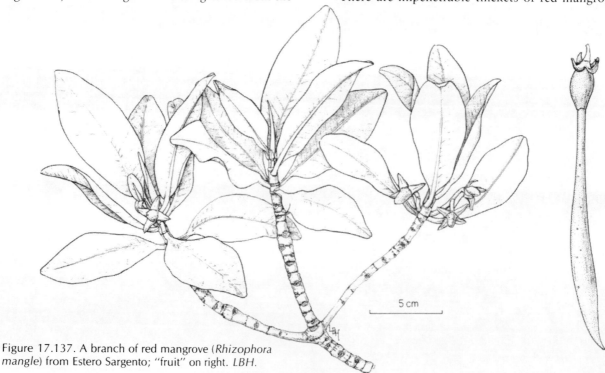

5 cm

Figure 17.137. A branch of red mangrove (*Rhizophora mangle*) from Estero Sargento; "fruit" on right. *LBH.*

intermingled with the other mangroves in shallow tidal waters of the esteros, quiet bays, and inlets. The northern limit is at the north end of the Infiernillo Channel in Estero Sargento.

Adornment: The embryo ("fruit") was cut into sections and strung as a necklace.

Dye: The roots were boiled and the decoction used for dying basketry splints black. It was used if *heep* from mesquite (*Prosopis*) or sea blite (*Suaeda moquinii*) was not available.

Firewood: Mangrove driftwood was sometimes used as firewood.

Food: Oysters (*Ostrea columbiensis*) were gathered from the stilt roots (see Figure 3.2). The Seri pointed out that oysters were found clinging to mangrove roots in large numbers in the esteros, but not at Sargento, the northern-most Sonoran locality for the red mangrove and the white mangrove (*Laguncularia*).

The "fruit" or enlarged embryo, called *mojépe pisj* 'sahuaro *pisj*' was eaten when green. It was cooked in ashes, the outside "shell" broken with a stone, and the inside eaten with sea turtle fat. Another method of preparation was to scrape the fruit with a knife to remove part of the outside "shell," wrap it in a cloth, and cook it in water with sea turtle fat. It was said that even the rank-smelling fat of a small sea turtle tasted good when eaten with mangrove "fruit." Some people said that only the "fruit" washed up on the beach was good eating (see *Simmondsia*).

Medicine: Dry "fruit" gathered from the beach, where it often washed ashore, was made into tea taken at the onset of dysentery.

Toys: Miniature boats were carved from the "fruit."

Rubiaceae—Madder Family

Randia thurberi S. Wats.
haalp
papache

Woody shrub, drought-deciduous foliage, and stiff spiny branches. The leaves are commonly 2 to 3 cm long and 1 to 2 cm wide, the blade orbicular or nearly so. The fruit is round, 1.5 to 2.5 cm in diameter, and mottled green and white. It bears fruit at least in spring. It is fairly common in and near the larger mountains on the mainland and Tiburón Island.

The Seri gave the following succinct and accurate information for this species: It is a spiny shrub about 1.5 m tall; the spines are large, and it does not have *zatx* (glochids or small spines). The plant is similar to *an icös* (*Phaulothamnus*) and *hahöj* (*Lycium* spp.), but the leaves are broader. The fruit has a hard shell, or outside, with a black ball inside containing seven or eight black seeds resembling watermelon seeds, but smaller. The seeds are like those of *Solanum hindsianum*. It occurs along most of the east side of Sierra Kunkaak, on Tiburón Island, and in the Sierra Seri range.

Facepaint: The black pulp of the fruit or the seeds were mashed in the palm of the hand and applied as facepaint.

Food: The black, sweet pulp, or inside of the fruit, was eaten fresh. In the old days, when the people were walking in the desert or going to Hermosillo or on some other long trip, they sometimes filled a turtle-stomach bag with the fruit and ate it while traveling. The children ate the fruit, but it was not considered to be a regular food.

Ruppiaceae—Ditch-grass Family

Ruppia maritima L.
zimö-taa 'when past be'
   "how did it happen?"
ditch-grass

Abundant undersea stands of ditch grass occur in shallow sea water in the Seri region from the north end of the Infiernillo Channel to Kino Bay and at Tastiota. Plant fragments are cast ashore in the beach drift in late spring and summer (Felger, Moser, and Moser 1980).

Ruppia is readily distinguished from *Zostera* (eelgrass) by its narrower and conspicuously shorter leaves, more slender stems and distinctive inflorescence and small seeds. However, young emerging shoots of the two species might be confused. The tiny fruit, borne on an elongated stalk, is about 2 x 1 mm and bears a single seed about 1 mm in diameter (Figure 17.138).

In 1974 many Seri over fifty or so years of age knew this plant by name and had some knowledge of its ecology and distribution, although the information was often not consistent. However, younger people generally did not know the plant by name and failed to distinguish it from eelgrass. The seeds were not eaten by the Seri, and there is no evidence that the people used the plant in any manner. Most of the elderly people did not classify it as *xpanáams*, the generic or distributive term for marine macro-algae.

In April, 1974, Sara Villalobos and María Luisa Chilión verified the identity of *zimö-taa* with fresh specimens. Sara said it was an old name. Both women told us it was a name given to it "at the beginning," when names were first given to the Seri (see Chapter 7).

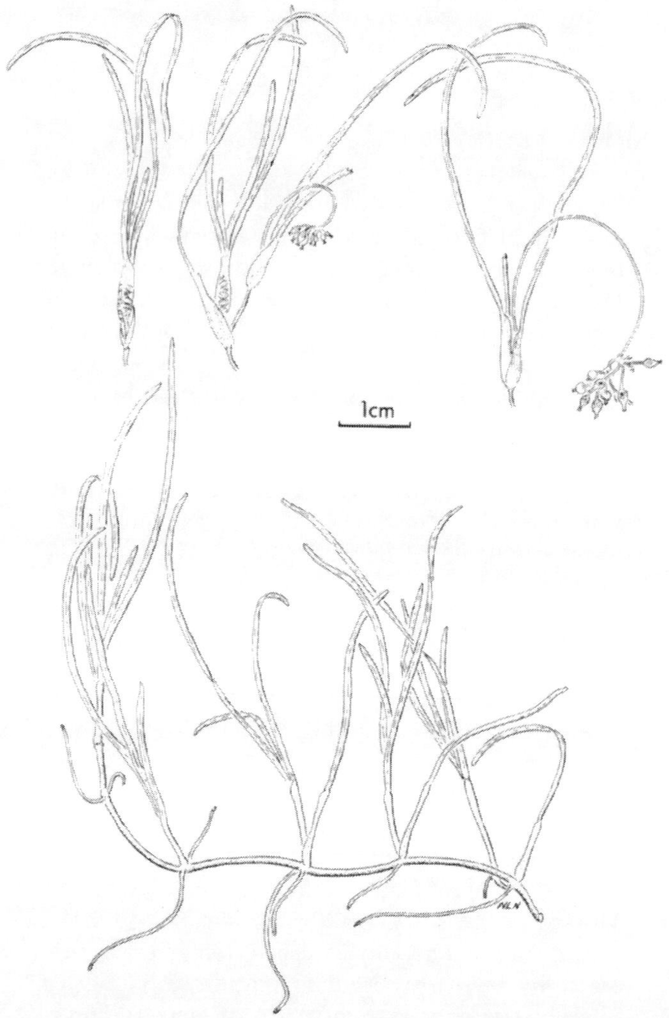

Concerning this plant Rosa Flores gave the following information:

> It is a kind of *xpanáams* (seaweed) and all marine plants are *xpanáams*. Sea turtles (*Chelonia*) eat it. It is like *eaz* (eelgrass) but the leaves are very narrow. It occurs in the Infiernillo and at Kino Bay, but not at El Desemboque. It does not have seeds. At Punta Chueca the sand is covered with it when the tide goes down. It is found in summer.

María Antonia Colosio and Roberto Herrera told us:

> It looks like eelgrass, but doesn't grow tall and has narrow strands. It does not have seeds—at least not within the long leaf. It is found in the stomachs of the *cooyam* [a certain kind of small green turtle] but not in the stomachs of larger sea turtles. The *cooyam* does not eat eelgrass (larger sea turtles do). The plant is not used by the Seri in any way. It does not wash ashore in large quantities like eelgrass does, and is mostly seen from boats when it is growing [rather than detached and floating like eelgrass].

Figure 17.138. Ditch grass (*Ruppia maritima*) from Kino Bay, April 1974. *NLN.*

Rutaceae—Citrus Family

* *Citrus aurantifolia* Swingle
limóon 'lime'
limón, lime

The Seri name is derived from Spanish.

* *Citrus limon* Burm.
sahmées ccapxl 'orange sour'
   "sour orange"
limón (*limón amarillo*), lemon

* *Citrus paradisi* Macf.
sahmées hamt caháacöl 'orange breast what-makes-large(plural)'
   "orange that enlarges the breast"
toronja, grapefruit

* *Citrus sinensis* Osbeck
sahmées
naranja, orange

Salicaceae—Willow Family

Salix gooddingii Ball
paij
sauce, willow

Due to the absence of perennial rivers or streams in the region, willows occur at only a few highly localized sites. Limited numbers of willow grow at the *Xapij* water hole, at the south end of Tiburón Island, and along irrigation ditches in the farming areas between Kino Bay and Hermosillo. At *Xapij* it is a shrub approximately 2 to 4 m tall. The term Sauzal, the local Spanish name for the *Xapij* water hole, indicates that willow was once more common there. In the 1960s salt cedar (*Tamarix ramosissima*) was abundant and seemingly was crowding out the willow.

In addition to being the specific term for willow, *paij* was also a generic word for driftwood. It was used for driftwood when one did not know or specify the name of the plant which produced the driftwood. However, there was also a certain kind of wood found only in the drift which the Seri thought was willow. This latter kind of *paij* became rare in the drift during the middle of the twentieth century. Per-

haps it originated from the great willow and cottonwood forests which once existed at the delta of the Colorado River. By the 1970s it was apparently no longer found in the beach drift in the Seri region.

Firewood: Although it was of inferior quality, the *paij* which was thought to be willow was used as firewood.

Hunting and Fishing: Temporary rafts were made of willow driftwood. The poles were lashed or tied together with rope made from mesquite (*Prosopis*) roots. The raft was used locally and then abandoned, although the mesquite cord was saved. These rafts were generally used for hunting sea turtles during the warmer months. The man hunting or fishing was called *paij iti caap* 'willow on who-stands' or "he who stands on willow." The raft itself had no special name.

Shelter: The wood was used for house posts and beams.

Transport: Yokes (*peen*) for carrying bundles of firewood were made from willow. These carrying yokes were said to be larger than those used for carrying water vessels.

Sapindaceae—Soapberry Family

Cardiospermum corindum L.
hax quipóin 'water what-is-closed'
tronador, balloon vine

This desert vine has hollow fruit resembling small paper lanterns (Figure 17.139). Each fruit bears three smooth, round seeds. It is common throughout the mainland and on Tiburón Island.

Adornment: While still green and soft the seeds were strung for necklaces.

Footwear: The long flexible stems were occasionally used to tie pads of creosotebush on one's feet as makeshift sandals (see *Larrea*).

Headpiece: The leafy vines were woven into a headband or wreath called *hehe yail it hapácatx* 'plant its-greenness on what-is-left-on.' It was worn during hot weather by men and women for shade and to keep the hair in place. Women and girls continued to use it in modern times (Bowen and Moser 1970a: 175).

Water Transport: As indicated by its name, this was one of several kinds of leafy plants stuffed into

the top of a water-filled olla or 20-liter can to prevent spillage.

Alfred Whiting (1957: #52) recorded the name of this plant as *ak^xpoi'en* and stated: "In carrying water it is necessary to put some green plant in the water to keep it from splashing. It must be a plant which will not make the water taste bad. This plant was commonly used for this purpose, as the name indicates *ax* water, *poi',en* shut."

Dodonaea viscosa Jacq.
casol caacöl 'casol large(plural)'
 "large *casol*"
hop bush

Plants on Tiburón Island were identified as *casol caacöl*, which was also the name for desert broom (*Baccharis sarothroides*). The term *casol* was associated with four genera in the composite family characterized by sticky and aromatic foliage. *Dodonaea* shares these characteristics.

2 cm

Figure 17.139. Balloon vine (*Cardiospermum corindum*) with inflated, papery fruit. *FR.*

This leafy shrub has elongate, simple leaves and papery, three-winged fruits. The young shoots are resinous. The sap is not milky. It occurs at higher elevations in the Sierra Seri and on Sierra Kunkaak.

When we brought some fresh leafy branches of this shrub to El Desemboque, Rosa Flores recognized the plant. She said that it comes from the mountain areas and has fruit like that of *Mascagnia*, but smaller. She did not know of a Seri name for it. Others confused it with *hierba de la flecha* (*Sapium*), which is considered to be a dangerous plant. One woman handled the branches and the next day became sick. She and her mother blamed the illness on the plant, because they thought it was *Sapium*.

Medicine: For a sore or aching leg, the leafy branches were put in hot water and a clean cloth was soaked in the hot solution and applied to the afflicted part as a compress.

Sapotaceae—Sapodilla Family

Bumelia occidentalis Hemsl.
paaza
hehe hatéen captax 'plant mouth perforated(plural)'
bebelama

Paaza is also the name for the Gila monster (*Heloderma suspectum*).

A large shrub or small tree, in remote places it develops into a tree up to 12 m tall (see Figure 2.16). The branches are stiff and the twigs often thorn tipped. The wood is very hard. The fruit is roundish—about 1.5 cm across—and fleshy, with a single seed. Seldom common, it occurs in arroyos and canyons and sometimes on rocky slopes and at waterholes on the mainland and Tiburón Island. Fruiting apparently occurs two or more times per year, but at various seasons.

The absence of large *Bumelia* trees in all except very remote regions indicates that they may have been cut for wood.

Boats: The wood was sometimes used for boat ribs.

Food: The fruit was eaten fresh. However, if one ate too much of it fresh from the tree, it was said to make the mouth sore. To get rid of this soreness or sting, one must put rattlesnake fat in the mouth. The name *hehe hatéen captax* derives from the sensation of the mouth feeling "punctured" or pricked when the fruit was eaten fresh. Women collected the fruit in large quantities and spread it in the sun. By evening, whatever danger existed from eating too much of it would be gone, and the fruit could then be eaten freely. It was also crushed in the fingers to remove the seeds. The dried fruit was said to be similar to raisins. It was not stored. However, the time of harvest was commonly extended by collecting it from pack rat nests.

Shelter: The wood was used for house posts and beams.

Sideroxylon leucophyllum S. Wats.
hehe pnaacoj 'plant mangrove'
    "mangrove plant"

Pnaacoj is the generic term for mangrove. Shrub or small tree, with hoary foliage superficially resembling that of black mangrove (*Avicennia*). It is primarily a Baja California species with populations on Angel de la Guarda and San Esteban islands.

Before we found this plant on San Esteban Island, we had been told it occurred there and was "like *pnaacoj-iscl*" (*Avicennia germinans*, black mangrove). Fresh, leafy branches were brought to El De-semboque, and the Seri immediately recognized and identified them as being from San Esteban Island.

Games: The leaves were mixed with cinnamon and held in the mouth to bring good luck in card games.

The Supernatural: Cutting the branches "made the wind blow."

Saururaceae—Lizard-tail Family

Anemopsis californica (Nutt.) Hook. & Arn.
comáanal
hierba de manso

This perennial herb is stoloniferous from a thick, creeping rootstock (Figure 17.140). The Seri knew of it at Tastiota and near ranches. It is restricted to wet, and usually alkaline, soils.

In the 1980s Rosa Flores obtained some *hierba de manso* plants from a local Mexican woman and planted them at her home at the El Desemboque well, along the Arroyo San Ignacio to the northeast of the village. She grew them in soil wet from seepage at the base of the water tank.

Medicine: A decoction made from the whole plant, including the roots, was said to be the best medicine to disinfect and cure sores. The sore was washed with the brew, or a cloth was soaked in the hot liquid and then placed on another cloth over the sore.

Tea from the root was held in the mouth over an aching tooth. A decoction made from the root, leaves of desert lavender (*Hyptis*), and the pith from the stems of jumping cholla (*Opuntia fulgida*) was another remedy that was held in the mouth to relieve a toothache.

Another medicinal decoction was made from *comáanal* and either *Hymenoclea salsola* or crucifixion thorn (*Koeberlinia*) twigs. The warm liquid was used to bathe the legs of one suffering from rheumatism. Tea made from *yerba buena* (cultivated mint, probably *Mentha arvensis* L.) and *comáanal* was taken to induce conception.

Hierba de manso was a much-used medicinal herb in northern Mexico and the southwestern United States (Bean and Saubel 1972:38–39, Curtin 1949:78, Uphof 1968:37).

Figure 17.140. *Hierba de manso* (*Anemopsis californica*) in Rosa Flores's garden. *RSF, April 1983.*

Scrophulariaceae—Snapdragon Family

Antirrhinum cyathiferum Benth.
hehe-monlc 'plant curly'
   "curly plant"
desert snapdragon

A very common and widespread ephemeral with tiny lavender snapdragon flowers (Figure 17.141). It appears at any time of year following adequate rainfall and ranges throughout the Seri region, including the islands.

Adornment: To bring good luck, pieces of the plant were put into small cloth bags which were tied on a cord necklace; sometimes the cloth bags were attached to a more elaborate necklace (see Figure 12.5 *A*).

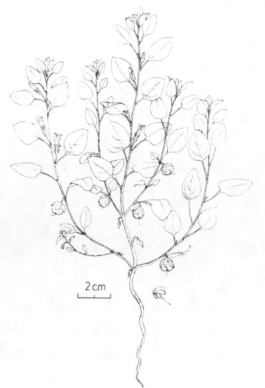

2 cm

Figure 17.141. Desert snapdragon (*Antirrhinum cyathiferum*). FR.

The plant was said to belong to girls and young women. After grinding the fresh, green plant, they mixed white clay with it and added water to make beads which they strung into necklaces. These were worn because of the fragrance.

Antirrhinum kingii S. Wats.
com-aacöl 'com-large(plural)'

A winter-spring ephemeral, the upper stems sometimes twining; seldom common.

Galvezia juncea (Benth.) Ball
noj-oopis caacöl 'hummingbird-what-it-sucks-out (*Justicia*) large(plural)'
"large *chuparosa*"

The same name is also given to *Nicotiana glauca* and *Penstemon* (see *Justicia* for etymology of the Seri name). This herbaceous perennial flowers at various times of the year. The bright red, tubular flowers are visited by hummingbirds. It is usually found at higher elevations and on north-facing rocky slopes.

Penstemon parryi (Gray) Gray
noj-oopis caacöl 'hummingbird-what-it-sucks-out (*Justicia*) large(plural)'
"large *chuparosa*"

This short-lived perennial occurs at higher elevations and on north-facing slopes. The attractive pink flowers appear in spring and are visited by hummingbirds.

Simaroubaceae—Quassia Family

Castela emoryi (Gray) Moran & Felger
   [= *Holacantha e.* Gray]
zazjc caacöl 'Castella-polyandra large(plural)'
   "large *Castella polyandra*"
corona de cristo, crucifixion thorn

Rosa Flores took us to a place in the desert about 15 km northeast of El Desemboque and more than 1 km inland from the road to the very place where there are several of these shrubs. We have not seen it elsewhere in the region. It is a large leafless shrub with stout, thorny branches.

Castela polyandra Moran & Felger
snaazx
zazjc

This low, spreading shrub is essentially leafless. The stiff, thorny branches form briar-like tangles. It is known only from the east side of Tiburón Island, the east side of Baja California, and several adjacent islands (Moran and Felger 1968).
Games: This was one of several kinds of plants carried on a person to insure good luck in betting.

Simmondsiaceae—Jojoba Family

Simmondsia chinensis (Link) Schneid.
pnaacöl
jojoba

Densely branched, many-stemmed, hemispherical shrub, with tardily drought-deciduous foliage of simple, ovate leaves (Figure 17.142). The wood is hard. The fruit is a nutlike, leathery capsule 1.5 to 2.0 cm long and has a single seed, which consists of approximately 50 percent "oil." The seeds are unique because the "oil" is a liquid wax rather than a fat. Common on the mainland and on Tiburón and San Esteban islands.

Adornment: The seeds were strung for necklaces (Figure 17.143).

Food: The seeds could be eaten fresh. However, the Seri regarded it as survival food, and not as something sustaining. Sara Villalobos said, "It's not food, you just eat it." Another person said, "It doesn't hurt you, but it isn't really food." Some said eating jojoba caused diarrhea. We were told of an occasion when some people were pursued by the military and they lived on mashed jojoba seeds mixed with prepared "fruit" of red mangrove (*Rhizophora*). Although edible, jojoba seeds are apparently not digestible.

Food Preparation: Two kinds of meat skewers were fashioned from the wood (see *Prosopis*, Food Preparation).

Hair Care: Shampoo made from the ground fruit continued to be popular in the early 1980s. The greener, young twigs of *Bursera microphylla*, with the bark removed, were sometimes added to the jojoba shampoo.

Ground jojoba seeds were rubbed into the hair to encourage luxuriant growth, and on the following day the hair was shampooed. If a man wanted to grow a beard, he roasted jojoba seeds in ashes. When blackened, the seeds were ground into a mash and rubbed on the chin. It was said to produce a better growth of beard after three days.

Medicine: The seeds were placed in hot ashes, removed before they burned, and crushed on a metate. The oil and blackened portions were applied to sores on the head. Three applications were said to be sufficient.

To relieve eye soreness the seeds were ground, put into a cloth, and the liquid squeezed into the eyes. Raw green seeds were chewed to deaden the pain of a sore throat.

As a remedy for a cold, the seeds were crushed and cooked in water for a short time. The resulting liquid was drunk, causing vomiting which was said to clear the throat. This same liquid was taken near the time of birth to aid in delivery. Alfred Whiting (1957: #50) noted that "the Seri chew the seeds . . . for cold, or pound and steep them and drink the liquid."

Figure 17.142. Jojoba (*Simmondsia chinensis*) from near Pozo Peña. *RSF, April 1983.*

2 cm

Figure 17.143. Necklace of jojoba seeds obtained in El Desemboque around 1957. *CMM.*

Solanaceae—Potato Family

Capsicum annuum L. var. *annuum*
coquée caacöl 'chile large(plural)'
  "large chiles"
chile rojo, red chile

The large chile when fully ripe and red.

coquée coil 'chile green'
  "green chile"

The same large chile as above, but still green (the long, pointed chile, not the bell pepper).

coquée quitajij 'chile pointed(plural)'
  "pointed chiles"

Small, hot chiles.

Capsicum annuum L. var. *aviculare* (Dierb.) D'Arcy & Eschb.
coquée quizil 'chile little(plural)'
  "little chiles"
chiltepín

Perennial, forming a small shrub. These tiny, round chiles are red and very hot.

The Seri knew of wild *chiltepines* in the mountains on the mainland. North of Rancho Guerreros, near Rancho Libertad, *chiltepines* were said to grow up through shrubs or trees.

Food: The fruit was used as seasoning, especially for meat.

Datura discolor Bernh.
hehe camóstim 'plant that-causes-grimacing (from being crazy)'
hehe carócot 'plant that-makes-one-crazy'
toloache, poisonous nightshade, Jimson weed

Use of *Datura* by various Indians for its hallucinatory properties is well known. Although these properties are indicated in the Seri names, it is not known to have been used by them for this purpose (see Kroeber 1931:14).

This species resembles other kinds of poisonous nightshade, or Jimson weed, but differs in having smaller flowers and in being generally less robust. It is an ephemeral. The fruit is covered with spiny projections and it has typical, black, nightshade seeds. It is widespread on the mainland and on Tiburón and San Esteban islands, and is a common weed near farms and ranches.

Medicine: Tea made from the dry, mashed seeds was taken for swollen throat because it deadened the pain. In 1975 Lolita Blanco and Carlota Colosio made tea with the dry, mashed seeds, adding cinnamon, sugar, and desert lavender (*Hyptis*) for treatment of sore throat. A poultice of the leaves was placed on boils. The latter remedy was given to a Seri by a Mexican.

The Supernatural: It was one of the first plants formed, and therefore many people would not touch it.

Lycium spp.

The various species of desert wolfberry (excluding *L. macrodon*) occurring in the Seri region were known as *hahöj*, and the Seri recognized five ethno-species. The green fruit was called *hahöj caxt* 'hahöj tender.' The tuning peg of the Seri violin was often whittled from a green branch of *hahöj* (Bowen & Moser 1970b).

Lycium andersonii Gray
hahöj-enej '*Lycium* empty'
  "empty *Lycium*"
hahöj ináil coopol '*Lycium* its-skin black'
  "black-barked *Lycium*"

The first name refers to the appearance of the inside of the fruit, which the Seri said was "empty," or "hollow." The Seri recognized these as two ethnotaxa of *hahöj*; these were indistinguishable to us, and both appeared to be *L. andersonii*.

It is a thorny shrub with relatively elongated, narrow leaves and flowers. The flowers are considerably smaller than those of *L. brevipes* and *L. fremontii*. The corolla is lavender and the small fruit is bright orange.

Food: The fruit was cooked in water. Eating the fruit was said to produce no adverse effects (see *L. fremontii*). It was not a major food resource.

Lycium brevipes Benth. var. *brevipes*
hahöj an quinelca '*Lycium* inside empty(plural)'
  "empty *Lycium*"

The Seri name refers to the appearance of the inside of the fruit, said to be "empty," or "hollow."

Stiff-branched, thorny shrub, widely distributed in the lowlands through the region, including Tiburón Island. Lavender corolla and fleshy, red fruit. Fruiting occurs at various times of the year, especially during spring.

Adornment: The green berries were strung for necklaces.

Food: The ripe fruit was eaten fresh or cooked in the same manner as that of *L. fremontii*. It was not a major food resource.

Lycium californicum Gray
hahöj-izij 'Lycium little(plural)'
   "small *Lycium*"

"Small" refers to the fruit, which is smaller than that of other *Lycium* species.

Shrub with very stiff, somewhat thick twigs, each ending in a sharp thorn. Common near the shore. The fruit is bitter and not eaten.

Child Care: When a baby cried excessively, the mother often passed it to another person, usually an older sister of the baby. Since the sister received no payment while tending the baby, she was said to be *hahöj-izij cöhaim* "thrown into the *Lycium* bush."

Lycium fremontii Gray
hahöj cacat 'Lycium tart'
   "tart *Lycium*"
frutilla

This desert wolfberry is widespread in the lowlands, and reaches maximum development on the floodplains of major arroyos in the northern part of the Seri region. It occurs on Tiburón Island but is not as abundant there as on the mainland. The corolla is lavender and the flowers are both long pedicelled and the largest of the several *hahöj* (Figure 17.144A). Each shrub commonly produces prodigious quantities of juicy, bright, orange-red berries averaging 12.5 mm long (range = 7 to 18 mm, n = 64, mean = 12.48). The fruit is ripe in March and April (Figure 17.144B).

Facepaint: During the cooking of the berries an orange-red foam develops. This foam was scraped off, placed in a large clamshell, and mixed with honey. When dry and sticky, the mixture was used as a facepaint.

Food: Of all of the various *hahöj*, the fruit of this species was the most extensively used. It was a favorite food, and in the 1980s women and girls continued to harvest substantial quantities of it (Figure 17.145).

The berries were picked by hand, collected in buckets or cans, and brought back to the camp or village. The fruit was also casually picked and eaten fresh in the desert.

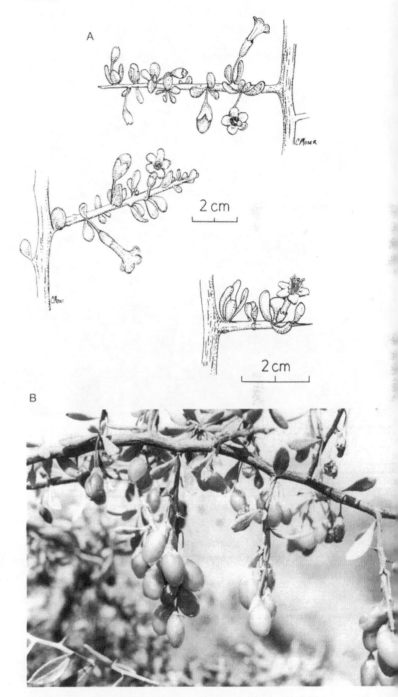

Figure 17.144. Desert wolfberry (*Lycium fremontii*). A) Details of flowering branches from a shrub at Arroyo San Ignacio. *CMM*. B) A branch with ripe berries near the base of *Hast Eemla. RSF, April 1983*.

Figure 17.145. Harvesting desert wolfberries (*Lycium fremontii*) in Arroyo San Ignacio. A) Angelita Torres picking the berries. B) This species produces large quantities of fruit in spring. The berries shown here are the results of about an hour of picking. *RSF, April 1983.*

Prior to the early part of the twentieth century, as the berries were picked, they probably would have been put in a special bowl called *ziix ito* 'thing its-eyes.' This distinctively shaped vessel was a squat, thick-walled bowl that was most commonly used as a container when women gathered berries (Bowen and Moser 1968:122).

The fruit usually was mashed in the hands and thoroughly mixed. A bit of sugar or honey was gener-

ally added, and the mixture cooked in water for about one-half hour. There is considerable variation in the tartness or sweetness of the fruit. Sugar or honey was not added to berries which were sufficiently sweet. Shrubs with very tart fruit were known as *hahöj xoácat* 'Lycium very-tart.' Some women added a bit of wheat flour to the berries, while others added barrel cactus flowers. It was said that if one ate too much of this mixture it could cause a temporary paralysis of the limbs.

In April, 1978, Rosa Flores gathered 756 g of the fresh berries in about forty-five minutes in Arroyo San Ignacio, about 4 km northeast of El Desemboque. She brought the berries home to El Desemboque, cooked them in water, added several tablespoons of honey and allowed the mixture to simmer for several minutes. The berries were then ready to eat.

The berries commonly harbor an insect larva. When shown the larvae, some women remarked, "Those aren't maggots, those are just live things (*ziix ccam*)." Fruit from any of the various species of *hahöj* may contain these larvae.

Hahöj fruit, probably mostly this species, was also collected from pack rat nests. The fruit was found packed together in slabs, which were broken up, cooked in water, and then eaten.

Music: The violin bow *heenj ica ihoom* 'violin with what-lies-on' was sometimes made from this shrub.

Toys: Boys used the slender sticks to fashion toy arrows.

Weapons: The foreshaft of the arrow was often made of this wood.

Lycium macrodon Gray var. *macrodon*
hehe iix coil 'plant its-liquid blue/green' "blue/green-sapped plant"

A thorny shrub, distinguished from other *Lycium* species in the region by its glaucous leaves, greenish corolla, and inedible fruit, which is conspicuously lobed. It is botanically distinct from the other local species of *Lycium*, and the Seri similarly distinguished it. They did not consider it to be a kind of *hahöj*. Common in the floodplain of Arroyo San Ignacio northeast of El Desemboque and elsewhere along major drainageways in the northeastern part of the Seri region.

Adornment: The relatively hard berries, while still green, were sometimes strung as necklaces.

Lycopersicon esculentum Mill.
xtoozp hapéc 'Physalis-crassifolia what-is-planted'
  "cultivated desert ground cherry"
tomate, tomato

The tomato was thought to resemble the fruit of *Physalis crassifolia*.

Nicotiana clevelandii Gray
xeezej islítx 'badger its-inner-ear'
  "badger's inner ear"
Cleveland's tobacco

This winter-spring ephemeral characteristically occurs on the desert floor rather than on rocky slopes (see *N. trigonophylla*). Mainland and Tiburón Island.
    The Seri distinguished it from *N. trigonophylla* by its smaller size and wider leaves.
    Smoking: The leaves were dried and smoked, although *N. trigonophylla* leaves were preferred. Both were apparently prepared in a similar manner. The dried leaves were smoked in pipes made of clay or fashioned from a thin section of *xapij* (reedgrass, *Phragmites*) or a large tube worm shell, *xtozaöj* (*Tripsycha tripsycha*). No other people are known to have smoked this species of tobacco (Volney Jones, personal communication, 1974).

Nicotiana glauca Graham
noj-oopis caacöl 'hummingbird-what-it-sucks-out (*Justicia*) large(plural)'
  "large *chuparosa*"
tree tobacco

The same name was also given to *Galvezia* and *Penstemon* (see *Justicia* for etymology of the Seri name).
    A sparsely branched open shrub or small tree. The tubular yellow flowers are frequented by hummingbirds. In the Seri region it occurs as a weed on farms and ranches and other disturbed sites. Tree tobacco, native to South America, is a widespread weed in western North America.

Nicotiana trigonophylla Dunal
hapis casa 'what-is-smoked putrid'
  "putrid tobacco"
tabaco del coyote, desert tobacco

Perennial or facultatively annual herb, with whitish tubular corolla. Widespread and common, it usually grows among rocks on north-facing slopes and along arroyos. Found on the mainland and Tiburón and San Esteban islands; also common on Alcatraz Island in Kino Bay.
    The Seri distinguished this species from *N. clevelandii* by its smaller and narrower leaves.
    Smoking: Although the leaves of any plant of this species might have been smoked, certain populations with superior smoking qualities were well known. The men made special trips to harvest this tobacco and brought it back to camp in quantity. Most of the leaves were allowed to dry naturally, but a small quantity was often parched, so that it would be immediately available for smoking. For this purpose the leaves were chopped up in a sea turtle carapace, then placed in a small, shallow clay bowl and parched with live coals. When used for this purpose, the bowl was called *hapis ihacáat* 'what-is-smoked(tobacco) to-toast' (Bowen and Moser 1968:122).
    Hapis casa was smoked in clay pipes, or pipes made from a large tube worm shell (*Tripsycha*) or a thin section of *xapij* (reedgrass, *Phragmites*). The leaves were said to smell bad, although they were better for smoking than the galls from creosotebush (*Larrea*). Clay pipes found in middens and old campsites were probably used primarily for smoking this tobacco.
    Special tobacco was found at *Hapis Ihoom* 'what-is-smoked where-it-is' or "where tobacco occurs." The site is near the mouth of a cave in the Cerros Los Mochos, about 20 km northeast of El Desemboque. This tobacco, said to be very potent, was specifically sought after. Plants from *Hapis Ihoom* do not seem to be morphologically distinguishable from other populations of *N. trigonophylla*.
    The San Esteban men collected a tobacco from San Lorenzo Island which seems to be this species. It was called *poora* (from the Spanish for cigar) and apparently had a superior flavor.
    The Supernatural: A Seri version of the story of Lola Casanova and Coyote Iguana relates that some men went to *Hapis Ihoom*, and tells of a magical appearance of *hapis* plants (see *Bursera microphylla*).

Physalis crassifolia Benth.
xtoozp
tomatillo del desierto, desert ground-cherry

There are two varieties in the Seri region: *P. crassi-*

folia var. *infundibularis* I.M. Johnston occurs on San Esteban Island, and var. *crassifolia* occurs on Tiburón Island and the mainland. There seems to be no difference in the fruit.

Annual or perennial herb, commonly found on rocky slopes and gravelly arroyo margins. The calyx grows over the small tomato-like fruit to form a paper lantern-like structure. Flowering and fruiting occur at various times of the year.

Food: The fruit was eaten fresh.

Physalis cf. *pubescens* L.
insáacaj
tomatillo, ground cherry

It was described by the Seri as follows: about 25 cm tall; small leaves like *ziim xat* (*Chenopodium murale*), but thinner; a "soft" plant (no thorns); many fruits on each plant; fruit like *xtoozp* (*Physalis crassifolia*)—but more pointed at the end—light green and fleshy like a grape, sweet, and without seeds; occurring in an arroyo on the south side of Tiburón Island, mostly under trees, and also near ranches.

Food: The fruit was eaten fresh, and said to have the consistency of a tomato, but sweeter.

Solanum hindsianum Benth.
hap itapxén 'mule-deer its-inner-canthus'
   "inner corner of mule-deer's eye"
mariola

A sparsely branched, spiny shrub with showy lavender corolla and yellow stamens. It is common nearly throughout the Gulf region, including the islands.

Medicine: Tea made from the flowers and roots was taken as a remedy for diarrhea.

*Solanum tuberosum L.
xoját hapéc '*Amoreuxia* what-is-planted'
   "cultivated *saiya*"
papa, potato

The common potato was thought to resemble the root of *saiya* (*Amoreuxia*).

Sterculiaceae—Cacao Family

Melochia tomentosa L. var. *tomentosa*
hehe coyóco 'plant dove'
   "dove plant"

Medium-size shrub, slender stems, and smallish pubescent leaves with deeply incised veins. Showy rose-lavender flowers appear at various times of the year, including dry seasons. The Seri said it often blooms when other plants are not in flower.

Dye: The root was occasionally used for a reddish brown dye. It was crushed on a metate and boiled. Basketry splints (see *Jatropha cuneata*) were steeped in the brew until stained the desired shade.

Games: Women used the sticks to roll hoops in the hoop rolling race called *hehe hahójoz cmaam* 'wood what-is-made-to-flee female' (see *Prosopis*).

Song: The following song was sung by the dove plant:

Hehe coox yapxöt hant impáailx
Isoj iháai hisoj conyáanomam
Coox com yapxotaj
Hant haa miifp
Hehe coox yapxotaj
Isoj iháai hisoj conyáanomam
Coox com yapxotaj.

All of the flowers have come,
you prepared me, covering me,
all have flowers.
The new year has arrived,
All plants have flowers.
You prepared me, covering me,
all have flowers.

The Supernatural: Young girls were cautioned not to touch the flowers or they would become promiscuous.

Tamaricaceae—Tamarisk Family

Tamarix aphylla (L.) Karst.
hocö hapéc 'wood what-is-planted'
　"cultivated tree"
pino, tamarisk

The Seri generally applied this name to any large cultivated or planted tree. Any extra-territorial wood was called *hocö*.

Tamarisk thrives in salt air and arid conditions, and provides dense shade (Figure 17.146). The first ones were planted at El Desemboque in about 1960, and within ten years they had developed into large trees. Most of them were planted by the Seri. The usual method of propagation was to cut off a branch, place it in water for a few days and then plant it in the ground. The Seri learned to grow the tree from their Mexican neighbors.

Tamarix ramosissima Ledeb.
salt cedar

Pink-flowered shrub tamarisk is a common farmland weed and has become established at the *Xapij* (Sauzal) water hole on Tiburón Island. The Seri had no name for it.

Figure 17.146. Salt cedar (*Tamarix aphylla*) trees have been planted by the Seri in El Desemboque for shade. *RSF, April 1983.*

Theophrastaceae—Theophrasta Family

Jacquinia pungens Gray
cof
San Juanico

The wind in the tree makes a rustling sound like that of the *coof*, the San Esteban chuckwalla (*Sauromalus varius*) dragging its body through the sand or across the ground.

This small tree has a thick trunk, and dense evergreen foliage of stiff, spine-tipped leaves (Figure 17.147). The orange-red corolla readily separates from the ovary and falls as a unit. The fruit is about 2 to 2.5 cm long, ovoid, and hard shelled, but fleshy or gelatinous inside. Common in arroyos and on plains from the vicinity of Kino Bay southward and on Tiburón Island. Flowering and fruiting occur at various times, with peak production in late spring.

Adornment: The freshly picked corollas were strung as necklaces (Figure 17.148; also see McGee 1898:171). These dried flowers reopen when placed in water, taking on a fresh and newly picked appearance, and may be revived many times. This necklace was called *cof yapxöt* 'cof its-flowering.' McGee

(1898:172) illustrated "nut pendants" which were made from the fruit of this tree.

Food: The fruit was cracked on a metate and the soft inside part eaten. However, if one ate too much it

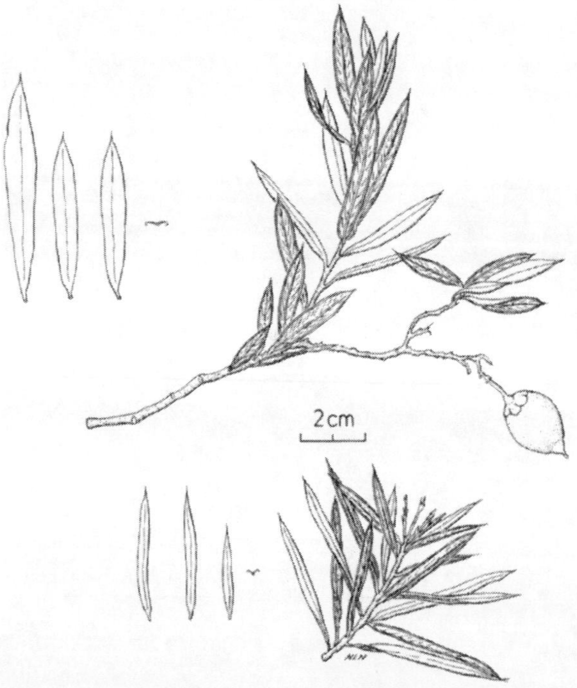

Figure 17.148. Necklace of *San Juanico* flowers (corollas). Made by Carlota Herrera in 1963, it is 70 cm in circumference. *CMM.*

Figure 17.147. Foliage and fruit of *San Juanico* (*Jacquinia pungens*) from near Kino Bay. The younger plants (*below*) have narrower leaves. *NLN.*

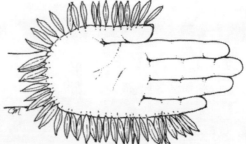

Figure 17.149. Girl's hand outlined with *San Juanico* leaves. Drawn from a photograph taken by MBM in 1975. *CMM.*

was said to cause dizziness and possible intestinal disorder.

Hunting: Great horned owls roosting in San Juanico trees were often killed with bow and arrow during the daytime.

Medicine: To cure dizziness an infusion of the flowers was used to wash the face. This liquid was also used for eardrops to cure an earache.

Play: Girls sometimes stuck the sharp leaf-tip just under the skin until the entire hand was outlined with leaves (Figure 17.149).

An occasion was recalled by an elderly man when a group of people was in the area of a *cof* tree. Breaking off branches with the sharp leaves, they had a free-for-all, slapping each other with branches, all in fun.

Typhaceae—Cattail Family

Typha domingensis Pers.
pat
tule, cattail

This species of cattail grows in shallow water at the *Xapij* water hole (Sauzal) at the south end of Tiburón Island. The Seri also knew of *pat* at the brackish water hole *Hax Cxana* 'water brackish' on Pico Johnson in the Sierra Seri. Both are permanent water places.

Clothing and Headpiece: The diary of one of the men on the ship Examiner in 1894 states that all the people except babies and the elderly wore "clouts of rushes and hats made of the same (Robinson and Flavell 1894:May 25)." They were at Tecomate, at the north end of Tiburón Island. Cattail from *Xapij* would be the most likely candidate, although we have found no other reference or evidence to support this statement.

Facepaint: The dry pollen, *pat yapxöt* "cattail's flower," bright orange-yellow in color, was applied in a narrow line across the face. It served as an accompaniment to a "brighter or stronger" color, such as red or blue.

Play: Boys made toy boats by coiling the long leaves and pinning the ends with thorns.

Ulmaceae—Elm Family

Celtis pallida Torr. subsp. *pallida*
ptaacal
garambullo, desert hackberry

Large, rambling, thorny shrub. The small, fleshy, orange fruit is often abundant in early fall. The wood is hard and flexible. It is common in the lowlands throughout the mainland and on Tiburón Island.

Cradleboard: The wood was used for cradleboard frames.

Food: The fruit was eaten fresh.

Weapons: It was said to be one of the best woods for making a bow, *haacni*.

Verbenaceae—Vervain Family

Aloysia lycioides Cham. var. *schulzae* (Standl.)
   Moldenke
hapsx iti icóocax
   "one who builds up her basket load better"

The name implies that the sticks of this shrub were used to build up the side of a basket load which a woman carried on her head (see Figure 15.24). The term *hapsx* implies "one who does something better," and *icóocax* derives from *coocax* 'build up a load in a basket.'

This slender shrub reaches 3 to 4 m in height and produces large quantities of tiny, whitish, and very fragrant flowers in spring. It grows along the larger drainageways on the inland side of the coastal mountains and in Arroyo San Ignacio in the northern part of the mainland region.

The women and girls esteemed it for its fragrance. In spring they picked bunches of the sweet-smelling flowering branches to take home.

Medicine: Tea made from the herbage (leafy branches) and cinnamon was taken as a remedy for a cold and flu (not stomach flu).

Avicennia germinans (L.) L.
pnaacoj-iscl 'mangrove drab'
  "drab mangrove"
mangle, black mangrove

Pnaacoj is the generic term for mangrove. The foliage is indeed drab as compared with the other mangroves in the Gulf (*Laguncularia* and *Rhizophora*). Black mangrove is a large shrub with dull, grayish foliage (Figure 17.150). It forms thickets in shallow tidal waters of esteros, quiet bays and inlets. It occurs on both sides of the Infiernillo Channel and southward through tropical America. The wood is hard.

Boat Building: The wood was considered excellent for the curved ribs of a boat.

Firewood: Mangrove driftwood was sometimes used as firewood.

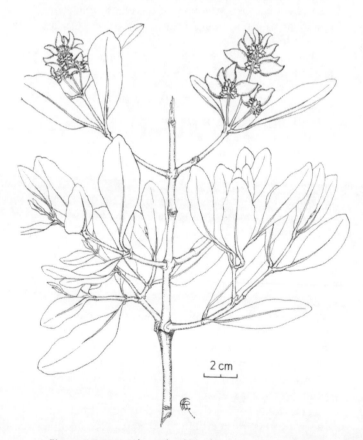

Figure 17.150. A branch of black mangrove (*Avicennia germinans*) from Estero Sargento. Note the fruit and the opposite, decussate branching. *FR.*

Shelter: The leafy branches were used extensively for shading and roofing of the brush house and ramada.

Lantana horrida H.B.K.
hamácj inoloj 'fires their-flames'
  "flames of fire"

It is a spindly shrub or bush, usually growing up through other shrubs. A dozen or more flowers are clustered into a round flower head. The corollas are bright yellow and fade to orange, which probably accounts for the poetic Seri name. The fruit is a cluster of small, round, fleshy drupes which turn black when ripe. It is seldom common and usually grows in larger arroyos and brushy places on north-facing mountain slopes.

Food: The ripe, black drupes ("berries") were eaten fresh.

Medicine: To cure dizziness, the head was washed with an infusion made from the fruit and leaves put in boiling water.

Lippia palmeri S. Wats.
xomcahíift
orégano

This shrub has thin, brittle stems and small leaves with deeply incised veins. The drought-deciduous foliage appears after each soaking rain. Flowering may occur several times per year. It is common in the desert throughout most of the region, including Tiburón and San Esteban islands.

Food: The dried, crushed leaves were used to flavor meat and fish. The flavor was similar to that of commercial oregano, although somewhat stronger. Ideally the leaves were collected after a heavy rain, for then the plant has tender foliage. When the shrub was in flower the leaves were sometimes not collected because they were said to be too strong. If harvested when in flower, then the small cone-like flower spikes were removed because of their bitter taste. It continued to be a much-used spice in the 1980s (Figure 17.151). Large bundles of the leafy branches were often stored on the roof tops.

When out in the desert, the people carried small cloth bags or bundles of the dried leaves mixed with salt, ready for flavoring meat. When drying meat, such as jerky, the crushed leaves could be used as a substitute for salt.

Food Gathering: *Hacozquíif*, the spike of the fruit-gathering pole for *pitaya agria* (*Stenocereus gummosus*) usually was made from wood of *Lippia* or creosotebush (*Larrea*).

Medicine: To cure dizziness, the head was washed with an infusion made from the fruit and leaves put in boiling water. The leaves were used to kill head lice (see *Bursera microphylla*).

Figure 17.151. Branches of *orégano* (*Lippia palmeri*) harvested near El Desemboque. *RSF, April 1983.*

Viscaceae—Mistletoe Family

Phoradendron californicum Nutt.
P. diguetianum Van Tieghem
aaxt
tojí, desert mistletoe

The Seri did not differentiate these species by name. *Phoradendron californicum* often festoons the common desert trees (Figure 17.152). The leafless green stems bear numerous small white or reddish translucent berries. *P. diguetianum* (Figure 17.153) has well-developed leaves and is not as widespread or common as *P. californicum*.

Mistletoe was classified according to the host plant as follows:

tis eaxt "catclaw's mistletoe" (*Acacia greggii*)
ziij eaxt "foothill palo verde's mistletoe" (*Cercidium microphyllum*)
comítin eaxt "ironwood's mistletoe" (*Olneya*)
heejac eaxt "*Pithecellobium confine*'s mistletoe"
haas eaxt "mesquite's mistletoe" (*Prosopis*)
cozi eaxt "*Condalia*'s mistletoe"
haaca eaxt "*Zizyphus*'s mistletoe"
haaxat eaxt "creosotebush's mistletoe" (*Larrea*)

Food: When growing on certain legumes (catclaw, ironwood, or mesquite) the fruit of *P. californicum* was eaten. However, it was said to be bitter when growing on other hosts, e.g., palo verde and condalia. The fruit was considered ripe and harvestable when it

Figure 17.152. Desert mistletoe (*Phoradendron californicum*) parasitic in a mesquite tree along Arroyo San Ignacio. *RSF, April 1983.*

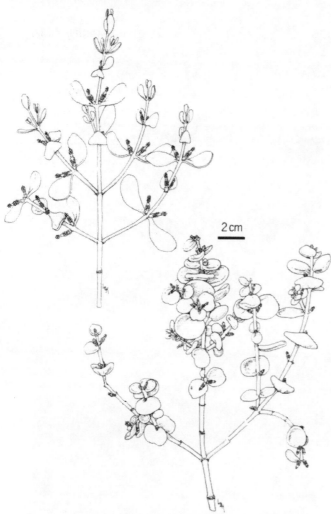

Figure 17.153. *Phoradendron diguetianum*. This mistletoe was parasitic in a *mangle dulce* (*Maytenus phyllanthoides*) along the Infiernillo Coast. *LBH*.

turned "glassy" (translucent). Women who went on special collecting trips for it were called *aaxt caaizi* 'mistletoe doers.' A common method of harvesting the fruit was to knock it off the tree onto a blanket. The berries were often fed to young children.

Alfred Whiting (1957: #10) recorded that "The white fruits are said to be sweet and are eaten by the Seri."

Medicine: Mistletoe tea was taken as a remedy for illness "inside the body." Tea made from the stems and leaves of mistletoe (*P. diguetianum*) growing on creosotebush was taken to cure diarrhea and to help a woman during childbirth.

Oral Tradition: Sometime in the mid–nineteenth century a group of Seri men went on a killing spree, involving both Seri and non-Seri victims. They then fled and lived in a cave on a mountain north of Hermosillo. They sometimes returned to raid their own people on the coast. On one such trip one of the men kidnapped a Seri girl, perhaps twelve or thirteen years old. He took her to the cave and tied her up when he went out. One of the other men released her, gave her water, and told her to go home. She made it back to the coast—a distance of some 100 km. She told her people that at night she was afraid to sleep even once on the ground or in a cave. Each day as evening drew near she looked for a large patch of mistletoe (it would have been *P. californicum*). She climbed up into it, worked her way into the mistletoe, and slept in it.

Zosteraceae—Eelgrass Family

Zostera marina L.
hatáam 'what is harvested'
eaz
zacate del mar, *trigo del mar*, eelgrass, wrack

While growing on the sea floor, eelgrass was called *hatáam*. The root of this word denotes harvesting or collecting long masses of plants—such as eelgrass or hanging branches—but not fruit or flowers. After the plant dislodges or washes ashore, mostly in spring, it was called *eaz*. The hard grain-like fruit (utricle) or seed was *xnois*. Comparable terms for seagrasses do not exist in English or Spanish. The following discussion of eelgrass is revised from three previous reports.*

Eelgrass occurs along the Atlantic and Pacific coasts of North America and Eurasia. In the Gulf of California it is known from El Desemboque southward to Sinaloa, and at various places off the Baja California peninsula. Eelgrass is usually perennial,

*Revised from R. S. Felger and M. B. Moser, "Eelgrass (*Zostera marina* L.) in the Gulf of California: discovery of its nutritional value by the Seri Indians," Science (1973) 131:355–356; T. E. Sheridan and R. S. Felger, "Indian utilization of eelgrass (*Zostera marina* L.) in northwestern Mexico: the Spanish colonial Record," The Kiva (1977) 43:89–92; R. S. Felger, M. B. Moser, and E. W. Moser, "Seagrasses in Seri Indian culture," pages 260–276, In R. C. Phillips and C. P. McRoy, eds., Handbook of seagrass biology: an ecosystem perspective (Garland STPM Press, 1980). Used with permission.

but in the Gulf of California annual eelgrass is prominent. It has a slender rootstock and long, thin leafy stems (Figure 17.154). The ribbon-like leaves are about 0.5 to 1.5 m long in deeper water, often much shorter in shallow water, and are 3.2 to 5.0 mm wide (Figure 17.155). The flask-shaped utricle, or fruit, bears a single seed about 1 × 3 mm.

In Sonora eelgrass generally grows at about 0.5 to 3 m below mean low tide. The long, pale green, ribbon-like shoots form undersea meadows in protected waters, from the vicinity of Kino Bay northward throughout the Infiernillo Channel. It does not occur off the outer or west coast of Tiburón Island nor off San Esteban Island. Eelgrass was an important component in the food web of the Gulf of California (see Figure 2.3).

Eelgrass in the Seri region is primarily a winter-spring plant. In October and early November, as the water temperatures begin to drop, vigorous and rapid-growing rhizomes and shoots begin to develop. In a matter of weeks these develop into vertical, ultimately reproductive (seed-bearing) shoots. Rapid growth continues through the mild winter. As water temperatures rise in late February and March, 100 percent of the eelgrass shoots become reproductive. Massive and essentially simultaneous ripening of the fruit occurs in the latter part of April and early May. Reproductive shoots, with ripe or rapidly ripening

Figure 17.154. A single mature shoot of eelgrass (*Zostera marina*) from shallow water in the vicinity of Campo Víboras, April 1974. The four lower spathes are pistillate (seed bearing) and the two upper ones are staminate. *NLN*.

1cm

Figure 17.155. Eelgrass along the Infiernillo Channel. A) Plants in shallow water at extreme low tide in Estero Santa Rosa. B) Plants in the beach drift. *RSF, April 1983.*

fruit, break off and float, often forming extensive rafts floating on the surface of the sea. Much of this floating eelgrass washes ashore in great quantity, accumulating in the beach drift (see Figure 2.2) where it was readily gathered by the Seri.

Eelgrass occupied a prominent position in Seri culture. The moon, or month, corresponding approximately to April was *xnois iháat iizax* "moon of the eelgrass harvest." A bird called *xnois cacáaso* "the foreteller of eelgrass seed," apparently the black brant, dives to feed on the plant. When seen diving in spring, this waterfowl was said to foretell the season or time of the eelgrass harvest. A small rock islet, Marito de Turner, off the southeast shore of Turners Island, was named *hast xnois*, "eelgrass-seed rock." *Hant xnois* 'land eelgrass-seed' is the term for trash, because when *xnois* is harvested it has shells, pieces of seaweed, sticks, and other unwanted objects in it—i.e. ". . . it is cluttered up."

The Seri had detailed knowledge of the distribution and ecology of eelgrass within their territory. They knew the precise locations of the eelgrass beds in Kino Bay and the Infiernillo Channel. Eelgrass is especially abundant near Campo Víboras on the mainland side of the Infiernillo Channel.

The bay at El Desemboque was said to have had an extensive growth of eelgrass which had disappeared by the 1940s. In spring 1973 several Seri pointed out where it had begun to reappear, but by 1983 it had not become common there. This site is relatively exposed, and the disappearance of the plant there may have been due to storm activity or other physical conditions. The Seri had no knowledge of any other extensive die-off of eelgrass.

A multiple burial discovered in 1974 on the mainland shore of the Infiernillo Channel contained eelgrass. Characteristic traditional Seri pottery, known as Tiburón Plain or "eggshell" ware (see Bowen 1976), was associated with the burial. Eelgrass from this site was dated by radiocarbon as being nearly two thousand years old (1940 ± 130 years B.P., University of Arizona #A-1640). This is probably a maximum age due to the marine effects, and the actual date might be a few hundred years younger. The burial consisted of two adults and one juvenile or child (Richard S. White, personal communication, 1974). There was evidence that the bodies were interred beneath layers of eelgrass. The eelgrass was well preserved, but no fruit, seeds, or flowering structures were evident among the several handfuls of it recovered from the site. This condition indicates that the burials probably took place in early spring (perhaps late February or early March), before reproductive structures would be evident, but late enough in the year that eelgrass was abundant in the beach drift.

Animal Food: The Seri knew that eelgrass was a favorite food of the green turtle and that it was extensively grazed upon by these animals. According to the Seri, turtles feeding on eelgrass had sweet, good-tasting meat, whereas those from the west coast of Tiburón Island—which ate *xpanáams*, or marine algae—were *cheemt* 'stinky.' Eelgrass was often found in the stomach of the green turtle.

Food: The seed of eelgrass was one of the most important foods of the Seri and was still occasionally harvested in the early 1980s. Eelgrass is the only marine seed plant in the Gulf of California which can be utilized as a major food resource. Earlier observers recognized its use among the Seri, but they either mistook it for *Ruppia* (Dawson 1944b:134) or seaweed or they did not botanically identify it (Kroeber 1931:11–12; McGee 1898:207; Quinn and Quinn 1965:164). Seri use of a "grain," or seed, from the sea is unique in the world.

Spanish colonial documents verify that a marine seed plant was an important component of the Seri diet in the seventeenth and eighteenth centuries. In August of 1729, Padre Nicolás de Perera of the Society of Jesus visited the mainland coast opposite Tiburón Island. He wrote that the Seri from the island told him it was

> . . . *sumamente estéril sin a ver en toda ella ni pedaso de tierra en que poder sembrar, y que solo ay algún sacate entre las peñas con que se mantienen tal qual bura, liebres, y ardillas, que es toda la cassa que ellos tienen para mantenerse juntando con ella una semilla que sacan de el mar. A tiempos padesen mucha hambre, y se ten obligados para que de el todo no les falta de el manteniento a mantenerse cada 24 horas con un puñado de la dha semilla lo qual savía yo aun antes de pasar a el mar* (Perera 1729).

> . . . extremely sterile with no hint of even a small piece of land that could be planted, and with only a little grass between cliffs which supports a few mule deer, jackrabbits, and squirrels; this is all the game they have to sustain themselves, together with a seed they gather from the sea. At times they suffer much hunger, and are obliged for their daily subsistence to exist on only a handful of this same seed, and this I knew about even before I arrived on the coast.

Padre Perera's brief but enlightening description indicates that this "seed from the sea" was one of the more significant food resources of the Seri. Furthermore, his account implies that the Seri subsisted on eelgrass seed when other foods were scarce. His comment that he knew about Seri use of eelgrass seed even before he journeyed to the coast indicates that exploitation of eelgrass was not an esoteric or sporadic practice. Instead, it was important enough to be noted by Jesuit missionaries as a major wild harvest even before they penetrated Seri territory.

Nearly a century before Perera's reconnaissance, Padre Andrés Pérez de Ribas, in his classic *Triunfos de Nuestra Santa Fe*, written in 1645, stated that Indians living along the coast of Sinaloa, which then incorporated Sonora, "*cogen una semilleja de hierba, que nace debajo del agua en el mar, que también les sirve de pan* . . . gather a small seed of a grass, which grows under the water in the sea, which also serves them for bread. (Pérez de Ribas 1944, I: 128)."

Pérez de Ribas was a precise and graceful writer, known for his careful choice of words (Ernest J. Burrus, S.J., personal communication). His use of the word *semilleja* reinforces the identification of this plant as eelgrass, since the seeds are about 3 mm in length. The phrase *les sirve de pan* does not necessarily imply that eelgrass was actually prepared as bread. It does suggest, however, that it was as important to the Indians as bread was to Europeans. Since Pérez de Ribas was among the first Europeans to colonize this region, his writings imply that eelgrass had been exploited since pre-Hispanic times.

Neither Pérez de Ribas nor his contemporaries had ever visited the Seri in the vicinity of Tiburón Island. Nevertheless, they may have been in contact with Spanish pearl fishermen familiar with the central Sonora coast or with occasional Seri from Tiburón, who came to trade with their more settled neighbors, the Yaqui, Opata, and Lower Pima Indians. Furthermore, it is possible that the use of eelgrass seed was not restricted to the Seri groups on or around Tiburón Island. Eelgrass meadows may exist in sufficient quantity for human exploitation as far south as the present-day border between Sonora and Sinaloa. Thus, the Guaymas—a southern Seri group missionized and reduced to village life in the late 1620s—might have harvested eelgrass seeds. It is also possible that coastal Yaqui and Mayo groups also exploited eelgrass.

The ripe fruit, *xnois*, was harvested in April or early May by both men and women. It was usually not harvested until the plants were floating loose in great masses close to the shore. At such times the people waded into the water, often to their chests, to pull in the long strands hand over hand. A person who harvested *eaz*, the floating plant, was called *capóee*. The weather was warm and pleasant. It was a time of happiness, with much shouting and laughter, and "everyone got wet." We have also observed fresh beach drift being gathered for harvest.

One who harvested *hatáam*, the growing plant, was called *cotáam*. *Hatáam* was generally harvested only when one was "in a hurry" to eat the seeds. However, *hatáam* was commonly harvested at *Cyazin*, a camp at Punta Perla on the northeast side of Tiburón Island. There it was possible to walk out at low tide into the shallow water and pull eelgrass loose from the sand.

On April 13, 1973, we were present when the fruit was just beginning to ripen. We were told that the eelgrass could not be successfully harvested just then because the south wind that was blowing was a damp wind that did not allow the seed-bearing strands to dry.

After the eelgrass was brought ashore, it was spread out along the beach to dry, substantially above the high tide line (Figure 17.156). From this point on the work was done by women.

After several days of favorable dry winds and after the morning dampness was gone, the job of separating the grain-like fruit began. The debris and marine macro-algae were picked out and thrown away. Freeing the grain-like fruit from the debris was called *caaptx*.

Women sat with large bunches of eelgrass placed before them on deer skins, or later on canvases, cloth, or large sheets of plastic. The dry masses of eelgrass were threshed with wooden clubs usually made from ocotillo (*Fouquieria splendens*) or mesquite (*Prosopis*). The fruit fell to the deer skin or canvas, from which it was collected. Bunches of eelgrass strands were then rolled between the palms to loosen any remaining fruit. Children often sat with their mothers and helped with the processing.

The fruit was placed in a shallow basket. It was then winnowed to remove the relatively large amount of remaining leaves, twigs, and debris (Figure 17.157). This type of winnowing technique was called *coospx*.

The fruit was toasted (parched) or left natural. The toasted and ground fruit, *xnois hapáha* "*xnois* that was ground up," was much preferred over the natural, untoasted fruit called *xnois hapánal* "fuzzy *xnois*." It was toasted in an open-mouthed pottery vessel, large sherd, or metal pan. The toasted grain

Figure 17.156. Aurora Colosio spreading out eelgrass gathered from beach drift to dry above the high tide line at Punta Chueca. *RSF, April 1974.*

Figure 17.157. Ramona Casanova using two methods to winnow eelgrass "grain" at El Desemboque, April 1972. *RSF.*

was then poured into a basket or turtle shell and pounded to break open the hard fruit, or utricle. The chaff was separated from the seed by a second *coospx* winnowing, also done by tossing the seed and chaff into the air. Finally the seeds were ground on a metate with a mano. The resulting flour was placed in a basket. The basket was tapped on a stick to bring the remaining chaff to the surface at the edge of the basket from where it was allowed to spill onto the ground, leaving the flour in the basket.

The flour was cooked in water and made into either a thick or thin gruel (*atole*). Since it was relatively bland, it was often eaten with other foods in nutritionally significant combinations. The preferred manner was to prepare it with sea turtle oil. The turtle fat was generally added after the gruel boiled and thickened. Eelgrass flour was also sometimes mixed with turtle oil and made into small dough balls which were added to a pot of cooking eelgrass gruel. The floating and bobbing of the dough balls in the gruel was called *hamápöjquim* 'what is floated.'

Xnois coinim 'eelgrass-seeds that-is-mixed' was made by grinding *cardón* (*Pachycereus*) seeds with eelgrass seeds. While preparing this mixture, the women enjoyed repeating a saying with a moral similar to "The pot calls the kettle black." The saying involves *cardón* seeds, which are shiny and black, and eelgrass seeds, which are dull in color. The *cardón* and eelgrass seeds are arguing:

He hin ntcmáinim hanso intóoscl hant zo
   commóii.
Me mos he hin ntcmáinim hanso intóopol hant zo
   commóii.

You weren't mixed with me,
   you're just so drab there where you are.
You, also, you weren't mixed with me,
   you're just so black there where you are.

Cardón seeds are rich in oil and were said to add a good flavor to the eelgrass. Eelgrass gruel was also sometimes flavored with honey. *Xnois* was often stored in sealed pottery vessels to be eaten at a later time, such as during the late summer-fall rainy period.

Although the fleshy roots and leaf bases of eelgrass have been eaten by various, distant peoples, such as in the Pacific Northwest (Turner and Bell 1973:274), there is no evidence for such use among the Seri. The Seri are the only people known for certain to have used the seed of a sea plant as a major food resource.

Two questions arise. Why have no other people used eelgrass seed? How might people learn to use a unique or "new" food?

In addressing the first question, it must be kept in mind that negative ethnological information is seldom conclusive. Much information was lost as native North American peoples were acculturated or vanquished. However, several other problems enter into this question. There is clinal increase in the percentage of reproductive shoots, or seed productivity, among eelgrass populations from north to south along the west coast of North America, with lowest values near the Arctic Circle in Alaska. The unique feature of the eelgrass in the Gulf of California is that 100 percent of the shoots become reproductive each year. Thus, it is not surprising that the Seri exploited this high seed production (Felger and McRoy 1975).

Another factor which might limit potential harvesting of eelgrass seed is the need to dry the plant quickly and early in the processing. For this reason the Seri did not harvest it when the moist south wind was blowing. Thus, high seed productivity and dry, warm, desert conditions seem to be needed.

Without direct evidence we can only speculate on the question of how people learned to exploit a new or unique resource. Nevertheless, there are some striking similarities between eelgrass and another major crop from the Gulf of California—Palmer saltgrass (*Distichlis palmeri*)—which may shed some light on the problem. Some Seri had opportunity to be in contact with the Cocopa, who lived in the delta region of the Colorado River and harvested Palmer saltgrass as one of their major food staples. Palmer saltgrass and eelgrass represent the only well-documented cases of seed plants growing with ocean water that have been used as major food resources (Castetter and Bell 1951; Felger and Nabhan 1978: 143).

Many of the procedures for the harvest and preparation of eelgrass and Palmer saltgrass are comparable. For example, both are harvested at the same time of year and under similar ecological conditions, and the growth-form and product are in many ways similar. Perhaps the Seri or their ancestors learned to harvest eelgrass after having been introduced to the harvest of Palmer saltgrass. Another factor to consider is that certain species, such as Palmer saltgrass or eelgrass, may have been more widely or differently distributed in earlier centuries or millennia. Furthermore, Palmer saltgrass is similar to commonly harvested terrestrial grasses.

Food Preparation: Since eelgrass was regarded as a "soft" plant and was readily available during winter and spring, it was often used to line a basket or sea turtle carapace to provide a clean bed on which meat was placed.

Hunting: Since eelgrass was a favorite food of the green turtle, hunters often sought their prey in the eelgrass meadows. Large sea turtles were known to hide by swimming into eelgrass thickets. When one was seen swimming into the eelgrass, a turtle hunter would stop his boat over the general area and stomp hard on the prow seat on which he was standing. This often caused the turtle to seek another clump of eelgrass and the hunter could pursue and try to harpoon it.

Figure 17.158. This eelgrass doll was made by Ramona Casanova at El Desemboque in April 1972. It is 31 cm long and bound with green and orange cloth. *HT, 1972; ASM-35448.*

Localized eelgrass meadows in the Infiernillo Channel were well known to the turtle hunters, and were readily located by them at sea, day or night. Different eelgrass beds, or areas, of the channel were named and could be located by aligning the boat with mountains and other topographic features on the mainland and Tiburón Island. More precise location of a given eelgrass stand, if not evident from the surface, was accomplished by placing one's ear against the end of a wooden paddle pushed into the water, blade first. Listening "through" the paddle, the hunter could hear the swish of the eelgrass as it moved in the current.

Medicine: A child suffering from diarrhea was said to recover if given gruel made from eelgrass flour.

Play: A deer or bighorn sheep scrotum was stuffed with dry eelgrass to make a ball for children to play with. Dolls were fashioned from small bundles of eelgrass (*eaz*) bound with strips of cloth (Figure 17.158). Like most Seri dolls and figures, it was faceless and of haunting simplicity. In the 1980s girls still played with these dolls, which they usually made for themselves.

Pottery Sealant: A scum that formed after boiling *xnois* flour was skimmed off and rubbed into the exterior of an olla in order to seal it for use as a water-carrying vessel.

Shelter: Eelgrass was spread over the frame of a brush house for shade and roofing. Later some temporary brush houses, built along beach dunes in winter, were covered with polyethylene sheeting, which in turn was covered with eelgrass. Pillows were made with cloth bags stuffed with eelgrass.

Zygophyllaceae—Creosotebush Family

Fagonia californica Benth.
F. pachyacantha Rydb.
xtamóosn-oohit 'desert-tortoise what-it-eats'
"what desert tortoises eat"

Fagonia californica and *F. pachyacantha* are sympatric, biologically different species, although the Seri did not recognize them as being distinct.

These low, spreading, perennial herbs have small, unpleasant spines and tri-foliate leaves. *Fagonia* occurs on dry rocky slopes and the desert floor of the mainland and Tiburón Island.

Facepaint: A yellow pigment was obtained by grinding the stem tips (probably with the leaves) with a liquid prepared by boiling the flowers of the *siml*, barrel cactus (*Ferocactus wislizenii*).

A bright green, or chartreuse, facepaint was obtained by crushing the plant on a metate, squeezing the juice through a white cloth into a large clamshell (e.g., *Laevicardium elatum*), and allowing it to dry. To use it, a tiny bit of water was added. When mixed with a small amount of sugar, it became brighter green.

Guaiacum coulteri Gray
mocni
guayacán

This large, woody shrub or small tree has very hard wood, and bright green, pinnate leaves. In May and June, at the height of the dry season, it produces spectacular masses of deep violet-blue flowers. The fruit ripens in summer and fall. *Mocni* grows in the interior of Tiburón Island and on the mainland coast from near Tastiota and southward; inland it ranges north of Hermosillo. This species extends southward to Oaxaca. *G. sanctum* and *G. officinale*, the lignum-vitae of commerce, are tropical relatives well known for their extremely hard wood and medicinal properties.

Hair Care: The roots were mashed and moistened, and then used as shampoo.

Medicine: The fruit was crushed and cooked in water, and the tea was taken as a cure for dysentery. In curing the sick, a shaman sometimes used Seri Blue for painting crosses on the patient's chest.

Pigment: A blue pigment, Seri Blue, was made with *guayacán* resin and several other ingredients, none of which is blue in its unaltered state. This blue pigment or paint was an important constituent in Seri aesthetic expression. It was used for facepaint, for coloring clay necklace beads, and for decorating pottery, clay figurines, fetishes, headdresses, arrows, wrist guards, game sticks, and a wide range of other objects. In fact, it was used for almost any decorated object other than basketry. The following discussion is modified from a previous report.*

In 1692 Gilg mentioned that the "cheeks are

---

*Modified from M. Moser, "Seri blue," The Kiva (1964) 30(2): 27–32. Used with permission.

painted blue (Di Peso and Matson 1965:53)," and many others have subsequently recorded blue as a prominent color in Seri facepainting. McGee (1898: 165–166) and Kroeber (1931:27) reported the mineral dumortierite as the source of the blue pigment used in Seri facepaint. The Coolidges (1939:200) described two blues—one a mixture of gypsum and a boiled root, the other a blue stone found on Tiburón Island. Xavier (1946) mentioned dumortierite and commercial bluing. She also referred to a third blue obtained from a plant on Tiburón Island. The Seri said that the only blues they ever used were commercial bluing (*añil*), called *icahóil* 'to cause to be blue,' and the highly prized *hantézj coil* 'clay blue' of their own manufacture, known as Seri Blue.

The components and the method of manufacture of *hantézj coil* were said to have been taught to the women long ago by *Hant Iha Quimx* "he who tells what there is on the land" (see Chapter 7). The actual color transformation was veiled in mystery to the Seri, who said that if the process of making the pigment was observed by a girl who had not passed the menarche, the blue would not appear. Appearance of the blue color, however, was not attributed to shamanistic power.

Seri Blue, *hantézj coil*, began to be in less demand around the middle of the twentieth century because commercial bluing became readily available. Nevertheless, *hantézj coil* was the blue preferred by those women who continued to practice facepainting.

The three ingredients used in making Seri Blue are the root bark of *xcoctz* (*Ambrosia dumosa*/spp.), *iqui icóizj* 'with what-one-grinds-with' (a mixed clay in which montmorillonitic layers predominate), and *mocn-ooxö* 'mocni excrement' (*guayacán* resin). The Seri said that the clay and *mocni* (*guayacán*) occur only on Tiburón Island. The root bark of canyon ragweed (*Ambrosia ambrosioides*) or *Stegnosperma* was sometimes used instead of the root bark of bur-sage (*Ambrosia dumosa*/spp.). Canyon ragweed root was said to impart a distinct aroma to the pigment. Pierce (1964) analyzed Seri Blue and determined that the essential ingredients were the inorganic clay, *guayacán* resin, and water.

Although the woman involved normally made the Seri Blue in complete privacy, on two occasions María Antonia Colosio agreed to let Mary Beck Moser observe the process. On the first occasion seven young men and women were near a fire outside María's house as she seated herself on the ground behind a water drum. As she began to work she cautioned them not to watch. (The restriction as stated by several Seri women pertained only to girls below the age of menarche. However, the young men and women, most of them in their twenties, were still not allowed to observe.) While on Tiburón Island in 1951, Alfred F. Whiting (1951:12) noted that a woman "has been preparing blue face paint, all very secret and hush hush."

María commenced by scraping the roots of several *xcoctz* (*Ambrosia dumosa*) plants to remove any soil and then moistened them with water. Using a mano and metate, she split the bark and removed it. The rest of the root was discarded. She then pounded the bark into a pulp, moistening it at the same time with more water. This pulp was put in a cup and set to one side.

Next she ground several small chunks of the clay, *iqui icóizj*, into a fine powder (Figure 17.159). Then she placed a piece of hard *guayacán* resin, *mocn-ooxö* 'mocni excrement,' the size of a marble, on the white powder. At this point she told her daughter, one of the two young women outside, to bring her a clean white rag and a blanket. After again warning the young people to stay away, she covered herself and her metate with the blanket, leaving only a small opening by her shoulder through which Moser could observe. The warning was heeded by all the youths

Figure 17.159. María Antonia Colosio grinding the white clay during the preparation of Seri Blue. *MBM, El Desemboque, 1963.*

except the daughter, who came over once to take a peek. The group remained interested, and several times someone asked, "Is it blue yet?"

Beneath the blanket, María ground the dry resin into the clay, thus mixing the powders thoroughly. After scraping the mixture to one side of the metate, she wrapped the bur-sage bark pulp in the white cloth and squeezed approximately a teaspoonful of liquid from it onto the center of the stone. Then she combined some of the powder with the liquid and ground the resultant paste for about ten minutes. As necessary, she added more powder or liquid to the mixture. Finally she daubed a streak of the mixture onto the back of her hand, put her hand into the sunlight, and remarked that it was beginning to turn blue. At that time it was grayish-blue. She continued to work for several more minutes and then tested the color again. This time she seemed satisfied that it was the right shade. It was now a bright, or as she put it, a *caixaj* "strong" blue. This was the *hantézj coil*, the prized Seri Blue pigment.

The making of the *hantézj coil* was followed immediately by the making of a second portion of clay mixture which is a lighter shade of blue, and is thus called *hahóoscl*, the root meaning of which is "drab" or "grayish." *Hahóoscl* was considered to be inferior to the *hantézj coil*. After María had scraped the finished *hantézj coil* from the metate and mano and transferred it to a saucer, she added more ground *iqui icóizj* clay to the small amount of remaining clay-resin powder mixture and combined this with the blue clay mixture remains on the metate. She then squeezed a few more drops of liquid from the cloth-covered bur-sage bark pulp onto the powder and resumed the grinding and mixing. After working several minutes she scraped up the second batch of clay mixture and placed it opposite the *hantézj coil* on the saucer. Only then did she drop the blanket and display the results of her work. The pastes were soft and sticky, and the difference in the shades of blue was noticeable. She said that on the following day she would form the pastes into the proper shapes.

The next day she kneaded and molded each of the two paste masses, which by that time had dried enough to be workable. She molded the *hantézj coil* into two capsule-shaped pieces, each about 2.5 cm long and tapered at both ends. She molded the *hahóoscl* into a disk about 1 cm thick and almost 2.5 cm in diameter. The different shape permitted it to be distinguished from the *hantézj coil*. Since both blues were made at a single sitting, the process was spoken

of only as "making *hantézj coil*." The blue "pencil" described and illustrated by McGee (1898:166 and facing 170) was apparently the *hantézj coil*.

To use the pellets of Seri Blue as paint, a bit of water was placed on a stone and the pellet was rubbed into it to make a paste.

The Supernatural: Seri Blue and powdered *palo zorrillo* wood was used for seeking good luck (see *Atamisquea*).

Kallstroemia grandiflora Torr.
hast ipénim 'rock what-it-is-splattered with' "splattered against rock"

Summer-fall ephemeral with showy orange flowers.

Larrea divaricata Cav. subsp. *tridentata* (DC.) Felger & Lowe
haaxat
"with smoke or steam"
gobernadora, hediondilla, creosotebush

The name *haaxat* may be related to the term *caaxat* 'with smoke or steam.' Creosotebush (Figure 17.160) is one of the most abundant shrubs in the deserts of North and South America. It ranges over most of the lowland deserts of western Sonora, including Tiburón Island. It is absent from San Esteban Island.

This multiple-stem shrub has very hard wood, although the stems seldom exceed 5 cm in diameter. Bright yellow-petaled flowers and the gummy foliage develop in response to rainfall at any time of the year. The resinous foliage produces a strong odor which is particularly evident after a rain. Its characteristic odor is reflected in the Spanish and English names. The branches are sometimes encrusted with *csipx*, a reddish brown lac from the excretions of a scale insect (*Tachardiella larreae*; Figure 17.161). Lac-bearing plants are particularly abundant in the Agua Dulce Valley on Tiburón Island and between Pozo Coyote and Puerto Libertad.

One evening shortly before sunset in late May, Carlota Comito said, "There will be dew tonight. I can smell the *haaxat*."

Adhesives and Sealants: Pieces of lac, *cspix*, were picked off the branches and placed in a sea turtle shell or other suitable receptacle. A hot coal was held near them, melting them into one piece, which was then formed into a rounded glob for carrying, called *csipx hapétij* 'lac circular-flat.' It could be kept for future use and melted when needed. Men carried a supply of *csipx* in their arrow quivers.

Creosotebush lac was used extensively as an all purpose glue and sealant. The lac is plastic when heated but hardens again on cooling, forming a strong bond akin to commercial sealing wax (Euler and Jones 1956; Bohrer 1962). It was used for hafting harpoon and arrow points, repairing holes and cracks in pottery vessels, and sealing clamshell, pottery, or stone lids on storage vessels. The lid was easily removed by severing the hard, dry lac with a heated knife blade.

Csipx was extensively used to repair ironwood carvings and to fill cracks, small holes, and crevices in the sculpture (see *Olneya*). In 1969 one man used the lac to build up the cylinder head to meet the gasket of his outboard motor.

The leafy branches were used to seal pottery vessels (see *Carnegiea*).

Boats: Creosotebush wood made into spikes called *icócaöj*, were stuck into the top sides of the balsa for fastening cargo (see *Phragmites*)

Containers: Boxes of soft wood, made from red elephant tree (*Bursera hindsiana*), were held together by pieces of creosotebush wood used as nails.

Fishing: The serrated points for both single- and double-pronged fish harpoons were made of creosotebush or catclaw (*Acacia greggii*). The point of the turtle harpoon, almost always made of catclaw before metal was available, was apparently sometimes made of creosotebush.

While harpooning fish in waist-deep water, a fisherman strung the fish on a line that used a creosotebush stick as the point of the stringer (see *Prosopis*, Cordage). The creosotebush stick was used because it does not soften in sea water.

Food Gathering: *Hactáapa*, the fruit-gathering pole for organ pipe (*Stenocereus thurberi*) fruit, had a tip called *hactáapa iti ihíip*, which used to be made of a sharpened spike of creosotebush wood. Later it was made of metal. *Hacozquíif*, the fruit-gathering pole for *pitaya agria* (*Stenocereus gummosus*), had a transverse hardwood spike, often made from creosotebush or *Lippia*.

Food Preparation: *Hamquée*, a stick used for removing meat from a cooking pot, was made of a short creosotebush branch. The two kinds of meat skewers were made from creosotebush (see *Prosopis*, Food Preparation). The branches were used to brush glochids and spines off the fruit of cholla and pricklypear (*Opuntia* spp.) before they were peeled and eaten.

Footwear: The people sometimes wrapped their feet with pads of leafy branches of creosotebush and tied them on with balloon vines (*Cardiospermum*). This practice was improvised if one was out in the desert without sandals in the summer, since the ground was often too hot to walk on barefoot.

Figure 17.160. Creosotebush (*Larrea divaricata*). A) Angelita Torres picking branches at Arroyo San Ignacio for medicinal use. B) Closeup of a branch with flowers and buds. *RSF, March 1983.*

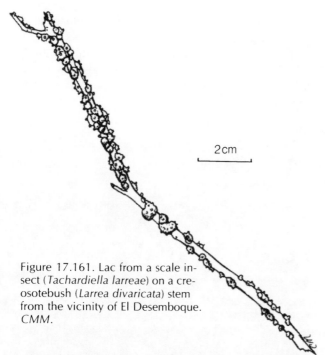

Figure 17.161. Lac from a scale insect (*Tachardiella larreae*) on a creosotebush (*Larrea divaricata*) stem from the vicinity of El Desemboque. *CMM.*

2cm

Medicine: Creosotebush was still much used for medicinal purposes in the 1980s. The leafy branches were often tied in a bundle and stored in the house until needed. Leafy branches heated in fire were used as a poultice to reduce swelling or to relieve abdominal pain after parturition. For aching feet, leafy branches were placed on hot coals, a cloth placed over the branches, and the tired feet rested on the cloth. For a sore leg, hand, or arm, a hole in the ground was filled with hot coals; then leafy creosotebush branches were added, the hole was covered with a cloth, and the afflicted limb was held into the smoke. For headache, hot leafy branches wrapped in a cloth were held to the head, or the head was washed in water in which the foliage had been cooked.

For difficulty in breathing, about one-half liter of *Larrea* tea was drunk. A feather was used to make the patient vomit, after which he could breathe. This was done on an empty stomach in the morning. One who drank it was called *hehe quisi* 'plant drinker.'

The roots were boiled and the liquid taken to cure dizziness. For a stingray wound, a tea was boiled and the afflicted part was soaked.

A contraceptive was made from the lac, *csipx*. A wad of lac was placed on a stick and heated in a flame. As it melted, drops were caught in a container of water held beneath it. A woman drank the water as a contraceptive.

A woman suffering from an aching back following childbirth sometimes lay on top of leafy creosotebush branches placed over a small pit filled with warm ashes. She also drank a lot of hot water, and a heated stone wrapped in a cloth was placed on her stomach.

Alfred Whiting (1957: #34) noted that it was "Used as medicine in several different ways. Fumes for colds. It may be boiled and applied to knees, chest, etc."

Smoking: A round insect gall found among the foliage twigs was smoked like tobacco. The gall was called *haaxat ihíix cocoj* 'creosotebush where-it-is-globular rounded.' It is produced by the creosotebush gall midge (*Asphondylia*).

The Supernatural: Children wore amulets of creosotebush lac (*csipx*) on a string or mesquite twine around their necks. Amulets of *csipx*, sometimes formed into an hourglass shape, were occasionally worn by adults to bring good luck.

Other Uses: The lac was used to decorate arrows. Lac was softened by holding a burning stick near it; it was then picked up in a glob, or wad, with another stick. Next, it was held over a container of water and again melted. As it melted and dripped into the water, it burned and turned black. The melted, blackened lac was called *hacánoj* 'what is flamed.' The blackened drops of *hacánoj* were taken from the water, reheated, and a single hair from a horse's tail was drawn through it. The horse hair, with the hot *hacánoj* on it, was used to make thin black rings near the end of the mainshaft. The rings were covered with transparent sinew, which bound the foreshaft and mainshaft. This decoration was done so that one could know whether or not an arrow was made by a Seri.

For stuffing and taxidermy of a mule deer or jackrabbit head, the eyes were represented with balls formed from *hacánoj*, the melted lac (*csipx*). The deer head was used in the deer dance or as a decoy for hunting; the jackrabbit head also served as a hunter's decoy or a girl's doll (Bowen and Moser 1970b: 173).

**Tribulus terrestris* L.
heen ilít 'cow its-head'
  "cow's head"
cözazni caacöl 'tangled (*Cenchrus*) large (plural)'
  "large sandbur"
hee inóosj 'jackrabbit its-claw'
  "jackrabbit's claw"

hehe cosyat 'plant spines'
  "spiny plant"
cosi cahóota 'spine that-causes-one-to-urinate
  (male speaking)'
puncture vine, bullhead, goathead

This Old World weed did not become established in El Desemboque until the early 1960s; it soon became abundant in the villages, although we have not seen it on Tiburón Island. The Seri names are descriptive and there is little constancy in their use; some people did not have a name for it. Elvira Valenzuela did not know its name, and said, "It is a new plant, it comes from the outside."

Animal Food: Rosa Flores and her husband, José Astorga, said that burros, cows, horses, and pigs eat the plant, but that the people do not use it.

Viscainoa geniculata (Kell.) Greene var. *geniculata*
xneejam is hayáa '*xneejam* its-seeds what-is-owned'
  "*xneejam* whose seeds are owned"

The name indicates this *xneejam* can be "owned" or used (as in necklaces), as contrasted with *xneejam-siictoj* (*Stegnosperma*), which has relatively little use.

Very common desert shrub on the mainland and islands. Relatively large, green capsules containing about eight black seeds 4 to 5 mm long. It is readily distinguished from *Stegnosperma* by its nonsucculent, puberulent leaves and nonfleshy fruit (Figure 17.162).

The Seri pointed out that it is particularly abundant in the vicinity of Tecomate on the north side of Tiburón Island, and on the mainland between Kino Bay and Punta Chueca, and near Campo *Oona*.

Adornment: The seeds were commonly strung in necklaces, often in combination with shells or with fish or snake vertebrae (see Figure 12.5A). Until about the mid-twentieth century the Seri made extensive use of these necklaces for their own adornment. In the 1970s and 1980s the people often went to Tecomate to collect the seeds because necklaces with these seeds commanded good prices. The seeds could be stored. While at Tecomate on Tiburón Island in June 1951, Alfred Whiting (1957:#35) recorded that the plant "produces black seeds which are used for beads."

The Supernatural: There was a belief that the plant could grow from a buried fingernail or toenail.

Figure 17.162. *Viscainoa geniculata* from Tiburón Island with flowers and mature capsules. *FR.*

Unidentified Plants

There were some plants for which the Seri had names and information, but we were unable to locate and identify these plants. Most of these were plants occurring at the periphery of the region, or plants known only to elderly people unable to travel into the field to locate them. In some cases information on these plants was inconsistent and conflicting, indicating erosion of information. Unidentified plants and Seri descriptions of them are given below.

cop quiinl '*cop* that-which-has-fingers'

This is a vine on Tiburón Island. The flower is blue and very pretty. It does not have a use. It may be an annual species of morning glory (*Ipomoea*).

coquéeen 'who walks with shoulders swinging'

This is a small plant like a small *hee xoját* (*Tiquilia palmeri*). The root is identical to that of *hee xoját*.

The outside of the root is black. The women cut the root into pieces, dried them and strung them as necklaces. It has a pleasant odor. Some said that there is much of it near Tastiota.

hatoj caihöj 'eyes to-make-red' "cause red eye"*

This plant was said to resemble *hant yax* (*Maximowiczia*). It has a large underground root which does not dry up (it is perennial). Some said the root is like a potato. It is a large plant and has leaves like those of a watermelon plant. The fruit is red and resembles that of *xtooxt* (*Neoevansia*) but is larger, the plant has 40 to 50 fruits, and the seed is red. It grows with the summer rains, at the same time as *casol heecto* (*Pectis papposa*). The fruit was eaten fresh like *pitaya* fruit, in October after the summer rains. It occurs near Pozo Coyote and elsewhere in the Arroyo San Ignacio, and at Rancho Estrella. This information indicates that it is probably a member of the gourd family.

hatoj ipápl 'eyes strung'

Some said it is like *hehe coozlil* (*Horsfordia alata*). However, others did not know of it, pointing out that *hatoj ipápl quiip* is a kind of bird. Still others had heard of it but did not know it.

hax oohit 'water what-it-eats' "what the water eats"

---

*Since this book was originally published, this plant has been identified as *Tumamoca macdougalii* Rose, a member of the Gourd Family (Cucurbitaceae).

Although some said this is a plant, others said that it is a kind of bird.

heetes

One woman identified *Abutilon incanum* as *heetes*. However, others assured us that *heetes* was not this shrub. It is a shrub, and has rough bark and a narrow leaf. One man said it is like jojoba (*Simmondsia*); another person said it is large, like *ptaact* (*Colubrina*). It does not have any use, although the wood is scented. It occurs at Rancho Estrella; some said it grows in the large arroyo (but we failed to find it there), while others said it occurs on mountain sides.

mahyan

The plant is about one meter or less in height, like a shrub (others said that it is small, like a tomato plant); it does not have spines, and the leaves are soft and resemble those of *ziim* (*Amaranthus palmeri*). The fruit is about 2 cm long, somewhat dark in color, resembling a grape but smaller. It is larger than the fruit of *tinóopa*, *Vallesia*). It is sweet and eaten fresh—in quantity when in season. It grows in summer after the summer rains begin. It was found at Rancho Costa Rica in the desert beneath trees; some said that it grows "below Pozo Peña" (at the south end of the Sierra Seri). It apparently has blue flowers (see *Amoreuxia*).

pnaacoj quistj 'mangrove that-which-has-leaves'

This was said to be a small plant, with leaves resembling those of *comáanal*, occurring at *Zozni Cmiipla* at the southeast side of Tiburón Island.

APPENDIX A

Seri Plant Names

SERI NAME	SCIENTIFIC NAME	FAMILY
aaxt	*Phoradendron californicum* *P. diguetianum*	Viscaceae
an icoquéetc 'in *icoquéetc*'	*Ambrosia deltoidea*/spp.	Compositae

an icös
   'inside its-thorns'
   "thorny inside"
	Phaulothamnus spinescens	Phytolaccaceae

caail iti śiml
   'dry-lake on barrel-cactus'
   "dry lake barrel cactus"
	Ferocactus covillei	Cactaceae

caail oocmoj
   'dry-lake what-it-wears-on-waist'
   "dry lake's waist cord"
	Juncus acutus	Juncaginaceae

caal oohit
   'companion-child what-he-eats'
   "what the companion-child eats"
	Triteleiopsis palmeri	Liliaceae

caal oohit caacöl
   'companion-child what-he-eats
   (*Triteleiopsis*) large(plural)'
   "large blue sand lily"
	Brodiaea pulchella	Liliaceae

caaöj	*Baccharis salicifolia*	Compositae

caatc ipápl
   'grasshopper what-it-is-strung-
   with(plural)'
   "what grasshoppers are strung with"
	Abutilon incanum
	A. palmeri
	Horsfordia alata
		Malvaceae
		
		Malvaceae

caay ixám
   'horse its-gourd'
   "horse's gourd"
	Cucurbita pepo	Cucurbitaceae

cacátajc 'what causes vomiting'	*Chaenactis carphoclinia* *Pectis papposa*	Compositae Compositae
caháahazxot 'what causes sneezing'	*Baileya multiradiata* *Dyssodia concinna*	Compositae Compositae

389

SERI NAME	SCIENTIFIC NAME	FAMILY
caháahazxot ctam 'what-causes-sneezing (*Baileya*) male' "male desert marigold"	*Phacelia ambigua*	Hydrophyllaceae
camótzila 'guamúchil'	**Pithecellobium dulce*	Leguminosae
cap	*Acacia willardiana*	Leguminosae
casol caacöl '*casol* large(plural)' "large *casol*"	*Baccharis sarothroides* *Dodonaea viscosa*	Compositae Sapindaceae
casol cacat '*casol* bitter' "bitter *casol*"	*Haplopappus sonorensis* *Hymenoclea salsola*	Compositae Compositae
casol coozlil '*casol* sticky' "sticky *casol*"	*Hymenoclea monogyra* *H. salsola*	Compositae
casol heecto '*casol* small' "small *casol*"	*Pectis papposa*	Compositae
casol heecto caacöl '*casol* small (*Pectis papposa*) large(plural)' "large cinchweed"	*Pectis palmeri*	Compositae
casol ihasíi tiipe '*casol* its-odor is-good' "fragrant *casol*"	*Pectis papposa*	Compositae
casol itac coosotoj '*casol* its-bone thin(plural)' "thin stemmed *casol*"	*Hymenoclea monogyra*	Compositae
casol ziix ic cöihíipe '*casol* thing with make-better' "medicinal *casol*"	*Haplopappus sonorensis* *Hymenoclea salsola*	Compositae Compositae
caztaz 'Castilian *Zostera*' "Castilian eelgrass"	**Triticum aestivum*	Gramineae
cmajíic ihásaquim 'women what-one-brushes-hair-with' "what women brush their hair with"	*Ambrosia deltoidea* *A. divaricata* *A. dumosa* *A. magdalenae* *Zinnia acerosa*	Compositae Compositae
cmapöjquij 'what bursts open' "bursts open"	*Caesalpinia palmeri* *Coursetia glandulosa*	Leguminosae Leguminosae
coap	*Cnidoscolus palmeri*	Euphorbiaceae
cocásjc	*Jouvea pilosa*	Gramineae
cocazn-ootizx 'rattlesnake what-it-peels-back' "rattlesnake's foreskin"	*Trixis californica*	Compositae

SERI NAME	SCIENTIFIC NAME	FAMILY
cocazn-ootizx caacöl 'rattlesnake what-it-peels-back (*Trixis*) large(plural)' "large *Trixis*"	*Verbesina palmeri*	Compositae
cocóol	*Cleome tenuis* *Descurainia pinnata* *Sisymbrium irio*	Capparaceae Cruciferae Cruciferae
cocóol cmaam '*cocóol* (*Descurainia*) female' "female tansy mustard"	*Draba cuneifolia*	Cruciferae
cof	*Jacquinia pungens*	Theophrastaceae
com-aacöl '*com*-large(plural)'	*Antirrhinim kingii* *Trianthema portulacastrum*	Scrophulariaceae Aizoaceae
comáanal	*Anemopsis californica*	Saururaceae
comíma	*Brickelia coulteri* *Eupatorium sagittatum*	Compositae Compositae
comítin	*Olneya tesota*	Leguminosae
com-ixaz '*com*-rattle'	*Cocculus diversifolius*	Menispermaceae
comot	*Matelea cordifolia*	Asclepiadaceae
conée caacöl 'grass large(plural)' "large grass"	unidentified	Gramineae
conée ccapxl 'grass sour' "sour grass"	*Brachiaria arizonica* *Eragrostis* sp. *Erioneuron pulchellum*	Gramineae Gramineae Gramineae
conée cosyat 'grass spines' "spiny grass"	*Erioneuron pulchellum*	Gramineae
conée csai 'grass hairbrush'	*Aristida californica*	Gramineae
coote	*Opuntia bigelovii*	Cactaceae
cop quiinl '*cop* that-which-has-fingers'	unidentified	
coptoj	*Agave angustifolia*	Agavaceae
coquée	*Lepidium lasiocarpum*	Cruciferae
coquée caacöl 'chile large(plural)' "large chiles"	*Capsicum annuum*	Solanaceae
coquée coil 'chile green' "green chile"	*Capsicum annuum*	Solanaceae
coquée coopol 'chile black' "black chile"	*Piper nigrum*	Piperaceae
coquéeen 'who walks with shoulders swinging'	unidentified	

SERI NAME	SCIENTIFIC NAME	FAMILY
coquée quitajij 'chile pointed(plural)' "pointed chiles"	*Capsicum annuum	Solanaceae
coquée quizil 'chile little(plural)' "little chiles"	*Capsicum annuum* var. *aviculare*	Solanaceae
cos	*Maytenus phyllanthoides*	Celastraceae
cosi cahóota 'thorn that-causes-one-to-urinate (male speaking)'	*Tribulus terrestris	Zygophyllaceae
cotéexet 'Opuntia bigelovii -eexet'	*Opuntia fulgida*	Cactaceae
cototax	*Fouquieria columnaris*	Fouquieriaceae
cotx 'acrid smell'	*Encelia farinosa*	Compositae
coxi ihéet 'dead-one his-gambling-sticks' "dead man's gambling sticks"	*Errazurizia megacarpa*	Leguminosae
coxi ixám 'dead-one his-squash' "dead man's squash"	*Citrullus lanatus	Cucurbitaceae
cozi	*Condalia globosa*	Rhamnaceae
cozi hax ihapóin 'Condalia water what-it-is-closed-with'	*Tephrosia palmeri*	Leguminosae
cöset	*Atamisquea emarginata*	Capparaceae
cötep	*Monanthochloe littoralis*	Gramineae
cözazni 'tangled'	*Cenchrus palmeri*	Gramineae
cözazni caacöl 'tangled (Cenchrus) large(plural)' "large sandbur"	*Tribulus terrestris *Xanthium strumarium	Zygophyllaceae Compositae
cpooj	*Digitaria californica* *Sporobolus cryptandrus* unidentified grass	Gramineae Gramineae Gramineae
eaz	*Zostera marina*	Zosteraceae
haaca	*Zizyphus obtusifolia*	Rhamnaceae
haacoz	*Melilotus indica*	Leguminosae
haalp	*Randia thurberi*	Rubiaceae
haamoja iháap 'pronghorn its-tepary' "pronghorn's tepary"	*Phaseolus filiformis*	Leguminosae
haamxö	*Agave subsimplex*	Agavaceae
haamxö caacöl 'Agave-subsimplex large(plural)' "large agave"	*Agave colorata* A. *fortiflora*	Agavaceae
haamxöii 'agave first' "trimmed agave"	*Hechtia montana* *Ananas comosus	Bromeliaceae Bromeliaceae

SERI NAME	SCIENTIFIC NAME	FAMILY
haap	*Phaseolus acutifolius*	Leguminosae
haas	*Prosopis glandulosa*	Leguminosae
haat	*Jatropha cuneata*	Euphorbiaceae
haaxat "with smoke or steam"	*Larrea divaricata*	Zygophyllaceae
haaxo 'ajo' (garlic)	**Allium sativum*	Liliaceae

hacáiin cooscl
  'what-is-gathered-for-windbreak drab'
  "drab windbreak"
	Croton californicus	Euphorbiaceae

hacx cahóit
  'somewhere what-causes-to-descend'
  "what wastes it" or "what causes to be lost"
	Tulostoma sp.	Fungus

hahöj an quinelca
  '*Lycium* inside empty(plural)'
  "empty *Lycium*"
	Lycium brevipes	Solanaceae

hahöj cacat
  '*Lycium* bitter'
  "bitter *Lycium*"
	Lycium fremontii	Solanaceae

hahöj ináil coopol
  '*Lycium* its-skin black'
  "black-barked *Lycium*"
	Lycium andersonii	Solanaceae

hahöj-enej
  '*Lycium* empty'
  "empty *Lycium*"
	Lycium andersonii	Solanaceae

hahöj-izij
  '*Lycium* little(plural)'
  "little *Lycium*"
	Lycium californicum	Solanaceae

halít an caascl
  'head on what-causes-dandruff'
  "causes dandruff"
	Tidestromia lanuginosa	Amaranthaceae

hamácj 'fires'	*Euphorbia misera*	Euphorbiaceae

hamácj inoloj
  'fires their-flames'
  "flames of fire"
	Lantana horrida	Verbenaceae

hamat	*Yucca arizonica*	Agavaceae
hamíp	*Boerhaavia coulteri*	Nyctaginaceae

hamíp caacöl
  '*Boerhaavia-coulteri* large(plural)'
  "large spiderling"
	Boerhaavia erecta	Nyctaginaceae

hamíp cmaam
  '*Boerhaavia-coulteri* female'
  "female spiderling"
	Allionia incarnata	Nyctaginaceae

hamísj	*Jatropha cinerea*	Euphorbiaceae
hamítom hant cocpétij (see *tomítom* . . .)		
hamoc	*Agave angustifolia*	Agavaceae

SERI NAME	SCIENTIFIC NAME	FAMILY
hamt ináil 'soil its-skin' "soil's skin"	unidentified soil algae	Soil algae
hamt inóosj 'soil its-claw' "soil's claw"	*Hofmeisteria laphamioides*	Compositae
hamt itóozj 'soil its-intestines' "soil's intestines"	*Cuscuta corymbosa* *C. leptantha*	Convolvulaceae
hamt yamása 'soil its-algae/moss' "soil's algae (or moss)"	unidentified soil algae or moss	Soil algae or moss
hanaj iit ixac 'raven its-lice their-eggs' "nits of the raven's lice"	*Marina parryi*	Leguminosae
hanaj itáamt 'raven its-sandals' "raven's sandals"	*Dalea mollis*	Leguminosae
hant caitoj 'land creeper'	*Brandegea bigelovii* *Vaseyanthus insularis*	Cucurbitaceae Cucurbitaceae
hant iipzx iteja 'land its-torn (place) its-bladder' "arroyo's bladder"	*Mammillaria microcarpa* *M. sheldonii* *M.* spp.	Cactaceae
hant iipzx iteja caacöl 'land its-torn (place) its-bladder (*Mammillaria*) large(plural)' "large fishhook cactus"	*Mammillaria estebanensis* *Echinocereus engelmannii* *E. fendleri* *E. grandis* *E. pectinatus*	Cactaceae Cactaceae
hant iit 'land its-lice' "land's lice"	*Mollugo cerviana*	Aizoaceae
hant ipásaquim 'land what-it-is-swept-with' "broom"	*Abutilon californica*	Malvaceae
hant ipépj 'land *ipépj*'	*Lotus salsuginosus* *L. tomentellus*	Leguminosae
hant istj 'land its-leaf' "land's leaf"	*Dithyrea californica*	Cruciferae
hant iteja 'land its-bladder' "land's bladder"	*Colpomenia tuberculata*	Marine alga
hant-oosinaj 'land *oosinaj*'	*Abronia villosa*	Nyctaginaceae
hant-oosinaj cooxp 'land-*oosinaj* (*Abronia villosa*) white' "white sand verbena"	*Oenothera californica*	Onagraceae
hant-oosinaj ctam 'land-*oosinaj* (*Abronia villosa*) male' "male sand verbena"	*Camissonia claviformis*	Onagraceae

SERI NAME	SCIENTIFIC NAME	FAMILY
hant ootizx 'land what-it-peels-back' "land's foreskin"	*Battarrea diguetii* *Podaxis pistillaris*	Fungus Fungus
hant otópl 'land what-it-sticks-to' "what the lands sticks to"	*Heliotropium* *curassavicum*	Boraginaceae
hant yapxöt 'land its-flowering' "land's flower"	*Achyronychia cooperi* *Tillaea erecta*	Caryophyllaceae Crassulaceae
hant yax 'land its-belly' "land's belly"	*Maximowiczia sonorae*	Cucurbitaceae
hap itapxén 'mule-deer its-inner-canthus' "inner corner of mule deer's eye"	*Solanum hindsianum*	Solanaceae
hap oacajam 'mule-deer what-it-hits' "what mule deer flay antlers on"	*Caesalpinia palmeri* *Echinopterys eglandulosa* *Thryallis angustifolia*	Leguminosae Malpighiaceae Malpighiaceae
hapats imóon 'Apache his-beans' "Apache's beans"	*Phaseolus lunatus*	Leguminosae
hapis casa 'what-is-smoked putrid' "putrid tobacco"	*Nicotiana trigonophylla*	Solanaceae
hapis coil 'what-is-smoked green' "green tobacco"	*Cannabis sativa*	Cannabaceae
hapsx iti icóocax "one who builds up her basket load better"	*Aloysia lycioides*	Verbenaceae
hapxöl	*Zea mays*	Gramineae
hasac	*Setaria macrostachya*	Gramineae
hasahcápöj '*has*-what is chewed' "chewed fruit"	*Lophocereus schottii*	Cactaceae
hasla an ihoom 'outer-ear in where-it-lies' "ear is its place"	*Abutilon incanum*	Malvaceae
hasoj an hehe 'river area plant'	*Andrachne ciliato-* *glandulosa* *Teucrium glandulosum* unidentified mustard	Euphorbiaceae Labiatae Cruciferae
hasot "narrow"	*Agave chrysoglossa*	Agavaceae
hast ipénim 'rock what-it-is-splattered-with' "splattered against rock"	*Camissonia californica* *Eschscholzia parishii* *Kallstroemia grandiflora*	Onagraceae Papaveraceae Zygophyllaceae
hast ipénim ctam 'rock what-it-is splattered-with male'	*Senecio douglasii*	Compositae

SERI NAME	SCIENTIFIC NAME	FAMILY
hast iti coocp 'rock on what-grows' "what grows on rock"	*Galaxaura fastigiata*	Marine alga
hast iti coteja 'rock on what-sways' "what sways on rock"	*Colpomenia phaeodactyla* *Cutleria hancockii*	Marine alga Marine alga
hast yamása 'rock lichen'	foliose and crustose lichens	
hast yapxöt 'rock its-flowering' "rock's flower"	*Dudleya arizonica* *Hofmeisteria laphamioides*	Crassulaceae Compositae
hatáaij 'what is spun (like a top)'	*Ipomoea* sp.	Convolvulaceae
hatáam 'what is harvested'	*Zostera marina*	Zosteraceae
hatáast an ihíih 'tooth in where-it-is' "what gets in the teeth"	*Aristolochia watsonii*	Aristolochiaceae
hataj-en 'vulva inside'	*Plantago insularis*	Plantaginaceae
hataj-ipol 'vulva black' "black vulva"	*Suaeda moquinii*	Chenopodiaceae
hataj-isijc 'vulva immature' "immature vulva"	*Atriplex canescens* *A. linearis*	Chenopodiaceae
hataj-ixp 'vulva white' "white vulva"	*Atriplex polycarpa* *A. linearis*	Chenopodiaceae
hatoj caihöj 'eyes to-make-red' "causes red eye"	unidentified	
hatoj ipápl 'eyes strung'	unidentified	
haxoj ano ihímz 'shoreline in its-fringes' "fringes of the shoreline"	*Spyridia filamentosa*	Marine algae
hax oohit 'water what-it-eats' "what the water eats"	unidentified	
hax quipóin 'water what-is-closed'	*Cardiospermum corindum*	Sapindaceae
haxz iiztim 'dog its-hipbone' "dog's hipbone"	*Calliandra eriophylla* *Hoffmanseggia intricata* *Krameria parvifolia*	Leguminosae Leguminosae Krameriaceae
haxz oocmoj 'dog what-it-wears-on-waist' "dog's waist cord"	*Mascagnia macroptera*	Malpighiaceae

SERI NAME	SCIENTIFIC NAME	FAMILY
hee imcát 'jackrabbit what-it-doesn't-bite-off' "what the jackrabbit doesn't bite off"	*Machaeranthera parviflora* *Perityle emoryi*	Compositae Compositae
hee imcát caacöl 'jackrabbit what-it-doesn't-bite-off (*Perityle emoryi*) large(plural)' "large rock daisy"	*Perityle leptoglossa*	Compositae
hee inóosj 'jackrabbit its-claw' "jackrabbit's claw"	*Lotus salsuginosus* *L. tomentellus* *Tribulus terrestris*	Leguminosae Zygophyllaceae
hee xoját 'jackrabbit *Amoreuxia*' "jackrabbit *saiya*"	*Tiquilia palmeri*	Boraginaceae
heecl	*Jatropha cardiophylla*	Euphorbiaceae
heecoj	*Roccella babingtonia*	Lichen
heejac	*Pithecellobium confine*	Leguminosae
heel	*Opuntia violacea*	Cactaceae
heel cocsar yaa 'prickly-pear Mexican his-belonging' "Mexican's prickly-pear"	*Opuntia ficus-indica*	Cactaceae
heel cooxp 'prickly-pear white' "white prickly-pear"	*Opuntia ficus-indica*	Cactaceae
heel hayéen ipáii 'prickly-pear face what-it-is-done-with' "prickly-pear used for face painting"	*Opuntia phaeacantha*	Cactaceae
heem	*Opuntia arbuscula*	Cactaceae
heemáa '*Opuntia arbuscula* true' "true pencil cholla"	*Opuntia arbuscula*	Cactaceae
heem icös cmasl '*Opuntia-abruscula* its-spines yellow(plural)' "yellow-spined pencil cholla"	*Opuntia arbuscula* *O. cf. burrageana*	Cactaceae
heem icös cmaxlilca '*Opuntia-arbuscula* its-spines stiffly- protruding(plural)' "stiff spined pencil cholla"	*Opuntia versicolor*	Cactaceae
heen ilít 'cow its-head' "cow's head"	*Tribulus terrestris*	Zygophyllaceae
heepol	*Krameria grayi*	Krameriaceae
heetes	unidentified	
hehe caacoj 'plant large' "large plant"	*Ricinus communis*	Euphorbiaceae
hehe cacátajc 'plant that-causes-vomiting'	*Machaeranthera parviflora*	Compositae

SERI NAME	SCIENTIFIC NAME	FAMILY
hehe camóstim 'plant that-causes-grimacing'	*Datura discolor*	Solanaceae
hehe carócot 'plant that-makes-one-crazy'	*Datura discolor*	Solanaceae
hehe casa 'plant putrid' "putrid plant"	*Desmanthus fruticosus*	Leguminosae
hehe ccon 'plant that-reeks'	**Allium cepa*	Liliaceae
hehe coanj 'plant poison' "poisonous plant"	*Sapium biloculare*	Euphorbiaceae
hehe cocóozxlim 'plant that-causes-rash'	*Tragia amblyodonta*	Euphorbiaceae
hehe coozlil 'plant sticky' "sticky plant"	*Horsfordia alata*	Malvaceae
hehe cosyat 'plant spines' "spiny plants"	**Tribulus terrestris*	Zygophyllaceae
hehe cotázita 'plant that-pinches'	*Mimosa laxifolia*	Leguminosae
hehe cotópl 'plant that-clings'	*Cryptantha angustifolia* *C. maritima* *C.* spp. *Mentzelia adhaerens* *M. involucrata* *Perityle emoryi*	Boraginaceae Loasaceae Compositae
hehe coyóco 'plant dove' "dove plant"	*Melochia tomentosa*	Sterculiaceae
hehe ctoozi 'plant resilient' "resilient plant"	*Coursetia glandulosa* *Lysiloma divaricata*	Leguminosae Leguminosae
hehe czatx 'plant stickery' "stickery plant"	*Cryptantha angustifolia* *C. maritima* *C.* spp. *Mentzelia adhaerens*	Boraginaceae Loasaceae
hehe czatx caacöl 'plant stickery large(plural)' "large stickery plant"	*Argythamnia neomexicana* *A. lanceolata*	Euphorbiaceae
hehe hatéen captax 'plant mouth punctured(plural)'	*Bumelia occidentalis*	Sapotaceae
hehe iix coil 'plant its-liquid blue/green' "blue/green-sapped plant"	*Lycium macrodon*	Solanaceae
hehe iix cooxp 'plant its-liquid white' "white-sapped plant"	*Euphorbia xanti*	Euphorbiaceae

SERI NAME	SCIENTIFIC NAME	FAMILY
hehe imixáa 'plant rootless' "rootless plant"	*Dithyrea californica* *Stephanomeria pauciflora*	Cruciferae Compositae
hehe imoz coopol 'plant its-heart black' "black-hearted plant"	*Viguiera deltoidea*	Compositae
hehe is quiixlc 'plant its-fruit globular(plural)' "round-fruited plant"	**Punica granatum*	Punicaceae
hehe is quisil 'plant its-fruit little' "small-fruited plant"	*Lophocereus schottii*	Cactaceae
hehe itac coozalc 'plant its-bones ribbed' "ribbed-stem plant"	*Teucrium cubense* *T. glandulosum*	Labiatae
hehe iti scahjíit 'plant on let's-fall' "let's fall on the plant"	*Brandegea bigelovii*	Cucurbitaceae
hehe iyas 'tree its-liver' "tree's liver"	*Ganoderma lucidum*	Fungus
hehe-monlc 'plant curly' "curly plant"	*Antirrhinum cyathiferum*	Scrophulariaceae
hehe pnaacoj 'plant mangrove' "mangrove plant"	*Croton magdalenae* *Sideroxylon leucophyllum*	Euphorbiaceae Sapindaceae
hehe quiijam 'plant that-curls-around-it'	*Ipomoea* sp. *Janusia californica* *J. gracilis*	Convolvulaceae Malphigiaceae
hehe quiinla 'plant that-rings'	*Cassia covesii*	Leguminosae
hehe quina 'plant with-hair' "hairy plant"	*Notholaena standleyi*	Fern
hehe quina caacöl 'plant with-hair (*Notholaena*) large(plural)' "large rock fern"	*Selaginella arizonica*	Fern relative
hehe yapxöt imóxi 'plant its-flowering what-doesn't-die' "plant whose flower doesn't die"	*Salvia columbariea*	Labiatae
hehet ináil coopl 'plants their-skin black(plural)' "black-barked plant"	*Cordia parvifolia*	Boraginaceae
heme	*Agave cerulata*	Agavaceae
hepem ihéem 'white-tailed-deer its-*Opuntia-arbuscula*' "white tailed deer's pencil cholla"	*Opuntia versicolor*	Cactaceae

SERI NAME	SCIENTIFIC NAME	FAMILY
hepem ijcóa 'white-tailed-deer its-*Sphaeralcea*' "white-tailed deer's globe-mallow"	*Hibiscus denudatus*	Malvaceae
hepem isla 'white-tailed-deer its-ear' "white-tailed-deer's ear"	*Mirabilis bigelovii*	Nyctaginaceae
hexe	*Sarcostemma cynanchoides*	Asclepiadaceae
hocö hapéc 'wood what-is-planted' "cultivated tree"	*Tamarix aphylla*	Tamaricaceae
hohr-oohit 'donkey what-it-eats' "what donkeys eat"	*Nama hispidum*	Hydrophyllaceae
hoinalca 'low hills'	*Holographis virgata* *Croton sonorae*	Acanthaceae Euphorbiaceae
icapánim "what one washes hair with"	*Agave schottii*	Agavaceae
iicj ano meróon 'sand from-in melon' "sand melon"	*Cucumis melo*	Cucurbitaceae
iicj yamása 'sand its-algae' "sand algae"	unidentified soil algae (on sandy soil)	Soil algae
iipxö	*Opuntia leptocaulis*	Cactaceae
iix casa insíi 'his-water putrid who-doesn't-smell' "who doesn't smell his putrid water"	*Astragalus magdalenae*	Leguminosae
iiz	*Cercidium floridum*	Leguminosae
impós	*Aristida adscensionis* *Muhlenbergia microsperma*	Gramineae Gramineae
insáacaj	*Physalis* cf. *pubescens*	Solanaceae
inyéeno 'faceless'	*Agave pelona*	Agavaceae
isnáap ic is 'its-side with its-fruit' "whose fruit is on one side"	*Bouteloua aristidoides* *B. barbata* *B. repens* *Lepidium lasiocarpum*	Gramineae Cruciferae
jcoa	*Sphaeralcea ambigua* *S. coulteri*	Malvaceae
jöene	*Passiflora palmeri*	Passifloraceae
limóon '*limón*' "lime"	*Citrus aurantifolia*	Rutaceae
maas	*Cercidium praecox*	Leguminosae
mahyan	unidentified	
matar (Papago man's name)	*Orobanche cooperi*	Orobanchaceae

SERI NAME	SCIENTIFIC NAME	FAMILY

meróon
    'melón'
    "melon"
	Cucumis melo	Cucurbitaceae

mocni	*Guaiacum coulteri*	Zygophyllaceae
mojépe	*Carnegiea gigantea*	Cactaceae

mojépe ihásaquim cmaam
    'sahuaro what-one-brushes-hair-with
    female'
    "female sahuaro hairbrush"
	Zinnia acerosa	Compositae

mojépe siml 'sahuaro barrel-cactus'	*Ferocactus acanthodes*	Cactaceae

mojet oohit
    'mountain-sheep what-it-eats'
    "what mountain sheep eat"
	Allium haematochiton	Liliaceae

mooj	*Gossypium* spp.	Malvaceae

moosni iha
    'sea-turtle its-possessions'
    "sea turtle's possessions"
	Dithyrea californica
	Palafoxia arida
	Rhodymenia hancockii
		Cruciferae
		Compositae
		Marine alga

moosníil ihaquéepe
    'blue-turtle what-it-likes'
    "what blue turtle likes"
	Asparagopsis taxiformis	Marine alga

moosni ipnáil
    'sea-turtle its-skirt'
    "sea turtle's skirt"
	Cryptonemia obovata
	Halymenia coccinea
	Padina durvillaei
	Rhodymenia divaricata
		Marine alga
		Marine alga
		Marine alga
		Marine alga

moosni iti hatépx
    'sea-turtle on what-is-rested-on'
    "what sea turtle meat rests on"
	Croton californica	Euphorbiaceae

moosni yazj
    'sea-turtle its-membrane'
    "sea turtle's membranes"
	Gracilaria textorii
	Padina durvillaei
		Marine alga
		Marine alga

moosn-oohit
    'sea-turtle what-it-eats'
    "what sea turtles eat"
	Asparagopsis taxiformis
	Palafoxia arida
		Marine alga
		Compositae

najcáazjc	*Asclepias albicans* *A. subulata*	Asclepiadaceae
najmís	*Phacelia ambigua*	Hydrophyllaceae
nas	*Matelea pringlei*	Asclepiadaceae
naz	*Cordia parvifolia*	Boraginaceae

noj-oopis
    'hummingbird what-it-sucks-out'
    "what hummingbirds suck out"
	Justicia californica	Acanthaceae

noj-oopis caacöl
    'hummingbird what-it-sucks-out
    (*Justicia*) large(plural)'
    "large *chuparosa*"
	Galvezia juncea
	Nicotiana glauca
	Penstemon parryi
		Scrophulariaceae
		Solanaceae
		Scrophulariaceae

nojóo ixpanáams
    'spotted-sand-bass its-seaweed'
    "spotted sand bass's seaweed"
	Galaxaura fastigiata	Marine alga

SERI NAME	SCIENTIFIC NAME	FAMILY
oeno-raama 'buena rama'	*Acacia constricta*	Leguminosae
ool	*Stenocereus thurberi*	Cactaceae
ool-axö '*Stenocereus-thurberi* its-excrement' "organ pipe cactus' excrement"	*Stenocereus gummosus*	Cactaceae
oot ijöéne 'coyote its-*Passiflora-palmeri*' "coyote's passion vine"	*Passiflora arida*	Passifloraceae
paar icomíhlc 'padre his-mesquite-seed' "padre's mesquite seed"	** Pisum sativum*	Leguminosae
paar icomítin 'padre his-ironwood-seed' "padre's ironwood seed"	** Cicer arietinum*	Leguminosae
paaza	*Bumelia occidentalis*	Sapindaceae
paij	*Salix gooddingii*	Salicaceae
pat	*Typha domingensis*	Typhaceae
paxáaza	*Ambrosia* cf. *confertifolia*	Compositae
paxóocsim (see *xpaxóocsim*)		
pnaacoj hacáaiz 'mangrove harpoon'	*Laguncularia racemosa*	Combretaceae
pnaacoj-iscl 'mangrove drab' "drab mangrove"	*Avicennia germinans*	Verbenaceae
pnaacoj quistj 'mangrove that-which-has-leaves'	unidentified	
pnaacoj-xnazolcam (*pnaazolcam*) 'mangrove crisscrossed' "crisscrossed mangrove"	*Rhizophora mangle*	Rhizophoraceae
pnaacöl	*Simmondsia chinensis*	Simmondsiaceae
poháas camoz 'future-mesquite thinker' "what thinks it's a mesquite"	*Acacia farnesiana* *Desmanthus covillei*	Leguminosae Leguminosae
ponás camoz 'future-*Matalea-pringlei* thinker' "what thinks it's a *Matalea pringlei*"	*Lyrocarpa coulteri*	Cruciferae
posapátx camoz 'future-*Bebbia* thinker' "what thinks it's a sweet-bush"	*Stephanomeria pauciflora*	Compositae
potács camoz 'future-*Allenrolfea* thinker' "what thinks it's an iodine bush"	*Heliotropium curassavicum*	Boraginaceae
ptaacal	*Celtis pallida*	Ulmaceae
ptaact	*Colubrina viridis*	Rhamnaceae
ptcamn iha 'lobster its-possessions' "lobster's possessions"	*Rhodymenia divaricata*	Marine alga

SERI NAME	SCIENTIFIC NAME	FAMILY
pteept	*Euphorbia eriantha*	Euphorbiaceae
queejam iti hacníix 'bearer-out-of-season on what-is-dumped' "what out-of-season fruit is dumped on"	*Acalypha californica*	Euphorbiaceae
queeto oohit 'Aldebaran what-it-eats' "what Aldebaran eats"	*Lepidium lasiocarpum*	Cruciferae
saapom	*Opuntia violacea*	Cactaceae
saapom ipémt '*Opuntia-violacea* what-it-is-rubbed-with' "what purple prickly-pear is rubbed with"	*Zinnia acerosa*	Compositae
sahmées	** Citrus sinensis*	Rutaceae
sahmées ccapxl 'orange sour' "sour orange"	** Citrus limon*	Rutaceae
sahmées hamt caháacöl 'orange breast what-makes-large(plural)' "orange that enlarges the breast"	** Citrus paradisi*	Rutaceae
sapátx	*Bebbia juncea*	Compositae
satóoml	*Ruellia californica*	Acanthaceae
sea	*Opuntia bigelovii*	Cactaceae
sea cotópl '*Opuntia-bigelovii* that-clings' "clinging teddybear cholla"	*Opuntia fulgida*	Cactaceae
sea icös cooxp '*Opuntia-bigelovii* its-spines white' "white-spined teddybear cholla"	*Opuntia fulgida*	Cactaceae
seepol	*Frankenia palmeri*	Frankeniaceae
siml	*Ferocactus wislizenii*	Cactaceae
siml caacöl '*Ferocactus-wislizenii* large(plural)' "large barrel cactus"	*Ferocactus covillei*	Cactaceae
siml cöquicöt '*Ferocactus-wislizenii* that-kills' "killer barrel cactus"	*Ferocactus covillei*	Cactaceae
siml yapxöt cheel '*Ferocactus-wislizenii* its-flowering red' "red-flowered barrel cactus"	*Ferocactus covillei*	Cactaceae
simláa 'true barrel cactus'	*Ferocactus wislizenii*	Cactaceae
sipöj yanéaax 'cardinal what-it-washes-its-hands-with' "what the cardinal washes its hands with"	*Suaeda esteroa*	Chenopodiaceae

SERI NAME	SCIENTIFIC NAME	FAMILY
snaazx	*Castela polyandra*	Simaroubaceae
snapxöl	**Parkinsonia aculeata*	Leguminosae
spitj	*Atriplex barclayana*	Chenopodiaceae
spitj caacöl '*Atriplex-barclayana* large(plural)' "large coastal saltbush"	*Sesuvium verrucosum*	Aizoaceae
spitj cmajíic '*Atriplex-barclayana* females' "female coastal saltbush"	*Abronia maritima*	Nyctaginaceae
spitj ctamcö '*Atriplex-barclayana* males' "male coastal saltbush"	*Sesuvium verrucosum*	Aizoaceae
taapt (see *pteept*)		
taca imas 'triggerfish its-body-hair' "triggerfish's body hair"	*Asparagopsis taxiformis* *Hofmeisteria fasciculata*	Marine alga Compositae
taca-noosc 'triggerfish its-roughness' "triggerfish's papillae"	*Eucheuma uncinatum*	Marine alga
taca oomas 'triggerfish what-it-twined' "the cord that the triggerfish twined"	*Codium simulans* *Galaxaura arborea*	Marine alga Marine alga
taca oomas cooxp 'triggerfish what-it-twined white' "white *Codium*"	*Codium amplivesticulatum*	Marine alga
tacj-anóosc 'porpoise its-roughness' "porpoise's papillae"	*Gigartina johnstonii* *G. pectinata*	Marine alga
tacj iha 'porpoise its-possessions' "porpoise's possessions"	*Gelidiopsis variabilis*	Marine alga
tacj oomas 'porpoise what-it-twined' "the cord that the porpoise twined"	*Codium simulans* *Digenia simplex*	Marine alga Marine alga
tacs	*Allenrolfea occidentalis*	Chenopodiaceae
tee 'té' "tea"	*Chorizanthe corrugata* *Eriogonum inflatum*	Polygonaceae Polygonaceae
tee caacöl '*Eriogonum-inflatum* large(plural)' "large desert trumpet"	*Senecio douglasii*	Compositae
tee cmaam '*Eriogonum-inflatum* female' "female desert trumpet"	*Eriogonum trichopes*	Polygonaceae
teepar 'tepary'	**Phaseolus acutifolius* (domesticated)	Leguminosae
teepar cmasol 'tepary yellow' "yellow tepary"	**Phaseolus acutifolius* (domesticated)	Leguminosae

SERI NAME	SCIENTIFIC NAME	FAMILY
teepar coopol 'tepary black' "black tepary"	*Phaseolus acutifolius* (domesticated)	Leguminosae
teepar coospoj 'tepary spotted' "spotted tepary"	*Phaseolus acutifolius* (domesticated)	Leguminosae
teepar cooxp 'tepary white' "white tepary"	*Phaseolus acutifolius* (domesticated)	Leguminosae
teept (see *pteept*)		
tincl	*Ambrosia ambrosioides*	Compositae
tinóopa (see *tonóopa*)		
tis	*Acacia greggii*	Leguminosae
tomáasa	*Oligomeris linifolia*	Resedaceae
tomítom hant cocpétij '*tomítom* land circular-flat' "prostrate *tomítom*"	*Euphorbia polycarpa*/spp. *Achyronychia cooperi*	Euphorbiaceae Caryophyllaceae
tomítom hant cocpétij caacöl '*tomítom* land circular-flat (*E. polycarpa*/spp.) large(plural)' "large spurge"	*Euphorbia tomentulosa* *Achyronychia cooperi*	Euphorbiaceae Caryophyllaceae
tonóopa	*Vallesia glabra*	Apocynaceae
tziino ixám 'Chinaman his-squash' "Chinaman's squash"	*Cucurbita mixta* *C. moschata*	Cucurbitaceae
xaasj	*Pachycereus pringlei*	Cactaceae
xam	*Cucurbita mixta* *C. moschata*	Cucurbitaceae
xamáasa (see *tomáasa*)		
xam coozalc 'squash ribbed' "ribbed squash"	*Cucurbita pepo*	Cucurbitaceae
xapij	*Phragmites australis* *Arundo donax*	Gramineae
xapij-aacöl '*Phragmites* large(plural)' "large reedgrass"	*Arundo donax*	Gramineae
xapij-aas 'reedgrass-*aas*'	*bamboo?	Gramineae
xapij coatöj '*Phragmites* sweet' "sweet reedgrass"	*Saccharum officinarum*	Gramineae
xasáacoj	*Stenocereus alamosensis*	Cactaceae
xat 'hail'	*Tillaea erecta*	Crassulaceae
xazácöz	*Argemone mexicana* *A. pleiacantha*	Papaveraceae
xazácöz caacöl '*Argemone* large(plural)' "large prickly poppy"	*Argemone subintegrifolia*	Papaveraceae

SERI NAME	SCIENTIFIC NAME	FAMILY
xcocoj	*Struthanthus haenkeanus*	Loranthaceae
xcoctz 'old'	*Ambrosia deltoidea* *A. divaricata* *A. dumosa* *A. ilicifolia* *A. magdalenae*	Compositae
xeescl	*Hyptis emoryi*	Labiatae
xeezej islítx 'badger its-inner-ear' "badger's inner ear"	*Nicotiana clevelandii*	Solanaceae
xepe an ihíms 'sea in its-fringe' "fringe of the sea"	*Hypnea valentiae* *Laurencia johnstonii* *Spyridia filamentosa*	Marine alga Marine alga Marine alga
xepe an impós 'sea in *Muhlenbergia* (or *Aristida-* *adscensionis*)' "small grass in the sea"	*Gelidiopsis variabilis* *Gymnogongrus johnstonii* *Hypnea valentiae* *Laurencia johnstonii* *Spyridia filamentosa*	Marine alga Marine alga Marine alga Marine alga Marine alga
xepe oil 'sea tuft' "sea fan"	*Galaxaura fastigiata*	Marine alga
xepe oil caitic 'sea tuft soft' "soft sea fan"	*Eucheuma uncinatum*	Marine alga
xepe oohit 'sea what-it-eats' "what the sea eats"	*Amphiroa beauvoisii* *A. van-bosseae* *Digenia simplex*	Marine alga Marine alga Marine alga
xepe yazj 'sea its-membrane'	*Cutleria hancockii*	Marine alga
xepe zatx 'sea glochid'	*Dictyota flabellata*	Marine alga
xica caacöl 'things large(plural)' "large things"	*Zea mays*	Gramineae
xica coosotoj 'things thin(plural)' "thin things"	*Oryza sativa*	Gramineae
xica ihíijim coopl 'things its-sight black(plural)' "black-eyed things"	*Vigna unguiculata*	Leguminosae
xica imám coopol 'things its-fruit black' "black fruited things"	*Ziziphus obtusifolia*	Rhamnaceae
xica is cheel 'things its-seeds red' "red seeded things"	*Phaseolus vulgaris*	Leguminosae
xica potáat cmis 'things maggots what-resemble' "what resemble maggots"	*Oryza sativa*	Gramineae

SERI NAME	SCIENTIFIC NAME	FAMILY
xica quiix 'things globular(plural)' "globular things"	*Setaria macrostachya*	Gramineae
xjii	** Lagenaria siceraria*	Cucurbitaceae
xloolcö	*Erythrina flabelliformis*	Leguminosae
xnaa caaa 'south-wind what-calls' "what calls for the south wind"	*Salicornia bigelovii*	Chenopodiaceae
xneejam is hayáa '*xneejam* its-seed what-is owned' "*xneejam* whose seeds are owned"	*Viscainoa geniculata*	Zygophyllaceae
xneejam-siictoj '*xneejam* red'	*Stegnosperma halimifolium*	Phytolaccaceae
xojásjc	*Sporobolus virginicus*	Gramineae
xoját	*Amoreuxia palmatifida*	Cochlosper- maceae
xoját hapéc '*Amoreuxia* what-is-planted' "cultivated *saiya*"	** Solanum tuberosum*	Solanaceae
xomáasa (see *tomáasa*)		
xomcahíift	*Lippia palmeri*	Verbenaceae
xomcahóij	*Opuntia marenae* *O. reflexispina*	Cactaceae
xomée	*Marsdenia edulis*	Asclepiadaceae
xométe	*Psorothamnus emoryi*	Leguminosae
xomítom hant cocpétij (see *tomítom* . . .)		
xomítom hant cocpétij caacöl (see *tomitom* . . .)		
xomxéziz	*Fouquieria splendens*	Fouquieriaceae
xomxéziz caacöl '*Fouquieria-splendens* large(plural)' "large ocotillo"	*Chorizanthe rigida* *Fouquieria diguetii*	Polygonaceae Fouquieriaceae
xonj	*Proboscidea altheifolia*	Martyniaceae
xonj caacöl '*Proboscidea-altheifolia* large(plural)' "large devil's claw"	*Proboscidea parviflora*	Martyniaceae
xonj itáast cmis '*Proboscidea* its-tooth what-resembles' "like devil's claw's tooth"	** Musa sapientum*	Musaceae
xooml	*Koeberlinia spinosa*	Koeberliniaceae
xoop	*Bursera microphylla*	Burseraceae
xoop caacöl '*Bursera-microphylla* large' "large elephant tree"	*Bursera laxiflora*	Burseraceae
xoop inl '*Bursera-microphylla* its-fingers' "elephant tree's fingers"	*Bursera hindsiana*	Burseraceae

SERI NAME	SCIENTIFIC NAME	FAMILY
xpaasni	*Ficus petiolaris*	Moraceae
xpacóocsim (see *xpaxóocsim*)		

xpanáams
    'sea in fringe'
    "fringe of the sea"
	Sargassum herporhizum
	S. sinicola
		Marine alga

xpanáams caacöl
    'seaweed large'
    "large seaweed"
	Sargassum sinicola	Marine alga

xpanáams caitic
    'seaweed soft'
    "soft seaweed"
	Dasya baillouviana	Marine alga

xpanáams ccapxl
    'seaweed bitter'
    "bitter seaweed"
	Amphiroa beauvoisii
	A. van-bosseae
	Galaxaura fastigiata
		Marine alga
		
		Marine alga

xpanáams cheel
    'seaweed red'
    "red seaweed"
	Rhodymenia divaricata	Marine alga

xpanáams coil
    'seaweed green'
    "green seaweed"
	Enteromorpha
	    acanthophora
		Marine alga

xpanáams coopol
    'seaweed black'
    "black seaweed"
	Digenia simplex	Marine alga

xpanáams coozlil
    'seaweed sticky'
    "sticky seaweed"
	Rhodymenia hancockii	Marine alga

xpanáams cöquihméel
    'seaweed purple'
    "purple seaweed"
	Rhodymenia hancockii	Marine alga

xpanáams cquihöj
    'seaweed red'
    "red seaweed"
	Lomentaria catenata
	Plocamium cartilagineum
		Marine alga
		Marine alga

xpanáams hasít 'seaweed earring'	unidentified	Marine alga

xpanáams isoj
    'seaweed true(genuine)'
    "real seaweed"
	Sargassum herporhizum
	S. sinicola
		Marine alga

xpanáams itojípz 'seaweed its-eyelashes'	*Lomentaria catenata*	Marine alga

xpanáams mojépe
    'seaweed sahuaro'
    "sahuaro seaweed"
	unidentified	Marine alga

xpanáams oaf 'seaweed waist-cord'	*Sargassum sinicola*	Marine alga

xpanáams ool
    'seaweed *Stenocereus-thurberi*'
    "organ pipe seaweed"
	Rhodymenia hancockii	Marine alga

xpanáams xaasj
    'seaweed *Pachycereus*'
    "*cardón* seaweed"
	unidentified	Marine alga

SERI NAME	SCIENTIFIC NAME	FAMILY
xpanéezj 'sea membrane'	*Enteromorpha* *acanthophora* unidentified	Marine alga Marine alga
xpanéezj cheel 'sea-membrane red' "red sea membrane"	unidentified	Marine alga
xpaxóocsim '*xpax*-chew and spit out'	*Batis maritima*	Bataceae
xpeetc 'sea scrotum'	*Colpomenia tuberculata*	Marine alga
xtamáaij-oohit 'mud-turtle what-it-eats' "what mud turtles eat"	*Nemocladus glanduliferus*	Campanulaceae
xtamóosn-oohit 'desert-tortoise what-it-eats' "what desert tortoises eat"	*Chaenactis carphoclinia* *Chorizanthe brevicornu* *Fagonia californica* *F. pachyacantha*	Compositae Polygonaceae Zygophyllaceae
xtisil	*Porophyllum gracile*	Compositae
xtooxt	*Neoevansia striata*	Cactaceae
xtoozp	*Physalis crassifolia*	Solanaceae
xtoozp hapéc '*Physalis-crassifolia* what-is-planted' "cultivated desert ground cherry"	**Lycopersicon esculentum*	Solanaceae
yori imóon 'Mexican his-beans' "Mexican's beans"	** Vigna unguiculata*	Leguminosae
zaah coocta 'sun watcher'	*Lupinus arizonica* *Machaeranthera parvifolia*	Leguminosae Compositae
zaaj iti cocái 'cliff on what-hangs' "cliff hanger"	*Eucnide rupestris*	Loasaceae
zaaj iti cocái cooxp 'cliff on what-hangs white' "white *Eucnide rupestris*"	*Mentzelia involucrata*	Loasaceae
zai	unidentified grass	Gramineae
zamij cmaam 'palm female' "female palm"	*Brahea aculeata*	Palmae
zamij ctam 'palm male' "male palm"	*Sabal uresana* and/or *Washingtonia robusta*	Palmae
zazjc	*Castela polyandra*	Simaroubaceae
zazjc caacöl '*Castela-polyandra* large(plural)' "large *Castela polyandra*"	*Castela emoryi*	Simaroubaceae
ziij	*Cercidium floridum*	Leguminosae
ziim caacöl '*ziim* large(plural)' "large *ziim*"	**Salsola kali*	Chenopodiaceae

SERI NAME	SCIENTIFIC NAME	FAMILY
ziim caitic '*ziim* soft' "soft *ziim*"	*Amaranthus fimbriatus*	Amaranthaceae
ziim quicös '*ziim* prickly' "prickly *ziim*"	*Amaranthus palmeri*	Amaranthaceae
ziim xat '*ziim* hail'	*Chenopodium murale*	Chenopodiaceae
ziipxöl	*Cercidium microphyllum*	Leguminosae
ziix atc casa insíi 'thing testicle putrid not-smell'	*Cressa truxillensis*	Convolvulaceae
ziix hant cpatj oohit 'thing land flat what-it-eats' "thing that the flounder eats"	*Amphiroa beauvoisii* *A. van-bosseae* *Asparagopsis taxiformis* *Dictyota flabellata* *Galaxaura arborea*	Marine alga Marine alga Marine alga Marine alga Marine alga
ziix is ccapxl 'thing its-fruit sour' "sour-fruited thing"	*Stenocereus gummosus*	Cactaceae
ziix is cmasol 'thing its-fruit yellow' "yellow-fruited thing"	*Cucurbita digitata*	Cucurbitaceae
ziix is coil 'thing its-fruit green' "green-fruited thing"	*Citrullus lanatus*	Cucurbitaceae
ziix is cquihöj 'thing its-seeds red' "red seeded thing"	*Phaseolus vulgaris*	Leguminosae
ziix is quicös 'thing its-fruit prickly' "prickly-fruited thing"	*Matelea pringlei*	Asclepiadaceae
ziix istj captalca 'thing its-leaves wide(plural)' "wide-leaved thing"	*Opuntia violacea*	Cactaceae
ziizil	*Muhlenbergia microsperma* *Setaria liebmannii*	Gramineae Gramineae
zimö-taa 'when past-be' "how did it happen!"	*Ruppia maritima*	Ruppiaceae
znaazj	*Castela polyandra*	Simaroubaceae

*Plant not native to the Seri region.

APPENDIX B

Seri People Named in the Text

AURORA ASTORGA, born about 1942; daughter of José Astorga and Rosa Flores; renowned for her ironwood sculpture.

JOSÉ ASTORGA, born about 1915; husband of Rosa Flores and founder of the Seri ironwood sculpture industry.

ROSA ÁVILA, died about 1944; granddaughter of Coyote Iguana and the mother of Ramona Casanova.

CHAPO BARNET, born in 1936; son of Miguel Barnet and María Victoria Astorga; well known for performing traditional dances at fiestas.

MIGUEL BARNET, born about 1915; a good boat maker in his early years; half-brother of Rosa Flores.

LOLITA BLANCO, born about 1929; youngest daughter of Santo Blanco, and half-sister of María Antonia Colosio; well known for her basketry and traditional knowledge of plants.

MATILDA BLANCO, born about 1928; daughter of Ramón Blanco and Chona.

RAMÓN BLANCO, died about 1929; made one of the last balsas in the 1920s.

SANTO BLANCO was born in the latter part of the nineteenth century and died in 1939. A well-known singer and shaman, he was featured by Coolidge and Coolidge (1939), Davis (Quinn and Quinn 1965), and others. He was said to be the first Seri to drive a motor vehicle—a truck used to haul water and food on periodic trips from the Pozo Peña camp to the coast in the 1930s.

AUGUSTINA BURGOS, born about 1908 and died in 1971; mother of Carmelita Burgos.

LOLA CASANOVA was a young Mexican woman who was captured by Coyote Iguana in 1850 and became his wife. The story became the basis of a romantic Mexican story, several novels, and a movie. However, the Seri oral history version seems to be the most accurate account (Lowell 1970). Many Seri trace their ancestry to Lola Casanova.

RAMONA CASANOVA was born about 1920 near a large ironwood tree at a camp called *Cyajoj* on the southern part of Tiburón Island. Her father was Chico Francisco, and her mother, Rosa Avila, was the granddaughter of Coyote Iguana (Jesús Avila) and Lola Casanova. Ramona married Roberto Herrera in 1935. They raised seven children.

MARÍA LUISA CHILIÓN was born around 1887 at Rancho Santa Ana and died in El Desemboque in 1978. She was known as *Comcáii Ilít Cooxp* "Old Woman of White Hair." She was one of the *Xica Xnai Ic Coii* "they who live on the side of the south wind"—the people from the Tastiota Region. She was already married when the boat-burning incident occurred at Punta Perla in 1904 (wooden boats which the Seri had made or stolen from Mexican fishermen were burned by authorities in retaliation for Seri actions). She married Urbano Sánchez and they had six children, two of whom were living in 1984. María Luisa was one of the last Seri potters.

AURORA COLOSIO, born about 1915; known as an excellent maker of baskets, dolls, and traditional Seri dresses; sister of María Antonia Colosio, and very knowledgeable about plants and the desert.

CARLOTA COLOSIO (COMITO), born about 1924; daughter of Pedro Colosio and Loreto Marcos; an industrious basketmaker and very knowledgeable about plants; sister of María Antonia Colosio.

MARÍA ANTONIA COLOSIO was born around 1904 at *Hast An Iicp* "The Side Toward the Mountains," near *Cyajoj* on Tiburón Island. Her father was Pedro Colosio and her mother was Loreto Marcos. María Antonia was known among the Seri as *Comcáii Mayóor* "Elder Old Woman." She was the oldest of seven living brothers and sisters; the youngest, a half-sister, was 53 years old in 1984. María Antonia was very knowledgeable about the entire Seri region, particularly Tiburón Island and the opposite mainland coast. She believed the time of her birth was approximately the year when the renegade Yaqui camping on the northwest side of Tiburón Island were killed by the Seri, and their hands and hats presented to the Mexican authorities. This incident occurred in 1904 (García y Alva 1905:19–33). Her husband was Jesús Morales, and they raised six children.

LUPE COMITO, born about 1913; daughter of Pedro Colosio and Loreto Marcos and wife of Nacho Morales; well known for her basketmaking skill.

PEDRO COMITO, born about 1910; brother of María Antonia Colosio and son of Pedro Colosio and Loreto Marcos. When he was a child his family moved to the interior of Tiburón Island where they lived for several years. He was the last boy to have a puberty fiesta.

PORFIRIO DÍAZ, deceased at an advanced age in 1959 at El Desemboque. His father was Quicilio and his mother was María Juana Necia. Porfirio had extensive traditional knowledge, some of which was recorded by Edward and Mary Beck Moser. Porfirio knew the distinctive "sing-song" dialect of the San Esteban people. Although he was not born on San Esteban, his parents went to live on the island when he was a boy. He was said to be the last San Esteban person, and was known as an *hast ano ctam* 'mountain from man,' or "a man from San Esteban." He and his wife Juana had at least seven children.

RAMONA DÍAZ died in 1958 at an advanced age.

MANUEL ENCINAS died around 1931. The grandfather of Roberto Herrera M., he was mentioned frequently by Edward H. Davis and the Coolidges.

MARÍA FÉLIX, born about 1939; married to Pancho Hofer. Her father, Rafael Félix, was the son of Ramón Blanco.

ROSA FLORES was born about 1916 at *Tis Cyeeno* (Rancho Costa Rica). Her parents were Juan Flores and Catalina Contreras. She married José Astorga and they raised eight children and had many grandchildren. Rosa had extensive knowledge of the northern part of the Seri territory and Tiburón Island. About 1980 Rosa and José moved to a small ranch between Pozo Coyote and El Desemboque, where they maintained the well which supplied water to El Desemboque.

ROBERTO HERRERA M., husband of Ramona Casanova, was born in 1916 at *Hatájc* (Pozo Coyote). His parents were Antonio Herrera and María Marcos, and he was baptized Roberto Thomson by Roberto Thomson Encinas of Hermosillo. He learned Spanish at an early age. He gave assistance to many researchers and was a primary consultant for Edward and Mary Beck Moser, helping them with translation and linguistic work. At various times he was Comisario in El Desemboque, Presidente de la Cooperativa de Pesca, and Jefe de la Tribu Seri. He was particularly knowledgeable about the sea, hunting, and animals in general. Over the years he was patient and meticulous in explaining detailed and often complex information and concepts.

RAQUEL HOFER, born in 1964; daughter of Pancho Hofer and María Félix.

CARLOS IBARRA, born in the early part of the twentieth century; brother-in-law of María Antonia Colosio. He died in a tragic boat accident in 1950 in which ten Seri perished.

JESÚS IBARRA was born in the late 1880s and died in 1973.

COYOTE IGUANA (JESÚS ÁVILA). In 1850 a group of Seri men, including Coyote Iguana, attacked a stage coach at La Pintada between Guaymas and Hermosillo. All of the passengers and crew were killed except Lola Casanova, who was kidnapped by Coyote Iguana and eventually became his wife. He was said to have extraordinary shamanistic and supernatural powers, and to have led an unusually colorful but violent life (which was not unusual for the times). He was killed by soldiers in the latter part of the nineteenth century (see Lowell 1970).

JUANA, the wife of Porfirio Díaz, died in 1965 at an advanced age.

ANITA LÓPEZ, born about 1946; sister of Evangelina Lopez and wife of Felipe Romero.

ANTONIO LÓPEZ, born about 1916; husband of Lola, a direct descendent of Coyote Iguana and Lola Casanova.

GUADALUPE LÓPEZ, born about 1930; very knowledgeable about sea turtles and one of the last great turtle hunters.

JUAN MARCOS (younger) died about 1919. He was the father of Antonio López.

LORETO MARCOS was born in the late 1800s and died in 1968. She knew many songs and was a fine basket maker.

JUAN MATA, brother of Ramón Blanco, was born in the late nineteenth century and died about 1943. He was orphaned at an early age. As a boy he stayed for a long time on ranches and received help from the Mexican ranchers. He made the first Seri wooden boat and was known for his shamanistic powers.

ROSITA MÉNDEZ, born in the early 1920s; known for her basketmaking skill.

ALBERTO MOLINO "VAQUERO" VILLALOBOS, born around 1906. When he was a boy he lived with his father, Manuel Molino (younger) between Kino Bay and Tastiota. Although blind in his later years, he carved ironwood sculptures.

MANUEL MOLINO (elder); grandfather of Alberto Molino Villalobos. In the mid-nineteenth century he planted dates and *guamúchil* trees at one or more camps on the coast south of Cerro San Nicolás.

NACHO MOLINO, born about 1924; known as a good boat maker and ironwood sculptor.

CAYETANO MONTAÑO was born about 1924. He and his family lived at Punta Chueca for several decades and were known for embracing traditional values.

JESÚS MORALES was born about 1904 at *Cösaacöla* on the Infiernillo coast of Tiburón Island and died in 1975 at El Desemboque. According to what was recounted to him, he was born nine days after his wife, María Antonia Colosio, was born. He was well known among the Seri for his traditional knowledge, including a vast repertoire of Seri songs. He was always sought after to sing at fiestas, and was a noted violinist. He was able to recall in detail many older customs and practices—including older artifacts and the processes involved in making them—most of which he had learned as a young man. He was an important consultant for Thomas Bowen, Edward Moser, and others. He made many artifacts, some of them replicas, which are in various museum collections, such as the Instituto Nacional de Antropología e Historia in Hermosillo, the Arizona State Museum, the Amerind Foundation, and the Anthropology Museum of the University of Michigan.

NACHO MORALES was one of the last practicing shamans with extensive traditional knowledge, and was widely acknowledged for his shamanistic powers. He was born about 1910 and died in 1969 in El Desemboque. He was married to Lupe Comito. They had five daughters, four of whom were living in 1984.

JUANA NECIA died in the late nineteenth century. She was the aunt of Miguel Barnet and the sister of Francisco Molino, who was Charles Sheldon's guide (Sheldon 1979).

GUILLERMO ORTEGA was born in the 1920s. His brother was Nacho Morales and his mother-in-law was Sara Villalobos.

ANDREA ROMERO, born about 1910; step-daughter of Chico Romero; known for her basketry skill.

CHICO ROMERO was born about 1888 at Rancho Costa Rica and died in El Desemboque in 1974. His parents were Juan Chávez and María Antonia. When Chico was about six years old his father was arrested and never seen again by the Seri. The incident was mentioned by McGee (1898:120): Chico's father was one of the seven warriors captured, "taken to Hermosillo, tried, and, according to oral accounts, banished," because of supposed involvement with the killing of two North Americans on Tiburón Island. Chico was a large man, over six feet tall, and was prominent in relationships with outsiders, often representing the Seri to various government officials. He served as a consultant to Alfred L. Kroeber in 1930 and assisted subsequent researchers and authors. During his later years he became blind.

TOLA lived in the nineteenth century. She was the wife of Manuel Molino (elder).

ANA TORRES, born in 1953; sister of Angelita Torres.

ANGELITA TORRES, born about 1941; daughter of Elvira Valenzuela and José Torres; well known for her excellent baskets and her skill as a seamstress of fine Seri skirts and blouses.

ARMANDO TORRES, born about 1944; brother of Angelita Torres. Although handicapped and unable to walk, he helped support his mother and sisters with his exquisite ironwood sculpture.

LUIS TORRES died in 1948 at an advanced age. He was a brother of Sara Villalobos.

LUIS TORRES (elder) died in 1931 at Rancho Santa Cruz during a flu epidemic. He was the father of Chavela, the wife of Cayetano Montaño.

ELVIRA VALENZUELA FÉLIX was born about 1915 at *Hastoj Isxoj* "Gray Mountains" near *Quipcö Quih an Icahéme* "Camp on the Thick Sand Dune," on the mainland side of the Infiernillo coast. Her parents were Ilario Covilla and Margarita. When she was a girl, her extended family roamed the desert north and east of El Desemboque. Her husband, José Torres, died in 1962. She raised six children. Elvira was particularly knowledgeable about the northern part of the Seri region. She said that when she was young she hardly knew the people on Tiburón Island and never went to Kino Bay.

SARA VILLALOBOS was born about 1900 at *Masánaj* near Punta Sargento, and died at El Desemboque in November 1982. Her parents were Jesús Félix* (Buro Alazán) and Rosa Castilla. Her ancestors came from the Tastiota region. Widowed at an early age, she struggled against great adversity to raise her family. Of her nine children, two sons drowned, one was killed, and by 1965 only her daughter Selsa Félix was still living. She had many descendents, including two great great grandchildren while she was still living.

She had extensive knowledge of the desert and traditional information, and added much to our understanding of Seri culture. She was well known for her outgoing, magnanimous personality, and visitors remembered her and looked for her on their return visits. She was much photographed and featured in a number of publications.

---

* Her husband was also called Jésus Félix (see Figure 9.13).

Literature Cited

Abbott, I. A., and E. H. Williamson
   1974   Limu: an ethnobotanical study of some edible Hawaiian seaweeds. Pacific Tropical Botanical Garden, Lawai, Kauai.

Alegre, F. J.
   1960   Historia de la Provincia de la Compañia de Jesús de Nueva España [1780]. E. J. Burrus and F. Zubillaga, eds., vol. 4. Institutum Historicum S. J., Rome.

Arizona Citizen
   1872   Article on bees brought from California. July 27, 3:2.

Arizona Daily Star
   1897   Article about first bees in Arizona. January 3, 4:4.

Ascher, R.
   1962   Ethnography for archaeology: a case from the Seri Indians. Ethnology 1(3):360–369.

Aschmann, H.
   1959   The Central Desert of Baja California: demography and ecology. Ibero-Americana 42. University of California Press, Berkeley.

Baegert, J. J.
   1952   Observations in Lower California. Translated by M. M. Brandenburg and C. L. Baumann. University of California Press, Berkeley.

Bahre, C. J.
   1967   Historic Seri residence, range, and sociopolitical structure. The Kiva 45(3):197–209.

Bandelier, A. F.
   1890   Contributions to the history of the southwestern portion of the United States. Papers of the Archaeological Institute of America, American Series, vol. 5, J. Wilson and Son, Cambridge. [Reprinted 1976 by Kraus Reprint Co., Millwood, New York.]

Beals, R. L.
   1932   The comparative ethnology of northern Mexico before 1750. Ibero-Americana 2. University of California Press, Berkeley.

Bean, L. J., and K. S. Saubel
   1972   Temalpakh. Malki Museum Press. Banning, California.

Bell, W. H., and E. F. Castetter
   1937   The utilization of mesquite and screwbean by the aborigines in the American Southwest. University of New Mexico Bulletin 314.

Berlin, B.
   1973   Folk systematics in relation to biological classification and nomenclature. Annual Review of Ecology and Systematics 4:259–271.

Berlin, B., D. E. Breedlove, and P. H. Raven
   1968   Covert categories and folk taxonomies. American Anthropologist 70:290–299.

Bohrer, V. L.
   1962   Nature and interpretation of ethnobotanical materials from Tonto National Monument. Pages 75–114, IN: L. R. Caywood, ed. Archaeological studies at Tonto National Monument, Arizona. Southwest Monuments Association Technical Series 2.

Bowen, T.
   1973   Seri basketry: a comparative view. The Kiva 38(3–4):141–172.

   1976   Seri prehistory. Anthropological Papers of the University of Arizona 27. University of Arizona Press, Tucson.

   1983   Seri. Pages 230–249; IN: A. Ortiz, ed. Handbook of North American Indians, vol. 10. Smithsonian Institution, Washington, D.C.

Bowen, T., and E. Moser
   1968   Seri pottery. The Kiva 33(3):89–132.

415

Bowen, T., and E. Moser (*cont.*)
  1970a Seri headpieces and hats. The Kiva 35(4): 168–177.
  1970b Material and functional aspects of Seri instrumental music. The Kiva 35(4):178–200.
Brusca, R.
  1980 Common intertidal invertebrates of the Gulf of California. Second edition, revised. University of Arizona Press, Tucson.
Burckhalter, D.
  1976 The Seris. University of Arizona Press, Tucson.
  1982 The power of Seri baskets, spirits, traditions and beauty. American West 19(1): 38–45.
Carmony, N. B., and D. E. Brown, eds.
  1983 Tales from Tiburon. Southwest Natural History Association, Phoenix.
Castetter, E. F., and W. H. Bell
  1937 The aboriginal utilization of the tall cacti in the American Southwest. University of New Mexico Bulletin 307.
  1951 Yuman Indian agriculture. University of New Mexico Press, Albuquerque.
Castetter, E. F., W. H. Bell, and A. D. Grove
  1938 The early utilization and the distribution of Agave in the American Southwest. University of New Mexico Bulletin 335.
Clavigero, F. J.
  1937 History of (Lower) California. Translated by S. E. Lake and A. A. Gray. Stanford University Press, Stanford.
Cliffton, K., D. O. Cornejo, and R. S. Felger
  1982 Sea turtles of the Pacific coast of Mexico. Pages 199–209, IN: K. Bjorndal, ed. Biological conservation of sea turtles. Smithsonian Institution Press, Washington, D.C.
Coolidge, D., and M. R. Coolidge
  1939 The last of the Seri. E. P. Dutton, New York. [Reprinted 1971 by Rio Grande Press, Glorieta, New Mexico.]
Cornejo, D. O., L. S. Leigh, R. S. Felger, and C. F. Hutchinson
  1982 Utilization of mesquite in the Sonoran Desert: past and future. Pages Q1–Q20, IN: H. W. Parker, ed. Mesquite utilization, proceedings of the symposium. Texas Tech University, Lubbock.
Culin, S.
  1907 Games of the North American Indians. Bureau of American Ethnology Annual Report 24, Washington, D.C.
Curtin, L. S. M.
  1949 By the prophet of the earth. San Vicente Foundation, Santa Fe, New Mexico

Davis, E. H.
  1922, 1924, 1926, 1929, 1934, 1936, and 1939. Field notes on file at the Museum of the American Indian, Heye Foundation, New York.
Dawson, E. Y.
  1944a Some ethnobotanical notes on the Seri Indians. Desert Plant Life 16(9):133–138.
  1944b The marine algae of the Gulf of California. Allan Hancock Pacific Expeditions 3:189–453. University of Southern California Press, Los Angeles.
  1966 Marine algae in the vicinity of Puerto Peñasco. Gulf of California Field Guide Series No. 1. The University of Arizona, Tucson.
del Barco, M.
  1973 Historia natural y crónica de la Antigua California. M. León-Portilla, ed. Universidad Nacional Autonoma de México, Mexico City. [Translated 1980 by F. Tiscareno. The natural history of Baja California. Dawson's Book Shop, Los Angeles.]
Densmore, F.
  1932 Yuman and Yaqui music. Bureau of American Ethnology, Bulletin 110.
Di Peso, C. C.
  1974 Casas Grandes, 3 vols. Northland Press, Flagstaff.
Di Peso, C. C., and D. S. Matson
  1965 The Seri Indians in 1692 as described by Adamo Gilg, S. J. Arizona and the West 7(1):33–56.
Euler, R. C., and V. H. Jones
  1956 Hermetic sealing as a technique of food preservation among the Indians of the American Southwest. Proceedings of the American Philosophical Society 100:87–99.
Felger, R. S.
  1965 *Xantusia vigilis* and its habitat in Sonora, Mexico. Herpetologica 21:146–147.
  1966 Ecology of the Islands and Gulf Coast of Sonora, Mexico. Ph.D. dissertation, University of Arizona.
  1976a Investigaciones ecologicas en Sonora y localidades adyacentes en Sinaloa—una perspectiva. Pages 21–62, IN: B. Braniff C. and R. S. Felger, eds. Sonora, Antropología del desierto. Instituto Nacional de Antropología e Historia, Colección Científica 27. Mexico City.
  1976b The Gulf of California: an ethno-ecological perspective. Natural Resources Journal 16(3):451–464.
  1977 Mesquite in Indian Cultures of Southwestern North America. Pages 150–176, IN: B.

B. Simpson, ed. Mesquite: its biology in two desert scrub ecosystems. Dowden, Hutchinson, and Ross, Stroudsburg, Pennsylvania.

1980   Vegetation and flora of the Gran Desierto, Sonora, Mexico. Desert Plants 2(2):87–114.

Felger, R. S., K. Cliffton, and P. J. Regal
1976   Winter dormancy in sea turtles: independent discovery and exploitation in the Gulf of California by two local cultures. Science 191:283–285.

Felger, R. S., and C. H. Lowe
1976   The island and coastal vegetation and flora of the northern part of the Gulf of California, Mexico. Natural History Museum of Los Angeles County, Contributions in Science 285.

Felger, R. S., and C. P. McRoy
1975   Seagrasses as potential food plants. Pages 62–74, IN: G. F. Somers, ed. Seedbearing halophytes as food plants. University of Delaware, Newark.

Felger, R. S., and M. B. Moser
1970   Seri use of Agave (century plant). The Kiva 35:(4):159–167.

1971   Seri use of mesquite (Prosopis glandulosa var. torreyana). The Kiva 37(3–4):53–60.

1973   Eelgrass (Zostera marina L.) in the Gulf of California: discovery of its nutritional value by the Seri Indians. Science 181:355–356.

1974a   Columnar cacti in Seri Indian culture. The Kiva 39(3–4):257–275.

1974b   Seri Indian pharmacopoeia. Economic Botany 28(4):414–436.

1976   Seri Indian food plants: desert subsistence without agriculture. Ecology of Food and Nutrition 5(1):13–27.

Felger, R. S., M. B. Moser, and E. W. Moser
1980   Seagrasses in Seri Indian culture. Pages 260–276; IN: R. C. Phillips and C. P. McRoy, eds. Handbook of seagrass biology, an ecosystem perspective. Garland STPM Press, New York.

1983   The desert tortoise in Seri Indian culture. Pages 113–120, IN: Desert Tortoise Council, proceedings of 1981 symposium. Desert Tortoise Council, 5319 Cerritos Avenue, Long Beach, California 90805.

Felger, R. S., and G. P. Nabhan
1978   Agroecosystem diversity: a model from the Sonoran Desert. Pages 128–149; IN: N. L. Gonzalez, ed. Social and technological management in dry lands. Westview Press, Boulder, Colorado.

Fenical, W.
1983   Investigation of benthic marine algae as a resource for new pharmaceuticals. Pages 497–521, IN: C. K. Tseng, ed. Proceedings of the joint China–U.S. Phycological Symposium. Science Press, Beijing, China.

Findley, L.
1971   Preliminary notes on the ethno-ichthyology of the Seri Indians. Unpublished manuscript.

Fontana, B. L., and H. Fontana
1983   A search for the Seris, 1895. Pages 23–55, IN: N. B. Carmony and D. E. Brown, eds. Tales from Tiburon. Southwestern Natural History Association, Phoenix.

Freeman, G. F.
1918   Southwestern beans and teparies. University of Arizona Agricultural Experiment Station Bulletin 68:1–55 (revised).

García, E.
1981   Modificaciones al sistema de clasificación climática de Köppen para adaptarlo a las condiciones de la República Méxicana. Third Edition. Instituto de Geografía, Universidad Autonomo de México, Mexico City.

García y Alva, F.
1905   Album—Directorio del Estado de Sonora. Hermosillo, Sonora.

Gastil, R. G., and D. Krummenacher
1977   Reconnaissance geology of coastal Sonora between Puerto Lobos and Bahia Kino. Geological Society of America Bulletin 88:189–198.

Gentry, H. S.
1942   Rio Mayo plants. Carnegie Institution of Washington 527. Washington, D.C.

1949   Land plants collected by the Vallero III, Allan Hancock Pacific Expeditions 1937–1951. Allan Hancock Pacific Expeditions 13. University of Southern California Press, Los Angeles.

1972   The agave family in Sonora. Agriculture Handbook 399. U.S. Department of Agriculture, Washington, D.C.

1978   The agaves of Baja California. California Academy of Sciences Occasional Paper 130, San Francisco.

1982   Agaves of continental North America. University of Arizona Press, Tucson.

Griffen, W. B.
1959   Notes on Seri Indian culture, Sonora, Mexico. Latin American Monographs 10. University of Florida Press, Gainesville.

1961   Seventeenth century Seri. The Kiva 27(2):12–21.

Hardy, R. W. H.
 1829   Travels in the Interior of Mexico in 1825,
        1826, 1827, and 1828. Colburn and Bently,
        London. [Reprinted 1977 by Rio Grande
        Press, Glorieta, New Mexico. Portion deal-
        ing with Tiburón Island reprinted in Car-
        mony and Brown 1983].
Harrington, G.—see G. Xavier
Hastings, J. R., and R. R. Humphrey, eds.
 1969   Climatological data and statistics for
        Sonora and northern Sinaloa. Technical
        Reports on the Meteorology and Clima-
        tology of Arid Regions 19. University of
        Arizona Institute for Atmospheric Physics,
        Tucson.
Hastings, J. R., R. M. Turner, and D. K. Warren
 1972   An atlas of some plant distributions in the
        Sonoran Desert. Technical Reports on the
        Meteorology and Climatology of Arid Re-
        gions 21. University of Arizona Institute of
        Atmospheric Physics, Tucson.
Haury, E. W.
 1976   The Hohokam. University of Arizona Press,
        Tucson.
Hayden, J. D.
 1942   Seri Indians on Tiburon Island. Arizona
        Highways 18(1):22–29, 40–41.
 1969   Gyratory crushers of the Sierra Pinacate,
        Sonora. American Antiquity 34(3):154–
        161.
 1972   Hohokam petroglyphs of the Sierra Pina-
        cate, Sonora, and the Hohokam shell expe-
        ditions. The Kiva 37(2):74–83.
Heizer, R.
 1945   Honey-dew "sugar" in western North
        America. The Masterkey 14:140–145.
Hills, J.
 1973   An ecological interpretation of prehistoric
        Seri settlement patterns in Sonora, Mex-
        ico. Master's thesis. Arizona State Univer-
        sity, Tempe.
 1977   The finishers. American Indian art 2(2):
        32–38.
Hinton, T.
 1955   A Seri girl's puberty ceremony at Desem-
        boque, Sonora. The Kiva 20(4):8–11.
Hole, F., F. V. Flannery, and J. A. Neeley
 1969   Prehistory land human ecology of the Deh
        Luran Plain: an early village sequence
        from Khuzistan, Iran. Memoir 1. Museum
        of Anthropology, University of Michigan.
Janzen, D. H., and P. S. Martin
 1982   Neotropical anachronisms: the fruit the
        gomphotheres ate. Science 215:19–27.

Johnson, B.
 1959   Seri Indian basketry. The Kiva 25(1):
        10–13.
Johnson G., D; R. Aguierre M.; L. Carillo M.; and
        F. Noriega V.
 1965   Vegetación del Estado de Sonora. COTE-
        COCA, Secretaría de Agricultura y Gana-
        dería, Hermosillo.
Johnston, B.
 1968   Seri ironwood carving. The Kiva 33(3):
        156–166.
 1969   Seri sculpture. Pacific Discovery 22(2):
        9–15.
 1970   The Seri Indians. University of Arizona
        Press, Tucson.
Johnston, I. M.
 1924   Expedition of the California Academy of
        Sciences to the Gulf of California in 1921,
        the botany (vascular plants). Proceedings
        of the California Academy of Sciences, 4th
        series, 12:951–1218.
Jones, C. E.
 1978   Pollinator constancy as a pre-pollination
        isolating mechanism between sympatric
        species of Cercidium. Evolution 32:189–
        198.
Jones, V. H.
 1945   The use of honey-dew as food by Indians.
        The Masterkey 14:145–149.
Keen, A. M.
 1971   Sea shells of tropical west America. Second
        edition. Stanford University Press,
        Stanford.
Kirk, D. R.
 1970   Wild edible plants of the western United
        States. Naturegraph Publishers, Happy
        Camp, California.
Kroeber, A.
 1931   The Seri. Southwest Museum Papers,
        number 6.
Langdon, M., and S. Silver, eds.
 1976   Hokan studies. Mouton, The Hague.
Lee, R. B., and I. DeVore, eds.
 1968   Man the hunter. Aldine Publishing Com-
        pany, Chicago.
Lowell, E. S.
 1970   A comparison of Mexican and Seri Indian
        versions of the legend of Lola Casanova.
        The Kiva 35(4):144–158.
Lubchenco, J. L., and J. E. Cubit
 1980   Effects of herbivors on heteromorphology
        in some marine algae. Ecology 61(3):676–
        687.

McConnell, O. J., and W. Fenical
   1976   Halogen chemistry. Phytochemistry 16: 367–368.

McGee, W. J.
   1894–1896   Diary, on file at U.S. Library of Congress. [Portion covering December 14 through December 28, 1895, in Fontana and Fontana 1983.]
   1898   The Seri Indians. Seventeenth Annual Report of the Bureau of American Ethnology, Part 1. Smithsonian Institution, Washington, D.C. [Reprinted 1971, as The Seri Indians of Bahia Kino and Sonora, Mexico. Rio Grande Press, Glorieta, New Mexico, with additional photographs and introduction, list of references cited, and index by B. L. Fontana.]

Mails, T. E.
   1974   The people called Apache. Prentice-Hall Inc., Englewood Cliffs, New Jersey.

Malkin, B.
   1962   Seri ethnozoology. Occasional Papers of the Idaho State College Museum 7.

Marlett, S. A.
   1981   The structure of Seri. Ph.D. dissertation, University of California, San Diego.

Mason, O. T.
   1894   North American bows, arrows, and quivers. Pages 631–680, IN: Annual Report of the Smithsonian Institution for 1893. Washington, D.C.

Massey, W. C.
   1966   Archaeology and ethnohistory of Lower California. Pages 38–58, IN: G. F. Eckholm and G. R. Willey, eds. Archaeological Frontiers and External Connections. Handbook of Middle American Indians, vol. 4. University of Texas Press, Austin.

Medsger, O. P.
   1939   Edible wild plants. The Macmillan Company, New York.

Meigs, P.
   1953   World distribution of arid and semi-arid homoclimates. Reviews of Research on Arid Zone Hydrology, Arid Zone Programme 1: 203–209. UNESCO, Paris.
   1966   Geography of coastal deserts. Arid Zone Research 28. UNESCO, Paris.

Moran, R., and R. S. Felger
   1968   *Castela polyandra*, a new species in a new section: union of *Holacantha* with *Castela* (Simaroubaceae). Transactions of the San Diego Society of Natural History 15(4): 31–40.

Moser, E. W.
   1963a   Seri bands. The Kiva 28(3): 14–27.
   1963b   The two brothers who went away angry: a Seri legend. Tlalocan 4(2): 157–160.
   1968   Two Seri myths. Tlalocan 5(4): 364–367.
   1973   Seri basketry. The Kiva 38(3–4): 105–140.
   1976   Guia bibliográfica de las fuentes para el estudio de la etnografía seri. Pages 365–375, IN: B. Braniff C. and R. S. Felger, eds. Sonora, Antropología del desierto. Instituto Nacional de Antropología e Historia, Colección Científica 27. Mexico City.

Moser, E., and R. S. White
   1968   Seri clay figurines. The Kiva 33(3): 133–154.

Moser, M. B.
   1964   Seri blue. The Kiva 30(2): 27–32.
   1970a   Seri: from conception through infancy. The Kiva 35(4): 201–210. [Revised 1982. Pages 221–232, IN: M. A. Kay, ed. Anthropology of human birth, F. A. Davis Company, Philadelphia].
   1970b   Seri elevated burials. Kiva 35(4): 211–216.

Nabhan, G. P., and R. S. Felger
   1978   Teparies in southwestern North America. Economic Botany 32(1): 2–19.

Nolasco, M.
   1967   Los Seris, desierto y mar. Anales del Instituto Nacional de Antropología e Historia 18: 125–194.

Norris, J. N.
   1978   Seri Indian seaweed knowledge: taxonomy, ecology and uses. Pages 205–206, IN: American Philosophical Society Yearbook, 1977. American Philosophical Society, Philadelphia.
   In prep.   Marine algae of the northern Gulf of California.

Norris, J. N., and K. E. Bucher
   1976   New records of marine algae from the 1974 R/V Dolphin Cruise to the Gulf of California. Smithsonian Contributions to Botany, No. 34.

Núñez Cabeza de Vaca, A.
   1555   La relacion y comentarios del gouernador Aluar nuñez cabeça de vaca, de lo acaescido en las dos jornadas que hizo a las Indias. Francisco fernandez de Cordoua, Valladolid.

Ogren, L., and C. McVea, Jr.
   1982   Apparent hibernation by sea turtles in North American waters. Pages 127–132,

Ogren, L., and C. McVea, Jr. (*cont.*)
   IN: K. Bjorndal, ed. Biology and conservation of sea turtles. Smithsonian Institution, Washington, D.C.

Oviedo y Valdés, G. de
1972   Joint report to the Audencia of Santo Domingo, translated by G. Theisen. Pages 161–271, IN: The narrative of Alvar Núñez Cabeza de Vaca. The Imprint Society, Barre, Massachusetts.

Pennington, C. W.
1963   The Tarahumar of Mexico. University of Utah Press, Salt Lake City.
1980   The Pima Bajo. University of Utah Press, Salt Lake City.

Perera, N. de
1729   Carta al Padre Visitador Nicolas de Oro (September 17, 1729). Archivo Histórico de Hacienda, Temporalidades 17, 57. Mexico City.

Pérez de Ribas, A.
1944   Historia de los triunfos de nuestra Santa Fé entre gentes las más bárbaras y fieras del Nuevo Orbe . . . , vols. 1 and 2. Editorial "Layac," Mexico City.

Pfefferkorn, I.
1949   Sonora: a description of the province. Translated by T. E. Treutlein. Coronado Cuarto Centennial Publications, 12. University of New Mexico Press, Albuquerque.

Pierce, H. W.
1964   Seri blue—an explanation. The Kiva 30 (2):33–39.

Quinn, C. R., and E. Quinn, eds.
1965   Edward H. Davis and the Indians of the southwest United States and northwest Mexico. Elena Quinn, Downey, California.

Raunkiaer, C.
1934   The life-forms of plants and statistical plant geography; being the collected papers of C. Raunkiaer. Clarendon Press, Oxford.

Record, S. J., and R. W. Hess
1943   Timbers of the New World. Yale University Press, New Haven.

Robinson, M. K.
1973   Atlas of monthly mean sea surface and subsurface temperatures in the Gulf of California, Mexico. San Diego Society of Natural History Memoir 5.

Robinson, R. E. L., and G. F. Flavell
1894   Cruise of the sloop Examiner in the Gulf of California from Yuma, Arizona to Guaymas, Mexico. On file at U.S. Park Service Archives, Grand Canyon National Park.

Roden, G. I.
1964   Oceanographic aspects of the Gulf of California. Pages 30–58, IN: T. H. van Andel and G. G. Shore, Jr., eds. Marine geology of the Gulf of California. American Association of Petroleum Geologists, Memoir 3., Tulsa.

Rosenthal, G. A.
1977   The biological effects and mode of action of L-canavanine, a structural analogue of L-arginine. Quarterly Review of Biology 52:155–178.

Runsk, G. A., and R. L. Fisher
1964   Structural history and evolution of the Gulf of California. Pages 144–156, IN: T. H. van Andel and G. G. Shore, Jr., eds. Marine geology of the Gulf of California. American Association of Petroleum Geologists, Memoir 3. Tulsa.

Russell, F.
1908   The Pima Indians. Annual Report of the Bureau of American Ethnology 26:3–389. Washington, D.C., Government Printing Office. [Reprinted 1975, University of Arizona Press, Tucson, with introduction, citation sources, and bibliography by B. L. Fontana].

Ryerson, S.
1976   Seri ironwood carving: an economic view. Pages 119–136, IN: N. H. H. Graburn, ed. Ethnic and tourist arts. University of California Press, Berkeley.

Schindler, S.
1981   The material culture and technoeconomic system of the Seri Indians: an experimental reconstruction. Ph.D. dissertation. Southern Illinois University, Carbondale.

Sheldon, C.
1979   The wilderness of desert bighorns and Seri Indians. The Arizona Desert Bighorn Sheep Society, Phoenix. [Section on the Seri reprinted in Carmony and Brown 1983].

Sheridan, T. E.
1979   Cross or arrow? The breakdown in Spanish-Seri relations 1729–1750. Arizona and the West 21(4):317–334.
1982   Seri bands in cross-cultural perspective. The Kiva 47(4):185–213.

Sheridan, T. E., and R. S. Felger
1977   Indian utilization of eelgrass (*Zostera marina* L.) in northwestern Mexico: the Spanish colonial record. The Kiva 43:89–92.

Shreve, F.
1951   Vegetation of the Sonoran Desert. Carnegie Institution of Washington, Publication 591.

Shreve, F., and I. L. Wiggins
1964   Vegetation and flora of the Sonoran Desert, vols. I and II. Stanford University Press, Stanford.

Simpson, B. B.
1977   Mesquite: its biology in two desert scrub ecosystems. Dowden, Hutchinson, and Ross, Stroudsburg, Pennsylvania.

Smith, W. N.
1959   Observations regarding Seri Indian basketry. The Kiva 25(1):14–17.
1974   The Seri Indians and the sea turtles. The Journal of Arizona History 15(2):139–158.

Spicer, E. H.
1962   Cycles of conquest. University of Arizona Press, Tucson.
1980   The Yaquis. University of Arizona Press, Tucson.

Standley, P.
1920–1926   Trees and shrubs of Mexico. Contributions from the United States National Herbarium, vol. 23. Smithsonian Institution, Washington, D.C.

Tanaka, T.
1976   Tanaka's cyclopedia of edible plants of the world. Keigaku Publishing Co., Tokyo.

Thomson, D. A., L. T. Findley, and A. N. Kerstitch
1979   Reef fishes of the Sea of Cortez. John Wiley & Sons, New York

Turner, N. C., and A. M. Bell
1973   The ethnobotany of the Southern Kwakiutl Indians of British Columbia. Economic Botany 27(3):257–310.

Uphof, J. C. T.
1968   Dictionary of economic plants. Second edition, revised. J. Cramer, Lehre, Germany.

United States Hydrographic Office
1880   The west coast of Mexico, from the boundary line between the United States and Mexico to Cape Corrientes, including the Gulf of California. Publications of the U. S. Hydrographic Office, Bureau of Navigation, number 56. Government Printing Office, Washington, D.C.

Velasco, J. F.
1850   Noticias estadisticas del estado de Sonora. Ignacio Cumplido, Mexico City.

Venegas, M.
1759   A natural and civil history of California. 2 vols. J. Rivington and J. Fletcher, London. [Reprinted 1966 by University Microfilms, Inc., Ann Arbor.]

Wagner, H. R.
1929   Spanish voyages to the northwest coast of America in the Sixteenth Century. California Historical Society, San Francisco.

West, J. A., and M. H. Hommersand
1981   Rhodophyta: life histories. Pages 133–193, IN: C. S. Lobban and M. J. Wynne, eds. The biology of seaweeds. Botanical Monographs 17. University of California Press, Berkeley.

Whiting, A. F.
1951   A Seri diary, June 11 to 24, 1951. On file at Arizona State Museum, University of Arizona, Tucson.
1957   Some preliminary notes on the ethnobotany of the Seri. On file at Arizona State Museum, University of Arizona, Tucson.

Wiggins, I. L.
1964   Flora of the Sonoran Desert. Part II, IN: F. Shreve and I. L. Wiggins, Flora and Vegetation of the Sonoran Desert. 2 vols. Stanford University Press, Stanford.
1980   Flora of Baja California. Stanford University Press, Stanford.

Woodward, A.
1938   Bees in California. Masterkey 12(3):112–115.

Xavier, G. Harrington
1941   Seri Indian drawings. On file at Arizona State Museum, University of Arizona, Tucson.
1946   Seri face painting. The Kiva 11(2):15–20.

Index

Plants are indexed under the scientific name, English common names, and Spanish common names; Seri names for plants are listed in alphabetical order in Appendix A, which gives the corresponding scientific name for each plant. Page numbers in **boldface** type under a scientific plant name indicate the location of the major entry for that plant in the Species Accounts.

423

Plant Genera and Families or Major Groups

Abronia Nyctaginaceae
Abutilon Malvaceae
Acacia Leguminosae
Acalypha Euphorbiaceae
Achyronychia Caryophyllaceae
Agave Agavaceae
Allenrolfea Chenopodiaceae
Allionia Nyctaginaceae
Allium Liliaceae
Aloysia Verbenaceae
Amaranthus Amaranthaceae
Ambrosia Compositae
Amoreuxia Cochlospermaceae
Amphiroa Marine alga
Ananas Bromeliaceae
Andrachne Euphorbiaceae
Anemopsis Saururaceae
Antirrhinum Scrophulariaceae
Argemone Papaveraceae
Argythamnia Euphorbiaceae
Aristida Gramineae
Aristolochia Aristolochiaceae
Arundo Gramineae
Asclepias Asclepiadaceae
Asparagopsis Marine alga
Astragalus Leguminosae
Atamisquea Capparaceae
Atriplex Chenopodiaceae
Avicennia Verbenaceae
Baccharis Compositae
Baileya Compositae
Batis Bataceae
Battarrea Fungus
Bebbia Compositae
Boerhaavia Nyctaginaceae
Bouteloua Gramineae
Brachiaria Gramineae
Brahea Palmae
Brandegea Cucurbitaceae
Brickelia Compositae
Brodiaea Liliaceae
Bumelia Sapotaceae
Bursera Burseraceae
Caesalpinia Leguminosae
Calliandra Leguminosae
Camissonia Onagraceae

Cannabis Cannabaceae
Capsicum Solanaceae
Cardiospermum Sapindaceae
Carnegiea Cactaceae
Cassia Leguminosae
Castela Simaroubaceae
Celtis Ulmaceae
Cenchrus Gramineae
Cercidium Leguminosae
Chaenactis Compositae
Chenopodium Chenopodiaceae
Chorizanthe Polygonaceae
Cicer Leguminosae
Citrullus Cucurbitaceae
Citrus Rutaceae
Cleome Capparaceae
Cnidoscolus Euphorbiaceae
Cocculus Menispermaceae
Codium Marine alga
Colpomenia Marine alga
Colubrina Rhamnaceae
Condalia Rhamnaceae
Cordia Boraginaceae
Coursetia Leguminosae
Cressa Convolvulaceae
Croton Euphorbiaceae
Cryptantha Boraginaceae
Cryptonemia Marine alga
Cucumis Cucurbitaceae
Cucurbita Cucurbitaceae
Cuscuta Convolvulaceae
Cutleria Marine alga
Dalea Leguminosae
Dasya Marine alga
Datura Solanaceae
Descurainia Cruciferae
Desmanthus Leguminosae
Dictyota Marine alga
Digenia Marine alga
Digitaria Gramineae
Dithyrea Cruciferae
Dodonaea Sapindaceae
Draba Cruciferae
Dudleya Crassulaceae
Dyssodia Compositae
Echinocereus Cactaceae

Echinopterys Malpighiaceae
Encelia Compositae
Enteromorpha Marine alga
Eragrostis Gramineae
Eriogonum Polygonaceae
Erioneuron Gramineae
Errazurizia Leguminosae
Erythrina Leguminosae
Eschscholzia Papaveraceae
Eucheuma Marine alga
Eucnide Loasaceae
Eupatorium Compositae
Euphorbia Euphorbiaceae
Fagonia Zygophyllaceae
Ferocactus Cactaceae
Ficus Moraceae
Fouquieria Fouquieriaceae
Frankenia Frankeniaceae
Galaxaura Marine alga
Galvezia Scrophulariaceae
Ganoderma Fungus
Guaiacum Zygophyllaceae
Gelidiopsis Marine alga
Gigartina Marine alga
Gossypium Malvaceae
Gracilaria Marine alga
Gymnogongrus Marine alga
Halymenia Marine alga
Haplopappus Compositae
Hechtia Bromeliaceae
Heliotropium Boraginaceae
Hibiscus Malvaceae
Hoffmanseggia Leguminosae
Hofmeisteria Compositae
Holographis Acanthaceae
Horsfordia Malvaceae
Hymenoclea Compositae
Hypnea Marine alga
Hyptis Labiatae
(Idria) Fouquieriaceae
Ipomoea Convolvulaceae
Jacquinia Theophrastaceae
Janusia Malpighiaceae
Jatropha Euphorbiaceae
Jouvea Gramineae

Juncus Juncaceae
Justicia Acanthaceae
Kallstroemia Zygophyllaceae
Koeberlinia Koeberliniaceae
Krameria Krameriaceae
Lagenaria Cucurbitaceae
Laguncularia Combretaceae
Lantana Verbenaceae
Larrea Zygophyllaceae
Laurencia Marine alga
Lepidium Cruciferae
Lippia Verbenaceae
Lomentaria Marine alga
Lophocereus Cactaceae
Lotus Leguminosae
Lupinus Leguminosae
Lycium Solanaceae
Lycopersicon Solanaceae
Lyrocarpa Cruciferae
Lysiloma Leguminosae
Machaeranthera Compositae
Malva Malvaceae
Mammillaria Cactaceae
Marina Leguminosae
Marsdenia Asclepiadaceae
Mascagnia Malpighiaceae
Matelea Asclepiadaceae
Maximowiczia Cucurbitaceae
Maytenus Celastraceae
Melilotus Leguminosae
Melochia Sterculiaceae
Mentzelia Loasaceae
Mimosa Leguminosae
Mirabilis Nyctaginaceae
Mollugo Aizoaceae
Monanthochloe Gramineae
Muhlenbergia Gramineae
Musa Musaceae
Nama Hydrophyllaceae
Nemocladus Campanulaceae
Neoevansia Cactaceae
Nicotiana Solanaceae
Notholaena Fern
Oenothera Onagraceae
Oligomeris Resedaceae

Olneya Leguminosae
Opuntia Cactaceae
Orobanche Orobanchaceae
Oryza Gramineae
Pachycereus Cactaceae
Padina Marine alga
Palafoxia Compositae
Parkinsonia Leguminosae
Passiflora Passifloraceae
Pectis Compositae
Penstemon Scrophulariaceae
Perityle Compositae
Phacelia Hydrophyllaceae
Phaseolus Leguminosae
Phaulothamnus Phytolaccaceae
Phoenix Palmae
Phoradendron Viscaceae
Phragmites Gramineae
Physalis Solanaceae
Pilostyles Rafflesiaceae
Piper Piperaceae
Pisum Leguminosae
Pithecellobium Leguminosae
Plantago Plantaginaceae
Plocamium Marine alga
Podaxis Fungus
Polysiphonia Marine alga
Porophyllum Compositae
Proboscidea Martyniaceae
Prosopis Leguminosae
Psorothamnus Leguminosae
Punica Punicaceae
Randia Rubiaceae
Rhizophora Rhizophoraceae
Rhodymenia Marine alga
Ricinus Euphorbiaceae
Roccella Lichen
Ruellia Acanthaceae
Ruppia Ruppiaceae
Sabal Palmae
Saccharum Gramineae
Salicornia Chenopodiaceae
Salix Salicaceae
Salsola Chenopodiaceae
Salvia Labiatae
Sarcostemma Asclepiadaceae

Sargassum Marine alga
Selaginella Fern relative
Senecio Compositae
Sesuvium Aizoaceae
Setaria Gramineae
Sideroxylon Sapindaceae
Simmondsia Simmondsiaceae
Sisymbrium Cruciferae
Solanum Solanaceae
Sphaeralcea Malvaceae
Sporobolus Gramineae
Spyridia Marine algae
Stegnosperma Phytolaccaceae
Stenocereus Cactaceae
Stephanomeria Compositae
Struthanthus Loranthaceae
Suaeda Chenopodiaceae
Tamarix Tamaricaceae
Tephrosia Leguminosae
Teucrium Labiatae
Thryallis Malpighiacea
Tidestromia Amaranthaceae
Tillaea Crassulaceae
Tiquilia Boraginaceae
Tragia Euphorbiaceae
Trianthema Aizoaceae
Tribulus Zygophyllaceae
Triteleiopsis Liliaceae
Triticum Gramineae
Trixis Compositae
Tulostoma Fungus
Typha Typhaceae
Vallesia Apocynaceae
Vaseyanthus Cucurbitaceae
Verbesina Compositae
Vigna Leguminosae
Viguiera Compositae
Viscainoa Zygophyllaceae
Washingtonia Palmae
Xanthium Compositae
Yucca Agavaceae
Zea Gramineae
Zinnia Compositae
Ziziphus Rhamnaceae
Zostera Zosteraceae